P9-ARH-227

College Algebra
Hybrid

ABOUT THE AUTHORS

JAMES **STEWART** received his MS from Stanford University and his PhD from the University of Toronto. He did research at the University of London and was influenced by the famous mathematician, George Polya, at Stanford University. Stewart is Professor Emeritus at McMaster University and is currently Professor of Mathematics at the University of Toronto. His research field is harmonic analysis and the connections between mathematics and music.

James Stewart is the author of a best-selling calculus textbook series published by Brooks/Cole, Cengage Learning, including *Calculus, Calculus: Early Transcendentals,* and *Calculus: Concepts and Contexts,* a series of precalculus texts, as well as a series of high-school mathematics textbooks.

LOTHAR **REDLIN** grew up on Vancouver Island, received a Bachelor of Science degree from the University of Victoria, and a PhD from McMaster University in 1978. He subsequently did research and taught at the University of Washington; the University of Waterloo; and California State University, Long Beach.

He is currently Professor of Mathematics at The Pennsylvania State University, Abington Campus. His research field is topology.

SALEEM **WATSON** received his Bachelor of Science degree from Andrews University in Michigan. He did graduate studies at Dalhousie University and McMaster University, where he received his PhD in 1978.

He subsequently did research at the Mathematics Institute of the University of Warsaw in Poland. He also taught at The Pennsylvania State University. He is currently Professor of Mathematics at California State University, Long Beach. His research field is functional analysis.

The authors have also published *Precalculus: Mathematics for Calculus, Algebra and Trigonometry,* and *Trigonometry.*

ABOUT THE COVER

The building portrayed on the cover is 30 St. Mary Axe in London, England. More commonly known as "the Gherkin," it was designed by the renowned architect Sir Norman Foster and completed in 2004. Although the building gives an overall curved appearance, its exterior actually contains only one curved piece of glass—the lens-shaped cap at the very top. In fact, the striking shape of this building hides a complex mathematical structure. Mathematical curves have been used in architecture throughout history, for structural reasons as well as for their intrinsic beauty.

College Algebra
Hybrid

James Stewart
McMaster University and University of Toronto

Lothar Redlin
The Pennsylvania State University, Abington Campus

Saleem Watson
California State University, Long Beach

Australia · Brazil · Japan · Korea · Mexico · Singapore · Spain · United Kingdom · United States

BROOKS/COLE
CENGAGE Learning™

College Algebra: Hybrid
James Stewart, Lothar Redlin, Saleem Watson
Acquisitions Editor: Gary Whalen
Developmental Editor: Stacy Green
Assistant Editor: Cynthia Ashton
Editorial Assistant: Guanglei Zhang
Media Editor: Lynh Pham
Marketing Manager: Myriah Fitzgibbon
Marketing Assistant: Angela Kim
Marketing Communications Manager: Katy
    Malatesta
Content Project Manager: Jennifer Risden
Creative Director: Rob Hugel
Art Director: Vernon Boes
Print Buyer: Judy Inouye
Rights Acquisitions Account Manager, Text:
    Roberta Broyer
Rights Acquisitions Account Manager, Image:
    Don Schlotman
Production Service: MPS Limited,
    a Macmillan Company
Text Designer: Lisa Henry
Art Editor: Martha Emry
Photo Researcher: PrePress PMG
Copy Editor: Barbara Willette, Richard Camp
Illustrator: Precision Graphics; Jade Myers, Matrix
Cover Designer: Lisa Henry
Cover Image: © Paul Hardy/Corbis
Compositor: MPS Limited, a Macmillan Company

© 2011 Brooks/Cole, Cengage Learning

ALL RIGHTS RESERVED. No part of this work covered by the copyright
herein may be reproduced, transmitted, stored, or used in any form or by
any means graphic, electronic, or mechanical, including but not limited to
photocopying, recording, scanning, digitizing, taping, Web distribution, information networks, or information storage and retrieval systems, except as
permitted under Section 107 or 108 of the 1976 United States Copyright Act,
without the prior written permission of the publisher.

For product information and technology assistance, contact us at
**Cengage Learning Customer & Sales Support,
1-800-354-9706.**

For permission to use material from this text or
product, submit all requests online at
cengage.com/permissions.
Further permissions questions can be e-mailed to
permissionrequest@cengage.com.

Library of Congress Control Number: 2009937732

ISBN-13: 978-0-538-74029-6
ISBN-10: 0-538-74029-9

Brooks/Cole
20 Davis Drive
Belmont, CA 94002-3098
USA

Cengage Learning is a leading provider of customized learning solutions
with office locations around the globe, including Singapore, the United
Kingdom, Australia, Mexico, Brazil, and Japan. Locate your local office at
www.cengage.com/global.

Cengage Learning products are represented in Canada by Nelson
Education, Ltd.

To learn more about Brooks/Cole, visit **www.cengage.com/brookscole**

Purchase any of our products at your local college store or at our preferred
online store **www.CengageBrain.com.**

Printed in the United States of America
1 2 3 4 5 6 7 13 12 11 10 09

CONTENTS

PREFACE

The art of teaching
is the art of assisting discovery.
MARK VAN DOREN

For many students a College Algebra course represents the first opportunity to discover the beauty and practical power of mathematics. Many traditional lecture-based College Algebra courses are evolving into lecture-lab courses or into courses in which all homework, even tests, is delivered online. Moreover, distance learning is growing rapidly. Thus instructors are increasingly faced with the dual challenge of integrating Internet technology and teaching the concepts and skills of algebra, while at the same time imparting a sense of its utility. In this hybrid text, as in the five editions of our hardcover *College Algebra* book, our aim is to provide instructors and students with tools they can use to meet this challenge.

What do we mean by *hybrid*? A hybrid text involves an integration of several products and can be used best in a course with a strong blend of lecture time and online course work.

In writing this hybrid text, one of our main goals was to encourage students to be active learners. So, for instance, each chapter opens with *Warm-Up Exercises*, which will check and reinforce the students' understanding of what they should know before beginning a new chapter. All end-of-section homework exercises (excluding Discovery-Discussion-Writing exercises) have been put into *Enhanced WebAssign®*—an easy-to-use online homework management system—which allows for repeated homework, quizzing, study, and review. Concept exercises, also included in *Enhanced WebAssign*, encourage students to work with the basic concepts of the section and to use algebra vocabulary appropriately.

We believe a slightly smaller book will be more manageable for the student who is spending his or her homework time on a computer. (So historical notes, projects, and extra modeling material found in *College Algebra*, Fifth Edition, have been removed for the hybrid.) Still, our focus in the exposition on concepts remains unchanged. Our premise continues to be that conceptual understanding, technical skill, and real-world applications all go hand in hand, each reinforcing the others.

Special Features for College Algebra: Hybrid

- All end-of-section homework exercises (excluding Discovery-Discussion-Writing exercises) are now included in *Enhanced WebAssign*. In any mathematics course, the most important way to foster conceptual understanding is through the problems that the instructor assigns. To that end we have provided a wide selection of exercises. Each exercise set is carefully graded, progressing from basic conceptual exercises and skill-development problems to more challenging problems requiring synthesis of previously learned material with new concepts. Concept exercises, found at the beginning of each exercise set, are included in *Enhanced WebAssign*. We have included substantial applications of algebra that we believe will capture the interest of students.

- Each chapter of the text opens with *Warm-Up Exercises* that allow the student to check their understanding of previously covered material, particularly material relevant to the upcoming chapter.

- *Enhanced WebAssign* directions, review material, and formulas are included in the text on perforated pages.
- *Focus on Modeling* and *Discovery Projects* have been removed for this hybrid text. If you would like to have this material included, it is available online at the book companion site and can also be customized into the text. We have decided to include the *Focus on Problem Solving: General Principles* section in this hybrid since the problem-solving approach described there is referenced throughout the text.

Additional Features

Study Aids   *Learning Objectives* are found at the beginning of each section and Review sections at the end of each chapter. The review includes a summary of the main *Properties and Formulas* of the chapter and a *Concept Summary* keyed to specific review exercises.

Real-World Applications   We have included substantial applications of algebra that we believe will capture the interest of students. These are integrated throughout the text and are seen in the exercises available in Enhanced WebAssign.

Graphing Calculators and Computers   Calculator and computer technology extends in a powerful way our ability to calculate and to visualize mathematics. We have integrated the use of the graphing calculator throughout the text—to graph and analyze functions, families of functions, and sequences; to calculate and graph regression curves; to perform matrix algebra; to graph linear inequalities; and other such powerful uses. We also exploit the programming capabilities of the graphing calculator to provide simple programs that model real-life situations. The graphing calculator sections, subsections, examples, and exercises, all marked with the special symbol ▦, are optional and may be omitted without loss of continuity.

Check Your Answer   The *Check Your Answer* feature is used, wherever possible, to emphasize the importance of looking back to check whether an answer is reasonable.

Review Sections and Chapter Tests   Each chapter ends with an extensive review section, including a *Chapter Test* designed to help students gauge their progress. Brief answers to odd-numbered review exercises and to all questions in the Chapter Tests are given in the back of the book. The review material in each chapter begins with a summary of the main *Properties and Formulas* and a *Concept Summary*. These two features provide a concise synopsis of the material in the chapter. *Cumulative Review Tests* follow Chapters 2, 5, and 8.

Ancillaries

College Algebra: Hybrid is supported by a complete set of ancillaries developed under our direction. Each piece has been designed to enhance student understanding and to facilitate creative instruction. **Enhanced WebAssign (EWA)**, our Web-based homework system, allows instructors to assign, collect, grade, and record homework assignments online, minimizing workload and streamlining the grading process. EWA also gives students the ability to stay organized with assignments and have up-to-date grade information. All end-of-section homework exercises (excluding Discovery-Discussion-Writing exercises) are available in Enhanced WebAssign.

Acknowledgments

We would like to acknowledge the following reviewers of our original *College Algebra* text, now in its fifth edition.

Beth-Allyn Osikiewicz, *Kent State University—Tuscarawas Campus*

Faiz Al-Rubaee, *University of North Florida*

Christian Barrientos, *Clayton State University*

Candace Blazek, *Anoka-Ramsey Community College*

Catherine May Bonan-Hamada, *Mesa State College*

Barry W. Brunson, *Western Kentucky University*

Richard Dodge, *Jackson Community College*

Floyd Downs, *Arizona State University at Tempe*

Gay Ellis, *Southwest Missouri State University*

Mohamed Elhamdadi, *University of South Florida*

José D. Flores, *University of South Dakota*

Carl L. Hensley, *Indian River Community College*

Linda B. Hutchison, *Belmont University*

Christy Leigh Jackson, *University of Arkansas, Little Rock*

George Johnson, *St. Phillips College*

Marjorie Kreienbrink, *University of Wisconsin*

Martha Ann Larkin, *Southern Utah University*

Scott Lewis, *Utah Valley State College*

Gene Majors, *Fullerton College*

Theresa McChesney, *Johnson County Community College*

O. Michael Melko, *Northern State University*

Franklin A. Michello, *Middle Tennessee State University*

Terry Nyman, *University of Wisconsin—Fox Valley*

Keith Oberlander, *Pasadena City College*

Christine Oxley, *Indiana University*

Christine Panoff, *University of Michigan at Flint*

David Rollins, *University of Central Florida*

George Rust, *West Virginia State University*

Alicia Serfaty de Markus, *Miami Dade College—Kendell Campus*

Randy Scott, *Santiago Canyon College*

Stanley Stascinsky, *Tarrant County College*

Fereja Tahir, *Illinois Central College*

Mary Ann Teel, *University of North Texas*

Arnold Volbach, *University of Houston, University Park*

Tom Walsh, *City College of San Francisco*

George Wang, *Whittier College*

Max Warshauer, *Southwest Texas State University*

David Watson, *Rutgers University*

Kathryn Wetzel, *Amarillo College*

Muserref Wiggins, *University of Akron*

Vassil Yorgov, *Fayetteville State University*

Naveed Zaman, *West Virginia State University*

Xiaohong Zhang, *West Virginia State University*

We have benefited greatly from the suggestions and comments of our colleagues who have used an edition of *College Algebra*. We extend special thanks in this regard to Larry Brownson, Linda Byun, Bruce Chaderjian, David Gau, Daniel Hernandez, YongHee Kim-Park, Daniel Martinez, David McKay, Robert Mena, Kent Merryfield, Viet Ngo, Marilyn Oba, Alan Safer, Robert Valentini, and Derming Wang, from California State University, Long Beach; to Karen Gold, Betsy Huttenlock, Cecilia McVoy, Mike McVoy, Samir Ouzomgi, and Ralph Rush, of The Pennsylvania State University, Abington College; and to Fred Safier, of the City College of San Francisco. We have learned much from our students; special thanks go to Devaki Shah and Ellen Newman for their helpful suggestions.

Ann Ostberg did a masterful job of checking the correctness of the examples and answers to exercises. We extend our heartfelt thanks for her timely and accurate work. We also thank Phyllis Panman-Watson for solving the exercises and checking the answer manuscript. We thank Doug Shaw for creating the Warm-Up Exercises and Melissa Pfohl for verifying their accuracy. We also thank Andy Bulman-Fleming and Doug Shaw for their inspired work in producing the supplemental study guide.

At Brooks/Cole, our thanks go to Developmental Editor Stacy Green, Assistant Editor Cynthia Ashton, Editorial Assistant Guanglei Zhang, Content Project Manager Jennifer Risden, and Marketing Manager Myriah Fitzgibbon. They have all done an outstanding job.

Finally, we thank our editor Gary Whalen for carefully and thoughtfully guiding this book through the writing and production process. His support and editorial insight played a crucial role in completing this edition.

Ancillaries for College Algebra

INSTRUCTOR RESOURCES

Enhanced WebAssign

Easy to assign. Easy to use. Easy to manage. Enhanced WebAssign allows you to assign, collect, grade, and record homework assignments via the web and includes links to text-specific content, video examples, and problem-specific tutorials. Enhanced WebAssign is more than a homework system; it is a complete learning system for your students.

Test Bank

ISBN-10: 0-495-56526-1; ISBN-13: 978-0-495-56526-0

The Test Bank includes six tests per chapter as well as three final exams. The tests are made up of a combination of multiple-choice, free-response, true/false, and fill-in-the-blank questions.

Instructor's Guide

ISBN-10: 0-495-56527-X; ISBN-13: 978-0-495-56527-7

The Instructor's Guide contains points to stress, suggested time to allot, text discussion topics, core materials for lecture, workshop/discussion suggestions, group work exercises in a form suitable for handout, and suggested homework problems.

Instructor's Guide Hybrid Course Update

ISBN-10: 0-538-74132-5; ISBN-13: 978-0-538-74132-3

The Instructor's Guide Hybrid Course Update includes teaching tips and materials targeted toward the challenges of teaching an online course.

ExamView® for Algorithmic Equations (Windows®/Macintosh®)

Create, deliver, and customize tests and study guides (both print and online) in minutes with this easy-to-use assessment and tutorial software on CD. Includes complete questions from the *College Algebra* Test Bank. Included on the PowerLecture™ CD.

PowerLecture™ CD

ISBN-10: 0-495-56529-6; ISBN-13: 978-0-495-56529-1

Contains PowerPoint® lecture outlines, a database of all the art in the text, ExamView, and electronic copies of the Instructor's Guide and Instructor's Solution Manual.

Text-Specific DVDs

ISBN-10: 0-495-56522-9; ISBN-13: 978-0-495-56522-2

This set of DVDs is available free upon adoption of the text. Each video offers one chapter of the text and is broken down into 10- to 20-minute problem-solving lessons that cover each section of the chapter. These videos can be used as additional explanations for each chapter of the text.

STUDENT RESOURCES

Study Guide

ISBN-10: 0-495-56523-7; ISBN-13: 978-0-495-56523-9

Reinforces student understanding with detailed explanations, worked-out examples, and practice problems. Lists key ideas to master and builds problem-solving skills. There is a section in the Study Guide corresponding to each section in the text.

TO THE STUDENT

This textbook was written for you to use as a guide to mastering College Algebra. Here are some suggestions to help you get the most out of your course.

First of all, you should read the appropriate section of text *before* you attempt your homework problems. Reading a mathematics text is quite different from reading a novel, a newspaper, or even another textbook. You may find that you have to reread a passage several times before you understand it. Pay special attention to the examples, and work them out yourself with pencil and paper as you read. With this kind of preparation you will be able to do your homework much more quickly and with more understanding.

Don't make the mistake of trying to memorize every single rule or fact you may come across. Mathematics doesn't consist simply of memorization. Mathematics is a *problem-solving art*, not just a collection of facts. To master the subject you must solve problems— lots of problems. Though you will likely be doing all your homework via the Enhanced WebAssign online program, the way you solve a problem should be no different than how you would solve it using pencil and paper. The difference is only in how you write your answer and submit it for grading. Be sure to write your solutions in a logical, step-by-step fashion. Don't give up on a problem if you can't solve it right away. Try to understand the problem more clearly—reread it thoughtfully and relate it to what you have learned from your teacher and from the examples in the text. Struggle with it until you solve it. Once you have done this a few times you will begin to understand what mathematics is really all about.

Answers to the odd-numbered chapter review exercises, as well as all the answers to each chapter test, appear at the back of the book. If your answer differs from the one given, don't immediately assume that you are wrong. There may be a calculation that connects the two answers and makes both correct. For example, if you get $1/(\sqrt{2} - 1)$ but the answer given is $1 + \sqrt{2}$, your answer *is* correct, because you can multiply both numerator and denominator of your answer by $\sqrt{2} + 1$ to change it to the given answer.

The symbol ⊘ is used to warn against committing an error. We have placed this symbol in the margin to point out situations where we have found that many of our students make the same mistake.

CALCULATORS
AND CALCULATIONS

Calculators are essential in most mathematics and science subjects. They free us from performing routine tasks, so we can focus more clearly on the concepts we are studying. Calculators are powerful tools but their results need to be interpreted with care. In what follows, we describe the features that a calculator suitable for a College Algebra course should have, and we give guidelines for interpreting the results of its calculations.

Scientific and Graphing Calculators

For this course you will need a *scientific* calculator—one that has, as a minimum, the usual arithmetic operations ($+$, $-$, $\times$, $\div$) as well as exponential and logarithmic functions (e^x, 10^x, $\ln x$, $\log x$). In addition, a memory and at least some degree of programmability will be useful.

Your instructor may recommend or require that you purchase a *graphing* calculator. This book has optional subsections and exercises that require the use of a graphing calculator or a computer with graphing software. These special subsections and exercises are indicated by the symbol 🖩. Besides graphing functions, graphing calculators can also be used to find functions that model real-life data, solve equations, perform matrix calculations (which are studied in Chapter 7), and help you perform other mathematical operations. All these uses are discussed in this book.

It is important to realize that, because of limited resolution, a graphing calculator gives only an *approximation* to the graph of a function. It plots only a finite number of points and then connects them to form a *representation* of the graph. In Section 2.3, we give guidelines for using a graphing calculator and interpreting the graphs that it produces.

Calculations and Significant Figures

Most of the applied examples and exercises in this book involve approximate values. For example, one exercise states that the moon has a radius of 1074 miles. This does not mean that the moon's radius is exactly 1074 miles but simply that this is the radius rounded to the nearest mile.

One simple method for specifying the accuracy of a number is to state how many **significant digits** it has. The significant digits in a number are the ones from the first nonzero digit to the last nonzero digit (reading from left to right). Thus, 1074 has four significant digits, 1070 has three, 1100 has two, and 1000 has one significant digit. This rule may sometimes lead to ambiguities. For example, if a distance is 200 km to the nearest kilometer, then the number 200 really has three significant digits, not just one. This ambiguity is avoided if we use scientific notation—that is, if we express the number as a multiple of a power of 10:

$$2.00 \times 10^2$$

When working with approximate values, students often make the mistake of giving a final answer with *more* significant digits than the original data. This is incorrect because you cannot "create" precision by using a calculator. The final result can be no more accurate than the measurements given in the problem. For example, suppose we are told that the two shorter sides of a right triangle are measured to be 1.25 and 2.33 inches long. By the Pythagorean Theorem, we find, using a calculator, that the hypotenuse has length

$$\sqrt{1.25^2 + 2.33^2} \approx 2.644125564 \text{ in.}$$

But since the given lengths were expressed to three significant digits, the answer cannot be any more accurate. We can therefore say only that the hypotenuse is 2.64 in. long, rounding to the nearest hundredth.

In general, the final answer should be expressed with the same accuracy as the *least-accurate* measurement given in the statement of the problem. The following rules make this principle more precise.

RULES FOR WORKING WITH APPROXIMATE DATA

1. When multiplying or dividing, round off the final result so that it has as many *significant digits* as the given value with the fewest number of significant digits.

2. When adding or subtracting, round off the final result so that it has its last significant digit in the *decimal place* in which the least-accurate given value has its last significant digit.

3. When taking powers or roots, round off the final result so that it has the same number of *significant digits* as the given value.

As an example, suppose that a rectangular table top is measured to be 122.64 in. by 37.3 in. We express its area and perimeter as follows:

Area = length × width = 122.64 × 37.3 ≈ 4570 in²   *Three significant digits*

Perimeter = 2(length + width) = 2(122.64 + 37.3) ≈ 319.9 in.   *Tenths digit*

 Note that in the formula for the perimeter, the value 2 is an exact value, not an approximate measurement. It therefore does not affect the accuracy of the final result. In general, if a problem involves only exact values, we may express the final answer with as many significant digits as we wish.

Note also that to make the final result as accurate as possible, *you should wait until the last step to round off your answer.* If necessary, use the memory feature of your calculator to retain the results of intermediate calculations.

ABBREVIATIONS

cm	centimeter	**mg**	milligram
dB	decibel	**MHz**	megahertz
F	farad	**mi**	mile
ft	foot	**min**	minute
g	gram	**mL**	milliliter
gal	gallon	**mm**	millimeter
h	hour	**N**	Newton
H	henry	**qt**	quart
Hz	Hertz	**oz**	ounce
in.	inch	**s**	second
J	Joule	**Ω**	ohm
kcal	kilocalorie	**V**	volt
kg	kilogram	**W**	watt
km	kilometer	**yd**	yard
kPa	kilopascal	**yr**	year
L	liter	**°C**	degree Celsius
lb	pound	**°F**	degree Fahrenheit
lm	lumen	**K**	Kelvin
M	mole of solute	$\Rightarrow$	implies
	per liter of solution	$\Leftrightarrow$	is equivalent to
m	meter		

College Algebra
Hybrid

Prerequisites

In this chapter we review some of the fundamental ideas of algebra that we'll use throughout this book.

The following Warm-Up Exercises will help you review the basic algebra skills that you will need for this chapter.

WARM-UP EXERCISES

1. Consider the following number line:

(a) Plot the values $\dfrac{5}{2}$ and -1.75 on the number line.

(b) What is the value of the point marked x?

2. Compute the following:

(a) 4×5 (b) -6.25×8 (c) $-9 \times \left(-\dfrac{4}{3}\right)$

3. Compute the following

(a) $52 - 25$ (b) $25 - 52$ (c) $-25 - 52$

4. Compute the following:

(a) $6 + 4 \times 3$ (b) $\dfrac{-3}{20 + 7}$ (c) $5(6 - 3)$

5. Compute the following:

(a) $\dfrac{1}{6} - \dfrac{2}{15}$ (b) $\dfrac{1 + \dfrac{1}{2}}{2 - \dfrac{1}{4}}$

6. Which of the following quantities are **undefined**?

(a) $\sqrt{25}$ (b) $\dfrac{3}{5 - 6}$ (c) $\dfrac{2 + 5}{6 - 6}$ (d) $\sqrt{8}$ (e) $\sqrt{-6}$

7. If $PV = nRT$, and we know that $R = 8.3$, $V = 10$, $n = 7.5$, and $T = 294$, compute P.

8. Make a table that gives the value of x^2, for each integer x from 1 through 6.

P.1 | Modeling the Real World with Algebra

LEARNING OBJECTIVES

After completing this section, you will be able to:

- Use an algebra model
- Make an algebra model

In algebra we use letters to stand for numbers. This allows us to describe patterns that we see in the real world.

For example, if we let N stand for the number of hours you work and W stand for your hourly wage, then the formula

$$P = NW$$

gives your pay P. The formula $P = NW$ is a description or *model* for pay. We can also call this formula an *algebra model*. We summarize the situation as follows:

Real World

You work for an hourly *wage*. You would like to know your *pay* for any *number* of hours worked.

Algebra Model

$P = NW$

The model $P = NW$ gives the pattern for finding the pay for *any* worker, with *any* hourly wage, working *any* number of hours. That's the power of algebra: By using letters to stand for numbers, we can write a single formula that describes many different situations.

We can now use the model $P = NW$ to answer questions such as "I make $10 an hour, and I worked 35 hours; how much do I get paid?" or "I make $8 an hour; how many hours do I need to work to get paid $1000?"

In general, a **model** is a mathematical representation (such as a formula) of a real-world situation. **Modeling** is the process of making mathematical models. Once a model has been made, it can be used to answer questions about the thing being modeled.

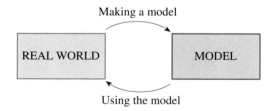

The examples we study in this section are simple, but the methods are far reaching.

■ Using Algebra Models

We begin our study of modeling by using models that are given to us. In the next subsection we learn how to make our own models.

▶ **EXAMPLE 1** | Using a Model for Pay

Use the model $P = NW$ to answer the following question: Aaron makes $10 an hour and worked 35 hours last week. How much did he get paid?

▼ **SOLUTION**   We know that $N = 35$ h and $W = \$10$. We substitute these values into the formula.

$$P = NW \qquad \text{Model}$$

$$= 35 \times 10 \qquad \text{Substitute } N = 35,\ W = 10$$

$$= 350 \qquad \text{Calculator}$$

So Aaron got paid $350.

▶ **EXAMPLE 2** | Using a Model for Pay

Use the model $P = NW$ to solve the following problem: Neil makes \$9.00 an hour tutoring mathematics in the Learning Center. He wants to earn enough money to buy a calculus text that costs \$126 (including tax). How many hours does he need to work to earn this amount?

▼ **SOLUTION**   We know that Neil's hourly wage is $W = \$9.00$ and the amount of pay he needs to buy the book is $P = \$126$. To find N, we substitute these values into the formula.

$$P = NW \qquad \text{Model}$$

$$126 = 9N \qquad \text{Substitute } P = 126,\ W = 9.00$$

$$\frac{126}{9} = N \qquad \text{Divide by 9}$$

$$N = 14 \qquad \text{Calculator}$$

So Neil must work 14 hours to buy this book.

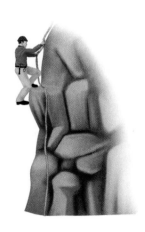

▶ **EXAMPLE 3** | Using an Elevation-Temperature Model

A mountain climber uses the model

$$T = 20 - 10h$$

to estimate the temperature T (in °C) at elevation h (in kilometers, km).

(a) Make a table that gives the temperature for each 1-km change in elevation, from elevation 0 km to elevation 5 km. How does temperature change as elevation increases?

(b) If the temperature is 5°C, what is the elevation?

▼ **SOLUTION**

(a) Let's use the model to find the temperature at elevation $h = 3$ km.

$$T = 20 - 10h \qquad \text{Model}$$

$$= 20 - 10(3) \qquad \text{Substitute } h = 3$$

$$= -10 \qquad \text{Calculator}$$

So at an elevation of 3 km the temperature is -10°C. The other entries in the following table are calculated similarly.

Elevation (km)	Temperature (°C)
0	20°
1	10°
2	0°
3	−10°
4	−20°
5	−30°

We see that temperature decreases as elevation increases.

(b) We substitute $T = 5°C$ in the model and solve for h.

$$T = 20 - 10h \qquad \text{Model}$$

$$5 = 20 - 10h \qquad \text{Substitute } T = 5$$

$$-15 = -10h \qquad \text{Subtract 20}$$

$$\frac{-15}{-10} = h \qquad \text{Divide by -10}$$

$$1.5 = h \qquad \text{Calculator}$$

The elevation is 1.5 km.   ▲

■ Making Algebra Models

In the next example we explore the process of making an algebra model for a real-life situation.

► **EXAMPLE 4** | Making a Model for Gas Mileage

The gas mileage of a car is the number of miles it can travel on one gallon of gas.

(a) Find a formula that models gas mileage in terms of the number of miles driven and the number of gallons of gasoline used.

(b) Henry's car used 10.5 gallons to drive 230 miles. Find its gas mileage.

12 mi/gal   40 mi/gal

Thinking About the Problem

Let's try a simple case. If a car uses 2 gallons to drive 100 miles, we easily see that

$$\text{gas mileage} = \frac{100}{2} = 50 \text{ mi/gal}$$

So gas mileage is the number of miles driven divided by the number of gallons used.

▼ **SOLUTION**

(a) To find the formula we want, we need to assign symbols to the quantities involved:

In Words	In Algebra
Number of miles driven	N
Number of gallons used	G
Gas mileage (mi/gal)	M

We can express the model as follows:

$$\text{gas mileage} = \frac{\text{number of miles driven}}{\text{number of gallons used}}$$

$$M = \frac{N}{G} \qquad \text{Model}$$

(b) To get the gas mileage, we substitute $N = 230$ and $G = 10.5$ in the formula.

$$M = \frac{N}{G} \qquad \text{Model}$$

$$= \frac{230}{10.5} \qquad \text{Substitute } N = 230, \, G = 10.5$$

$$\approx 21.9 \qquad \text{Calculator}$$

The gas mileage for Henry's car is 21.9 mi/gal.

<table><tr><td>**P.2**</td><td># Real Numbers and Their Properties</td></tr></table>

LEARNING OBJECTIVES

After completing this section, you will be able to:

- Classify real numbers
- Use properties of real numbers
- Use properties of negatives
- Add, subtract, multiply, and divide fractions

■ Types of Real Numbers

Let's review the types of numbers that make up the real number system. We start with the **natural numbers**:

$$1, 2, 3, 4, \ldots$$

The different types of real numbers were invented to meet specific needs. For example, natural numbers are needed for counting, negative numbers for describing debt or below-zero temperatures, rational numbers for concepts such as "half a gallon of milk," and irrational numbers for measuring certain distances, such as the diagonal of a square.

The **integers** consist of the natural numbers together with their negatives and 0:

$$\ldots, -3, -2, -1, 0, 1, 2, 3, 4, \ldots$$

We construct the **rational numbers** by taking ratios of integers. Thus, any rational number r can be expressed as

$$r = \frac{m}{n}$$

where m and n are integers and $n \neq 0$. Examples are

$$\tfrac{1}{2} \qquad -\tfrac{3}{7} \qquad 46 = \tfrac{46}{1} \qquad 0.17 = \tfrac{17}{100}$$

(Recall that division by 0 is always ruled out, so expressions such as $\frac{3}{0}$ and $\frac{0}{0}$ are undefined.) There are also real numbers, such as $\sqrt{2}$, that cannot be expressed as a ratio of integers and are therefore called **irrational numbers**. It can be shown, with varying degrees of difficulty, that these numbers are also irrational:

$$\sqrt{3} \qquad \sqrt{5} \qquad \sqrt[3]{2} \qquad \pi \qquad \frac{3}{\pi^2}$$

The set of all real numbers is usually denoted by the symbol $\mathbb{R}$. When we use the word *number* without qualification, we will mean "real number." Figure 1 is a diagram of the types of real numbers that we work with in this book.

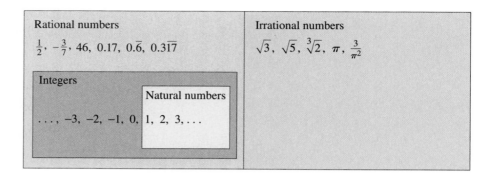

FIGURE 1 The real number system

Every real number has a decimal representation. If the number is rational, then its corresponding decimal is repeating. For example,

$$\tfrac{1}{2} = 0.5000\ldots = 0.5\overline{0} \qquad\qquad \tfrac{2}{3} = 0.66666\ldots = 0.\overline{6}$$

$$\tfrac{157}{495} = 0.3171717\ldots = 0.3\overline{17} \qquad \tfrac{9}{7} = 1.285714285714\ldots = 1.\overline{285714}$$

(The bar indicates that the sequence of digits repeats forever.) If the number is irrational, the decimal representation is nonrepeating:

$$\sqrt{2} = 1.414213562373095\ldots \qquad \pi = 3.141592653589793\ldots$$

If we stop the decimal expansion of any number at a certain place, we get an approximation to the number. For instance, we can write

$$\pi \approx 3.14159265$$

where the symbol $\approx$ is read "is approximately equal to." The more decimal places we retain, the better our approximation.

A repeating decimal such as
$$x = 3.5474747\ldots$$
is a rational number. To convert it to a ratio of two integers, we write

$$1000x = 3547.47474747\ldots$$
$$10x = 35.47474747\ldots$$
$$\overline{990x = 3512.0}$$

Thus, $x = \tfrac{3512}{990}$. (The idea is to multiply x by appropriate powers of 10 and then subtract to eliminate the repeating part.)

▶ **EXAMPLE 1** | Classifying Real Numbers

Determine whether each given real number is a natural number, an integer, a rational number, or an irrational number.

(a) 999   **(b)** $-\tfrac{6}{5}$ **(c)** $-\tfrac{6}{3}$ **(d)** $\sqrt{25}$ **(e)** $\sqrt{3}$

▼ **SOLUTION**

(a) 999 is a positive whole number, so it is a natural number.

(b) $-\tfrac{6}{5}$ is a ratio of two integers, so it is a rational number.

(c) $-\tfrac{6}{3}$ equals -2, so it is an integer.

(d) $\sqrt{25}$ equals 5, so it is a natural number.

(e) $\sqrt{3}$ is a nonrepeating decimal (approximately 1.7320508075689), so it is an irrational number. ▲

■ Operations on Real Numbers

Real numbers can be combined using the familiar operations of addition, subtraction, multiplication, and division. When evaluating arithmetic expressions that contain several of these operations, we use the following conventions to determine the order in which the operations are performed:

1. Perform operations inside parentheses first, beginning with the innermost pair. In dividing two expressions, the numerator and denominator of the quotient are treated as if they are within parentheses.

2. Perform all multiplications and divisions, working from left to right.

3. Perform all additions and subtractions, working from left to right.

▶ **EXAMPLE 2** | Evaluating an Arithmetic Expression

Find the value of the expression

$$3\left(\frac{8 + 10}{2 \cdot 3} + 4\right) - 2(5 + 9)$$

▼ **SOLUTION** First we evaluate the numerator and denominator of the quotient, since these are treated as if they are inside parentheses:

$$3\left(\frac{8 + 10}{2 \cdot 3} + 4\right) - 2(5 + 9) = 3\left(\frac{18}{6} + 4\right) - 2(5 + 9) \qquad \text{Evaluate numerator and denominator}$$

$$= 3(3 + 4) - 2(5 + 9) \qquad \text{Evaluate quotient}$$

$$= 3 \cdot 7 - 2 \cdot 14 \qquad \text{Evaluate parentheses}$$

$$= 21 - 28 \qquad \text{Evaluate products}$$

$$= -7 \qquad \text{Evaluate difference} \quad \blacktriangle$$

■ Properties of Real Numbers

We all know that $2 + 3 = 3 + 2$. We also know that $5 + 7 = 7 + 5$, $513 + 87 = 87 + 513$, and so on. In algebra we express all these (infinitely many) facts by writing

$$a + b = b + a$$

where a and b stand for any two numbers. In other words, "$a + b = b + a$" is a concise way of saying that "when we add two numbers, the order of addition doesn't matter." This fact is called the *Commutative Property of Addition*. From our experience with numbers we know that the properties in the following box are also valid.

PROPERTIES OF REAL NUMBERS

Property	Example	Description
Commutative Properties		
$a + b = b + a$	$7 + 3 = 3 + 7$	When we add two numbers, order doesn't matter.
$ab = ba$	$3 \cdot 5 = 5 \cdot 3$	When we multiply two numbers, order doesn't matter.
Associative Properties		
$(a + b) + c = a + (b + c)$	$(2 + 4) + 7 = 2 + (4 + 7)$	When we add three numbers, it doesn't matter which two we add first.
$(ab)c = a(bc)$	$(3 \cdot 7) \cdot 5 = 3 \cdot (7 \cdot 5)$	When we multiply three numbers, it doesn't matter which two we multiply first.
Distributive Property		
$a(b + c) = ab + ac$	$2 \cdot (3 + 5) = 2 \cdot 3 + 2 \cdot 5$	When we multiply a number by a sum of two numbers,
$(b + c)a = ab + ac$	$(3 + 5) \cdot 2 = 2 \cdot 3 + 2 \cdot 5$	we get the same result as multiplying the number by each of the terms and then adding the results.

The Distributive Property is crucial because it describes the way addition and multiplication interact with each other.

The Distributive Property applies whenever we multiply a number by a sum. Figure 2 explains why this property works for the case in which all the numbers are positive integers, but the property is true for any real numbers a, b, and c.

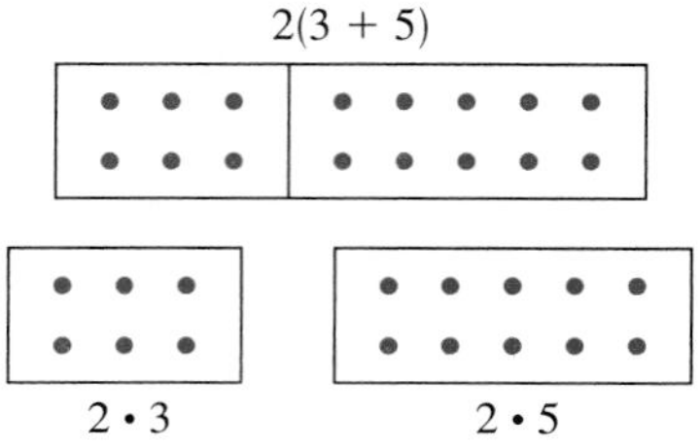

$$2(3 + 5)$$

$$2 \cdot 3 \qquad 2 \cdot 5$$

FIGURE 2 The Distributive Property

▶ **EXAMPLE 3** | Using the Properties of Real Numbers

(a) $2 + (3 + 7) = 2 + (7 + 3)$ Commutative Property of Addition

                 $= (2 + 7) + 3$ Associative Property of Addition

(b) $2(x + 3) = 2 \cdot x + 2 \cdot 3$ Distributive Property

           $= 2x + 6$ Simplify

(c) $(a + b)(x + y) = (a + b)x + (a + b)y$ Distributive Property

                    $= (ax + bx) + (ay + by)$ Distributive Property

                    $= ax + bx + ay + by$ Associative Property of Addition

In the last step we removed the parentheses because, according to the Associative Property, the order of addition doesn't matter. ▲

■ Addition and Subtraction

The number 0 is special for addition; it is called the **additive identity** because $a + 0 = a$ for any real number a. Every real number a has a **negative**, $-a$, that satisfies $a + (-a) = 0$. **Subtraction** is the operation that undoes addition; to subtract a number from another, we simply add the negative of that number. By definition

$$a - b = a + (-b)$$

⊘ Don't assume that $-a$ is a negative number. Whether $-a$ is negative or positive depends on the value of a. For example, if $a = 5$, then $-a = -5$, a negative number, but if $a = -5$, then $-a = -(-5) = 5$ (Property 2), a positive number.

To combine real numbers involving negatives, we use the following properties.

PROPERTIES OF NEGATIVES	
Property	**Example**
1. $(-1)a = -a$	$(-1)5 = -5$
2. $-(-a) = a$	$-(-5) = 5$
3. $(-a)b = a(-b) = -(ab)$	$(-5)7 = 5(-7) = -(5 \cdot 7)$
4. $(-a)(-b) = ab$	$(-4)(-3) = 4 \cdot 3$
5. $-(a + b) = -a - b$	$-(3 + 5) = -3 - 5$
6. $-(a - b) = b - a$	$-(5 - 8) = 8 - 5$

Property 6 states the intuitive fact that $a - b$ and $b - a$ are negatives of each other. Property 5 is often used with more than two terms:

$$-(a + b + c) = -a - b - c$$

► **EXAMPLE 4** | Using Properties of Negatives

Let x, y, and z be real numbers.

(a) $-(3 + 2) = -3 - 2$ Property 5: $-(a + b) = -a - b$

(b) $-(x + 2) = -x - 2$ Property 5: $-(a + b) = -a - b$

(c) $-(x + y - z) = -x - y - (-z)$ Property 5: $-(a + b) = -a - b$

$ = -x - y + z$ Property 2: $-(-a) = a$ ▲

> The word **algebra** comes from the ninth-century Arabic book *Hisâb al-Jabr w'al-Muqabala*, written by al-Khowarizmi. The title refers to transposing and combining terms, two processes that are used in solving equations. In Latin translations the title was shortened to *Aljabr*, from which we get the word *algebra*. The author's name itself made its way into the English language in the form of our word *algorithm*.

■ Multiplication and Division

The number 1 is special for multiplication; it is called the **multiplicative identity** because $a \cdot 1 = a$ for any real number a. Every nonzero real number a has an **inverse**, $1/a$, that satisfies $a \cdot (1/a) = 1$. **Division** is the operation that undoes multiplication; to divide by a number, we multiply by the inverse of that number. If $b \neq 0$, then by definition

$$a \div b = a \cdot \frac{1}{b}$$

We write $a \cdot (1/b)$ as simply a/b. We refer to a/b as the **quotient** of a and b or as the **fraction** a over b; a is the **numerator**, and b is the **denominator** (or **divisor**). To combine real numbers using the operation of division, we use the following properties.

PROPERTIES OF FRACTIONS

Property	Example	Description
1. $\dfrac{a}{b} \cdot \dfrac{c}{d} = \dfrac{ac}{bd}$	$\dfrac{2}{3} \cdot \dfrac{5}{7} = \dfrac{2 \cdot 5}{3 \cdot 7} = \dfrac{10}{21}$	When **multiplying fractions**, multiply numerators and denominators.
2. $\dfrac{a}{b} \div \dfrac{c}{d} = \dfrac{a}{b} \cdot \dfrac{d}{c}$	$\dfrac{2}{3} \div \dfrac{5}{7} = \dfrac{2}{3} \cdot \dfrac{7}{5} = \dfrac{14}{15}$	When **dividing fractions**, invert the divisor and multiply.
3. $\dfrac{a}{c} + \dfrac{b}{c} = \dfrac{a + b}{c}$	$\dfrac{2}{5} + \dfrac{7}{5} = \dfrac{2 + 7}{5} = \dfrac{9}{5}$	When **adding fractions** with the **same denominator**, add the numerators.
4. $\dfrac{a}{b} + \dfrac{c}{d} = \dfrac{ad + bc}{bd}$	$\dfrac{2}{5} + \dfrac{3}{7} = \dfrac{2 \cdot 7 + 3 \cdot 5}{35} = \dfrac{29}{35}$	When **adding fractions** with **different denominators**, find a common denominator. Then add the numerators.
5. $\dfrac{ac}{bc} = \dfrac{a}{b}$	$\dfrac{2 \cdot 5}{3 \cdot 5} = \dfrac{2}{3}$	**Cancel** numbers that are **common factors** in the numerator and denominator.
6. If $\dfrac{a}{b} = \dfrac{c}{d}$, then $ad = bc$	$\dfrac{2}{3} = \dfrac{6}{9}$, so $2 \cdot 9 = 3 \cdot 6$	**Cross multiply**.

When adding fractions with different denominators, we don't usually use Property 4. Instead, we rewrite the fractions so that they have the smallest possible common denominator (often smaller than the product of the denominators), and then we use Property 3. This denominator is the **L**east **C**ommon **D**enominator (LCD) described in the next example.

► **EXAMPLE 5** | Using the LCD to Add Fractions

Evaluate: $\dfrac{5}{36} + \dfrac{7}{120}$

▼ **SOLUTION**   Factoring each denominator into prime factors gives

$$36 = 2^2 \cdot 3^2 \quad \text{and} \quad 120 = 2^3 \cdot 3 \cdot 5$$

We find the least common denominator (LCD) by forming the product of all the factors that occur in these factorizations, using the highest power of each factor. Thus, the LCD is $2^3 \cdot 3^2 \cdot 5 = 360$. So

$$\frac{5}{36} + \frac{7}{120} = \frac{5 \cdot 10}{36 \cdot 10} + \frac{7 \cdot 3}{120 \cdot 3} \qquad \text{Use common denominator}$$

$$= \frac{50}{360} + \frac{21}{360} = \frac{71}{360} \qquad \text{Property 3: Adding fractions with the same denominator} \qquad \blacktriangle$$

P.3 | # The Real Number Line and Order

LEARNING OBJECTIVES
After completing this section, you will be able to:

- Graph numbers on the real line
- Use the order symbols $<, \leq, >, \geq$
- Work with set and interval notation
- Find and use absolute values of real numbers
- Find distances on the real line

■ The Real Line

The real numbers can be represented by points on a line, as shown in Figure 1. The positive direction (toward the right) is indicated by an arrow. We choose an arbitrary reference point O, called the **origin**, which corresponds to the real number 0. Given any convenient unit of measurement, each positive number x is represented by the point on the line a distance of x units to the right of the origin, and each negative number $-x$ is represented by the point x units to the left of the origin. Thus, every real number is represented by a point on the line, and every point P on the line corresponds to exactly one real number. The number associated with the point P is called the coordinate of P, and the line is then called a **coordinate line**, or a **real number line**, or simply a **real line**. Often we identify the point with its coordinate and think of a number as being a point on the real line.

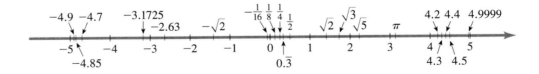

FIGURE 1 The real line

■ Order on the Real Line

The real numbers are *ordered*. We say that **a is less than b** and write $a < b$ if $b - a$ is a positive number. Geometrically, this means that a lies to the left of b on the number line. (Equivalently, we can say that **b is greater than a** and write $b > a$.) The symbol $a \leq b$ (or $b \geq a$) means that either $a < b$ or $a = b$ and is read "a is less than or equal to b." For instance, the following are true inequalities (see Figure 2):

$$-\pi < -3 \qquad \sqrt{2} < 2 \qquad 2 \leq 2 \qquad 7 < 7.4 < 7.5$$

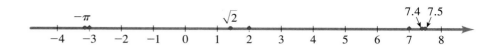

FIGURE 2

▶ **EXAMPLE 1** | Graphing Inequalities

(a) On the real line, graph all the numbers x that satisfy the inequality $x < 3$.

(b) On the real line, graph all the numbers x that satisfy the inequality $x \geq -2$.

▼ **SOLUTION**

(a) We must graph the real numbers that are smaller than 3—those that lie to the left of 3 on the real line. The graph is shown in Figure 3. Note that the number 3 is indicated with an open dot on the real line, since it does not satisfy the inequality.

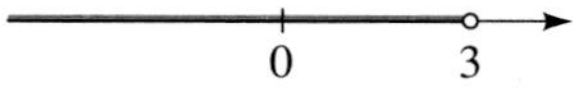

(b) We must graph the real numbers that are greater than or equal to -2: those that lie to the right of -2 on the real line, including the number -2 itself. The graph is shown in Figure 4. Note that the number -2 is indicated with a solid dot on the real line, since it satisfies the inequality.   ▲

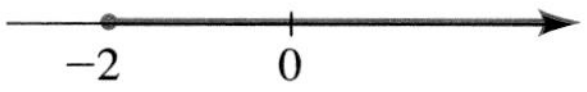

FIGURE 3

FIGURE 4

◼ Sets and Intervals

A **set** is a collection of objects, and these objects are called the **elements** of the set. If S is a set, the notation $a \in S$ means that a is an element of S, and $b \notin S$ means that b is not an element of S. For example, if Z represents the set of integers, then $-3 \in Z$ but $\pi \notin Z$.

Some sets can be described by listing their elements within braces. For instance, the set A that consists of all positive integers less than 7 can be written as

$$A = \{1, 2, 3, 4, 5, 6\}$$

We could also write A in **set-builder notation** as

$$A = \{x \mid x \text{ is an integer and } 0 < x < 7\}$$

which is read "A is the set of all x such that x is an integer and $0 < x < 7$."

If S and T are sets, then their **union** $S \cup T$ is the set that consists of all elements that are in S or T (or in both). The **intersection** of S and T is the set $S \cap T$ consisting of all elements that are in both S and T. In other words, $S \cap T$ is the common part of S and T. The **empty set**, denoted by $\varnothing$, is the set that contains no element.

▶ **EXAMPLE 2** | Union and Intersection of Sets

If $S = \{1, 2, 3, 4, 5\}$, $T = \{4, 5, 6, 7\}$, and $V = \{6, 7, 8\}$, find the sets $S \cup T$, $S \cap T$, and $S \cap V$.

▼ **SOLUTION**

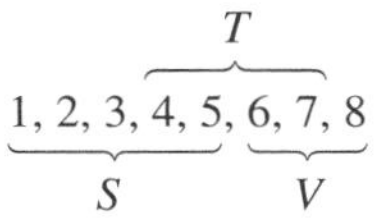

$$S \cup T = \{1, 2, 3, 4, 5, 6, 7\} \qquad \text{All elements in } S \text{ or } T$$

$$S \cap T = \{4, 5\} \qquad \text{Elements common to both } S \text{ and } T$$

$$S \cap V = \varnothing \qquad S \text{ and } V \text{ have no element in common}$$ ▲

Certain sets of real numbers, called **intervals**, occur frequently in calculus and correspond geometrically to line segments. For example, if $a < b$, then the **open interval** from

FIGURE 5 The open interval (a, b)

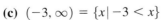

FIGURE 6 The closed interval $[a, b]$

a to b consists of all numbers between a and b and is denoted by the symbol (a, b). Using set-builder notation, we can write

$$(a, b) = \{x \mid a < x < b\}$$

Note that the endpoints, a and b, are excluded from this interval. This fact is indicated by the parentheses ( ) in the interval notation and the open circles on the graph of the interval in Figure 5.

The **closed interval** from a to b is the set

$$[a, b] = \{x \mid a \leq x \leq b\}$$

Here the endpoints of the interval are included. This is indicated by the square brackets [ ] in the interval notation and the solid circles on the graph of the interval in Figure 6. It is also possible to include only one endpoint in an interval, as shown in the table of intervals below.

We also need to consider infinite intervals, such as

$$(a, \infty) = \{x \mid a < x\}$$

This does not mean that ∞ ("infinity") is a number. The notation (a, ∞) stands for the set of all numbers that are greater than a, so the symbol ∞ simply indicates that the interval extends indefinitely far in the positive direction.

The following table lists the nine possible types of intervals. When these intervals are discussed, we will always assume that $a < b$.

Notation	Set description	Graph
(a, b)	$\{x \mid a < x < b\}$	
$[a, b]$	$\{x \mid a \leq x \leq b\}$	
$[a, b)$	$\{x \mid a \leq x < b\}$	
$(a, b]$	$\{x \mid a < x \leq b\}$	
(a, ∞)	$\{x \mid a < x\}$	
$[a, \infty)$	$\{x \mid a \leq x\}$	
$(-\infty, b)$	$\{x \mid x < b\}$	
$(-\infty, b]$	$\{x \mid x \leq b\}$	
$(-\infty, \infty)$	$\mathbb{R}$ (set of all real numbers)	

▶ **EXAMPLE 3** | Graphing Intervals

Express each interval in terms of inequalities, and then graph the interval.

(a) $[-1, 2) = \{x \mid -1 \leq x < 2\}$

(b) $[1.5, 4] = \{x \mid 1.5 \leq x \leq 4\}$

(c) $(-3, \infty) = \{x \mid -3 < x\}$

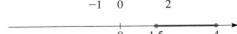

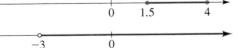

▲

▶ **EXAMPLE 4** | Finding Unions and Intersections of Intervals

Graph each set.

(a) $(1, 3) \cap [2, 7]$ **(b)** $(1, 3) \cup [2, 7]$

No Smallest or Largest Number in an Open Interval

Any interval contains infinitely many numbers—every point on the graph of an interval corresponds to a real number. In the closed interval $[0, 1]$, the smallest number is 0 and the largest is 1, but the open interval $(0, 1)$ contains no smallest or largest number. To see this, note that 0.01 is close to zero but 0.001 is closer, 0.0001 closer yet, and so on. So we can always find a number in the interval $(0, 1)$ closer to zero than any given number. Since 0 itself is not in the interval, the interval contains no smallest number. Similarly, 0.99 is close to 1, but 0.999 is closer, 0.9999 closer yet, and so on. Since 1 itself is not in the interval, the interval has no largest number.

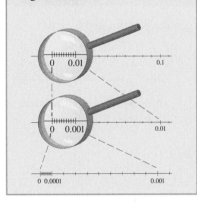

▼ **SOLUTION**

(a) The intersection of two intervals consists of the numbers that are in both intervals. Therefore,

$$(1, 3) \cap [2, 7] = \{x \mid 1 < x < 3 \text{ and } 2 \le x \le 7\}$$
$$= \{x \mid 2 \le x < 3\} = [2, 3)$$

This set is illustrated in Figure 7.

(b) The union of two intervals consists of the numbers that are in either one interval or the other (or both). Therefore,

$$(1, 3) \cup [2, 7] = \{x \mid 1 < x < 3 \text{ or } 2 \le x \le 7\}$$
$$= \{x \mid 1 < x \le 7\} = (1, 7]$$

This set is illustrated in Figure 8.

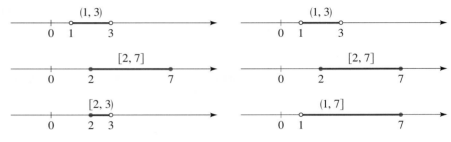

FIGURE 7 $(1, 3) \cap [2, 7]$ **FIGURE 8** $(1, 3) \cup [2, 7]$ ▲

■ Absolute Value and Distance

The **absolute value** of a number a, denoted by $|a|$, is the distance from a to 0 on the real number line (see Figure 9). Distance is always positive or zero, so we have $|a| \ge 0$ for every number a. Remembering that $-a$ is positive when a is negative, we have the following definition.

$|-3| = 3$ $|5| = 5$

FIGURE 9

DEFINITION OF ABSOLUTE VALUE

If a is a real number, then the **absolute value** of a is

$$|a| = \begin{cases} a & \text{if } a \ge 0 \\ -a & \text{if } a < 0 \end{cases}$$

▶ **EXAMPLE 5** | Evaluating Absolute Values of Numbers

(a) $|3| = 3$

(b) $|-3| = -(-3) = 3$

(c) $|0| = 0$

(d) $|\sqrt{2} - 1| = \sqrt{2} - 1$ (since $\sqrt{2} > 1 \Rightarrow \sqrt{2} - 1 > 0$)

(e) $|3 - \pi| = -(3 - \pi) = \pi - 3$ (since $\pi > 3 \Rightarrow 3 - \pi < 0$)   ▲

When working with absolute values, we use the following properties.

PROPERTIES OF ABSOLUTE VALUE

Property	Example	Description
1. $\|a\| \geq 0$	$\|-3\| = 3 \geq 0$	The absolute value of a number is always positive or zero.
2. $\|a\| = \|-a\|$	$\|5\| = \|-5\|$	A number and its negative have the same absolute value.
3. $\|ab\| = \|a\|\|b\|$	$\|-2 \cdot 5\| = \|-2\|\|5\|$	The absolute value of a product is the product of the absolute values.
4. $\left\|\dfrac{a}{b}\right\| = \dfrac{\|a\|}{\|b\|}$	$\left\|\dfrac{12}{-3}\right\| = \dfrac{\|12\|}{\|-3\|}$	The absolute value of a quotient is the quotient of the absolute values.

FIGURE 10

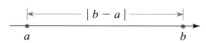

FIGURE 11 Length of a line segment is $\|b - a\|$

What is the distance on the real line between the numbers -2 and 11? From Figure 10 we see that the distance is 13. We arrive at this by finding either $\|11 - (-2)\| = 13$ or $\|(-2) - 11\| = 13$. From this observation we make the following definition (see Figure 11).

DISTANCE BETWEEN POINTS ON THE REAL LINE

If a and b are real numbers, then the **distance** between the points a and b on the real line is

$$d(a, b) = |b - a|$$

From Property 6 of negatives it follows that $|b - a| = |a - b|$. This confirms that, as we would expect, the distance from a to b is the same as the distance from b to a.

▶ **EXAMPLE 6** | Distance Between Points on the Real Line

The distance between the numbers -8 and 2 is

$$d(a, b) = |-8 - 2| = |-10| = 10$$

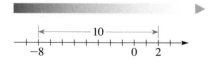

FIGURE 12

We can check this calculation geometrically, as shown in Figure 12.   ▲

P.4 | Integer Exponents

LEARNING OBJECTIVES

After completing this section, you will be able to:

▨ Use exponential notation
▨ Simplify expressions using the Laws of Exponents
▨ Write numbers in scientific notation

In this section we review the rules for working with exponent notation. We also see how exponents can be used to represent very large and very small numbers.

▨ **Exponential Notation**

A product of identical numbers is usually written in exponential notation. For example, $5 \cdot 5 \cdot 5$ is written as 5^3. In general, we have the following definition.

EXPONENTIAL NOTATION

If a is any real number and n is a positive integer, then the **nth power** of a is

$$a^n = \underbrace{a \cdot a \cdots a}_{n \text{ factors}}$$

The number a is called the **base**, and n is called the **exponent**.

▶ **EXAMPLE 1** | Exponential Notation

(a) $\left(\frac{1}{2}\right)^5 = \left(\frac{1}{2}\right)\left(\frac{1}{2}\right)\left(\frac{1}{2}\right)\left(\frac{1}{2}\right)\left(\frac{1}{2}\right) = \frac{1}{32}$

(b) $(-3)^4 = (-3) \cdot (-3) \cdot (-3) \cdot (-3) = 81$

(c) $-3^4 = -(3 \cdot 3 \cdot 3 \cdot 3) = -81$ ▲

⊘ Note the distinction between $(-3)^4$ and -3^4. In $(-3)^4$ the exponent applies to -3, but in -3^4 the exponent applies only to 3.

We can state several useful rules for working with exponential notation. To discover the rule for multiplication, we multiply 5^4 by 5^2:

$$5^4 \cdot 5^2 = \underbrace{(5 \cdot 5 \cdot 5 \cdot 5)}_{4 \text{ factors}}\underbrace{(5 \cdot 5)}_{2 \text{ factors}} = \underbrace{5 \cdot 5 \cdot 5 \cdot 5 \cdot 5 \cdot 5}_{6 \text{ factors}} = 5^6 = 5^{4+2}$$

It appears that *to multiply two powers of the same base, we add their exponents*. In general, for any real number a and any positive integers m and n we have

$$a^m a^n = \underbrace{(a \cdot a \cdots a)}_{m \text{ factors}}\underbrace{(a \cdot a \cdots a)}_{n \text{ factors}} = \underbrace{a \cdot a \cdot a \cdots a}_{m + n \text{ factors}} = a^{m+n}$$

Thus, $a^m a^n = a^{m+n}$.

We would like this rule to be true even when m and n are 0 or negative integers. For instance, we must have

$$2^0 \cdot 2^3 = 2^{0+3} = 2^3$$

But this can happen only if $2^0 = 1$. Likewise, we want to have

$$5^4 \cdot 5^{-4} = 5^{4+(-4)} = 5^{4-4} = 5^0 = 1$$

and this will be true if $5^{-4} = 1/5^4$. These observations lead to the following definition.

ZERO AND NEGATIVE EXPONENTS

If $a \neq 0$ is any real number and n is a positive integer, then

$$a^0 = 1 \qquad \text{and} \qquad a^{-n} = \frac{1}{a^n}$$

▶ **EXAMPLE 2** | Zero and Negative Exponents

(a) $\left(\frac{4}{7}\right)^0 = 1$

(b) $x^{-1} = \frac{1}{x^1} = \frac{1}{x}$

(c) $(-2)^{-3} = \frac{1}{(-2)^3} = \frac{1}{-8} = -\frac{1}{8}$

■ Rules for Working with Exponents

Familiarity with the following rules is essential for our work with exponents and bases. In the table the bases a and b are real numbers, and the exponents m and n are integers.

LAWS OF EXPONENTS

Law	Example	Description
1. $a^m a^n = a^{m+n}$	$3^2 \cdot 3^5 = 3^{2+5} = 3^7$	To multiply two powers of the same number, add the exponents.
2. $\dfrac{a^m}{a^n} = a^{m-n}$	$\dfrac{3^5}{3^2} = 3^{5-2} = 3^3$	To divide two powers of the same number, subtract the exponents.
3. $(a^m)^n = a^{mn}$	$(3^2)^5 = 3^{2\cdot5} = 3^{10}$	To raise a power to a new power, multiply the exponents.
4. $(ab)^n = a^n b^n$	$(3\cdot4)^2 = 3^2 \cdot 4^2$	To raise a product to a power, raise each factor to the power.
5. $\left(\dfrac{a}{b}\right)^n = \dfrac{a^n}{b^n}$	$\left(\dfrac{3}{4}\right)^2 = \dfrac{3^2}{4^2}$	To raise a quotient to a power, raise both numerator and denominator to the power.

▼ **PROOF OF LAW 3**   If m and n are positive integers, we have

$$(a^m)^n = \underbrace{(a \cdot a \cdots \cdots a)}_{m \text{ factors}}{}^n$$

$$= \underbrace{\underbrace{(a \cdot a \cdots \cdots a)}_{m \text{ factors}}\underbrace{(a \cdot a \cdots \cdots a)}_{m \text{ factors}} \cdots \underbrace{(a \cdot a \cdots \cdots a)}_{m \text{ factors}}}_{n \text{ groups of factors}}$$

$$= \underbrace{a \cdot a \cdots \cdots a}_{mn \text{ factors}} = a^{mn}$$

The cases for which $m \le 0$ or $n \le 0$ can be proved by using the definition of negative exponents.   ▲

▼ **PROOF OF LAW 4**   If n is a positive integer, we have

$$(ab)^n = \underbrace{(ab)(ab) \cdots (ab)}_{n \text{ factors}} = \underbrace{(a \cdot a \cdots \cdots a)}_{n \text{ factors}} \cdot \underbrace{(b \cdot b \cdots \cdots b)}_{n \text{ factors}} = a^n b^n$$

Here we have used the Commutative and Associative Properties repeatedly. If $n \le 0$, Law 4 can be proved by using the definition of negative exponents.   ▲

▶ **EXAMPLE 3** | Using Laws of Exponents

(a) $x^4 x^7 = x^{4+7} = x^{11}$ Law 1: $a^m a^n = a^{m+n}$

(b) $y^4 y^{-7} = y^{4-7} = y^{-3} = \dfrac{1}{y^3}$ Law 1: $a^m a^n = a^{m+n}$

(c) $\dfrac{c^9}{c^5} = c^{9-5} = c^4$ Law 2: $a^m/a^n = a^{m-n}$

(d) $(b^4)^5 = b^{4\cdot5} = b^{20}$ Law 3: $(a^m)^n = a^{mn}$

(e) $(3x)^3 = 3^3 x^3 = 27x^3$ Law 4: $(ab)^n = a^n b^n$

(f) $\left(\dfrac{x}{2}\right)^5 = \dfrac{x^5}{2^5} = \dfrac{x^5}{32}$ Law 5: $(a/b)^n = a^n/b^n$ ▲

▶ **EXAMPLE 4** | Simplifying Expressions with Exponents

Simplify: **(a)** $(2a^3b^2)(3ab^4)^3$ **(b)** $\left(\dfrac{x}{y}\right)^3\left(\dfrac{y^2x}{z}\right)^4$

▼ **SOLUTION**

(a)
$$
\begin{aligned}
(2a^3b^2)(3ab^4)^3 &= (2a^3b^2)[3^3a^3(b^4)^3] && \text{Law 4: } (ab)^n = a^n b^n \\
&= (2a^3b^2)(27a^3b^{12}) && \text{Law 3: } (a^m)^n = a^{mn} \\
&= (2)(27)a^3a^3b^2b^{12} && \text{Group factors with the same base} \\
&= 54\,a^6b^{14} && \text{Law 1: } a^m a^n = a^{m+n}
\end{aligned}
$$

(b)
$$
\begin{aligned}
\left(\frac{x}{y}\right)^3\left(\frac{y^2x}{z}\right)^4 &= \frac{x^3}{y^3}\frac{(y^2)^4x^4}{z^4} && \text{Laws 5 and 4} \\
&= \frac{x^3}{y^3}\frac{y^8x^4}{z^4} && \text{Law 3} \\
&= (x^3x^4)\left(\frac{y^8}{y^3}\right)\frac{1}{z^4} && \text{Group factors with the same base} \\
&= \frac{x^7y^5}{z^4} && \text{Laws 1 and 2}
\end{aligned}
$$
▲

When simplifying an expression, you will find that many different methods will lead to the same result; you should feel free to use any of the rules of exponents to arrive at your own method. We now give two additional laws that are useful in simplifying expressions with negative exponents.

LAWS OF EXPONENTS

Law	Example	Description
6. $\left(\dfrac{a}{b}\right)^{-n} = \left(\dfrac{b}{a}\right)^n$	$\left(\dfrac{3}{4}\right)^{-2} = \left(\dfrac{4}{3}\right)^2$	To raise a fraction to a negative power, invert the fraction and change the sign of the exponent.
7. $\dfrac{a^{-n}}{b^{-m}} = \dfrac{b^m}{a^n}$	$\dfrac{3^{-2}}{4^{-5}} = \dfrac{4^5}{3^2}$	To move a number raised to a power from numerator to denominator or from denominator to numerator, change the sign of the exponent.

▼ **PROOF OF LAW 7**   Using the definition of negative exponents and then Property 2 of fractions (page 9), we have

$$
\frac{a^{-n}}{b^{-m}} = \frac{1/a^n}{1/b^m} = \frac{1}{a^n}\cdot\frac{b^m}{1} = \frac{b^m}{a^n}
$$
▲

▶ **EXAMPLE 5** | Simplifying Expressions with Negative Exponents

Eliminate negative exponents and simplify each expression.

(a) $\dfrac{6st^{-4}}{2s^{-2}t^2}$ **(b)** $\left(\dfrac{y}{3z^3}\right)^{-2}$

▼ **SOLUTION**

(a) We use Law 7, which allows us to move a number raised to a power from the numerator to the denominator (or vice versa) by changing the sign of the exponent.

> t^{-4} moves to denominator and becomes t^4.

$$\frac{6st^{-4}}{2s^{-2}t^2} = \frac{6s\,s^2}{2t^2\,t^4} \qquad \text{Law 7}$$

> s^{-2} moves to numerator and becomes s^2.

$$= \frac{3s^3}{t^6} \qquad \text{Law 1}$$

(b) We use Law 6, which allows us to change the sign of the exponent of a fraction by inverting the fraction.

$$\left(\frac{y}{3z^3}\right)^{-2} = \left(\frac{3z^3}{y}\right)^2 \qquad \text{Law 6}$$

$$= \frac{9z^6}{y^2} \qquad \text{Laws 5 and 4}$$

▲

■ Scientific Notation

Exponential notation is used by scientists as a compact way of writing very large numbers and very small numbers. For example, the nearest star beyond the sun, Proxima Centauri, is approximately 40,000,000,000,000 km away. The mass of a hydrogen atom is about 0.00000000000000000000000166 g. Such numbers are difficult to read and to write, so scientists usually express them in *scientific notation*.

SCIENTIFIC NOTATION

A positive number x is said to be written in **scientific notation** if it is expressed as follows:

$$x = a \times 10^n \qquad \text{where } 1 \le a < 10 \text{ and } n \text{ is an integer}$$

For instance, when we state that the distance to the star Proxima Centauri is 4×10^{13} km, the positive exponent 13 indicates that the decimal point should be moved 13 places to the *right*:

$$4 \times 10^{13} = 40{,}000{,}000{,}000{,}000$$

Move decimal point 13 places to the right.

When we state that the mass of a hydrogen atom is 1.66×10^{-24} g, the exponent -24 indicates that the decimal point should be moved 24 places to the *left*:

$$1.66 \times 10^{-24} = 0.00000000000000000000000166$$

Move decimal point 24 places to the left.

▶ **EXAMPLE 6** | Changing from Decimal to Scientific Notation

Write each number in scientific notation.

(a) 56,920   (b) 0.000093

▼ **SOLUTION**

(a) $\underbrace{56{,}920}_{4 \text{ places}} = 5.692 \times 10^4$ (b) $\underbrace{0.000093}_{5 \text{ places}} = 9.3 \times 10^{-5}$

▶ **EXAMPLE 7** | Changing from Decimal to Scientific Notation

Write each number in decimal notation.

(a) 6.97×10^9 (b) 4.6271×10^{-6}

▼ **SOLUTION**

(a) $6.97 \times 10^9 = \underbrace{6{,}970{,}000{,}000}_{9 \text{ places}}$ *Move decimal 9 places to the right*

(b) $4.6271 \times 10^{-6} = \underbrace{0.0000046271}_{6 \text{ places}}$ *Move decimal 6 places to the left*

To use scientific notation on a calculator, use a key labeled $\boxed{\text{EE}}$ or $\boxed{\text{EXP}}$ or $\boxed{\text{EEX}}$ to enter the exponent. For example, to enter the number 3.629×10^{15} on a TI-83 calculator, we enter

3.629 $\boxed{\text{2ND}}$ $\boxed{\text{EE}}$ 15

and the display reads

$3.629\text{E}15$

Scientific notation is often used on a calculator to display a very large or very small number. For instance, if we use a calculator to square the number 1,111,111, the display panel may show (depending on the calculator model) the approximation

$\boxed{\texttt{1.234568   12}}$ or $\boxed{\texttt{1.234568   E12}}$

Here the final digits indicate the power of 10, and we interpret the result as

$$1.234568 \times 10^{12}$$

▶ **EXAMPLE 8** | Calculating with Scientific Notation

If $a \approx 0.00046$, $b \approx 1.697 \times 10^{22}$, and $c \approx 2.91 \times 10^{-18}$, use a calculator to approximate the quotient ab/c.

▼ **SOLUTION**   We could enter the data using scientific notation, or we could use laws of exponents as follows:

$$\frac{ab}{c} \approx \frac{(4.6 \times 10^{-4})(1.697 \times 10^{22})}{2.91 \times 10^{-18}}$$

$$= \frac{(4.6)(1.697)}{2.91} \times 10^{-4+22+18}$$

$$\approx 2.7 \times 10^{36}$$

We state the answer correct to two significant figures because the least accurate of the given numbers is stated to two significant figures.

P.5 | Rational Exponents and Radicals

LEARNING OBJECTIVES:

After completing this section, you will be able to:

- Simplify expressions involving radicals
- Simplify expressions involving rational exponents
- Express radicals using rational exponents
- Rationalize a denominator

In this section we learn to work with expressions that contain radicals or rational exponents.

■ Radicals

We know what 2^n means whenever n is an integer. To give meaning to a power, such as $2^{4/5}$, whose exponent is a rational number, we need to discuss radicals.

The symbol $\sqrt{}$ means "the positive square root of." Thus,

$$\sqrt{a} = b \quad \text{means} \quad b^2 = a \quad \text{and} \quad b \geq 0$$

It is true that the number 9 has two square roots, 3 and -3, but the notation $\sqrt{9}$ is reserved for the *positive* square root of 9 (sometimes called the *principal square root* of 9). If we want the negative root, we must write $-\sqrt{9}$, which is -3.

Since $a = b^2 \geq 0$, the symbol $\sqrt{a}$ makes sense only when $a \geq 0$. For instance,

$$\sqrt{9} = 3 \quad \text{because} \quad 3^2 = 9 \quad \text{and} \quad 3 \geq 0$$

Square roots are special cases of nth roots. The nth root of x is the number that, when raised to the nth power, gives x.

> ### DEFINITION OF nth ROOT
>
> If n is any positive integer, then the **principal nth root** of a is defined as follows:
> $$\sqrt[n]{a} = b \quad \text{means} \quad b^n = a$$
> If n is even, we must have $a \geq 0$ and $b \geq 0$.

Thus

$$\sqrt[4]{81} = 3 \quad \text{because} \quad 3^4 = 81 \quad \text{and} \quad 3 \geq 0$$
$$\sqrt[3]{-8} = -2 \quad \text{because} \quad (-2)^3 = -8$$

But $\sqrt{-8}$, $\sqrt[4]{-8}$, and $\sqrt[6]{-8}$ are not defined. (For instance, $\sqrt{-8}$ is not defined because the square of every real number is nonnegative.)

Notice that

$$\sqrt{4^2} = \sqrt{16} = 4 \quad \text{but} \quad \sqrt{(-4)^2} = \sqrt{16} = 4 = |-4|$$

So the equation $\sqrt{a^2} = a$ is not always true; it is true only when $a \geq 0$. However, we can always write $\sqrt{a^2} = |a|$. This last equation is true not only for square roots, but for any even root. This and other rules used in working with nth roots are listed in the following box. In each property we assume that all the given roots exist.

> ### PROPERTIES OF nth ROOTS
>
> | Property | Example |
> |---|---|
> | 1. $\sqrt[n]{ab} = \sqrt[n]{a}\sqrt[n]{b}$ | $\sqrt[3]{-8 \cdot 27} = \sqrt[3]{-8}\sqrt[3]{27} = (-2)(3) = -6$ |
> | 2. $\sqrt[n]{\dfrac{a}{b}} = \dfrac{\sqrt[n]{a}}{\sqrt[n]{b}}$ | $\sqrt[4]{\dfrac{16}{81}} = \dfrac{\sqrt[4]{16}}{\sqrt[4]{81}} = \dfrac{2}{3}$ |
> | 3. $\sqrt[m]{\sqrt[n]{a}} = \sqrt[mn]{a}$ | $\sqrt{\sqrt[3]{729}} = \sqrt[6]{729} = 3$ |
> | 4. $\sqrt[n]{a^n} = a \quad$ if n is odd | $\sqrt[3]{(-5)^3} = -5, \sqrt[5]{2^5} = 2$ |
> | 5. $\sqrt[n]{a^n} = |a| \quad$ if n is even | $\sqrt[4]{(-3)^4} = |-3| = 3$ |

▶ **EXAMPLE 1** | Simplifying Expressions Involving nth Roots

(a) $\sqrt[3]{x^4} = \sqrt[3]{x^3 x}$ *Factor out the largest cube*

$\qquad\quad = \sqrt[3]{x^3}\sqrt[3]{x}$ *Property 1: $\sqrt[3]{ab} = \sqrt[3]{a}\sqrt[3]{b}$*

$\qquad\quad = x\sqrt[3]{x}$ *Property 4: $\sqrt[3]{a^3} = a$*

(b) $\sqrt[4]{81x^8y^4} = \sqrt[4]{81}\,\sqrt[4]{x^8}\,\sqrt[4]{y^4}$ Property 1: $\sqrt[4]{abc} = \sqrt[4]{a}\,\sqrt[4]{b}\,\sqrt[4]{c}$

$= 3\sqrt[4]{(x^2)^4}\,|y|$ Property 5: $\sqrt[4]{a^4} = |a|$

$= 3x^2\,|y|$ Property 5: $\sqrt[4]{a^4} = |a|,\ |x^2| = x^2$ ▲

It is frequently useful to combine like radicals in an expression such as $2\sqrt{3} + 5\sqrt{3}$. This can be done by using the Distributive Property. Thus,

$$2\sqrt{3} + 5\sqrt{3} = (2 + 5)\sqrt{3} = 7\sqrt{3}$$

The next example further illustrates this process.

▶ **EXAMPLE 2** | Combining Radicals

⊘ Avoid making the following error:

$\sqrt{a + b} \ \cancel{\ }\ \sqrt{a} + \sqrt{b}$

For instance, if we let $a = 9$ and $b = 16$, then we see the error:

$\sqrt{9 + 16} \overset{?}{=} \sqrt{9} + \sqrt{16}$

$\sqrt{25} \overset{?}{=} 3 + 4$

$5 \overset{?}{=} 7$ Wrong!

$\sqrt{32} + \sqrt{200} = \sqrt{16 \cdot 2} + \sqrt{100 \cdot 2}$ Factor out the largest squares

$= \sqrt{16}\sqrt{2} + \sqrt{100}\sqrt{2}$ Property 1

$= 4\sqrt{2} + 10\sqrt{2} = 14\sqrt{2}$ Distributive Property ▲

■ **Rational Exponents**

To define what is meant by a *rational exponent* or, equivalently, a *fractional exponent* such as $a^{1/3}$, we need to use radicals. To give meaning to the symbol $a^{1/n}$ in a way that is consistent with the Laws of Exponents, we would have to have

$$(a^{1/n})^n = a^{(1/n)n} = a^1 = a$$

So by the definition of nth root,

$$\boxed{a^{1/n} = \sqrt[n]{a}}$$

In general, we define rational exponents as follows.

DEFINITION OF RATIONAL EXPONENTS

For any rational exponent m/n in lowest terms, where m and n are integers and $n > 0$, we define

$$a^{m/n} = \left(\sqrt[n]{a}\right)^m \qquad \text{or, equivalently,} \qquad a^{m/n} = \sqrt[n]{a^m}$$

If n is even, then we require that $a \geq 0$.

With this definition it can be proved that *the Laws of Exponents also hold for rational exponents*.

▶ **EXAMPLE 3** | Using the Definition of Rational Exponents

(a) $4^{1/2} = \sqrt{4} = 2$

(b) $8^{2/3} = \left(\sqrt[3]{8}\right)^2 = 2^2 = 4$ Alternative solution: $8^{2/3} = \sqrt[3]{8^2} = \sqrt[3]{64} = 4$

(c) $(125)^{-1/3} = \dfrac{1}{125^{1/3}} = \dfrac{1}{\sqrt[3]{125}} = \dfrac{1}{5}$

(d) $\dfrac{1}{\sqrt[3]{x^4}} = \dfrac{1}{x^{4/3}} = x^{-4/3}$

▲

▶ **EXAMPLE 4** | Using the Laws of Exponents with Rational Exponents

(a) $a^{1/3}a^{7/3} = a^{8/3}$ Law 1: $a^m a^n = a^{m+n}$

(b) $\dfrac{a^{2/5}a^{7/5}}{a^{3/5}} = a^{2/5+7/5-3/5} = a^{6/5}$ Law 1, Law 2: $\dfrac{a^m}{a^n} = a^{m-n}$

(c) $(2a^3b^4)^{3/2} = 2^{3/2}(a^3)^{3/2}(b^4)^{3/2}$ Law 4: $(abc)^n = a^n b^n c^n$

$\qquad = (\sqrt{2})^3 a^{3(3/2)}b^{4(3/2)}$ Law 3: $(a^m)^n = a^{mn}$

$\qquad = 2\sqrt{2}\,a^{9/2}b^6$

(d) $\left(\dfrac{2x^{3/4}}{y^{1/3}}\right)^3 \left(\dfrac{y^4}{x^{-1/2}}\right) = \dfrac{2^3(x^{3/4})^3}{(y^{1/3})^3}\cdot(y^4 x^{1/2})$ Laws 5, 4, and 7

$\qquad = \dfrac{8x^{9/4}}{y}\cdot y^4 x^{1/2}$ Law 3

$\qquad = 8x^{11/4}y^3$ Laws 1 and 2

▲

▶ **EXAMPLE 5** | Simplifying by Writing Radicals as Rational Exponents

(a) $(2\sqrt{x})(3\sqrt[3]{x}) = (2x^{1/2})(3x^{1/3})$ Definition of rational exponents

$\qquad = 6x^{1/2+1/3} = 6x^{5/6}$ Law 1

(b) $\sqrt{x\sqrt{x}} = (xx^{1/2})^{1/2}$ Definition of rational exponents

$\qquad = (x^{3/2})^{1/2}$ Law 1

$\qquad = x^{3/4}$ Law 3

▲

■ Rationalizing the Denominator

It is often useful to eliminate the radical in a denominator by multiplying both numerator and denominator by an appropriate expression. This procedure is called **rationalizing the denominator**. If the denominator is of the form $\sqrt{a}$, we multiply numerator and denominator by $\sqrt{a}$. In doing this, we multiply the given quantity by 1, so we do not change its value. For instance,

$$\frac{1}{\sqrt{a}} = \frac{1}{\sqrt{a}}\cdot 1 = \frac{1}{\sqrt{a}}\cdot\frac{\sqrt{a}}{\sqrt{a}} = \frac{\sqrt{a}}{a}$$

Note that the denominator in the last fraction contains no radical. In general, if the denominator is of the form $\sqrt[n]{a^m}$ with $m < n$, then multiplying the numerator and denominator by $\sqrt[n]{a^{n-m}}$ will rationalize the denominator, because (for $a > 0$)

$$\sqrt[n]{a^m}\sqrt[n]{a^{n-m}} = \sqrt[n]{a^{m+n-m}} = \sqrt[n]{a^n} = a$$

▶ **EXAMPLE 6** | Rationalizing Denominators

Rationalize the denominator in each fraction.

(a) $\dfrac{2}{\sqrt{3}}$

(b) $\dfrac{1}{\sqrt[3]{5}}$

(c) $\dfrac{1}{\sqrt[5]{5x^2}}$

▼ **SOLUTION**   This equals 1.

(a) $\dfrac{2}{\sqrt{3}} = \dfrac{2}{\sqrt{3}} \cdot \dfrac{\sqrt{3}}{\sqrt{3}}$ Multiply by $\dfrac{\sqrt{3}}{\sqrt{3}}$

$\qquad = \dfrac{2\sqrt{3}}{3}$ $\sqrt{3} \cdot \sqrt{3} = 3$

(b) $\dfrac{1}{\sqrt[3]{5}} = \dfrac{1}{\sqrt[3]{5}} \cdot \dfrac{\sqrt[3]{5^2}}{\sqrt[3]{5^2}}$ Multiply by $\dfrac{\sqrt[3]{5^2}}{\sqrt[3]{5^2}}$

$\qquad = \dfrac{\sqrt[3]{25}}{5}$ $\sqrt[3]{5} \cdot \sqrt[3]{5^2} = \sqrt[3]{5^3} = 5$

(c) $\dfrac{1}{\sqrt[5]{x^2}} = \dfrac{1}{\sqrt[5]{x^2}} \cdot \dfrac{\sqrt[5]{x^3}}{\sqrt[5]{x^3}}$ Multiply by $\dfrac{\sqrt[5]{x^3}}{\sqrt[5]{x^3}}$

$\qquad = \dfrac{\sqrt[5]{x^3}}{x}$ $\sqrt[5]{x^2} \cdot \sqrt[5]{x^3} = \sqrt[5]{x^5} = x$ ▲

P.6 | Algebraic Expressions

LEARNING OBJECTIVES
After completing this section, you will be able to:

▨ Add and subtract polynomials
▨ Multiply algebraic expressions
▨ Use the Special Product Formulas

A **variable** is a letter that can represent any number from a given set of numbers. If we start with variables such as x, y, and z and some real numbers and we combine them using addition, subtraction, multiplication, division, powers, and roots, we obtain an **algebraic expression.** Here are some examples:

$$2x^2 - 3x + 4 \qquad \sqrt{x} + 10 \qquad \dfrac{y - 2z}{y^2 + 4}$$

A **monomial** is an expression of the form ax^k, where a is a real number and k is a nonnegative integer. A **binomial** is a sum of two monomials, and a **trinomial** is a sum of three monomials. In general, a sum of monomials is called a *polynomial.* For example, the first expression listed above is a polynomial, but the other two are not.

> **POLYNOMIALS**
>
> A **polynominal** in the variable x is an expression of the form
> $$a_n x^n + a_{n-1}x^{n-1} + \cdots + a_1 x + a_0$$
> where $a_0, a_1, \ldots, a_n$ are real numbers and n is a nonnegative integer. If $a_n \neq 0$, then the polynomial has **degree n**. The monomials $a_k x^k$ that make up the polynomial are called the **terms** of the polynomial.

Note that the degree of a polynomial is the highest power of the variable that appears in the polynomial.

Polynomial	Type	Terms	Degree
$2x^2 - 3x + 4$	trinomial	$2x^2, -3x, 4$	2
$x^8 + 5x$	binomial	$x^8, 5x$	8
$8 - x + x^2 - \frac{1}{2}x^3$	four terms	$-\frac{1}{2}x^3, x^2, -x, 8$	3
$5x + 1$	binomial	$5x, 1$	1
$9x^5$	monomial	$9x^5$	5
6	monomial	6	0

■ Adding and Subtracting Polynomials

We **add** and **subtract** polynomials using the properties of real numbers that were discussed in Section P.2. The idea is to combine **like terms** (that is, terms with the same variables raised to the same powers) using the Distributive Property. For instance,

Distributive Property
$ac + bc = (a + b)c$

$$5x^7 + 3x^7 = (5 + 3)x^7 = 8x^7$$

In subtracting polynomials, we have to remember that if a minus sign precedes an expression in parentheses, then the sign of every term within the parentheses is changed when we remove the parentheses:

$$-(b + c) = -b - c$$

[This is simply a case of the Distributive Property, $a(b + c) = ab + ac$, with $a = -1$.]

▶ **EXAMPLE 1** | Adding and Subtracting Polynomials

(a) Find the sum $(x^3 - 6x^2 + 2x + 4) + (x^3 + 5x^2 - 7x)$.

(b) Find the difference $(x^3 - 6x^2 + 2x + 4) - (x^3 + 5x^2 - 7x)$.

▼ **SOLUTION**

(a) $(x^3 - 6x^2 + 2x + 4) + (x^3 + 5x^2 - 7x)$
$= (x^3 + x^3) + (-6x^2 + 5x^2) + (2x - 7x) + 4$ Group like terms
$= 2x^3 - x^2 - 5x + 4$ Combine like terms

(b) $(x^3 - 6x^2 + 2x + 4) - (x^3 + 5x^2 - 7x)$
$= x^3 - 6x^2 + 2x + 4 - x^3 - 5x^2 + 7x$ Distributive Property
$= (x^3 - x^3) + (-6x^2 - 5x^2) + (2x + 7x) + 4$ Group like terms
$= -11x^2 + 9x + 4$ Combine like terms ▲

■ Multiplying Algebraic Expressions

To find the **product** of polynomials or other algebraic expressions, we need to use the Distributive Property repeatedly. In particular, using it three times on the product of two binomials, we get

$$(a + b)(c + d) = a(c + d) + b(c + d) = ac + ad + bc + bd$$

This says that we multiply the two factors by multiplying each term in one factor by each term in the other factor and adding these products. Schematically, we have

The acronym **FOIL** helps us to remember that the product of two binomials is the sum of the products of the **F**irst terms, the **O**uter terms, the **I**nner terms, and the **L**ast terms.

$$(a + b)(c + d) = ac + ad + bc + bd$$
$$\quad\quad\quad\quad\quad \uparrow \quad\ \uparrow \quad\ \uparrow \quad\ \uparrow$$
$$\quad\quad\quad\quad\quad F \quad\ O \quad\ I \quad\ L$$

In general, we can multiply two algebraic expressions by using the Distributive Property and the Laws of Exponents.

▶ **EXAMPLE 2** | Multiplying Binomials Using FOIL

$$(2x + 1)(3x - 5) = 6x^2 - 10x + 3x - 5 \quad \text{Distributive Property}$$
$$\quad\quad\quad\quad\quad\quad\ \uparrow \quad\quad \uparrow \quad\ \uparrow \quad\ \uparrow$$
$$\quad\quad\quad\quad\quad\quad\ F \quad\quad O \quad\ I \quad\ L$$
$$\quad\quad\quad\quad\ = 6x^2 - 7x - 5 \quad\quad\quad\quad \text{Combine like terms} \quad ▲$$

When we multiply trinomials or other polynomials with more terms, we use the Distributive Property. It is also helpful to arrange our work in table form. The next example illustrates both methods.

▶ **EXAMPLE 3** | Multiplying Polynomials

Find the product: $(2x + 3)(x^2 - 5x + 4)$

▼ **SOLUTION 1:** Using the Distributive Property

$$(2x + 3)(x^2 - 5x + 4) = 2x(x^2 - 5x + 4) + 3(x^2 - 5x + 4) \quad\quad \text{Distributive Property}$$
$$= (2x \cdot x^2 - 2x \cdot 5x + 2x \cdot 4) + (3 \cdot x^2 - 3 \cdot 5x + 3 \cdot 4) \quad\quad \text{Distributive Property}$$
$$= (2x^3 - 10x^2 + 8x) + (3x^2 - 15x + 12) \quad\quad \text{Laws of Exponents}$$
$$= 2x^3 - 7x^2 - 7x + 12 \quad\quad \text{Combine like terms}$$

▼ **SOLUTION 2:** Using Table Form

$x^2 - \ 5x + 4$	First factor
$2x + 3$	Second factor
$3x^2 - 15x + 12$	Multiply $x^2 - 5x + 4$ by 3
$2x^3 - 10x^2 + \ 8x$	Multiply $x^2 - 5x + 4$ by 2x
$2x^3 - \ 7x^2 - \ 7x + 12$	Add like terms

▲

■ Special Product Formulas

Certain types of products occur so frequently that you should memorize them. You can verify the following formulas by performing the multiplications.

> **SPECIAL PRODUCT FORMULAS**
>
> If A and B are any real numbers or algebraic expressions, then
>
> **1.** $(A + B)(A - B) = A^2 - B^2$ *Sum and difference of same terms*
> **2.** $(A + B)^2 = A^2 + 2AB + B^2$ *Square of a sum*
> **3.** $(A - B)^2 = A^2 - 2AB + B^2$ *Square of a difference*
> **4.** $(A + B)^3 = A^3 + 3A^2B + 3AB^2 + B^3$ *Cube of a sum*
> **5.** $(A - B)^3 = A^3 - 3A^2B + 3AB^2 - B^3$ *Cube of a difference*

The key idea in using these formulas (or any other formula in algebra) is the **Principle of Substitution**: We may substitute any algebraic expression for any letter in a formula. For example, to find $(x^2 + y^3)^2$, we use Product Formula 2, substituting x^2 for A and y^3 for B, to get

$$(x^2 + y^3)^2 = (x^2)^2 + 2(x^2)(y^3) + (y^3)^2$$

$$(A + B)^2 \ = \ A^2 \ + \ 2AB \ + \ B^2$$

▶ **EXAMPLE 4** | Using the Special Product Formulas

Use a Special Product Formula to find each product.

(a) $(3x + 5)^2$

(b) $(x^2 - 2)^3$

▼ **SOLUTION**

(a) Substituting $A = 3x$ and $B = 5$ in Product Formula 2, we get

$$(3x + 5)^2 = (3x)^2 + 2(3x)(5) + 5^2 = 9x^2 + 30x + 25$$

(b) Substituting $A = x^2$ and $B = 2$ in Product Formula 5, we get

$$(x^2 - 2)^3 = (x^2)^3 - 3(x^2)^2(2) + 3(x^2)(2)^2 - 2^3$$
$$= x^6 - 6x^4 + 12x^2 - 8 \qquad ▲$$

▶ **EXAMPLE 5** | Using the Special Product Formulas

Find each product.

(a) $(2x - \sqrt{y})(2x + \sqrt{y})$

(b) $(x + y - 1)(x + y + 1)$

▼ **SOLUTION**

(a) Substituting $A = 2x$ and $B = \sqrt{y}$ in Product Formula 1, we get

$$(2x - \sqrt{y})(2x + \sqrt{y}) = (2x)^2 - (\sqrt{y})^2 = 4x^2 - y$$

(b) If we group $x + y$ together and think of this as one algebraic expression, we can use Product Formula 1 with $A = x + y$ and $B = 1$.

$$(x + y - 1)(x + y + 1) = [(x + y) - 1][(x + y) + 1]$$
$$= (x + y)^2 - 1^2 \qquad \text{Product Formula 1}$$
$$= x^2 + 2xy + y^2 - 1 \qquad \text{Product Formula 2} \quad ▲$$

P.7 | Factoring

LEARNING OBJECTIVES
After completing this section, you will be able to:

- Factor out common factors
- Factor trinomials by trial and error
- Use the Special Factoring Formulas
- Factor algebraic expressions completely
- Factor by grouping terms

We use the Distributive Property to expand algebraic expressions. We sometimes need to reverse this process (again using the Distributive Property) by **factoring** an expression as a product of simpler ones. For example, we can write

$$x^2 - 4 = (x - 2)(x + 2)$$

We say that $x - 2$ and $x + 2$ are **factors** of $x^2 - 4$.

■ **Common Factors**

The easiest type of factoring occurs when the terms have a common factor.

▶ **EXAMPLE 1** | Factoring Out Common Factors

Factor each expression.

(a) $3x^2 - 6x$

(b) $8x^4y^2 + 6x^3y^3 - 2xy^4$

▼ **SOLUTION**

(a) The greatest common factor of the terms $3x^2$ and $-6x$ is $3x$, so we have

$$3x^2 - 6x = 3x(x - 2)$$

(b) We note that

8, 6, and -2 have the greatest common factor 2

x^4, x^3, and x have the greatest common factor x

y^2, y^3, and y^4 have the greatest common factor y^2

So the greatest common factor of the three terms in the polynomial is $2xy^2$, and we have

$$8x^4y^2 + 6x^3y^3 - 2xy^4 = (2xy^2)(4x^3) + (2xy^2)(3x^2y) + (2xy^2)(-y^2)$$
$$= 2xy^2(4x^3 + 3x^2y - y^2)$$ ▲

Check Your Answer

(a) Multiplying gives

$3x(x - 2) = 3x^2 - 6x$ ✔

(b) Multiplying gives

$2xy^2(4x^3 + 3x^2y - y^2) =$

$\qquad 8x^4y^2 + 6x^3y^3 - 2xy^4$ ✔

▶ **EXAMPLE 2** | Factoring Out a Common Factor

Factor: $(2x + 4)(x - 3) - 5(x - 3)$

▼ **SOLUTION** The two terms have the common factor $x - 3$.

$$(2x + 4)(x - 3) - 5(x - 3) = [(2x + 4) - 5](x - 3) \quad \text{Distributive Property}$$
$$= (2x - 1)(x - 3) \quad \text{Simplify} \quad \blacktriangle$$

■ Factoring Trinomials

To factor a trinomial of the form $x^2 + bx + c$, we note that

$$(x + r)(x + s) = x^2 + (r + s)x + rs$$

so we need to choose numbers r and s so that $r + s = b$ and $rs = c$.

▶ **EXAMPLE 3** | Factoring $x^2 + bx + c$ by Trial and Error

Factor: $x^2 + 7x + 12$

▼ **SOLUTION** We need to find two integers whose product is 12 and whose sum is 7. By trial and error we find that the two integers are 3 and 4. Thus, the factorization is

$$x^2 + 7x + 12 = (x + 3)(x + 4)$$

$$\underset{\text{factors of 12}}{\uparrow\qquad\uparrow}$$

$\blacktriangle$

Check Your Answer

Multiplying gives

$(x + 3)(x + 4) = x^2 + 7x + 12$ ✔

$$\overset{\text{factors of } a}{\overset{\downarrow\qquad\downarrow}{ax^2 + bx + c = (px + r)(qx + s)}}$$
$$\underset{\text{factors of } c}{\uparrow\qquad\uparrow}$$

To factor a trinomial of the form $ax^2 + bx + c$ with $a \neq 1$, we look for factors of the form $px + r$ and $qx + s$:

$$ax^2 + bx + c = (px + r)(qx + s) = pqx^2 + (ps + qr)x + rs$$

Therefore, we try to find numbers p, q, r, and s such that $pq = a$, $rs = c$, $ps + qr = b$. If these numbers are all integers, then we will have a limited number of possibilities to try for p, q, r, and s.

▶ **EXAMPLE 4** | Factoring $ax^2 + bx + c$ by Trial and Error

Factor: $6x^2 + 7x - 5$

▼ **SOLUTION** We can factor 6 as $6 \cdot 1$ or $3 \cdot 2$, and -5 as $-5 \cdot 1$ or $5 \cdot (-1)$. By trying these possibilities, we arrive at the factorization

$$\overset{\text{factors of 6}}{\overset{\downarrow\qquad\downarrow}{6x^2 + 7x - 5 = (3x + 5)(2x - 1)}}$$

$$\underset{\text{factors of } -5}{\uparrow\qquad\uparrow}$$

$\blacktriangle$

Check Your Answer

Multiplying gives

$(3x + 5)(2x - 1) = 6x^2 + 7x - 5$ ✔

▶ **EXAMPLE 5** | Recognizing the Form of an Expression

Factor each expression.

(a) $x^2 - 2x - 3$

(b) $(5a + 1)^2 - 2(5a + 1) - 3$

▼ **SOLUTION**

(a) $x^2 - 2x - 3 = (x - 3)(x + 1)$ Trial and error

(b) This expression is of the form

$$\square^2 - 2\,\square - 3$$

where ███ represents $5a + 1$. This is the same form as the expression in part (a), so it will factor as $(\;\blacksquare\; - 3)(\;\blacksquare\; + 1)$.

$$(5a + 1)^2 - 2(5a + 1) - 3 = [(5a + 1) - 3][(5a + 1) + 1]$$
$$= (5a - 2)(5a + 2)$$ ▲

■ Special Factoring Formulas

Some special algebraic expressions can be factored by using the following formulas. The first three are simply Special Product Formulas written backward.

FACTORING FORMULAS

Formula	Name
1. $A^2 - B^2 = (A - B)(A + B)$	Difference of squares
2. $A^2 + 2AB + B^2 = (A + B)^2$	Perfect square
3. $A^2 - 2AB + B^2 = (A - B)^2$	Perfect square
4. $A^3 - B^3 = (A - B)(A^2 + AB + B^2)$	Difference of cubes
5. $A^3 + B^3 = (A + B)(A^2 - AB + B^2)$	Sum of cubes

▶ **EXAMPLE 6** | Factoring Differences of Squares

Factor each polynomial:
(a) $4x^2 - 25$ (b) $(x + y)^2 - z^2$

▼ **SOLUTION**

(a) Using the Difference of Squares Formula with $A = 2x$ and $B = 5$, we have

$$4x^2 - 25 = (2x)^2 - 5^2 = (2x - 5)(2x + 5)$$

$$A^2 \;-\; B^2 \;=\; (A \;-\; B)\,(A \;+\; B)$$

(b) We use the Difference of Squares Formula with $A = x + y$ and $B = z$.

$$(x + y)^2 - z^2 = (x + y - z)(x + y + z)$$ ▲

▶ **EXAMPLE 7** | Factoring Differences and Sums of Cubes

Factor each polynomial:
(a) $27x^3 - 1$ (b) $x^6 + 8$

▼ **SOLUTION**

(a) Using the Difference of Cubes Formula with $A = 3x$ and $B = 1$, we get

$$27x^3 - 1 = (3x)^3 - 1^3 = (3x - 1)[(3x)^2 + (3x)(1) + 1^2]$$
$$= (3x - 1)(9x^2 + 3x + 1)$$

(b) Using the Sum of Cubes Formula with $A = x^2$ and $B = 2$, we have

$$x^6 + 8 = (x^2)^3 + 2^3 = (x^2 + 2)(x^4 - 2x^2 + 4)$$ ▲

A trinomial is a perfect square if it is of the form

$$A^2 + 2AB + B^2 \qquad \text{or} \qquad A^2 - 2AB + B^2$$

So we **recognize a perfect square** if the middle term ($2AB$ or $-2AB$) is plus or minus twice the product of the square roots of the outer two terms.

▶ **EXAMPLE 8** | Recognizing Perfect Squares

Factor each trinomial:
(a) $x^2 + 6x + 9$ (b) $4x^2 - 4xy + y^2$

▼ **SOLUTION**

(a) Here $A = x$ and $B = 3$, so $2AB = 2 \cdot x \cdot 3 = 6x$. Since the middle term is $6x$, the trinomial is a perfect square. By the Perfect Square Formula, we have

$$x^2 + 6x + 9 = (x + 3)^2$$

(b) Here $A = 2x$ and $B = y$, so $2AB = 2 \cdot 2x \cdot y = 4xy$. Since the middle term is $-4xy$, the trinomial is a perfect square. By the Perfect Square Formula, we have

$$4x^2 - 4xy + y^2 = (2x - y)^2$$ ▲

■ Factoring an Expression Completely

When we factor an expression, the result can sometimes be factored further. In general, we first factor out common factors, then inspect the result to see whether it can be factored by any of the other methods of this section. We repeat this process until we have factored the expression completely.

▶ **EXAMPLE 9** | Factoring an Expression Completely

Factor each expression completely.
(a) $2x^4 - 8x^2$ (b) $x^5y^2 - xy^6$

▼ **SOLUTION**

(a) We first factor out the power of x with the smallest exponent.

$$2x^4 - 8x^2 = 2x^2(x^2 - 4) \qquad \text{Common factor is } 2x^2$$
$$= 2x^2(x - 2)(x + 2) \qquad \text{Factor } x^2 - 4 \text{ as a difference of squares}$$

(b) We first factor out the powers of x and y with the smallest exponents.

$$x^5y^2 - xy^6 = xy^2(x^4 - y^4) \qquad \text{Common factor is } xy^2$$
$$= xy^2(x^2 + y^2)(x^2 - y^2) \qquad \text{Factor } x^4 - y^4 \text{ as a difference of squares}$$
$$= xy^2(x^2 + y^2)(x + y)(x - y) \qquad \text{Factor } x^2 - y^2 \text{ as a difference of squares}$$ ▲

In the next example we factor out variables with fractional exponents. This type of factoring occurs in calculus.

▶ **EXAMPLE 10** | Factoring Expressions with Fractional Exponents

Factor each expression.
(a) $3x^{3/2} - 9x^{1/2} + 6x^{-1/2}$ (b) $(2 + x)^{-2/3}x + (2 + x)^{1/3}$

▼ **SOLUTION**

(a) Factor out the power of x with the *smallest exponent*, that is, $x^{-1/2}$.

$$3x^{3/2} - 9x^{1/2} + 6x^{-1/2} = 3x^{-1/2}(x^2 - 3x + 2) \qquad \text{Factor out } 3x^{-1/2}$$
$$= 3x^{-1/2}(x - 1)(x - 2) \qquad \text{Factor the quadratic } x^2 - 3x + 2$$

(b) Factor out the power of $2 + x$ with the *smallest exponent*, that is, $(2 + x)^{-2/3}$.

$$(2 + x)^{-2/3}x + (2 + x)^{1/3} = (2 + x)^{-2/3}[x + (2 + x)] \qquad \text{Factor out } (2 + x)^{-2/3}$$
$$= (2 + x)^{-2/3}(2 + 2x) \qquad \text{Simplify}$$
$$= 2(2 + x)^{-2/3}(1 + x) \qquad \text{Factor out 2} \qquad \blacktriangle$$

Check Your Answer

To see that you have factored correctly, multiply using the Laws of Exponents.

(a) $3x^{-1/2}(x^2 - 3x + 2)$

$\quad = 3x^{3/2} - 9x^{1/2} + 6x^{-1/2} \qquad$ ✔

(b) $(2 + x)^{-2/3}[x + (2 + x)]$

$\quad = (2 + x)^{-2/3}x + (2 + x)^{1/3} \qquad$ ✔

■ Factoring by Grouping Terms

Polynomials with at least four terms can sometimes be factored by grouping terms. The following example illustrates the idea.

▶ **EXAMPLE 11** | Factoring by Grouping

Factor each polynomial.

(a) $x^3 + x^2 + 4x + 4$

(b) $x^3 - 2x^2 - 3x + 6$

▼ **SOLUTION**

(a) $x^3 + x^2 + 4x + 4 = (x^3 + x^2) + (4x + 4) \qquad$ Group terms

$\qquad\qquad\qquad\quad = x^2(x + 1) + 4(x + 1) \qquad$ Factor out common factors

$\qquad\qquad\qquad\quad = (x^2 + 4)(x + 1) \qquad$ Factor out x + 1 from each term

(b) $x^3 - 2x^2 - 3x + 6 = (x^3 - 2x^2) - (3x - 6) \qquad$ Group terms

$\qquad\qquad\qquad\qquad\; = x^2(x - 2) - 3(x - 2) \qquad$ Factor out common factors

$\qquad\qquad\qquad\qquad\; = (x^2 - 3)(x - 2) \qquad$ Factor out x − 2 from each term $\quad \blacktriangle$

P.8	Rational Expressions

LEARNING OBJECTIVES

After completing this section, you will be able to:

- Find the domain of an algebraic expression
- Simplify rational expressions
- Add and subtract rational expressions
- Multiply and divide rational expressions
- Simplify compound fractions
- Rationalize a denominator or numerator
- Avoid common errors

A quotient of two algebraic expressions is called a **fractional expression**. Here are some examples:

$$\frac{2x}{x - 1} \qquad \frac{\sqrt{x} + 3}{x + 1} \qquad \frac{y - 2}{y^2 + 4}$$

A **rational expression** is a fractional expression in which both the numerator and denominator are polynomials. For example, the following are rational expressions:

$$\frac{2x}{x - 1} \qquad \frac{x}{x^2 + 1} \qquad \frac{x^3 - x}{x^2 - 5x + 6}$$

Expression	Domain
$\dfrac{1}{x}$	$\{x \mid x \neq 0\}$
$\sqrt{x}$	$\{x \mid x \geq 0\}$
$\dfrac{1}{\sqrt{x}}$	$\{x \mid x > 0\}$

In this section we learn how to perform algebraic operations on rational expressions.

■ The Domain of an Algebraic Expression

In general, an algebraic expression might not be defined for all values of the variable. The **domain** of an algebraic expression is the set of real numbers that the variable is permitted to have. The table in the margin gives some basic expressions and their domains.

▶ **EXAMPLE 1** | Finding the Domain of an Expression

Consider the expression

$$\frac{2x + 4}{x - 3}$$

(a) Find the value of the expression for $x = 2$.

(b) Find the domain of the expression.

▼ **SOLUTION**

(a) We find the value by substituting 2 for x in the expression:

$$\frac{2(2) + 4}{2 - 3} = -8 \qquad \text{Substitute } x = 2$$

(b) The denominator is zero when $x = 3$. Since division by zero is not defined, we must have $x \neq 3$. Thus, the domain is all real numbers except 3. We can write this in set notation as

$$\{x \mid x \neq 3\}$$ ▲

▶ **EXAMPLE 2** | Finding the Domain of an Expression

Find the domains of the following expressions.

(a) $2x^2 + 3x - 1$ 　　　**(b)** $\dfrac{x}{x^2 - 5x + 6}$ 　　　**(c)** $\dfrac{\sqrt{x}}{x - 5}$

▼ **SOLUTION**

(a) This polynomial is defined for every x. Thus, the domain is the set $\mathbb{R}$ of real numbers.

(b) We first factor the denominator.

$$\frac{x}{x^2 - 5x + 6} = \frac{x}{(x - 2)(x - 3)}$$

> Denominator would be 0 if $x = 2$ or $x = 3$.

Since the denominator is zero when $x = 2$ or 3, the expression is not defined for these numbers. The domain is $\{x \mid x \neq 2 \text{ and } x \neq 3\}$.

(c) For the numerator to be defined, we must have $x \geq 0$. Also, we cannot divide by zero, so $x \neq 5$.

> Must have $x \geq 0$ to take square root.　　$\dfrac{\sqrt{x}}{x - 5}$　　Denominator would be 0 if $x = 5$.

Thus, the domain is $\{x \mid x \geq 0 \text{ and } x \neq 5\}$.   ▲

◼ Simplifying Rational Expressions

To **simplify rational expressions**, we factor both numerator and denominator and use the following property of fractions:

$$\frac{AC}{BC} = \frac{A}{B}$$

This allows us to **cancel** common factors from the numerator and denominator.

▶ **EXAMPLE 3** | Simplifying Rational Expressions by Cancellation

Simplify: $\dfrac{x^2 - 1}{x^2 + x - 2}$

▼ **SOLUTION**

$$\frac{x^2 - 1}{x^2 + x - 2} = \frac{(x-1)(x+1)}{(x-1)(x+2)} \qquad \text{Factor}$$

$$= \frac{x+1}{x+2} \qquad \text{Cancel common factors} \quad ▲$$

⊘ We can't cancel the x^2's in
$\dfrac{x^2 - 1}{x^2 + x - 2}$ because x^2 is not a factor.

◼ Multiplying and Dividing Rational Expressions

To **multiply rational expressions**, we use the following property of fractions:

$$\frac{A}{B} \cdot \frac{C}{D} = \frac{AC}{BD}$$

This says that to multiply two fractions, we multiply their numerators and multiply their denominators.

▶ **EXAMPLE 4** | Multiplying Rational Expressions

Perform the indicated multiplication and simplify:

$$\frac{x^2 + 2x - 3}{x^2 + 8x + 16} \cdot \frac{3x + 12}{x - 1}$$

▼ **SOLUTION**   We first factor.

$$\frac{x^2 + 2x - 3}{x^2 + 8x + 16} \cdot \frac{3x + 12}{x - 1} = \frac{(x-1)(x+3)}{(x+4)^2} \cdot \frac{3(x+4)}{x-1} \qquad \text{Factor}$$

$$= \frac{3(x-1)(x+3)(x+4)}{(x-1)(x+4)^2} \qquad \text{Property of fractions}$$

$$= \frac{3(x+3)}{x+4} \qquad \text{Cancel common factors} \quad ▲$$

To **divide rational expressions**, we use the following property of fractions:

$$\frac{A}{B} \div \frac{C}{D} = \frac{A}{B} \cdot \frac{D}{C}$$

This says that to divide a fraction by another fraction we invert the divisor and multiply.

▶ **EXAMPLE 5** | Dividing Rational Expressions

Perform the indicated division and simplify: $\dfrac{x-4}{x^2-4} \div \dfrac{x^2-3x-4}{x^2+5x+6}$

▼ **SOLUTION**

$$\frac{x-4}{x^2-4} \div \frac{x^2-3x-4}{x^2+5x+6} = \frac{x-4}{x^2-4} \cdot \frac{x^2+5x+6}{x^2-3x-4} \qquad \text{Invert divisor and multiply}$$

$$= \frac{(x-4)(x+2)(x+3)}{(x-2)(x+2)(x-4)(x+1)} \qquad \text{Factor}$$

$$= \frac{x+3}{(x-2)(x+1)} \qquad \text{Cancel common factors} \quad \blacktriangle$$

🚫 Avoid making the following error:

$$\frac{A}{B+C} \neq \frac{A}{B} + \frac{A}{C}$$

For instance, if we let $A = 2$, $B = 1$, and $C = 1$, then we see the error:

$$\frac{2}{1+1} \overset{?}{=} \frac{2}{1} + \frac{2}{1}$$

$$\frac{2}{2} \overset{?}{=} 2 + 2$$

$$1 \overset{?}{=} 4 \quad \text{Wrong!}$$

▪ Adding and Subtracting Rational Expressions

To **add or subtract rational expressions**, we first find a common denominator and then use the following property of fractions:

$$\frac{A}{C} + \frac{B}{C} = \frac{A+B}{C}$$

Although any common denominator will work, it is best to use the **least common denominator** (LCD) as explained in Section P.2. The LCD is found by factoring each denominator and taking the product of the distinct factors, using the highest power that appears in any of the factors.

▶ **EXAMPLE 6** | Adding and Subtracting Rational Expressions

Perform the indicated operations and simplify.

(a) $\dfrac{3}{x-1} + \dfrac{x}{x+2}$

(b) $\dfrac{1}{x^2-1} - \dfrac{2}{(x+1)^2}$

▼ **SOLUTION**

(a) Here the LCD is simply the product $(x-1)(x+2)$.

$$\frac{3}{x-1} + \frac{x}{x+2} = \frac{3(x+2)}{(x-1)(x+2)} + \frac{x(x-1)}{(x-1)(x+2)} \qquad \begin{array}{l}\text{Write fractions}\\\text{using LCD}\end{array}$$

$$= \frac{3x+6+x^2-x}{(x-1)(x+2)} \qquad \text{Add fractions}$$

$$= \frac{x^2+2x+6}{(x-1)(x+2)} \qquad \begin{array}{l}\text{Combine terms in}\\\text{numerator}\end{array}$$

(b) The LCD of $x^2 - 1 = (x-1)(x+1)$ and $(x+1)^2$ is $(x-1)(x+1)^2$.

$$\frac{1}{x^2-1} - \frac{2}{(x+1)^2} = \frac{1}{(x-1)(x+1)} - \frac{2}{(x+1)^2} \qquad \text{Factor}$$

$$= \frac{(x+1) - 2(x-1)}{(x-1)(x+1)^2}$$ Combine fractions using LCD

$$= \frac{x+1-2x+2}{(x-1)(x+1)^2}$$ Distributive Property

$$= \frac{3-x}{(x-1)(x+1)^2}$$ Combine terms in numerator ▲

■ Compound Fractions

A **compound fraction** is a fraction in which the numerator, the denominator, or both are themselves fractional expressions.

▶ **EXAMPLE 7** | Simplifying a Compound Fraction

Simplify: $\dfrac{\dfrac{x}{y} + 1}{1 - \dfrac{y}{x}}$

▼ **SOLUTION 1** We combine the terms in the numerator into a single fraction. We do the same in the denominator. Then we invert and multiply.

$$\frac{\dfrac{x}{y} + 1}{1 - \dfrac{y}{x}} = \frac{\dfrac{x+y}{y}}{\dfrac{x-y}{x}}$$

$$= \frac{x+y}{y} \cdot \frac{x}{x-y}$$

$$= \frac{x(x+y)}{y(x-y)}$$

▼ **SOLUTION 2** We find the LCD of all the fractions in the expression, then multiply the numerator and denominator by it. In this example the LCD of all the fractions is xy. Thus,

$$\frac{\dfrac{x}{y} + 1}{1 - \dfrac{y}{x}} = \frac{\dfrac{x}{y} + 1}{1 - \dfrac{y}{x}} \cdot \frac{xy}{xy}$$ Multiply numerator and denominator by xy

$$= \frac{x^2 + xy}{xy - y^2}$$ Simplify

$$= \frac{x(x+y)}{y(x-y)}$$ Factor ▲

The next two examples show situations in calculus that require the ability to work with fractional expressions.

▶ **EXAMPLE 8** | Simplifying a Compound Fraction

Simplify: $\dfrac{\dfrac{1}{a+h} - \dfrac{1}{a}}{h}$

▼ **SOLUTION**   We begin by combining the fractions in the numerator using a common denominator.

$$\frac{\dfrac{1}{a+h} - \dfrac{1}{a}}{h} = \frac{\dfrac{a-(a+h)}{a(a+h)}}{h}$$ Combine fractions in the numerator

$$= \frac{a-(a+h)}{a(a+h)} \cdot \frac{1}{h}$$ Invert divisor and multiply

$$= \frac{a-a-h}{a(a+h)} \cdot \frac{1}{h}$$ Distributive Property

$$= \frac{-h}{a(a+h)} \cdot \frac{1}{h}$$ Simplify

$$= \frac{-1}{a(a+h)}$$ Cancel common factors   ▲

▶ **EXAMPLE 9** | Simplifying a Compound Fraction

Simplify: $\dfrac{(1+x^2)^{1/2} - x^2(1+x^2)^{-1/2}}{1+x^2}$

▼ **SOLUTION 1**   Factor $(1+x^2)^{-1/2}$ from the numerator.

Factor out the power of $1+x^2$ with the *smallest* exponent, in this case $(1+x^2)^{-1/2}$.

$$\frac{(1+x^2)^{1/2} - x^2(1+x^2)^{-1/2}}{1+x^2} = \frac{(1+x^2)^{-1/2}\left[(1+x^2) - x^2\right]}{1+x^2}$$

$$= \frac{(1+x^2)^{-1/2}}{1+x^2} = \frac{1}{(1+x^2)^{3/2}}$$

▼ **SOLUTION 2**   Since $(1+x^2)^{-1/2} = 1/(1+x^2)^{1/2}$ is a fraction, we can clear all fractions by multiplying numerator and denominator by $(1+x^2)^{1/2}$.

$$\frac{(1+x^2)^{1/2} - x^2(1+x^2)^{-1/2}}{1+x^2} = \frac{(1+x^2)^{1/2} - x^2(1+x^2)^{-1/2}}{1+x^2} \cdot \frac{(1+x^2)^{1/2}}{(1+x^2)^{1/2}}$$

$$= \frac{(1+x^2) - x^2}{(1+x^2)^{3/2}} = \frac{1}{(1+x^2)^{3/2}}$$ ▲

▮ Rationalizing the Denominator or the Numerator

If a fraction has a denominator of the form $A + B\sqrt{C}$, we may rationalize the denominator by multiplying numerator and denominator by the **conjugate radical** $A - B\sqrt{C}$. This is effective because, by Product Formula 1 in Section P.6, the product of the denominator and its conjugate radical does not contain a radical:

$$(A + B\sqrt{C})(A - B\sqrt{C}) = A^2 - B^2C$$

► **EXAMPLE 10** | Rationalizing the Denominator

Rationalize the denominator: $\dfrac{1}{1 + \sqrt{2}}$

▼ **SOLUTION** We multiply both the numerator and the denominator by the conjugate radical of $1 + \sqrt{2}$, which is $1 - \sqrt{2}$.

$$\frac{1}{1 + \sqrt{2}} = \frac{1}{1 + \sqrt{2}} \cdot \frac{1 - \sqrt{2}}{1 - \sqrt{2}} \qquad \text{\textit{Multiply numerator and denominator by the conjugate radical}}$$

$$= \frac{1 - \sqrt{2}}{1^2 - (\sqrt{2})^2} \qquad \text{\textit{Product Formula 1: } } (a+b)(a-b) = a^2 - b^2$$

$$= \frac{1 - \sqrt{2}}{1 - 2} = \frac{1 - \sqrt{2}}{-1} = \sqrt{2} - 1 \qquad \blacktriangle$$

► **EXAMPLE 11** | Rationalizing the Numerator

Rationalize the numerator: $\dfrac{\sqrt{4 + h} - 2}{h}$

▼ **SOLUTION** We multiply numerator and denominator by the conjugate radical $\sqrt{4 + h} + 2$.

$$\frac{\sqrt{4 + h} - 2}{h} = \frac{\sqrt{4 + h} - 2}{h} \cdot \frac{\sqrt{4 + h} + 2}{\sqrt{4 + h} + 2} \qquad \text{\textit{Multiply numerator and denominator by the conjugate radical}}$$

$$= \frac{(\sqrt{4 + h})^2 - 2^2}{h(\sqrt{4 + h} + 2)} \qquad \text{\textit{Product Formula 1:}}$$
$$\text{\textit{(}a+b\text{)(}a-b\text{)} = a^2 - b^2}$$

$$= \frac{4 + h - 4}{h(\sqrt{4 + h} + 2)}$$

$$= \frac{h}{h(\sqrt{4 + h} + 2)} = \frac{1}{\sqrt{4 + h} + 2} \qquad \text{\textit{Cancel common factors}} \qquad \blacktriangle$$

■ Avoiding Common Errors

Don't make the mistake of applying properties of multiplication to the operation of addition. Many of the common errors in algebra involve doing just that. The following table states several properties of multiplication and illustrates the error in applying them to addition.

Correct multiplication property	Common error with addition
$(a \cdot b)^2 = a^2 \cdot b^2$	$(a + b)^2 \neq a^2 + b^2$
$\sqrt{a \cdot b} = \sqrt{a}\sqrt{b} \quad (a, b \geq 0)$	$\sqrt{a + b} \neq \sqrt{a} + \sqrt{b}$
$\sqrt{a^2 \cdot b^2} = a \cdot b \quad (a, b \geq 0)$	$\sqrt{a^2 + b^2} \neq a + b$
$\dfrac{1}{a} \cdot \dfrac{1}{b} = \dfrac{1}{a \cdot b}$	$\dfrac{1}{a} + \dfrac{1}{b} \neq \dfrac{1}{a + b}$
$\dfrac{ab}{a} = b$	$\dfrac{a + b}{a} \neq b$
$a^{-1} \cdot b^{-1} = (a \cdot b)^{-1}$	$a^{-1} + b^{-1} \neq (a + b)^{-1}$

To verify that the equations in the right-hand column are wrong, simply substitute numbers for a and b and calculate each side. For example, if we take $a = 2$ and $b = 2$ in the fourth error, we find that the left-hand side is

$$\frac{1}{a} + \frac{1}{b} = \frac{1}{2} + \frac{1}{2} = 1$$

whereas the right-hand side is

$$\frac{1}{a + b} = \frac{1}{2 + 2} = \frac{1}{4}$$

Since $1 \neq \frac{1}{4}$, the stated equation is wrong. You should similarly convince yourself of the error in each of the other equations.

▶ CHAPTER P | REVIEW

▼ PROPERTIES AND FORMULAS

Properties of Real Numbers (p. 7)

Commutative: $a + b = b + a$
$\qquad\qquad\quad ab = ba$
Associative: $(a + b) + c = a + (b + c)$
$\qquad\qquad\quad (ab)c = a(bc)$
Distributive: $a(b + c) = ab + ac$

Absolute Value (p. 13)

$$|a| = \begin{cases} a & \text{if } a \geq 0 \\ -a & \text{if } a < 0 \end{cases}$$

$$|ab| = |a||b|$$

$$\left|\frac{a}{b}\right| = \frac{|a|}{|b|}$$

Distance between a and b is

$$d(a, b) = |b - a|$$

Exponents (p. 16)

$a^m a^n = a^{m+n}$

$\dfrac{a^m}{a^n} = a^{m-n}$

$(a^m)^n = a^{mn}$

$(ab)^n = a^n b^n$

$\left(\dfrac{a}{b}\right)^n = \dfrac{a^n}{b^n}$

Radicals (p. 20)

$\sqrt[n]{a} = b$ means $b^n = a$

$\sqrt[n]{ab} = \sqrt[n]{a}\sqrt[n]{b}$

$\sqrt[n]{\dfrac{a}{b}} = \dfrac{\sqrt[n]{a}}{\sqrt[n]{b}}$

$\sqrt[m]{\sqrt[n]{a}} = \sqrt[mn]{a}$

If n is odd, then $\sqrt[n]{a^n} = a$

If n is even, then $\sqrt[n]{a^n} = |a|$

$a^{m/n} = \sqrt[n]{a^m}$

Special Product Formulas (p. 26)

Sum and difference of same terms:

$$(A + B)(A - B) = A^2 - B^2$$

Square of a sum or difference:

$$(A + B)^2 = A^2 + 2AB + B^2$$
$$(A - B)^2 = A^2 - 2AB + B^2$$

Cube of a sum or difference:

$$(A + B)^3 = A^3 + 3A^2B + 3AB^2 + B^3$$
$$(A - B)^3 = A^3 - 3A^2B + 3AB^2 - B^3$$

Factoring Formulas (p. 29)

Difference of squares:

$$A^2 - B^2 = (A + B)(A - B)$$

Perfect squares:

$$A^2 + 2AB + B^2 = (A + B)^2$$
$$A^2 - 2AB + B^2 = (A - B)^2$$

Sum or difference of cubes:

$$A^3 - B^3 = (A - B)(A^2 + AB + B^2)$$
$$A^3 + B^3 = (A + B)(A^2 - AB + B^2)$$

Rational Expressions (pp. 33–34)

We can cancel common factors:

$$\frac{AC}{BC} = \frac{A}{B}$$

To multiply two fractions, we multiply their numerators together and their denominators together:

$$\frac{A}{B} \times \frac{C}{D} = \frac{AC}{BD}$$

To divide fractions, we invert the divisor and multiply:

$$\frac{A}{B} \div \frac{C}{D} = \frac{A}{B} \times \frac{D}{C}$$

To add fractions, we find a common denominator:

$$\frac{A}{C} + \frac{B}{C} = \frac{A + B}{C}$$

▼ CONCEPT SUMMARY

	Review Exercises
Section P.1	
▪ Make and use algebra models	1–2
Section P.2	
▪ Classify real numbers	3–4
▪ Use properties of real numbers	5–8
▪ Add, subtract, multiply, and divide fractions	9–12

▼ EXERCISES

1–2 ■ Make and use an algebra model to solve the problem.

1. Elena regularly takes a multivitamin and mineral supplement. She purchases a bottle of 250 tablets and takes two tablets every day.
   (a) Find a formula for the number of tablets T that are left in the bottle after she has been taking the tablets for x days.
   (b) How many tablets are left after 30 days?
   (c) How many days will it take for her to run out of tablets?

2. Alonzo's Delivery is having a sale on calzones. Each calzone costs $2, and there is a $3 delivery charge for phone-in orders.
   (a) Find a formula for the total cost C of ordering x calzones for delivery.
   (b) How much would it cost to have 4 calzones delivered?
   (c) If you have $15, how many calzones can you order?

3–4 ■ Determine whether each number is rational or irrational. If it is rational, determine whether it is a natural number, an integer, or neither.

3. (a) 16 (b) -16 (c) $\sqrt{16}$ (d) $\sqrt{2}$
   (e) $\frac{8}{3}$ (f) $-\frac{8}{2}$

4. (a) -5 (b) $-\frac{25}{6}$ (c) $\sqrt{25}$ (d) 3π
   (e) $\frac{24}{16}$ (f) 10^{20}

5–8 ■ State the property of real numbers being used.

5. $3 + 2x = 2x + 3$

6. $(a + b)(a - b) = (a - b)(a + b)$

7. $A(x + y) = Ax + Ay$

8. $(A + 1)(x + y) = (A + 1)x + (A + 1)y$

9–12 ▪ Evaluate each expression. Express your answer as a fraction in lowest terms.

9. (a) $\dfrac{5}{6} + \dfrac{2}{3}$ (b) $\dfrac{5}{6} - \dfrac{2}{3}$

10. (a) $\dfrac{7}{10} - \dfrac{11}{15}$ (b) $\dfrac{7}{10} + \dfrac{11}{15}$

11. (a) $\dfrac{15}{8} \cdot \dfrac{12}{5}$ (b) $\dfrac{15}{8} \div \dfrac{12}{5}$

12. (a) $\dfrac{30}{7} \div \dfrac{12}{35}$ (b) $\dfrac{30}{7} \cdot \dfrac{12}{35}$

13–16 ▪ Express the interval in terms of inequalities, and then graph the interval.

13. $[-2, 6)$ **14.** $(0, 10]$

15. $(-\infty, 4]$ **16.** $[-2, \infty)$

17–20 ▪ Express the inequality in interval notation, and then graph the corresponding interval.

17. $x \geq 5$ **18.** $x < -3$

19. $-1 < x \leq 5$ **20.** $0 \leq x \leq \frac{1}{2}$

21–24 ▪ The sets A, B, C, and D are defined as follows:

$A = \{-1, 0, 1, 2, 3\}$ $B = \{\frac{1}{2}, 1, 4\}$
$C = \{x \mid 0 < x \leq 2\}$ $D = (-1, 1]$

Find each of the following sets.

21. (a) $A \cup B$ (b) $A \cap B$

22. (a) $C \cup D$ (b) $C \cap D$

23. (a) $A \cap C$ (b) $B \cap D$

24. (a) $A \cap D$ (b) $B \cap C$

25–36 ▪ Evaluate the expression.

25. $|7 - 10|$ **26.** $\left|-\frac{3}{2} - 5\right|$

27. $\big|3 - |-9|\big|$ **28.** $1 - \big|1 - |-1|\big|$

29. $2^{-3} - 3^{-2}$ **30.** $2^{1/2} 8^{1/2}$

31. $216^{-1/3}$ **32.** $64^{2/3}$

33. $\dfrac{\sqrt{242}}{\sqrt{2}}$ **34.** $\sqrt[4]{4}\,\sqrt[4]{324}$

35. $\sqrt[3]{-125}$ **36.** $\sqrt{2}\sqrt{50}$

37–38 ▪ Express the distance between the given numbers on the real line using an absolute value. Then evaluate this distance.

37. (a) 3 and 5 (b) 3 and -5

38. (a) 0 and -4 (b) 4 and -4

39–46 ▪ Write the expression as a power of x.

39. $\dfrac{1}{x^2}$ **40.** $x\sqrt{x}$

41. $x^2 x^m (x^3)^m$ **42.** $((x^m)^2)^n$

43. $x^a x^b x^c$ **44.** $((x^a)^b)^c$

45. $x^{c+1}(x^{2c-1})^2$ **46.** $\dfrac{(x^2)^n x^5}{x^n}$

47–50 ▪ Express the radical as a power with a rational exponent.

47. (a) $\sqrt[3]{7}$ (b) $\sqrt[5]{7^4}$ **48.** (a) $\sqrt[3]{5^7}$ (b) $(\sqrt[4]{5})^3$

49. (a) $\sqrt[6]{x^5}$ (b) $(\sqrt{x})^9$ **50.** (a) $\sqrt{y^3}$ (b) $(\sqrt[8]{y})^2$

51–60 ▪ Simplify the expression.

51. $(2x^3 y)^2 (3x^{-1} y^2)$ **52.** $(a^2)^{-3}(a^3 b)^2 (b^3)^4$

53. $\dfrac{x^4 (3x)^2}{x^3}$ **54.** $\left(\dfrac{r^2 s^{4/3}}{r^{1/3} s}\right)^6$

55. $\sqrt[3]{(x^3 y)^2 y^4}$ **56.** $\sqrt{x^2 y^4}$

57. $\dfrac{x}{2 + \sqrt{x}}$ **58.** $\dfrac{\sqrt{x} + 1}{\sqrt{x} - 1}$

59. $\dfrac{8r^{1/2} s^{-3}}{2r^{-2} s^4}$ **60.** $\left(\dfrac{ab^2 c^{-3}}{2a^2 b^{-4}}\right)^{-2}$

61. Write the number 78,250,000,000 in scientific notation.

62. Write the number 2.08×10^{-8} in decimal notation.

63. If $a \approx 0.00000293$, $b \approx 1.582 \times 10^{-14}$, and $c \approx 2.8064 \times 10^{12}$, use a calculator to approximate the number ab/c.

64. If your heart beats 80 times per minute and you live to be 90 years old, estimate the number of times your heart beats during your lifetime. State your answer in scientific notation.

65–86 ▪ Factor the expression.

65. $2x^2 y - 6xy^2$ **66.** $12x^2 y^4 - 3xy^5 + 9x^3 y^2$

67. $x^2 - 9x + 18$ **68.** $x^2 + 3x - 10$

69. $3x^2 - 2x - 1$ **70.** $6x^2 + x - 12$

71. $4t^2 - 13t - 12$ **72.** $x^4 - 2x^2 + 1$

73. $25 - 16t^2$ **74.** $2y^6 - 32y^2$

75. $x^6 - 1$ **76.** $y^3 - 2y^2 - y + 2$

77. $x^{-1/2} - 2x^{1/2} + x^{3/2}$ **78.** $a^4 b^2 + ab^5$

79. $4x^3 - 8x^2 + 3x - 6$ **80.** $8x^3 + y^6$

81. $(x^2 + 2)^{5/2} + 2x(x^2 + 2)^{3/2} + x^2 \sqrt{x^2 + 2}$

82. $3x^3 - 2x^2 + 18x - 12$ **83.** $a^2 y - b^2 y$

84. $ax^2 + bx^2 - a - b$ **85.** $(x + 1)^2 - 2(x + 1) + 1$

86. $(a + b)^2 + 2(a + b) - 15$

87–110 ▪ Perform the indicated operations.

87. $(2x + 1)(3x - 2) - 5(4x - 1)$

88. $(2y - 7)(2y + 7)$ **89.** $(2a^2 - b)^2$

90. $(1 + x)(2 - x) - (3 - x)(3 + x)$

91. $(x - 1)(x - 2)(x - 3)$ **92.** $(2x + 1)^3$

93. $\sqrt{x}(\sqrt{x} + 1)(2\sqrt{x} - 1)$

94. $x^3(x - 6)^2 + x^4(x - 6)$ **95.** $x^2(x - 2) + x(x - 2)^2$

96. $\dfrac{x^3 + 2x^2 + 3x}{x}$ **97.** $\dfrac{x^2 - 2x - 3}{2x^2 + 5x + 3}$

98. $\dfrac{t^3 - 1}{t^2 - 1}$ **99.** $\dfrac{x^2 + 2x - 3}{x^2 + 8x + 16} \cdot \dfrac{3x + 12}{x - 1}$

100. $\dfrac{x^3/(x - 1)}{x^2/(x^3 - 1)}$

101. $\dfrac{x^2 - 2x - 15}{x^2 - 6x + 5} \div \dfrac{x^2 - x - 12}{x^2 - 1}$

102. $x - \dfrac{1}{x + 1}$ **103.** $\dfrac{1}{x - 1} - \dfrac{x}{x^2 + 1}$

104. $\dfrac{2}{x} + \dfrac{1}{x - 2} + \dfrac{3}{(x - 2)^2}$ **105.** $\dfrac{1}{x - 1} - \dfrac{2}{x^2 - 1}$

106. $\dfrac{1}{x + 2} + \dfrac{1}{x^2 - 4} - \dfrac{2}{x^2 - x - 2}$

107. $\dfrac{\dfrac{1}{x} - \dfrac{1}{2}}{x - 2}$ **108.** $\dfrac{\dfrac{1}{x} - \dfrac{1}{x + 1}}{\dfrac{1}{x} + \dfrac{1}{x + 1}}$

109. $\dfrac{3(x + h)^2 - 5(x + h) - (3x^2 - 5x)}{h}$

110. $\dfrac{\sqrt{x + h} - \sqrt{x}}{h}$ (rationalize the numerator)

111–114 ▪ Find the domain of the algebraic expression.

111. $\dfrac{x + 5}{x + 10}$ **112.** $\dfrac{2x}{x^2 - 9}$

113. $\dfrac{\sqrt{x}}{x^2 - 3x - 4}$ **114.** $\dfrac{\sqrt{x - 3}}{x^2 - 4x + 4}$

115–122 ▪ State whether the given equation is true for all values of the variables. (Disregard any value that makes a denominator 0.)

115. $(x + y)^3 = x^3 + y^3$ **116.** $\dfrac{1 + \sqrt{a}}{1 - a} = \dfrac{1}{1 - \sqrt{a}}$

117. $\dfrac{12 + y}{y} = \dfrac{12}{y} + 1$ **118.** $\sqrt[3]{a + b} = \sqrt[3]{a} + \sqrt[3]{b}$

119. $\sqrt{a^2} = a$ **120.** $\dfrac{1}{x + 4} = \dfrac{1}{x} + \dfrac{1}{4}$

121. $x^3 + y^3 = (x + y)(x^2 + xy + y^2)$

122. $\dfrac{x^2 + 1}{x^2 + 2x + 1} = \dfrac{1}{2x + 1}$

123. If $m > n > 0$ and $a = 2mn$, $b = m^2 - n^2$, $c = m^2 + n^2$, show that $a^2 + b^2 = c^2$.

124. If $t = \dfrac{1}{2}\left(x^3 - \dfrac{1}{x^3}\right)$ and $x > 0$, show that

$$\sqrt{1 + t^2} = \dfrac{1}{2}\left(x^3 + \dfrac{1}{x^3}\right)$$

1. A pizzeria charges \$9 for a medium plain cheese pizza, plus \$1.50 for each extra topping.
   - (a) Find a formula that models the cost C of a medium pizza with x toppings.
   - (b) Use your model from part (a) to find the cost of a medium pizza with the following extra toppings: anchovies, ham, sausage, and pineapple.

2. Determine whether each number is rational or irrational. If it is rational, determine whether it is a natural number, an integer, or neither.
   - (a) 5
   - (b) $\sqrt{5}$
   - (c) $-\frac{9}{3}$
   - (d) $-1,000,000$

3. Let $A = \{-2, 0, 1, 3, 5\}$ and $B = \{0, \frac{1}{2}, 1, 5, 7\}$. Find each of the following sets.
   - (a) $A \cap B$
   - (b) $A \cup B$

4. (a) Graph the intervals $[-4, 2)$ and $[0, 3]$ on a real line.
   - (b) Find the intersection and the union of the intervals in part (a), and graph each of them on a real line.
   - (c) Use an absolute value to express the distance between -4 and 2 on the real line, and then evaluate this distance.

5. Evaluate each expression:
   - (a) -2^6
   - (b) $(-2)^6$
   - (c) 2^{-6}
   - (d) $\dfrac{7^{10}}{7^{12}}$
   - (e) $\left(\dfrac{3}{2}\right)^{-2}$
   - (f) $\dfrac{\sqrt[5]{32}}{\sqrt{16}}$
   - (g) $\sqrt[4]{\dfrac{3^8}{2^{16}}}$
   - (h) $81^{-3/4}$

6. Write each number in scientific notation.
   - (a) 186,000,000,000
   - (b) 0.0000003965

7. Simplify each expression. Write your final answer without negative exponents.
   - (a) $\sqrt{200} - \sqrt{32}$
   - (b) $(3a^3b^3)(4ab^2)^2$
   - (c) $\sqrt[3]{\dfrac{125}{x^{-9}}}$
   - (d) $\left(\dfrac{3x^{3/2}y^3}{x^2y^{-1/2}}\right)^{-2}$

8. Perform the indicated operations and simplify.
   - (a) $3(x + 6) + 4(2x - 5)$
   - (b) $(x + 3)(4x - 5)$
   - (c) $(\sqrt{a} + \sqrt{b})(\sqrt{a} - \sqrt{b})$
   - (d) $(2x + 3)^2$
   - (e) $(x + 2)^3$
   - (f) $x^2(x - 3)(x + 3)$

9. Factor each expression completely.
   - (a) $4x^2 - 25$
   - (b) $2x^2 + 5x - 12$
   - (c) $x^3 - 3x^2 - 4x + 12$
   - (d) $x^4 + 27x$
   - (e) $3x^{3/2} - 9x^{1/2} + 6x^{-1/2}$
   - (f) $x^3y - 4xy$

10. Simplify the rational expression.
    - (a) $\dfrac{x^2 + 3x + 2}{x^2 - x - 2}$
    - (b) $\dfrac{2x^2 - x - 1}{x^2 - 9} \cdot \dfrac{x + 3}{2x + 1}$
    - (c) $\dfrac{x^2}{x^2 - 4} - \dfrac{x + 1}{x + 2}$
    - (d) $\dfrac{\dfrac{y}{x} - \dfrac{x}{y}}{\dfrac{1}{y} - \dfrac{1}{x}}$

11. Rationalize the denominator and simplify.
    - (a) $\dfrac{6}{\sqrt[3]{4}}$
    - (b) $\dfrac{\sqrt{6}}{2 + \sqrt{3}}$

GENERAL PRINCIPLES

George Polya (1887–1985) is famous among mathematicians for his ideas on problem solving. His lectures on problem solving at Stanford University attracted overflow crowds whom he held on the edges of their seats, leading them to discover solutions for themselves. He was able to do this because of his deep insight into the psychology of problem solving. His well-known book *How To Solve It* has been translated into 15 languages. He said that Euler was unique among great mathematicians because he explained *how* he found his results. Polya often said to his students and colleagues, "Yes, I see that your proof is correct, but how did you discover it?" In the preface to *How To Solve It*, Polya writes, "A great discovery solves a great problem but there is a grain of discovery in the solution of any problem. Your problem may be modest; but if it challenges your curiosity and brings into play your inventive faculties, and if you solve it by your own means, you may experience the tension and enjoy the triumph of discovery."

There are no hard and fast rules that will ensure success in solving problems. However, it is possible to outline some general steps in the problem-solving process and to give principles that are useful in solving certain problems. These steps and principles are just common sense made explicit. They have been adapted from George Polya's insightful book *How To Solve It*.

1. Understand the Problem

The first step is to read the problem and make sure that you understand it. Ask yourself the following questions:

What is the unknown?

What are the given quantities?

What are the given conditions?

For many problems it is useful to

draw a diagram

and identify the given and required quantities on the diagram.

Usually, it is necessary to

introduce suitable notation

In choosing symbols for the unknown quantities, we often use letters such as a, b, c, m, n, x, and y, but in some cases it helps to use initials as suggestive symbols, for instance, V for volume or t for time.

2. Think of a Plan

Find a connection between the given information and the unknown that enables you to calculate the unknown. It often helps to ask yourself explicitly: "How can I relate the given to the unknown?" If you don't see a connection immediately, the following ideas may be helpful in devising a plan.

■ TRY TO RECOGNIZE SOMETHING FAMILIAR

Relate the given situation to previous knowledge. Look at the unknown and try to recall a more familiar problem that has a similar unknown.

■ TRY TO RECOGNIZE PATTERNS

Certain problems are solved by recognizing that some kind of pattern is occurring. The pattern could be geometric, numerical, or algebraic. If you can see regularity or repetition in a problem, then you might be able to guess what the pattern is and then prove it.

■ USE ANALOGY

Try to think of an analogous problem, that is, a similar or related problem but one that is easier than the original. If you can solve the similar, simpler problem, then it might give you the clues you need to solve the original, more difficult one. For instance, if a problem involves very large numbers, you could first try a similar problem with smaller numbers. Or if the problem is in three-dimensional geometry, you could look for something similar in two-dimensional geometry. Or if the problem you start with is a general one, you could first try a special case.

■ INTRODUCE SOMETHING EXTRA

You might sometimes need to introduce something new—an auxiliary aid—to make the connection between the given and the unknown. For instance, in a problem for which

Stanford University News Service

a diagram is useful, the auxiliary aid could be a new line drawn in the diagram. In a more algebraic problem the aid could be a new unknown that relates to the original unknown.

■ TAKE CASES

You might sometimes have to split a problem into several cases and give a different argument for each case. For instance, we often have to use this strategy in dealing with absolute value.

■ WORK BACKWARD

Sometimes it is useful to imagine that your problem is solved and work backward, step by step, until you arrive at the given data. Then you might be able to reverse your steps and thereby construct a solution to the original problem. This procedure is commonly used in solving equations. For instance, in solving the equation $3x - 5 = 7$, we suppose that x is a number that satisfies $3x - 5 = 7$ and work backward. We add 5 to each side of the equation and then divide each side by 3 to get $x = 4$. Since each of these steps can be reversed, we have solved the problem.

■ ESTABLISH SUBGOALS

In a complex problem it is often useful to set subgoals (in which the desired situation is only partially fulfilled). If you can attain or accomplish these subgoals, then you might be able to build on them to reach your final goal.

■ INDIRECT REASONING

Sometimes it is appropriate to attack a problem indirectly. In using **proof by contradiction** to prove that P implies Q, we assume that P is true and Q is false and try to see why this cannot happen. Somehow we have to use this information and arrive at a contradiction to what we absolutely know is true.

■ MATHEMATICAL INDUCTION

In proving statements that involve a positive integer n, it is frequently helpful to use the Principle of Mathematical Induction, which is discussed in Section 9.4.

3. Carry Out the Plan

In Step 2, a plan was devised. In carrying out that plan, you must check each stage of the plan and write the details that prove that each stage is correct.

4. Look Back

Having completed your solution, it is wise to look back over it, partly to see whether any errors have been made and partly to see whether you can discover an easier way to solve the problem. Looking back also familiarizes you with the method of solution, which may be useful for solving a future problem. Descartes said, "Every problem that I solved became a rule which served afterwards to solve other problems."

We illustrate some of these principles of problem solving with an example.

▶ **PROBLEM** | Average Speed

A driver sets out on a journey. For the first half of the distance, she drives at the leisurely pace of 30 mi/h; during the second half she drives 60 mi/h. What is her average speed on this trip?

Thinking About the Problem

It is tempting to take the average of the speeds and say that the average speed for the entire trip is

$$\frac{30 + 60}{2} = 45 \text{ mi/h}$$

But is this simple-minded approach really correct?

◀ *Try a special case*   Let's look at an easily calculated special case. Suppose that the total distance traveled is 120 mi. Since the first 60 mi is traveled at 30 mi/h, it takes 2 h. The second 60 mi is traveled at 60 mi/h, so it takes one hour. Thus, the total time is $2 + 1 = 3$ hours and the average speed is

$$\frac{120}{3} = 40 \text{ mi/h}$$

So our guess of 45 mi/h was wrong.

▼ **SOLUTION**

Understand the problem ▶ We need to look more carefully at the meaning of average speed. It is defined as

$$\text{average speed} = \frac{\text{distance traveled}}{\text{time elapsed}}$$

Introduce notation ▶ Let d be the distance traveled on each half of the trip. Let t_1 and t_2 be the times taken for the first and second halves of the trip. Now we can write down the information we have

State what is given ▶ been given. For the first half of the trip we have

$$30 = \frac{d}{t_1}$$

and for the second half we have

$$60 = \frac{d}{t_2}$$

Identify the unknown ▶ Now we identify the quantity that we are asked to find:

$$\text{average speed for entire trip} = \frac{\text{total distance}}{\text{total time}} = \frac{2d}{t_1 + t_2}$$

Connect the given ▶ To calculate this quantity, we need to know t_1 and t_2, so we solve the above equations for *with the unknown* these times:

$$t_1 = \frac{d}{30} \qquad t_2 = \frac{d}{60}$$

Now we have the ingredients needed to calculate the desired quantity:

$$\text{average speed} = \frac{2d}{t_1 + t_2} = \frac{2d}{\dfrac{d}{30} + \dfrac{d}{60}}$$

$$= \frac{60(2d)}{60\left(\dfrac{d}{30} + \dfrac{d}{60}\right)} \qquad \text{Multiply numerator and denominator by 60}$$

$$= \frac{120d}{2d + d} = \frac{120d}{3d} = 40$$

So the average speed for the entire trip is 40 mi/h.

Problems

1. **Distance, Time, and Speed**   A man drives from home to work at a speed of 50 mi/h. The return trip from work to home is traveled at the more leisurely pace of 30 mi/h. What is the man's average speed for the round-trip?

2. **Distance, Time, and Speed**   An old car has to travel a 2-mile route, uphill and down. Because it is so old, the car can climb the first mile—the ascent—no faster than an average speed of 15 mi/h. How fast does the car have to travel the second mile—on the descent it can go faster, of course—to achieve an average speed of 30 mi/h for the trip?

3. **A Speeding Fly**   A car and a van are parked 120 mi apart on a straight road. The drivers start driving toward each other at noon, each at a speed of 40 mi/h. A fly starts from the front bumper of the van at noon and flies to the bumper of the car, then immediately back to the bumper of the van, back to the car, and so on, until the car and the van meet. If the fly flies at a speed of 100 mi/h, what is the total distance it travels?

4. **Comparing Discounts**   Which price is better for the buyer, a 40% discount or two successive discounts of 20%?

5. **Cutting up a Wire**   A piece of wire is bent as shown in the figure. You can see that one cut through the wire produces four pieces and two parallel cuts produce seven pieces. How many pieces will be produced by 142 parallel cuts? Write a formula for the number of pieces produced by n parallel cuts.

6. **Amoeba Propagation**   An amoeba propagates by simple division; each split takes 3 minutes to complete. When such an amoeba is put into a glass container with a nutrient fluid, the container is full of amoebas in one hour. How long would it take for the container to be filled if we start with not one amoeba, but two?

7. **Running Laps**   Two runners start running laps at the same time, from the same starting position. George runs a lap in 50 s; Sue runs a lap in 30 s. When will the runners next be side by side?

8. **Batting Averages**   Player A has a higher batting average than player B for the first half of the baseball season. Player A also has a higher batting average than player B for the second half of the season. Is it necessarily true that player A has a higher batting average than player B for the entire season?

9. **Coffee and Cream**   A spoonful of cream is taken from a pitcher of cream and put into a cup of coffee. The coffee is stirred. Then a spoonful of this mixture is put into the pitcher of cream. Is there now more cream in the coffee cup or more coffee in the pitcher of cream?

10. **A Melting Ice Cube**   An ice cube is floating in a cup of water, full to the brim, as shown in the sketch. As the ice melts, what happens? Does the cup overflow, or does the water level drop, or does it remain the same? (You need to know Archimedes' Principle: A floating object displaces a volume of water whose weight equals the weight of the object.)

11. **Wrapping the World**   A ribbon is tied tightly around the earth at the equator. How much more ribbon would you need if you raised the ribbon 1 ft above the equator everywhere? (You don't need to know the radius of the earth to solve this problem.)

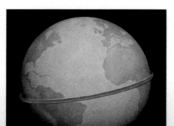

12. Irrational Powers   Prove that it's possible to raise an irrational number to an irrational power and get a rational result.   [*Hint:* The number $a = \sqrt{2}^{\sqrt{2}}$ is either rational or irrational. If a is rational, you are done. If a is irrational, consider $a^{\sqrt{2}}$.]

13. A Perfect Cube   Show that if you multiply three consecutive integers and then add the middle integer to the result, you get a perfect cube.

14. Number Patterns   Find the last digit in the number 3^{459}.   [*Hint:* Calculate the first few powers of 3, and look for a pattern.]

15. Number Patterns   Use the techniques of solving a simpler problem and looking for a pattern to evaluate the number

$$3999999999999^2$$

16. Ending Up Where You Started   A woman starts at a point P on the earth's surface and walks 1 mi south, then 1 mi east, then 1 mi north, and finds herself back at P, the starting point. Describe all points P for which this is possible (there are infinitely many).

17. Volume of a Truncated Pyramid   The ancient Egyptians, as a result of their pyramid building, knew that the volume of a pyramid with height h and square base of side length a is $V = \frac{1}{3}ha^2$. They were able to use this fact to prove that the volume of a truncated pyramid is $V = \frac{1}{3}h(a^2 + ab + b^2)$, where h is the height and b and a are the lengths of the sides of the square top and bottom, as shown in the figure. Prove the truncated pyramid volume formula.

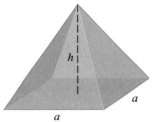
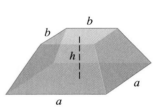

18. Area of a Ring   Find the area of the region between the two concentric circles shown in the figure.

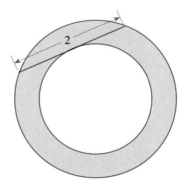

19. Bhaskara's Proof   The Indian mathematician Bhaskara sketched the two figures shown here and wrote below them, "Behold!" Explain how his sketches prove the Pythagorean Theorem.

20. Simple Numbers

(a) Use a calculator to find the value of the expression

$$\sqrt{3 + 2\sqrt{2}} - \sqrt{3 - 2\sqrt{2}}$$

The number looks very simple. Show that the calculated value is correct.

(b) Use a calculator to evaluate

$$\frac{\sqrt{2} + \sqrt{6}}{\sqrt{2} + \sqrt{3}}$$

Show that the calculated value is correct.

21. The Impossible Museum Tour   A museum is in the shape of a square with six rooms to a side; the entrance and exit are at diagonally opposite corners, as shown in the figure. Each pair of adjacent rooms is joined by a door. Some very efficient tourists would like to tour the museum by visiting each room *exactly* once. Can you find a path for such a tour? Here are examples of attempts that failed.

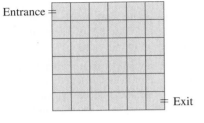

Entrance

Exit

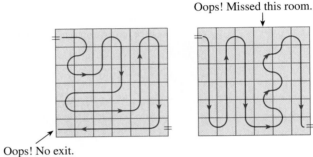

Oops! Missed this room.

Oops! No exit.

Here is how you can prove that the museum tour is not possible. Imagine that the rooms are colored black and white like a checkerboard.

(a) Show that the room colors alternate between white and black as the tourists walk through the museum.

(b) Use part (a) and the fact that there are an even number of rooms in the museum to conclude that the tour cannot end at the exit.

Equations and Inequalities

In this chapter we study equations and inequalities. We learn to solve linear, quadratic, power, and other equations using the rules of algebra. Solving equations that arise in the real world was the original motivation for inventing algebra. In this chapter we consider a variety of applications of equations and inequalities.

The following Warm-Up Exercises will help you review the basic algebra skills that you will need for this chapter.

WARM-UP EXERCISES

1. Simplify the following quantities:

    (a) $3x^2 - 4x + 2 + 5x + x^2$ (b) $(3y - 3) - (2y + 1)$ (c) $(2r + 1)(r - 4)$

2. Find the lowest common denominator (LCD) of the following fractions:

    (a) $\dfrac{1}{60}$ and $\dfrac{1}{90}$ (b) $\dfrac{1}{(2x + 1)}$ and $\dfrac{1}{x + 2}$

3. Compute the following:

    (a) $\sqrt[3]{64}$ (b) $\sqrt[4]{81}$ (c) $\sqrt[3]{27x^6}$ (d) $\sqrt{(2x + 1)^2}$

4. Compute 17% of 247.

5. (a) If a runner travels $\dfrac{5}{4}$ miles in 15 minutes, how fast is she traveling in miles per hour?

    (b) If a runner is traveling at 6 miles per hour, how long does it take her to run 1.5 miles?

6. If $a = 2$, $b = -7$, and $c = 3$, compute $\dfrac{-b - \sqrt{b^2 - 4ac}}{2a}$.

7. Factor the following expression: $x^3 + 2x^2 - 9x - 18$.

8. Compute the following:

    (a) $|63|$ (b) $|-\sqrt{2}|$ (c) $|2x + 1|$ if $x > 10$ (d) $|x - 2|$ if $x < 1$

1.1	Basic Equations

LEARNING OBJECTIVES

After completing this section, you will be able to:

■ Solve linear equations
■ Solve power equations
■ Solve for one variable in terms of others

Equations are the basic mathematical tool for solving real-world problems. In this chapter we learn how to solve equations, as well as how to construct equations that model real-life situations.

An equation is a statement that two mathematical expressions are equal. For example,

$$3 + 5 = 8$$

is an equation. Most equations that we study in algebra contain variables, which are symbols (usually letters) that stand for numbers. In the equation

$$4x + 7 = 19$$

the letter x is the variable. We think of x as the "unknown" in the equation, and our goal is to find the value of x that makes the equation true. The values of the unknown that make the equation true are called the **solutions** or **roots** of the equation, and the process of finding the solutions is called **solving the equation**.

Two equations with exactly the same solutions are called **equivalent equations**. To solve an equation, we try to find a simpler, equivalent equation in which the variable stands alone on one side of the "equal" sign. Here are the properties that we use to solve an equation. (In these properties, A, B, and C stand for any algebraic expressions, and the symbol $\Leftrightarrow$ means "is equivalent to.")

$x = 3$ is a solution of the equation $4x + 7 = 19$, because substituting $x = 3$ makes the equation true:

$$x = 3$$

$$4(3) + 7 = 19$$

PROPERTIES OF EQUALITY

Property	Description
1. $A = B \Leftrightarrow A + C = B + C$	Adding the same quantity to both sides of an equation gives an equivalent equation.
2. $A = B \Leftrightarrow CA = CB$ ($C \neq 0$)	Multiplying both sides of an equation by the same nonzero quantity gives an equivalent equation.

These properties require that you *perform the same operation on both sides of an equation* when solving it. Thus, if we say "*add* -7" when solving an equation, that is just a short way of saying "*add* -7 to each side of the equation."

■ **Solving Linear Equations**

The simplest type of equation is a *linear equation*, or first-degree equation, which is an equation in which each term is either a constant or a nonzero multiple of the variable.

LINEAR EQUATIONS

A **linear equation** in one variable is an equation that is equivalent to one of the form

$$ax + b = 0$$

where a and b are real numbers and x is the variable.

Here are some examples that illustrate the difference between linear and nonlinear equations.

Linear equations	Nonlinear equations	
$4x - 5 = 3$	$x^2 + 2x = 8$	Not linear; contains the square of the variable
$2x = \frac{1}{2}x - 7$	$\sqrt{x} - 6x = 0$	Not linear; contains the square root of the variable
$x - 6 = \frac{x}{3}$	$\frac{3}{x} - 2x = 1$	Not linear; contains the reciprocal of the variable

▶ **EXAMPLE 1** | Solving a Linear Equation

Solve the equation $7x - 4 = 3x + 8$.

▼ **SOLUTION** We solve this by changing it to an equivalent equation with all terms that have the variable x on one side and all constant terms on the other.

$$7x - 4 = 3x + 8 \qquad \text{Given equation}$$
$$(7x - 4) + 4 = (3x + 8) + 4 \qquad \text{Add 4}$$
$$7x = 3x + 12 \qquad \text{Simplify}$$
$$7x - 3x = (3x + 12) - 3x \qquad \text{Subtract } 3x$$
$$4x = 12 \qquad \text{Simplify}$$
$$\tfrac{1}{4} \cdot 4x = \tfrac{1}{4} \cdot 12 \qquad \text{Multiply by } \tfrac{1}{4}$$
$$x = 3 \qquad \text{Simplify}$$

Because it is important to CHECK YOUR ANSWER, we do this in many of our examples. In these checks, LHS stands for "left-hand side" and RHS stands for "right-hand side" of the original equation.

Check Your Answer	$x = 3$	$x = 3$
$x = 3$:	LHS $= 7(3) - 4$	RHS $= 3(3) + 8$
	$= 17$	$= 17$
LHS = RHS ✔		

▲

When a linear equation involves fractions, solving the equation is usually easier if we first multiply each side by the lowest common denominator (LCD) of the fractions, as we see in the following examples.

▶ **EXAMPLE 2** | Solving an Equation That Involves Fractions

Solve the equation $\dfrac{x}{6} + \dfrac{2}{3} = \dfrac{3}{4}x$.

▼ **SOLUTION** The LCD of the denominators 6, 3, and 4 is 12, so we first multiply each side of the equation by 12 to clear the denominators.

$$12 \cdot \left(\frac{x}{6} + \frac{2}{3} \right) = 12 \cdot \frac{3}{4}x \qquad \text{Multiply by LCD}$$
$$2x + 8 = 9x \qquad \text{Distributive Property}$$
$$8 = 7x \qquad \text{Subtract } 2x$$
$$\frac{8}{7} = x \qquad \text{Divide by 7}$$

The solution is $x = \frac{8}{7}$.

▲

In the next example we solve an equation that doesn't look like a linear equation, but it simplifies to one when we multiply by the LCD.

▶ **EXAMPLE 3** | An Equation Involving Fractional Expressions

Solve the equation $\dfrac{1}{x+1} + \dfrac{1}{x-2} = \dfrac{x+3}{x^2-x-2}$.

▼ **SOLUTION**   The LCD of the fractional expressions is $(x+1)(x-2) = x^2-x-2$. So as long as $x \neq -1$ and $x \neq 2$, we can multiply both sides of the equation by the LCD to get

$$(x+1)(x-2)\left(\frac{1}{x+1} + \frac{1}{x-2}\right) = (x+1)(x-2)\left(\frac{x+3}{x^2-x-2}\right) \quad \text{Mulitiply by LCD}$$

$$(x-2) + (x+1) = x+3 \qquad \text{Expand}$$

$$2x - 1 = x+3 \qquad \text{Simplify}$$

$$x = 4 \qquad \text{Solve}$$

The solution is $x = 4$.

Check Your Answer

$x = 4$:

$$\text{LHS} = \frac{1}{4+1} + \frac{1}{4-2} \qquad \text{RHS} = \frac{4+3}{4^2-4-2} = \frac{7}{10}$$

$$= \frac{1}{5} + \frac{1}{2} = \frac{7}{10}$$

LHS = RHS   ✔

▲

   It is always important to check your answer, even if you never make a mistake in your calculations. This is because you sometimes end up with **extraneous solutions**, which are potential solutions that do not satisfy the original equation. The next example shows how this can happen.

▶ **EXAMPLE 4** | An Equation with No Solution

Solve the equation $2 + \dfrac{5}{x-4} = \dfrac{x+1}{x-4}$.

▼ **SOLUTION**   First, we multiply each side by the common denominator, which is $x-4$.

$$(x-4)\left(2 + \frac{5}{x-4}\right) = (x-4)\left(\frac{x+1}{x-4}\right) \qquad \text{Multiply by } x-4$$

$$2(x-4) + 5 = x+1 \qquad \text{Expand}$$

$$2x - 8 + 5 = x+1 \qquad \text{Distributive Property}$$

$$2x - 3 = x+1 \qquad \text{Simplify}$$

$$2x = x+4 \qquad \text{Add 3}$$

$$x = 4 \qquad \text{Subtract } x$$

Check Your Answer

$x = 4$:

$$\text{LHS} = 2 + \frac{5}{4-4} = 2 + \frac{5}{0}$$

$$\text{RHS} = \frac{4+1}{4-4} = \frac{5}{0}$$

Impossible—can't divide by 0. LHS and RHS are undefined, so $x = 4$ is not a solution.   ✗

But now if we try to substitute $x = 4$ back into the original equation, we would be dividing by 0, which is impossible. So this equation has *no solution*.   ▲

The first step in the preceding solution, multiplying by $x-4$, had the effect of multiplying by 0. (Do you see why?) Multiplying each side of an equation by an expression that contains the variable may introduce extraneous solutions. That is why it is important to check every answer.

■ Solving Power Equations

Linear equations have variables only to the first power. Now let's consider some equations that involve squares, cubes, and other powers of the variable. Such equations will be studied more extensively in Sections 1.3 and 1.5. Here we just consider basic equations that can be simplified into the form $X^n = a$. Equations of this form are called **power equations** and can be solved by taking radicals of both sides of the equation.

"Algebra is a merry science," Uncle Jakob would say. "We go hunting for a little animal whose name we don't know, so we call it x. When we bag our game we pounce on it and give it its right name."

ALBERT EINSTEIN

SOLVING A POWER EQUATION

The power equation $X^n = a$ has the solution

$$X = \sqrt[n]{a} \qquad \text{if } n \text{ is odd}$$
$$X = \pm\sqrt[n]{a} \qquad \text{if } n \text{ is even and } a \geq 0$$

If n is even and $a < 0$, the equation has no real solution.

Here are some examples of solving power equations.

The equation $x^5 = 32$ has only one real solution: $x = \sqrt[5]{32} = 2$.

The equation $x^4 = 16$ has two real solutions: $x = \pm\sqrt[4]{16} = \pm 2$.

The equation $x^5 = -32$ has only one real solution: $x = \sqrt[5]{-32} = -2$.

The equation $x^4 = -16$ has no real solutions because $\sqrt[4]{-16}$ does not exist.

▶ **EXAMPLE 5** | Solving Power Equations

Solve each equation.

(a) $x^2 - 5 = 0$

(b) $(x - 4)^2 = 5$

▼ **SOLUTION**

(a) $x^2 - 5 = 0$

$$x^2 = 5 \qquad \text{Add 5}$$
$$x = \pm\sqrt{5} \qquad \text{Take the square root}$$

The solutions are $x = \sqrt{5}$ and $x = -\sqrt{5}$.

(b) We can take the square root of each side of this equation as well.

$$(x - 4)^2 = 5$$
$$x - 4 = \pm\sqrt{5} \qquad \text{Take the square root}$$
$$x = 4 \pm \sqrt{5} \qquad \text{Add 4}$$

The solutions are $x = 4 + \sqrt{5}$ and $x = 4 - \sqrt{5}$.

Be sure to check that each answer satisfies the original equation. ▲

We will revisit equations like the ones in Example 5 in Section 1.3.

▶ **EXAMPLE 6** | Solving Power Equations

Find all real solutions for each equation.

(a) $x^3 = -8$

(b) $16x^4 = 81$

▼ **SOLUTION**

(a) Since every real number has exactly one real cube root, we can solve this equation by taking the cube root of each side.

$$(x^3)^{1/3} = (-8)^{1/3}$$
$$x = -2$$

(b) Here we must remember that if n is even, then every positive real number has *two* real nth roots, a positive one and a negative one.

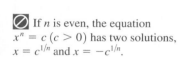 If n is even, the equation $x^n = c$ $(c > 0)$ has two solutions, $x = c^{1/n}$ and $x = -c^{1/n}$.

$$x^4 = \frac{81}{16} \qquad \text{Divide by 16}$$
$$(x^4)^{1/4} = \pm\left(\tfrac{81}{16}\right)^{1/4} \qquad \text{Take the fourth root}$$
$$x = \pm\tfrac{3}{2}$$

▲

The next example shows how to solve an equation that involves a fractional power of the variable.

▶ **EXAMPLE 7** | Solving an Equation with a Fractional Power

Solve the equation $5x^{2/3} - 2 = 43$.

▼ **SOLUTION**   The idea is to first isolate the term with the fractional exponent, then raise both sides of the equation to the *reciprocal* of that exponent.

If n is even, the equation $x^{n/m} = c$ has two solutions, $x = c^{m/n}$ and $x = -c^{m/n}$.

$$5x^{2/3} - 2 = 43$$
$$5x^{2/3} = 45 \qquad \text{Add 2}$$
$$x^{2/3} = 9 \qquad \text{Divide by 5}$$
$$x = \pm 9^{3/2} \qquad \text{Raise both sides to } \tfrac{3}{2} \text{ power}$$
$$x = \pm 27 \qquad \text{Simplify}$$

The solutions are $x = 27$ and $x = -27$.

Check Your Answers

$x = 27$:

LHS $= 5(27)^{2/3} - 2$

$= 5(9) - 2$

$= 43$

RHS $= 43$

LHS = RHS   ✔

$x = -27$:

LHS $= 5(-27)^{2/3} - 2$

$= 5(9) - 2$

$= 43$

RHS $= 43$

LHS = RHS   ✔

▲

■ Solving for One Variable in Terms of Others

Many formulas in the sciences involve several variables, and it is often necessary to express one of the variables in terms of the others. In the next example we solve for a variable in Newton's Law of Gravity.

▶ **EXAMPLE 8** | Solving for One Variable in Terms of Others

Solve for the variable M in the equation

$$F = G\frac{mM}{r^2}$$

This is Newton's Law of Gravity. It gives the gravitational force F between two masses m and M that are a distance r apart. The constant G is the universal gravitational constant.

▼ **SOLUTION**   Although this equation involves more than one variable, we solve it as usual by isolating M on one side and treating the other variables as we would numbers.

$$F = \left(\frac{Gm}{r^2}\right)M \qquad \text{Factor M from RHS}$$

$$\left(\frac{r^2}{Gm}\right)F = \left(\frac{r^2}{Gm}\right)\left(\frac{Gm}{r^2}\right)M \qquad \text{Multiply by reciprocal of } \frac{Gm}{r^2}$$

$$\frac{r^2 F}{Gm} = M \qquad \text{Simplify}$$

The solution is $M = \dfrac{r^2 F}{Gm}$.   ▲

▶ **EXAMPLE 9** | Solving for One Variable in Terms of Others

The surface area A of the closed rectangular box shown in Figure 1 can be calculated from the length l, the width w, and the height h according to the formula

$$A = 2lw + 2wh + 2lh$$

Solve for w in terms of the other variables in this equation.

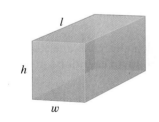

FIGURE 1 A closed rectangular box

▼ **SOLUTION**   Although this equation involves more than one variable, we solve it as usual by isolating w on one side, treating the other variables as we would numbers.

$$A = (2lw + 2wh) + 2lh \qquad \text{Collect terms involving } w$$

$$A - 2lh = 2lw + 2wh \qquad \text{Subtract } 2lh$$

$$A - 2lh = (2l + 2h)w \qquad \text{Factor } w \text{ from RHS}$$

$$\frac{A - 2lh}{2l + 2h} = w \qquad \text{Divide by } 2l + 2h$$

The solution is $w = \dfrac{A - 2lh}{2l + 2h}$.   ▲

1.2 | Modeling with Equations

LEARNING OBJECTIVES

After completing this section, you will be able to:

▪ Make equations that model real-world situations
▪ Solve problems about interest
▪ Solve problems about areas and lengths
▪ Solve problems about mixtures and concentrations
▪ Solve problems about the time needed to do a job
▪ Solve problems about distance, speed, and time

Many problems in the sciences, economics, finance, medicine, and numerous other fields can be translated into algebra problems; this is one reason that algebra is so useful. In this section we use equations as mathematical models to solve real-life problems.

▪ Making and Using Models

We will use the following guidelines to help us set up equations that model situations described in words. To show how the guidelines can help you to set up equations, we note them as we work each example in this section.

GUIDELINES FOR MODELING WITH EQUATIONS

1. **Identify the Variable.**   Identify the quantity that the problem asks you to find. This quantity can usually be determined by a careful reading of the question that is posed at the end of the problem. Then **introduce notation** for the variable (call it x or some other letter). ⊦

2. **Translate from Words to Algebra.**   Read each sentence in the problem again, and express all the quantities mentioned in the problem in terms of the variable you defined in Step 1. To organize this information, it is sometimes helpful to **draw a diagram** or **make a table**.

3. **Set Up the Model.**   Find the crucial fact in the problem that gives a relationship between the expressions you listed in Step 2. **Set up an equation** (or **model**) that expresses this relationship.

4. **Solve the Equation and Check Your Answer.**   Solve the equation, check your answer, and express it as a sentence that answers the question posed in the problem.

The following example illustrates how these guidelines are used to translate a "word problem" into the language of algebra.

▶ **EXAMPLE 1** | Renting a Car

A car rental company charges $30 a day and 15¢ a mile for renting a car. Helen rents a car for two days, and her bill comes to $108. How many miles did she drive?

▼ **SOLUTION**

Identify the variable.   We are asked to find the number of miles Helen has driven. So we let

$$x = \text{number of miles driven}$$

Translate from words to algebra.   Now we translate all the information given in the problem into the language of algebra.

In Words	In Algebra
Number of miles driven	x
Mileage cost (at $0.15 per mile)	$0.15x$
Daily cost (at $30 per day)	$2(30)$

Set up the model.   Now we set up the model.

$$\text{mileage cost} \quad + \quad \text{daily cost} \quad = \quad \text{total cost}$$

$$0.15x + 2(30) = 108$$

Solve.   Now we solve for x.

Check Your Answer

total cost = mileage cost + daily cost
   = 0.15(320) + 2(30)
   = 108   ✔

$$0.15x = 48 \qquad \text{Subtract 60}$$

$$x = \frac{48}{0.15} \qquad \text{Divide by 0.15}$$

$$x = 320 \qquad \text{Calculator}$$

Helen drove her rental car 320 miles.   ▲

In the examples and exercises that follow, we construct equations that model problems in many different real-life situations.

■ Problems About Interest

When you borrow money from a bank or when a bank "borrows" your money by keeping it for you in a savings account, the borrower in each case must pay for the privilege of using the money. The fee that is paid is called **interest**. The most basic type of interest is **simple interest**, which is just an annual percentage of the total amount borrowed or deposited. The amount of a loan or deposit is called the **principal** P. The annual percentage paid for the use of this money is the **interest rate** r. We will use the variable t to stand for the number of years that the money is on deposit and the variable I to stand for the total interest earned. The following **simple interest formula** gives the amount of interest I earned when a principal P is deposited for t years at an interest rate r.

$$\boxed{I = Prt}$$

Ø   When using this formula, remember to convert r from a percentage to a decimal. For example, in decimal form, 5% is 0.05. So at an interest rate of 5%, the interest paid on a $1000 deposit over a 3-year period is $I = Prt = 1000(0.05)(3) = \$150$.

▶ **EXAMPLE 2** | Interest on an Investment

Mary inherits $100,000 and invests it in two certificates of deposit. One certificate pays 6% and the other pays $4\frac{1}{2}\%$ simple interest annually. If Mary's total interest is $5025 per year, how much money is invested at each rate?

▼ **SOLUTION**

Identify the variable.   The problem asks for the amount she has invested at each rate. So we let

$$x = \text{the amount invested at } 6\%$$

Translate from words to algebra.   Since Mary's total inheritance is \$100,000, it follows that she invested $100,000 - x$ at $4\frac{1}{2}\%$. We translate all the information given into the language of algebra.

In Words	In Algebra
Amount invested at 6%	x
Amount invested at $4\frac{1}{2}\%$	$100,000 - x$
Interest earned at 6%	$0.06x$
Interest earned at $4\frac{1}{2}\%$	$0.045(100,000 - x)$

Set up the model.   We use the fact that Mary's total interest is \$5025 to set up the model.

$$\boxed{\text{interest at } 6\%} + \boxed{\text{interest at } 4\tfrac{1}{2}\%} = \boxed{\text{total interest}}$$

$$0.06x + 0.045(100,000 - x) = 5025$$

Solve.   Now we solve for x.

$$0.06x + 4500 - 0.045x = 5025 \qquad \text{Multiply}$$
$$0.015x + 4500 = 5025 \qquad \text{Combine the x-terms}$$
$$0.015x = 525 \qquad \text{Subtract 4500}$$
$$x = \frac{525}{0.015} = 35,000 \qquad \text{Divide by 0.015}$$

So Mary has invested \$35,000 at 6% and the remaining \$65,000 at $4\frac{1}{2}\%$.

Check Your Answer

$$\text{total interest} = 6\% \text{ of } \$35,000 + 4\tfrac{1}{2}\% \text{ of } \$65,000$$
$$= \$2100 + \$2925 = \$5025 \quad ✔$$

▲

■ **Problems About Area or Length**

When we use algebra to model a physical situation, we must sometimes use basic formulas from geometry. For example, we may need a formula for an area or a perimeter, or the formula that relates the sides of similar triangles, or the Pythagorean Theorem. Most of these formulas are listed on cards at the back of this book. The next two examples use these geometric formulas to solve some real-world problems.

▶ **EXAMPLE 3** | Dimensions of a Garden

A square garden has a walkway 3 ft wide around its outer edge, as shown in Figure 1. If the area of the entire garden, including the walkway, is 18,000 ft², what are the dimensions of the planted area?

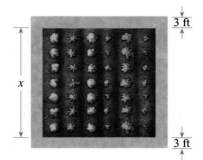

FIGURE 1

▼ **SOLUTION**

Identify the variable.   We are asked to find the length and width of the planted area. So we let

$$x = \text{the length of the planted area}$$

Translate from words to algebra.   Next, translate the information from Figure 1 into the language of algebra.

In Words	In Algebra
Length of planted area	x
Length of entire garden	$x + 6$
Area of entire garden	$(x + 6)^2$

Set up the model.   We now set up the model.

$$\text{area of entire garden} = 18{,}000 \text{ ft}^2$$

$$(x + 6)^2 = 18{,}000$$

Solve.   Now we solve for x.

$$x + 6 = \sqrt{18{,}000} \qquad \text{Take square roots}$$

$$x = \sqrt{18{,}000} - 6 \qquad \text{Subtract 6}$$

$$x \approx 128$$

The planted area of the garden is about 128 ft by 128 ft.   ▲

▶ **EXAMPLE 4** | Determining the Height of a Building Using Similar Triangles

A man 6 ft tall wishes to find the height of a certain four-story building. He measures its shadow and finds it to be 28 ft long, while his own shadow is $3\frac{1}{2}$ ft long. How tall is the building?

▼ **SOLUTION**

Identify the variable.   The problem asks for the height of the building. So let

$$h = \text{the height of the building}$$

Translate from words to algebra.   We use the fact that the triangles in Figure 2 are similar. Recall that for any pair of similar triangles the ratios of corresponding sides are equal. Now we translate these observations into the language of algebra.

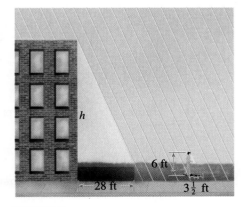

FIGURE 2

In Words	In Algebra
Height of building	h
Ratio of height to base in large triangle	$\frac{h}{28}$
Ratio of height to base in small triangle	$\frac{6}{3.5}$

Set up the model.   Since the large and small triangles are similar, we get the equation

$$\boxed{\begin{array}{c}\text{ratio of height to}\\ \text{base in large triangle}\end{array}} = \boxed{\begin{array}{c}\text{ratio of height to}\\ \text{base in small triangle}\end{array}}$$

$$\frac{h}{28} = \frac{6}{3.5}$$

Solve.   Now we solve for h.

$$h = \frac{6 \cdot 28}{3.5} = 48 \qquad \text{Multiply by 28}$$

So the building is 48 ft tall.   ▲

■ Problems About Mixtures

Many real-world problems involve mixing different types of substances. For example, construction workers may mix cement, gravel, and sand; fruit juice from concentrate may involve mixing different types of juices. Problems involving mixtures and concentrations make use of the fact that if an amount x of a substance is dissolved in a solution with volume V, then the concentration C of the substance is given by

$$C = \frac{x}{V}$$

So if 10 g of sugar is dissolved in 5 L of water, then the sugar concentration is $C = 10/5 = 2$ g/L. Solving a mixture problem usually requires us to analyze the amount x of the substance that is in the solution. When we solve for x in this equation, we see that $x = CV$. Note that in many mixture problems the concentration C is expressed as a percentage, as in the next example.

▶ **EXAMPLE 5** | Mixtures and Concentration

A manufacturer of soft drinks advertises their orange soda as "naturally flavored," although it contains only 5% orange juice. A new federal regulation stipulates that to be called "natural," a drink must contain at least 10% fruit juice. How much pure orange juice must this manufacturer add to 900 gal of orange soda to conform to the new regulation?

▼ **SOLUTION**

Identify the variable.   The problem asks for the amount of pure orange juice to be added. So let

$$x = \text{the amount (in gallons) of pure orange juice to be added}$$

Translate from words to algebra.   In any problem of this type—in which two different substances are to be mixed—drawing a diagram helps us to organize the given information (see Figure 3).

FIGURE 3

We now translate the information in the figure into the language of algebra.

In Words	In Algebra
Amount of orange juice to be added	x
Amount of the mixture	$900 + x$
Amount of orange juice in the first vat	$0.05(900) = 45$
Amount of orange juice in the second vat	$1 \cdot x = x$
Amount of orange juice in the mixture	$0.10(900 + x)$

Set up the model.   To set up the model, we use the fact that the total amount of orange juice in the mixture is equal to the orange juice in the first two vats.

$$\boxed{\begin{array}{c}\text{amount of} \\ \text{orange juice} \\ \text{in first vat}\end{array}} + \boxed{\begin{array}{c}\text{amount of} \\ \text{orange juice} \\ \text{in second vat}\end{array}} = \boxed{\begin{array}{c}\text{amount of} \\ \text{orange juice} \\ \text{in mixture}\end{array}}$$

$$45 + x = 0.1(900 + x) \qquad \text{From Figure 3}$$

Solve.   Now we solve for x.

$$45 + x = 90 + 0.1x \qquad \text{Distributive Property}$$

$$0.9x = 45 \qquad \text{Subtract 0.1x and 45}$$

$$x = \frac{45}{0.9} = 50 \qquad \text{Divide by 0.9}$$

The manufacturer should add 50 gal of pure orange juice to the soda.

Check Your Answer

$$\text{amount of juice before mixing} = 5\% \text{ of } 900 \text{ gal} + 50 \text{ gal pure juice}$$

$$= 45 \text{ gal} + 50 \text{ gal} = 95 \text{ gal}$$

$$\text{amount of juice after mixing} = 10\% \text{ of } 950 \text{ gal} = 95 \text{ gal}$$

Amounts are equal.   ✔

■ Problems About the Time Needed to Do a Job

When solving a problem that involves determining how long it takes several workers to complete a job, we use the fact that if a person or machine takes H time units to complete the task, then in one time unit the fraction of the task that has been completed is $1/H$. For example, if a worker takes 5 hours to mow a lawn, then in 1 hour the worker will mow 1/5 of the lawn.

▶ **EXAMPLE 6** | Time Needed to Do a Job

Because of an anticipated heavy rainstorm, the water level in a reservoir must be lowered by 1 ft. Opening spillway A lowers the level by this amount in 4 hours, whereas opening the smaller spillway B does the job in 6 hours. How long will it take to lower the water level by 1 ft if both spillways are opened?

▼ **SOLUTION**

Identify the variable. We are asked to find the time needed to lower the level by 1 ft if both spillways are open. So let

$$x = \text{the time (in hours) it takes to lower the water level}$$
$$\text{by 1 ft if both spillways are open}$$

Translate from words to algebra. Finding an equation relating x to the other quantities in this problem is not easy. Certainly x is not simply $4 + 6$, because that would mean that together the two spillways require longer to lower the water level than either spillway alone. Instead, *we look at the fraction of the job that can be done in 1 hour by each spillway.*

In Words	In Algebra
Time it takes to lower level 1 ft with A and B together	x h
Distance A lowers level in 1 h	$\frac{1}{4}$ ft
Distance B lowers level in 1 h	$\frac{1}{6}$ ft
Distance A and B together lower levels in 1 h	$\frac{1}{x}$ ft

Set up the model. Now we set up the model.

$$\text{fraction done by A} \; + \; \text{fraction done by B} \; = \; \text{fraction done by both}$$

$$\frac{1}{4} + \frac{1}{6} = \frac{1}{x}$$

Solve. Now we solve for x.

$$3x + 2x = 12 \quad \text{Multiply by the LCD, 12x}$$

$$5x = 12 \quad \text{Add}$$

$$x = \frac{12}{5} \quad \text{Divide by 5}$$

It will take $2\frac{2}{5}$ hours, or 2 h 24 min, to lower the water level by 1 ft if both spillways are open. ▲

■ Problems About Distance, Rate, and Time

The next example deals with distance, rate (speed), and time. The formula to keep in mind here is

$$\boxed{\text{distance} = \text{rate} \times \text{time}}$$

where the rate is either the constant speed or average speed of a moving object. For example, driving at 60 mi/h for 4 hours takes you a distance of $60 \cdot 4 = 240$ mi.

▶ **EXAMPLE 7** | Distance, Speed, and Time

Bill left his house at 2:00 P.M. and rode his bicycle down Main Street at a speed of 12 mi/h. When his friend Mary arrived at his house at 2:10 P.M., Bill's mother told her the direction in which Bill had gone, and Mary cycled after him at a speed of 16 mi/h. At what time did Mary catch up with Bill?

▼ **SOLUTION**

Identify the variable.   We are asked to find the time that it took Mary to catch up with Bill. Let

$$t = \text{the time (in hours) it took Mary to catch up with Bill}$$

Translate from words to algebra.   In problems involving motion, it is often helpful to organize the information in a table, using the formula distance = rate × time. First we fill in the "Speed" column in the table, since we are told the speeds at which Mary and Bill cycled. Then we fill in the "Time" column. (Because Bill had a 10-minute, or $\frac{1}{6}$-hour head start, he cycled for $t + \frac{1}{6}$ hours.) Finally, we multiply these columns to calculate the entries in the "Distance" column.

	Distance (mi)	Speed (mi/h)	Time (h)
Mary	$16t$	16	t
Bill	$12\left(t + \frac{1}{6}\right)$	12	$t + \frac{1}{6}$

Set up the model.   At the instant when Mary caught up with Bill, they had both cycled the same distance. We use this fact to set up the model for this problem.

$$\text{distance traveled by Mary} = \text{distance traveled by Bill}$$

$$16t = 12\left(t + \tfrac{1}{6}\right) \qquad \text{From table}$$

Solve.   Now we solve for t.

$$16t = 12t + 2 \qquad \text{Distributive Property}$$

$$4t = 2 \qquad \text{Subtract } 12t$$

$$t = \tfrac{1}{2} \qquad \text{Divide by 4}$$

Mary caught up with Bill after cycling for half an hour, that is, at 2:40 P.M.

Check Your Answer

Bill traveled for $\frac{1}{2} + \frac{1}{6} = \frac{2}{3}$ h, so

$$\text{distance Bill traveled} = 12 \text{ mi/h} \times \tfrac{2}{3} \text{h} = 8 \text{ mi}$$

$$\text{distance Mary traveled} = 16 \text{ mi/h} \times \tfrac{1}{2} \text{h} = 8 \text{ mi}$$

Distances are equal.   ✔

▲

1.3 | Quadratic Equations

LEARNING OBJECTIVES

After completing this section, you will be able to:

- Solve quadratic equations by factoring
- Solve quadratic equations by completing the square
- Solve quadratic equations using the Quadratic Formula
- Model with quadratic equations

Linear Equations

$$4x = -7$$

$$6x - 8 = 21$$

$$2 + 3x = \tfrac{1}{2} - \tfrac{3}{4}x$$

Quadratic Equations

$$x^2 - 2x - 8 = 0$$

$$3x + 10 = 4x^2$$

$$\tfrac{1}{2}x^2 + \tfrac{1}{3}x - \tfrac{1}{6} = 0$$

In Section 1.1 we learned how to solve linear equations, which are first-degree equations such as $2x + 1 = 5$ or $4 - 3x = 2$. In this section we learn how to solve quadratic equations, which are second-degree equations such as $x^2 + 2x - 3 = 0$ or $2x^2 + 3 = 5x$. We will also see that many real-life problems can be modeled using quadratic equations.

QUADRATIC EQUATIONS

A **quadratic equation** is an equation of the form

$$ax^2 + bx + c = 0$$

where a, b, and c are real numbers with $a \neq 0$.

■ Solving Quadratic Equations by Factoring

Some quadratic equations can be solved by factoring and using the following basic property of real numbers.

ZERO-PRODUCT PROPERTY

$$AB = 0 \quad \text{if and only if} \quad A = 0 \quad \text{or} \quad B = 0$$

This means that if we can factor the left-hand side of a quadratic (or other) equation, then we can solve it by setting each factor equal to 0 in turn. This method works only when the right-hand side of the equation is 0.

▶ **EXAMPLE 1** | Solving a Quadratic Equation by Factoring

Solve the equation $x^2 + 5x = 24$.

▼ **SOLUTION**   We must first rewrite the equation so that the right-hand side is 0.

$$x^2 + 5x = 24 \qquad \text{Given equation}$$

$$x^2 + 5x - 24 = 0 \qquad \text{Subtract 24}$$

$$(x - 3)(x + 8) = 0 \qquad \text{Factor}$$

$$x - 3 = 0 \quad \text{or} \quad x + 8 = 0 \qquad \text{Zero-Product Property}$$

$$x = 3 \qquad\qquad x = -8 \qquad \text{Solve}$$

The solutions are $x = 3$ and $x = -8$.

Check Your Answers

$x = 3$:

$\quad (3)^2 + 5(3) = 9 + 15 = 24 \quad ✔$

$x = -8$:

$\quad (-8)^2 + 5(-8) = 64 - 40 = 24$
$\qquad\qquad\qquad\qquad\qquad\qquad ✔$

Do you see why one side of the equation must be 0 in Example 1? Factoring the equation as $x(x + 5) = 24$ does not help us find the solutions, since 24 can be factored in infinitely many ways, such as $6 \cdot 4, \frac{1}{2} \cdot 48, \left(-\frac{2}{5}\right) \cdot (-60)$, and so on.

■ Solving Quadratic Equations by Completing the Square

As we saw in Section 1.1, Example 5(b), if a quadratic equation is of the form $(x \pm a)^2 = c$, then we can solve it by taking the square root of each side. In an equation of this form the left-hand side is a *perfect square*: the square of a linear expression in x. So if a quadratic equation does not factor readily, then we can solve it by **completing the square.**

Completing the Square
Area of blue region is

$$x^2 + 2\left(\frac{b}{2}\right)x = x^2 + bx$$

Add a small square of area $(b/2)^2$ to "complete" the square.

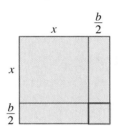

COMPLETING THE SQUARE

To make $x^2 + bx$ a perfect square, add $\left(\dfrac{b}{2}\right)^2$, the square of half the coefficient of x. This gives the perfect square

$$x^2 + bx + \left(\frac{b}{2}\right)^2 = \left(x + \frac{b}{2}\right)^2$$

To complete the square, we add a constant to a quadratic expression to make it a perfect square. For example, to make

$$x^2 + 6x$$

a perfect square, we must add $\left(\frac{6}{2}\right)^2 = 9$. Then

$$x^2 + 6x + 9 = (x + 3)^2$$

is a perfect square. The table gives some more examples of completing the square.

Expression	Add	Complete the square
$x^2 + 8x$	$\left(\dfrac{8}{2}\right)^2 = 16$	$x^2 + 8x + 16 = (x + 4)^2$
$x^2 - 12x$	$\left(-\dfrac{12}{2}\right)^2 = 36$	$x^2 - 12x + 36 = (x - 6)^2$
$x^2 + 3x$	$\left(\dfrac{3}{2}\right)^2 = \dfrac{9}{4}$	$x^2 + 3x + \dfrac{9}{4} = \left(x + \dfrac{3}{2}\right)^2$
$x^2 - \sqrt{3}x$	$\left(-\dfrac{\sqrt{3}}{2}\right)^2 = \dfrac{3}{4}$	$x^2 - \sqrt{3}x + \dfrac{3}{4} = \left(x - \dfrac{\sqrt{3}}{2}\right)^2$

▶ **EXAMPLE 2** | Solving Quadratic Equations by Completing the Square

Solve each equation.

(a) $x^2 - 8x + 13 = 0$ **(b)** $3x^2 - 12x + 6 = 0$

▼ **SOLUTION**

(a)

$x^2 - 8x + 13 = 0$	Given equation
$x^2 - 8x = -13$	Subtract 13
$x^2 - 8x + 16 = -13 + 16$	Complete the square: add $\left(\dfrac{-8}{2}\right)^2 = 16$
$(x - 4)^2 = 3$	Perfect square
$x - 4 = \pm\sqrt{3}$	Take square root
$x = 4 \pm \sqrt{3}$	Add 4

⊘ When completing the square, make sure the coefficient of x^2 is 1. If it isn't, you must factor this coefficient from both terms that contain x:

$$ax^2 + bx = a\left(x^2 + \frac{b}{a}x\right)$$

Then complete the square inside the parentheses. Remember that the term added inside the parentheses is multiplied by a.

(b) After subtracting 6 from each side of the equation, we must factor the coefficient of x^2 (the 3) from the left side to put the equation in the correct form for completing the square.

$$
\begin{aligned}
3x^2 - 12x + 6 &= 0 && \text{Given equation}\\
3x^2 - 12x &= -6 && \text{Subtract 6}\\
3(x^2 - 4x) &= -6 && \text{Factor 3 from LHS}
\end{aligned}
$$

Now we complete the square by adding $(-2)^2 = 4$ *inside* the parentheses. Since everything inside the parentheses is multiplied by 3, this means that we are actually adding $3 \cdot 4 = 12$ to the left side of the equation. Thus, we must add 12 to the right side as well.

$$
\begin{aligned}
3(x^2 - 4x + 4) &= -6 + 3 \cdot 4 && \text{Complete the square: add 4}\\
3(x - 2)^2 &= 6 && \text{Perfect square}\\
(x - 2)^2 &= 2 && \text{Divide by 3}\\
x - 2 &= \pm\sqrt{2} && \text{Take square root}\\
x &= 2 \pm \sqrt{2} && \text{Add 2}
\end{aligned}
$$

▲

■ The Quadratic Formula

We can use the technique of completing the square to derive a formula for the roots of the general quadratic equation $ax^2 + bx + c = 0$.

THE QUADRATIC FORMULA

The roots of the quadratic equation $ax^2 + bx + c = 0$, where $a \neq 0$, are

$$x = \frac{-b \pm \sqrt{b^2 - 4ac}}{2a}$$

▼ **PROOF**   First, we divide each side of the equation by a and move the constant to the right side, giving

$$x^2 + \frac{b}{a}x = -\frac{c}{a} \qquad \text{Divide by } a$$

We now complete the square by adding $(b/2a)^2$ to each side of the equation:

$$
\begin{aligned}
x^2 + \frac{b}{a}x + \left(\frac{b}{2a}\right)^2 &= -\frac{c}{a} + \left(\frac{b}{2a}\right)^2 && \text{Complete the square: Add } \left(\frac{b}{2a}\right)^2\\
\left(x + \frac{b}{2a}\right)^2 &= \frac{-4ac + b^2}{4a^2} && \text{Perfect square}\\
x + \frac{b}{2a} &= \pm\frac{\sqrt{b^2 - 4ac}}{2a} && \text{Take square root}\\
x &= \frac{-b \pm \sqrt{b^2 - 4ac}}{2a} && \text{Subtract } \frac{b}{2a}
\end{aligned}
$$

▲

The Quadratic Formula could be used to solve the equations in Examples 1 and 2. You should carry out the details of these calculations.

▶ **EXAMPLE 3** | Using the Quadratic Formula

Find all solutions of each equation.

(a) $3x^2 - 5x - 1 = 0$ (b) $4x^2 + 12x + 9 = 0$ (c) $x^2 + 2x + 2 = 0$

▼ **SOLUTION**

(a) In this quadratic equation $a = 3$, $b = -5$, and $c = -1$.

$$b = -5$$
$$3x^2 - 5x - 1 = 0$$
$$a = 3 \qquad c = -1$$

By the Quadratic Formula,

$$x = \frac{-(-5) \pm \sqrt{(-5)^2 - 4(3)(-1)}}{2(3)} = \frac{5 \pm \sqrt{37}}{6}$$

Another Method

$4x^2 + 12x + 9 = 0$

$(2x + 3)^2 = 0$

$2x + 3 = 0$

$x = -\frac{3}{2}$

If approximations are desired, we can use a calculator to obtain

$$x = \frac{5 + \sqrt{37}}{6} \approx 1.8471 \qquad \text{and} \qquad x = \frac{5 - \sqrt{37}}{6} \approx -0.1805$$

(b) Using the Quadratic Formula with $a = 4$, $b = 12$, and $c = 9$ gives

$$x = \frac{-12 \pm \sqrt{(12)^2 - 4 \cdot 4 \cdot 9}}{2 \cdot 4} = \frac{-12 \pm 0}{8} = -\frac{3}{2}$$

This equation has only one solution, $x = -\frac{3}{2}$.

(c) Using the Quadratic Formula with $a = 1$, $b = 2$, and $c = 2$ gives

$$x = \frac{-2 \pm \sqrt{2^2 - 4 \cdot 2}}{2} = \frac{-2 \pm \sqrt{-4}}{2} = \frac{-2 \pm 2\sqrt{-1}}{2} = -1 \pm \sqrt{-1}$$

Since the square of any real number is nonnegative, $\sqrt{-1}$ is undefined in the real number system. The equation has no real solution.   ▲

In the next section we study the complex number system, in which the square roots of negative numbers do exist. The equation in Example 3(c) does have solutions in the complex number system.

■ The Discriminant

The quantity $b^2 - 4ac$ that appears under the square root sign in the Quadratic Formula is called the *discriminant* of the equation $ax^2 + bx + c = 0$ and is given the symbol D. If $D < 0$, then $\sqrt{b^2 - 4ac}$ is undefined, and the quadratic equation has no real solution, as in Example 3(c). If $D = 0$, then the equation has only one real solution, as in Example 3(b). Finally, if $D > 0$, then the equation has two distinct real solutions, as in Example 3(a). The following box summarizes these observations.

> **THE DISCRIMINANT**
>
> The **discriminant** of the general quadratic $ax^2 + bx + c = 0$ $(a \neq 0)$ is $D = b^2 - 4ac$.
>
> **1.** If $D > 0$, then the equation has two distinct real solutions.
>
> **2.** If $D = 0$, then the equation has exactly one real solution.
>
> **3.** If $D < 0$, then the equation has no real solution.

▶ **EXAMPLE 4** | Using the Discriminant

Use the discriminant to determine how many real solutions each equation has.

(a) $x^2 + 4x - 1 = 0$ **(b)** $4x^2 - 12x + 9 = 0$ **(c)** $\frac{1}{3}x^2 - 2x + 4 = 0$

▼ **SOLUTION**

(a) The discriminant is $D = 4^2 - 4(1)(-1) = 20 > 0$, so the equation has two distinct real solutions.

(b) The discriminant is $D = (-12)^2 - 4 \cdot 4 \cdot 9 = 0$, so the equation has exactly one real solution.

(c) The discriminant is $D = (-2)^2 - 4\left(\frac{1}{3}\right)4 = -\frac{4}{3} < 0$, so the equation has no real solution.

▲

■ Modeling with Quadratic Equations

Let's look at some real-life problems that can be modeled by quadratic equations. The principles discussed in Section 1.2 for setting up equations as models are useful here as well.

▶ **EXAMPLE 5** | Dimensions of a Building Lot

A rectangular building lot is 8 ft longer than it is wide and has an area of 2900 ft². Find the dimensions of the lot.

▼ **SOLUTION**   We are asked to find the width and length of the lot. So let

Identify the variable ▶

$$w = \text{width of lot}$$

Then we translate the information given in the problem into the language of algebra (see Figure 1).

In Words	In Algebra
Width of lot	w
Length of lot	$w + 8$

Translate from words to algebra ▶

$w + 8$

FIGURE 1

Now we set up the model.

Set up the model ▶

$$\text{width of lot} \times \text{length of lot} = \text{area of lot}$$

Solve ▶

$$w(w + 8) = 2900$$
$$w^2 + 8w = 2900 \qquad \text{Expand}$$
$$w^2 + 8w - 2900 = 0 \qquad \text{Subtract 2900}$$
$$(w - 50)(w + 58) = 0 \qquad \text{Factor}$$
$$w = 50 \quad \text{or} \quad w = -58 \qquad \text{Zero-Product Property}$$

Since the width of the lot must be a positive number, we conclude that $w = 50$ ft. The length of the lot is $w + 8 = 50 + 8 = 58$ ft. ▲

▶ **EXAMPLE 6** | A Distance-Speed-Time Problem

A jet flew from New York to Los Angeles, a distance of 4200 km. The speed for the return trip was 100 km/h faster than the outbound speed. If the total trip took 13 hours, what was the jet's speed from New York to Los Angeles?

▼ **SOLUTION** We are asked for the speed of the jet from New York to Los Angeles. So let

Identify the variable ▶

$$s = \text{speed from New York to Los Angeles}$$

Then $\qquad s + 100 = \text{speed from Los Angeles to New York}$

Now we organize the information in a table. We fill in the "Distance" column first, since we know that the cities are 4200 km apart. Then we fill in the "Speed" column, since we have expressed both speeds (rates) in terms of the variable s. Finally, we calculate the entries for the "Time" column, using

$$\text{time} = \frac{\text{distance}}{\text{rate}}$$

Translate from words to algebra ▶

	Distance (km)	Speed (km/h)	Time (h)
N.Y. to L.A.	4200	s	$\dfrac{4200}{s}$
L.A. to N.Y.	4200	$s + 100$	$\dfrac{4200}{s + 100}$

The total trip took 13 hours, so we have the model

Set up the model ▶

$$\text{time from N.Y. to L.A.} + \text{time from L.A. to N.Y.} = \text{total time}$$

$$\frac{4200}{s} + \frac{4200}{s + 100} = 13$$

Multiplying by the common denominator, $s(s + 100)$, we get

$$4200(s + 100) + 4200s = 13s(s + 100)$$
$$8400s + 420{,}000 = 13s^2 + 1300s$$
$$0 = 13s^2 - 7100s - 420{,}000$$

Although this equation does factor, with numbers this large it is probably quicker to use the Quadratic Formula and a calculator.

Solve ▶

$$s = \frac{7100 \pm \sqrt{(-7100)^2 - 4(13)(-420,000)}}{2(13)}$$

$$= \frac{7100 \pm 8500}{26}$$

$$s = 600 \quad \text{or} \quad s = \frac{-1400}{26} \approx -53.8$$

Since s represents speed, we reject the negative answer and conclude that the jet's speed from New York to Los Angeles was 600 km/h. ▲

EXAMPLE 7 | The Path of a Projectile

This formula depends on the fact that acceleration due to gravity is constant near the earth's surface. Here, we neglect the effect of air resistance.

An object thrown or fired straight upward at an initial speed of v_0 ft/s will reach a height of h feet after t seconds, where h and t are related by the formula

$$h = -16t^2 + v_0 t$$

Suppose that a bullet is shot straight upward with an initial speed of 800 ft/s. Its path is shown in Figure 2.

(a) When does the bullet fall back to ground level?

(b) When does it reach a height of 6400 ft?

(c) When does it reach a height of 2 mi?

(d) How high is the highest point the bullet reaches?

▼ **SOLUTION**   Since the initial speed in this case is $v_0 = 800$ ft/s, the formula is

$$h = -16t^2 + 800t$$

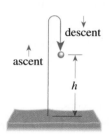

FIGURE 2

(a) Ground level corresponds to $h = 0$, so we must solve the equation

$$0 = -16t^2 + 800t \qquad \text{Set } h = 0$$
$$0 = -16t(t - 50) \qquad \text{Factor}$$

Thus, $t = 0$ or $t = 50$. This means the bullet starts ($t = 0$) at ground level and returns to ground level after 50 s.

(b) Setting $h = 6400$ gives the equation

$$6400 = -16t^2 + 800t \qquad \text{Set } h = 6400$$
$$16t^2 - 800t + 6400 = 0 \qquad \text{All terms to LHS}$$
$$t^2 - 50t + 400 = 0 \qquad \text{Divide by 16}$$
$$(t - 10)(t - 40) = 0 \qquad \text{Factor}$$
$$t = 10 \quad \text{or} \quad t = 40 \qquad \text{Solve}$$

The bullet reaches 6400 ft after 10 s (on its ascent) and again after 40 s (on its descent to earth).

(c) Two miles is $2 \times 5280 = 10,560$ ft.

$$10,560 = -16t^2 + 800t \qquad \text{Set } h = 10,560$$
$$16t^2 - 800t + 10,560 = 0 \qquad \text{All terms to LHS}$$
$$t^2 - 50t + 660 = 0 \qquad \text{Divide by 16}$$

The discriminant of this equation is $D = (-50)^2 - 4(660) = -140$, which is negative. Thus, the equation has no real solution. The bullet never reaches a height of 2 mi.

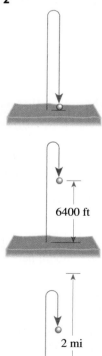

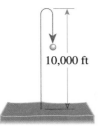

10,000 ft

(d) Each height that the bullet reaches is attained twice—once on its ascent and once on its descent. The only exception is the highest point of its path, which is reached only once. This means that for the highest value of h, the following equation has only one solution for t:

$$h = -16t^2 + 800t$$

$$16t^2 - 800t + h = 0 \qquad \text{All terms to LHS}$$

This in turn means that the discriminant D of the equation is 0, so

$$D = (-800)^2 - 4(16)h = 0$$

$$640,000 - 64h = 0$$

$$h = 10,000$$

The maximum height reached is 10,000 ft. ▲

1.4	Complex Numbers

LEARNING OBJECTIVES
After completing this section, you will be able to:

- Add and subtract complex numbers
- Multiply and divide complex numbers
- Simplify expressions with roots of negative numbers
- Find complex roots of quadratic equations

In Section 1.3 we saw that if the discriminant of a quadratic equation is negative, the equation has no real solution. For example, the equation

$$x^2 + 4 = 0$$

has no real solution. If we try to solve this equation, we get $x^2 = -4$, so

$$x = \pm\sqrt{-4}$$

But this is impossible, since the square of any real number is positive. [For example, $(-2)^2 = 4$, a positive number.] Thus, negative numbers don't have real square roots.

To make it possible to solve *all* quadratic equations, mathematicians invented an expanded number system, called the *complex number system*. First they defined the new number

$$i = \sqrt{-1}$$

This means that $i^2 = -1$. A complex number is then a number of the form $a + bi$, where a and b are real numbers.

DEFINITION OF COMPLEX NUMBERS

A **complex number** is an expression of the form

$$a + bi$$

where a and b are real numbers and $i^2 = -1$. The **real part** of this complex number is a and the **imaginary part** is b. Two complex numbers are **equal** if and only if their real parts are equal and their imaginary parts are equal.

Note that both the real and imaginary parts of a complex number are real numbers.

▶ **EXAMPLE 1** | Complex Numbers

The following are examples of complex numbers.

$3 + 4i$ Real part 3, imaginary part 4

$\frac{1}{2} - \frac{2}{3}i$ Real part $\frac{1}{2}$, imaginary part $-\frac{2}{3}$

$6i$ Real part 0, imaginary part 6

-7 Real part -7, imaginary part 0   ▲

A number such as $6i$, which has real part 0, is called a **pure imaginary number**. A real number like -7 can be thought of as a complex number with imaginary part 0.

In the complex number system every quadratic equation has solutions. The numbers $2i$ and $-2i$ are solutions of $x^2 = -4$ because

$$(2i)^2 = 2^2 i^2 = 4(-1) = -4 \qquad \text{and} \qquad (-2i)^2 = (-2)^2 i^2 = 4(-1) = -4$$

Although we use the term *imaginary* in this context, imaginary numbers should not be thought of as any less "real" (in the ordinary rather than the mathematical sense of that word) than negative numbers or irrational numbers. All numbers (except possibly the positive integers) are creations of the human mind—the numbers -1 and $\sqrt{2}$ as well as the number i. We study complex numbers because they complete, in a useful and elegant fashion, our study of the solutions of equations. In fact, imaginary numbers are useful not only in algebra and mathematics, but in the other sciences as well. To give just one example, in electrical theory the *reactance* of a circuit is a quantity whose measure is an imaginary number.

■ Arithmetic Operations on Complex Numbers

Complex numbers are added, subtracted, multiplied, and divided just as we would any number of the form $a + b\sqrt{c}$. The only difference that we need to keep in mind is that $i^2 = -1$. Thus, the following calculations are valid.

$$(a + bi)(c + di) = ac + (ad + bc)i + bdi^2 \qquad \text{Multiply and collect like terms}$$
$$= ac + (ad + bc)i + bd(-1) \qquad i^2 = -1$$
$$= (ac - bd) + (ad + bc)i \qquad \text{Combine real and imaginary parts}$$

We therefore define the sum, difference, and product of complex numbers as follows.

ADDING, SUBTRACTING, AND MULTIPLYING COMPLEX NUMBERS

Definition

Addition

$(a + bi) + (c + di) = (a + c) + (b + d)i$

Subtraction

$(a + bi) - (c + di) = (a - c) + (b - d)i$

Multiplication

$(a + bi) \cdot (c + di) = (ac - bd) + (ad + bc)i$

Description

To add complex numbers, add the real parts and the imaginary parts.

To subtract complex numbers, subtract the real parts and the imaginary parts.

Multiply complex numbers like binomials, using $i^2 = -1$.

Graphing calculators can perform arithmetic operations on complex numbers.

```
(3+5i)+(4-2i)
              7+3i
(3+5i)*(4-2i)
              22+14i
```

▶ **EXAMPLE 2** | Adding, Subtracting, and Multiplying Complex Numbers

Express the following in the form $a + bi$.

(a) $(3 + 5i) + (4 - 2i)$ **(b)** $(3 + 5i) - (4 - 2i)$

(c) $(3 + 5i)(4 - 2i)$ **(d)** i^{23}

▼ **SOLUTION**

(a) According to the definition, we add the real parts and we add the imaginary parts.

$$(3 + 5i) + (4 - 2i) = (3 + 4) + (5 - 2)i = 7 + 3i$$

(b) $(3 + 5i) - (4 - 2i) = (3 - 4) + [5 - (-2)]i = -1 + 7i$

(c) $(3 + 5i)(4 - 2i) = [3 \cdot 4 - 5(-2)] + [3(-2) + 5 \cdot 4]i = 22 + 14i$

(d) $i^{23} = i^{22+1} = (i^2)^{11} i = (-1)^{11} i = (-1)i = -i$ ▲

Complex Conjugates

Number	Conjugate
$3 + 2i$	$3 - 2i$
$1 - i$	$1 + i$
$4i$	$-4i$
5	5

Division of complex numbers is much like rationalizing the denominator of a radical expression, which we considered in Section P.8. For the complex number $z = a + bi$ we define its **complex conjugate** to be $\bar{z} = a - bi$. Note that

$$z \cdot \bar{z} = (a + bi)(a - bi) = a^2 + b^2$$

So the product of a complex number and its conjugate is always a nonnegative real number. We use this property to divide complex numbers.

DIVIDING COMPLEX NUMBERS

To simplify the quotient $\dfrac{a + bi}{c + di}$, multiply the numerator and the denominator by the complex conjugate of the denominator:

$$\frac{a + bi}{c + di} = \left(\frac{a + bi}{c + di}\right)\left(\frac{c - di}{c - di}\right) = \frac{(ac + bd) + (bc - ad)i}{c^2 + d^2}$$

Rather than memorizing this entire formula, it is easier to just remember the first step and then multiply out the numerator and the denominator as usual.

▶ **EXAMPLE 3** | Dividing Complex Numbers

Express the following in the form $a + bi$.

(a) $\dfrac{3 + 5i}{1 - 2i}$ **(b)** $\dfrac{7 + 3i}{4i}$

▼ **SOLUTION** We multiply both the numerator and denominator by the complex conjugate of the denominator to make the new denominator a real number.

(a) The complex conjugate of $1 - 2i$ is $\overline{1 - 2i} = 1 + 2i$.

$$\frac{3 + 5i}{1 - 2i} = \left(\frac{3 + 5i}{1 - 2i}\right)\left(\frac{1 + 2i}{1 + 2i}\right) = \frac{-7 + 11i}{5} = -\frac{7}{5} + \frac{11}{5}i$$

(b) The complex conjugate of $4i$ is $-4i$. Therefore,

$$\frac{7 + 3i}{4i} = \left(\frac{7 + 3i}{4i}\right)\left(\frac{-4i}{-4i}\right) = \frac{12 - 28i}{16} = \frac{3}{4} - \frac{7}{4}i$$ ▲

■ Square Roots of Negative Numbers

Just as every positive real number r has two square roots ($\sqrt{r}$ and $-\sqrt{r}$), every negative number has two square roots as well. If $-r$ is a negative number, then its square roots are $\pm i\sqrt{r}$, because $(i\sqrt{r})^2 = i^2r = -r$ and $(-i\sqrt{r})^2 = (-1)^2i^2r = -r$.

SQUARE ROOTS OF NEGATIVE NUMBERS

If $-r$ is negative, then the **principal square root** of $-r$ is

$$\sqrt{-r} = i\sqrt{r}$$

The two square roots of $-r$ are $i\sqrt{r}$ and $-i\sqrt{r}$.

We usually write $i\sqrt{b}$ instead of $\sqrt{b}i$ to avoid confusion with $\sqrt{bi}$.

▶ **EXAMPLE 4** | Square Roots of Negative Numbers

(a) $\sqrt{-1} = i\sqrt{1} = i$ **(b)** $\sqrt{-16} = i\sqrt{16} = 4i$ **(c)** $\sqrt{-3} = i\sqrt{3}$ ▲

Special care must be taken in performing calculations that involve square roots of negative numbers. Although $\sqrt{a} \cdot \sqrt{b} = \sqrt{ab}$ when a and b are positive, this is *not* true when both are negative. For example,

$$\sqrt{-2} \cdot \sqrt{-3} = i\sqrt{2} \cdot i\sqrt{3} = i^2\sqrt{6} = -\sqrt{6}$$

but

$$\sqrt{(-2)(-3)} = \sqrt{6}$$

so

$$\sqrt{-2} \cdot \sqrt{-3} \neq \sqrt{(-2)(-3)}$$

 When multiplying radicals of negative numbers, express them first in the form $i\sqrt{r}$ (where $r > 0$) to avoid possible errors of this type.

▶ **EXAMPLE 5** | Using Square Roots of Negative Numbers

Evaluate $(\sqrt{12} - \sqrt{-3})(3 + \sqrt{-4})$ and express in the form $a + bi$.

▼ **SOLUTION**

$$(\sqrt{12} - \sqrt{-3})(3 + \sqrt{-4}) = (\sqrt{12} - i\sqrt{3})(3 + i\sqrt{4})$$
$$= (2\sqrt{3} - i\sqrt{3})(3 + 2i)$$
$$= (6\sqrt{3} + 2\sqrt{3}) + i(2 \cdot 2\sqrt{3} - 3\sqrt{3})$$
$$= 8\sqrt{3} + i\sqrt{3}$$

▲

■ Complex Solutions of Quadratic Equations

We have already seen that if $a \neq 0$, then the solutions of the quadratic equation $ax^2 + bx + c = 0$ are

$$x = \frac{-b \pm \sqrt{b^2 - 4ac}}{2a}$$

If $b^2 - 4ac < 0$, then the equation has no real solution. But in the complex number system, this equation will always have solutions, because negative numbers have square roots in this expanded setting.

▶ **EXAMPLE 6** | Quadratic Equations with Complex Solutions

Solve each equation.

(a) $x^2 + 9 = 0$ (b) $x^2 + 4x + 5 = 0$

▼ **SOLUTION**

(a) The equation $x^2 + 9 = 0$ means $x^2 = -9$, so

$$x = \pm\sqrt{-9} = \pm i\sqrt{9} = \pm 3i$$

The solutions are therefore $3i$ and $-3i$.

(b) By the Quadratic Formula we have

$$x = \frac{-4 \pm \sqrt{4^2 - 4 \cdot 5}}{2}$$

$$= \frac{-4 \pm \sqrt{-4}}{2}$$

$$= \frac{-4 \pm 2i}{2} = \frac{2(-2 \pm i)}{2} = -2 \pm i$$

So the solutions are $-2 + i$ and $-2 - i$.   ▲

We see from Example 6 that if a quadratic equation with real coefficients has complex solutions, then these solutions are complex conjugates of each other. So if $a + bi$ is a solution, then $a - bi$ is also a solution.

▶ **EXAMPLE 7** | Complex Conjugates as Solutions of a Quadratic

Show that the solutions of the equation

$$4x^2 - 24x + 37 = 0$$

are complex conjugates of each other.

▼ **SOLUTION**   We use the Quadratic Formula to get

$$x = \frac{24 \pm \sqrt{(24)^2 - 4(4)(37)}}{2(4)}$$

$$= \frac{24 \pm \sqrt{-16}}{8} = \frac{24 \pm 4i}{8} = 3 \pm \frac{1}{2}i$$

So the solutions are $3 + \frac{1}{2}i$ and $3 - \frac{1}{2}i$, and these are complex conjugates.   ▲

1.5 | Other Types of Equations

LEARNING OBJECTIVES

After completing this section, you will be able to:

- Solve basic polynomial equations
- Solve equations involving radicals
- Solve equations of quadratic type
- Model with equations

So far, we have learned how to solve linear and quadratic equations. In this section we study other types of equations, including those that involve higher powers, fractional expressions, and radicals.

■ Polynomial Equations

Some equations can be solved by factoring and using the Zero-Product Property, which says that if a product equals 0, then at least one of the factors must equal 0.

▶ **EXAMPLE 1** | Solving an Equation by Factoring

Solve the equation $x^5 = 9x^3$.

▼ **SOLUTION**   We bring all terms to one side and then factor.

$$x^5 - 9x^3 = 0 \qquad \text{Subtract } 9x^3$$

$$x^3(x^2 - 9) = 0 \qquad \text{Factor } x^3$$

$$x^3(x - 3)(x + 3) = 0 \qquad \text{Difference of squares}$$

$$x^3 = 0 \quad \text{or} \quad x - 3 = 0 \quad \text{or} \quad x + 3 = 0 \qquad \text{Zero-Product Property}$$

$$x = 0 \qquad\qquad x = 3 \qquad\qquad x = -3 \qquad \text{Solve}$$

The solutions are $x = 0$, $x = 3$, and $x = -3$. You should check that each of these satisfies the original equation.   ▲

To divide each side of the equation in Example 1 by the common factor x^3 would be wrong, because in doing so, we would lose the solution $x = 0$. Never divide both sides of an equation by an expression that contains the variable unless you know that the expression cannot equal 0.

▶ **EXAMPLE 2** | Factoring by Grouping

Solve the equation $x^3 + 3x^2 - 4x - 12 = 0$.

▼ **SOLUTION**   The left-hand side of the equation can be factored by grouping the terms in pairs.

$$(x^3 + 3x^2) - (4x + 12) = 0 \qquad \text{Group terms}$$

$$x^2(x + 3) - 4(x + 3) = 0 \qquad \text{Factor } x^2 \text{ and } 4$$

$$(x^2 - 4)(x + 3) = 0 \qquad \text{Factor } x + 3$$

$$(x - 2)(x + 2)(x + 3) = 0 \qquad \text{Difference of squares}$$

$$x - 2 = 0 \quad \text{or} \quad x + 2 = 0 \quad \text{or} \quad x + 3 = 0 \qquad \text{Zero-Product Property}$$

$$x = 2 \qquad\qquad x = -2 \qquad\qquad x = -3 \qquad \text{Solve}$$

The solutions are $x = 2$, -2, and -3.   ▲

▶ **EXAMPLE 3** | An Equation Involving Fractional Expressions

Solve the equation $\dfrac{3}{x} + \dfrac{5}{x + 2} = 2$.

▼ **SOLUTION**   To simplify the equation, we multiply each side by the common denominator.

$$\left(\frac{3}{x} + \frac{5}{x+2}\right)x(x+2) = 2x(x+2) \qquad \text{Multiply by LCD } x(x+2)$$

$$3(x+2) + 5x = 2x^2 + 4x \qquad \text{Expand}$$

$$8x + 6 = 2x^2 + 4x \qquad \text{Expand LHS}$$

$$0 = 2x^2 - 4x - 6 \qquad \text{Subtract } 8x + 6$$

$$0 = x^2 - 2x - 3 \qquad \text{Divide both sides by 2}$$

$$0 = (x-3)(x+1) \qquad \text{Factor}$$

$$x - 3 = 0 \quad \text{or} \quad x + 1 = 0 \qquad \text{Zero-Product Property}$$

$$x = 3 \qquad\qquad x = -1 \qquad \text{Solve}$$

We must check our answers because multiplying by an expression that contains the variable can introduce extraneous solutions (see the *Warning* on pages 52–53). From *Check Your Answers* we see that the solutions are $x = 3$ and -1.   ▲

Check Your Answers

$x = 3$:

$$\text{LHS} = \frac{3}{3} + \frac{5}{3+2} = 2$$

$$\text{RHS} = 2$$

$$\text{LHS} = \text{RHS} \qquad ✔$$

$x = -1$:

$$\text{LHS} = \frac{3}{-1} + \frac{5}{-1+2} = 2$$

$$\text{RHS} = 2$$

$$\text{LHS} = \text{RHS} \qquad ✔$$

■ Equations Involving Radicals

When you solve an equation that involves radicals, you must be especially careful to check your final answers. The next example demonstrates why.

▶ **EXAMPLE 4** | An Equation Involving a Radical

Solve the equation $2x = 1 - \sqrt{2 - x}$.

▼ **SOLUTION**   To eliminate the square root, we first isolate it on one side of the equal sign, then square.

$$2x - 1 = -\sqrt{2-x} \qquad \text{Subtract 1}$$

$$(2x-1)^2 = 2 - x \qquad \text{Square each side}$$

$$4x^2 - 4x + 1 = 2 - x \qquad \text{Expand LHS}$$

$$4x^2 - 3x - 1 = 0 \qquad \text{Add } -2 + x$$

$$(4x+1)(x-1) = 0 \qquad \text{Factor}$$

$$4x + 1 = 0 \quad \text{or} \quad x - 1 = 0 \qquad \text{Zero-Product Property}$$

$$x = -\tfrac{1}{4} \qquad\qquad x = 1 \qquad \text{Solve}$$

The values $x = -\frac{1}{4}$ and $x = 1$ are only potential solutions. We must check them to see whether they satisfy the original equation. From *Check Your Answers* we see that $x = -\frac{1}{4}$ is a solution but $x = 1$ is not. The only solution is $x = -\frac{1}{4}$.   ▲

Check Your Answers

$x = -\frac{1}{4}$:

$$\text{LHS} = 2\left(-\tfrac{1}{4}\right) = -\tfrac{1}{2}$$

$$\text{RHS} = 1 - \sqrt{2 - \left(-\tfrac{1}{4}\right)}$$

$$= 1 - \sqrt{\tfrac{9}{4}}$$

$$= 1 - \tfrac{3}{2} = -\tfrac{1}{2}$$

$$\text{LHS} = \text{RHS} \qquad ✔$$

$x = 1$:

$$\text{LHS} = 2(1) = 2$$

$$\text{RHS} = 1 - \sqrt{2 - 1}$$

$$= 1 - 1 = 0$$

$$\text{LHS} \neq \text{RHS} \qquad ✘$$

When we solve an equation, we may end up with one or more **extraneous solutions**, that is, potential solutions that do not satisfy the original equation. In Example 4 the value $x = 1$ is an extraneous solution. Extraneous solutions may be introduced when we square each side of an equation because the operation of squaring can turn a false equation into a true one. For example, $-1 \neq 1$, but $(-1)^2 = 1^2$. Thus, the squared equation may be true for

⊘ more values of the variable than the original equation. That is why you must always check your answers to make sure that each satisfies the original equation.

■ Equations of Quadratic Type

An equation of the form $aW^2 + bW + c = 0$, where W is an algebraic expression, is an equation of **quadratic type**. We solve equations of quadratic type by substituting for the algebraic expression, as we see in the next two examples.

▶ **EXAMPLE 5** | An Equation of Quadratic Type

Solve the equation $\left(1 + \dfrac{1}{x}\right)^2 - 6\left(1 + \dfrac{1}{x}\right) + 8 = 0$.

▼ **SOLUTION** We could solve this equation by multiplying it out first. But it's easier to think of the expression $1 + \frac{1}{x}$ as the unknown in this equation and give it a new name W. This turns the equation into a quadratic equation in the new variable W.

$$\left(1 + \frac{1}{x}\right)^2 - 6\left(1 + \frac{1}{x}\right) + 8 = 0 \qquad \text{Given equation}$$

$$W^2 - 6W + 8 = 0 \qquad \text{Let } W = 1 + \frac{1}{x}$$

$$(W - 4)(W - 2) = 0 \qquad \text{Factor}$$

$$W - 4 = 0 \quad \text{or} \quad W - 2 = 0 \qquad \text{Zero-Product Property}$$

$$W = 4 \qquad\qquad W = 2 \qquad \text{Solve}$$

Now we change these values of W back into the corresponding values of x.

$$1 + \frac{1}{x} = 4 \qquad\quad 1 + \frac{1}{x} = 2 \qquad W = 1 + \frac{1}{x}$$

$$\frac{1}{x} = 3 \qquad\qquad \frac{1}{x} = 1 \qquad \text{Subtract 1}$$

$$x = \frac{1}{3} \qquad\qquad x = 1 \qquad \text{Take reciprocals}$$

The solutions are $x = \frac{1}{3}$ and $x = 1$.   ▲

▶ **EXAMPLE 6** | A Fourth-Degree Equation of Quadratic Type

Find all solutions of the equation $x^4 - 8x^2 + 8 = 0$.

▼ **SOLUTION** If we set $W = x^2$, then we get a quadratic equation in the new variable W:

$$(x^2)^2 - 8x^2 + 8 = 0 \qquad \text{Write } x^4 \text{ as } (x^2)^2$$

$$W^2 - 8W + 8 = 0 \qquad \text{Let } W = x^2$$

$$W = \frac{-(-8) \pm \sqrt{(-8)^2 - 4 \cdot 8}}{2} = 4 \pm 2\sqrt{2} \qquad \text{Quadratic Formula}$$

$$x^2 = 4 \pm 2\sqrt{2} \qquad\qquad\qquad W = x^2$$

$$x = \pm\sqrt{4 \pm 2\sqrt{2}} \qquad\qquad\qquad \text{Take square roots}$$

So, there are four solutions:

$$\sqrt{4 + 2\sqrt{2}}, \qquad \sqrt{4 - 2\sqrt{2}}, \qquad -\sqrt{4 + 2\sqrt{2}}, \qquad -\sqrt{4 - 2\sqrt{2}}$$

Using a calculator, we obtain the approximations $x \approx 2.61,\ 1.08,\ -2.61,\ -1.08$.   ▲

▶ **EXAMPLE 7** | An Equation Involving Fractional Powers

Find all solutions of the equation $x^{1/3} + x^{1/6} - 2 = 0$.

▼ **SOLUTION** This equation is of quadratic type because if we let $W = x^{1/6}$, then $W^2 = (x^{1/6})^2 = x^{1/3}$.

$$x^{1/3} + x^{1/6} - 2 = 0 \qquad \text{Given equation}$$

$$W^2 + W - 2 = 0 \qquad \text{Let } W = x^{1/6}$$

$$(W - 1)(W + 2) = 0 \qquad \text{Factor}$$

$$W - 1 = 0 \quad \text{or} \quad W + 2 = 0 \qquad \text{Zero-Product Property}$$

$$W = 1 \qquad\qquad W = -2 \qquad \text{Solve}$$

$$x^{1/6} = 1 \qquad\qquad x^{1/6} = -2 \qquad W = x^{1/6}$$

$$x = 1^6 = 1 \qquad\qquad x = (-2)^6 = 64 \qquad \text{Take the 6th power}$$

From *Check Your Answers* we see that $x = 1$ is a solution but $x = 64$ is not. The only solution is $x = 1$.

Check Your Answers

$x = 1$:
$$\text{LHS} = 1^{1/3} + 1^{1/6} - 2 = 0$$
$$\text{RHS} = 0$$
$$\text{LHS} = \text{RHS} \quad ✔$$

$x = 64$:
$$\text{LHS} = 64^{1/3} + 64^{1/6} - 2$$
$$= 4 + 2 - 2 = 4$$
$$\text{RHS} = 0$$
$$\text{LHS} \neq \text{RHS} \quad ✗$$

▲

■ Applications

Many real-life problems can be modeled with the types of equations that we have studied in this section.

▶ **EXAMPLE 8** | Dividing a Lottery Jackpot

A group of people come forward to claim a $1,000,000 lottery jackpot, which the winners are to share equally. Before the jackpot is divided, three more winning ticket holders show up. As a result, each person's share is reduced by $75,000. How many winners were in the original group?

▼ **SOLUTION** We are asked for the number of people in the original group. So let

Identify the variable ▶

$$x = \text{number of winners in the original group}$$

We translate the information in the problem as follows:

In Words	In Algebra
Number of winners in original group	x
Number of winners in final group	$x + 3$
Winnings per person, originally	$\dfrac{1{,}000{,}000}{x}$
Winnings per person, finally	$\dfrac{1{,}000{,}000}{x + 3}$

Translate from words to algebra ▶

Now we set up the model.

Set up the model ▶

$$\boxed{\begin{array}{c}\text{winnings per}\\\text{person, originally}\end{array}} - \boxed{\$75{,}000} = \boxed{\begin{array}{c}\text{winnings per}\\\text{person, finally}\end{array}}$$

$$\frac{1{,}000{,}000}{x} - 75{,}000 = \frac{1{,}000{,}000}{x + 3}$$

Solve ▶

$1{,}000{,}000(x + 3) - 75{,}000x(x + 3) = 1{,}000{,}000x$	Multiply by LCD $x(x + 3)$
$40(x + 3) - 3x(x + 3) = 40x$	Divide by 25,000
$x^2 + 3x - 40 = 0$	Expand, simplify, and divide by 3
$(x + 8)(x - 5) = 0$	Factor
$x + 8 = 0 \quad$ or $\quad x - 5 = 0$	Zero-Product Property
$x = -8 \qquad\qquad x = 5$	Solve

Since we can't have a negative number of people, we conclude that there were five winners in the original group.

Check Your Answer

$$\text{winnings per person, originally} = \frac{\$1{,}000{,}000}{5} = \$200{,}000$$

$$\text{winnings per person, finally} = \frac{\$1{,}000{,}000}{8} = \$125{,}000$$

$$\$200{,}000 - \$75{,}000 = \$125{,}000 \qquad ✔$$

▲

▶ **EXAMPLE 9** | Energy Expended in Bird Flight

Ornithologists have determined that some species of birds tend to avoid flights over large bodies of water during daylight hours, because air generally rises over land and falls over water in the daytime, so flying over water requires more energy. A bird is released from point A on an island, 5 mi from B, the nearest point on a straight shoreline. The bird flies to a point C on the shoreline and then flies along the shoreline to its nesting area D, as shown in Figure 1. Suppose the bird has 170 kcal of energy reserves. It uses 10 kcal/mi flying over land and 14 kcal/mi flying over water.

(a) Where should the point C be located so that the bird uses exactly 170 kcal of energy during its flight?

(b) Does the bird have enough energy reserves to fly directly from A to D?

▼ **SOLUTION**

(a) We are asked to find the location of C. So let

$$x = \text{distance from } B \text{ to } C$$

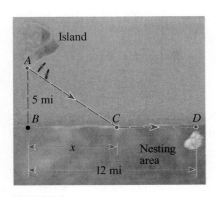

FIGURE 1

Identify the variable ▶

From the figure, and from the fact that

$$\text{energy used} = \text{energy per mile} \times \text{miles flown}$$

we determine the following:

In Words	In Algebra	
Distance from B to C	x	
Distance flown over water (from A to C)	$\sqrt{x^2 + 25}$	Pythagorean Theorem
Distance flown over land (from C to D)	$12 - x$	
Energy used over water	$14\sqrt{x^2 + 25}$	
Energy used over land	$10(12 - x)$	

Translate from words to algebra ▶

Now we set up the model.

Set up the model ▶

$$\text{total energy used} = \text{energy used over water} + \text{energy used over land}$$

$$170 = 14\sqrt{x^2 + 25} + 10(12 - x)$$

To solve this equation, we eliminate the square root by first bringing all other terms to the left of the equal sign and then squaring each side.

Solve ▶

$$170 - 10(12 - x) = 14\sqrt{x^2 + 25} \qquad \text{Isolate square root term on RHS}$$

$$50 + 10x = 14\sqrt{x^2 + 25} \qquad \text{Simplify LHS}$$

$$(50 + 10x)^2 = (14)^2(x^2 + 25) \qquad \text{Square each side}$$

$$2500 + 1000x + 100x^2 = 196x^2 + 4900 \qquad \text{Expand}$$

$$0 = 96x^2 - 1000x + 2400 \qquad \text{All terms to RHS}$$

This equation could be factored, but because the numbers are so large, it is easier to use the Quadratic Formula and a calculator:

$$x = \frac{1000 \pm \sqrt{(-1000)^2 - 4(96)(2400)}}{2(96)} = \frac{1000 \pm 280}{192}$$

$$x = 6\tfrac{2}{3} \qquad \text{or} \qquad x = 3\tfrac{3}{4}$$

Point C should be either $6\tfrac{2}{3}$ mi or $3\tfrac{3}{4}$ mi from B so that the bird uses exactly 170 kcal of energy during its flight.

(b) By the Pythagorean Theorem, the length of the route directly from A to D is $\sqrt{5^2 + 12^2} = 13$ mi, so the energy the bird requires for that route is $14 \times 13 = 182$ kcal. This is more energy than the bird has available, so it can't use this route.   ▲

1.6 | Inequalities

LEARNING OBJECTIVES
After completing this section, you will be able to:

- Solve linear inequalities
- Solve nonlinear inequalities
- Model with inequalities

Some problems in algebra lead to **inequalities** instead of equations. An inequality looks just like an equation, except that in the place of the equal sign is one of the symbols $<$, $>$, $\leq$, or $\geq$. Here is an example of an inequality:

$$4x + 7 \leq 19$$

x	$4x + 7 \leq 19$
1	$11 \leq 19$ ✔
2	$15 \leq 19$ ✔
3	$19 \leq 19$ ✔
4	$23 \leq 19$ ✗
5	$27 \leq 19$ ✗

The table in the margin shows that some numbers satisfy the inequality and some numbers do not.

To **solve** an inequality that contains a variable means to find all values of the variable that make the inequality true. Unlike an equation, an inequality generally has infinitely many solutions, which form an interval or a union of intervals on the real line. The following illustration shows how an inequality differs from its corresponding equation:

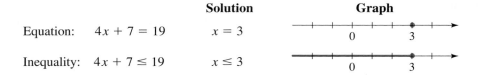

		Solution	**Graph**
Equation:	$4x + 7 = 19$	$x = 3$	
Inequality:	$4x + 7 \leq 19$	$x \leq 3$	

To solve inequalities, we use the following rules to isolate the variable on one side of the inequality sign. These rules tell us when two inequalities are *equivalent* (the symbol $\Leftrightarrow$ means "is equivalent to"). In these rules the symbols A, B, and C stand for real numbers or algebraic expressions. Here we state the rules for inequalities involving the symbol $\leq$, but they apply to all four inequality symbols.

RULES FOR INEQUALITIES

Rule	**Description**
1. $A \leq B \quad \Leftrightarrow \quad A + C \leq B + C$	**Adding** the same quantity to each side of an inequality gives an equivalent inequality.
2. $A \leq B \quad \Leftrightarrow \quad A - C \leq B - C$	**Subtracting** the same quantity from each side of an inequality gives an equivalent inequality.
3. If $C > 0$, then $A \leq B \quad \Leftrightarrow \quad CA \leq CB$	**Multiplying** each side of an inequality by the same *positive* quantity gives an equivalent inequality.
4. If $C < 0$, then $A \leq B \quad \Leftrightarrow \quad CA \geq CB$	**Multiplying** each side of an inequality by the same *negative* quantity *reverses the direction* of the inequality.
5. If $A > 0$ and $B > 0$, then $A \leq B \quad \Leftrightarrow \quad \dfrac{1}{A} \geq \dfrac{1}{B}$	**Taking reciprocals** of each side of an inequality involving *positive* quantities *reverses the direction* of the inequality.
6. If $A \leq B$ and $C \leq D$, then $A + C \leq B + D$	Inequalities can be added.

 Pay special attention to Rules 3 and 4. Rule 3 says that we can multiply (or divide) each side of an inequality by a *positive* number, but Rule 4 says that if we multiply each side of an inequality by a *negative* number, then we reverse the direction of the inequality. For example, if we start with the inequality

$$3 < 5$$

and multiply by 2, we get

$$6 < 10$$

but if we multiply by -2, we get

$$-6 > -10$$

Solving Linear Inequalities

An inequality is **linear** if each term is constant or a multiple of the variable. To solve a linear inequality, we isolate the variable on one side of the inequality sign.

▶ **EXAMPLE 1** | Solving a Linear Inequality

Solve the inequality $3x < 9x + 4$ and sketch the solution set.

▼ **SOLUTION**

$$3x < 9x + 4 \qquad \text{Given inequality}$$
$$3x - 9x < 9x + 4 - 9x \qquad \text{Subtract } 9x$$
$$-6x < 4 \qquad \text{Simplify}$$
$$\left(-\tfrac{1}{6}\right)(-6x) > \left(-\tfrac{1}{6}\right)(4) \qquad \text{Multiply by } -\tfrac{1}{6} \text{ (or divide by } -6)$$
$$x > -\tfrac{2}{3} \qquad \text{Simplify}$$

Multiplying by the negative number $-\tfrac{1}{6}$ *reverses* the direction of the inequality.

FIGURE 1

The solution set consists of all numbers greater than $-\tfrac{2}{3}$. In other words, the solution of the inequality is the interval $\left(-\tfrac{2}{3}, \infty\right)$. It is graphed in Figure 1.   ▲

▶ **EXAMPLE 2** | Solving a Pair of Simultaneous Inequalities

Solve the inequalities $4 \leq 3x - 2 < 13$.

▼ **SOLUTION**   The solution set consists of all values of x that satisfy both of the inequalities $4 \leq 3x - 2$ and $3x - 2 < 13$. Using Rules 1 and 3, we see that the following inequalities are equivalent:

$$4 \leq 3x - 2 < 13 \qquad \text{Given inequality}$$
$$6 \leq 3x < 15 \qquad \text{Add 2}$$
$$2 \leq x < 5 \qquad \text{Divide by 3}$$

FIGURE 2

Therefore, the solution set is $[2, 5)$, as shown in Figure 2.   ▲

Solving Nonlinear Inequalities

To solve inequalities involving squares and other powers of the variable, we use factoring, together with the following principle.

THE SIGN OF A PRODUCT OR QUOTIENT

If a product or a quotient has an *even* number of *negative* factors, then its value is *positive*.

If a product or a quotient has an *odd* number of *negative* factors, then its value is *negative*.

For example, to solve the inequality $x^2 - 5x \leq -6$, we first move all terms to the left-hand side and factor to get

$$(x - 2)(x - 3) \leq 0$$

This form of the inequality says that the product $(x - 2)(x - 3)$ must be negative or zero, so to solve the inequality, we must determine where each factor is negative or positive (because the sign of a product depends on the sign of the factors). The details are explained in Example 3, in which we use the following guidelines.

GUIDELINES FOR SOLVING NONLINEAR INEQUALITIES

1. **Move All Terms to One Side.**   If necessary, rewrite the inequality so that all nonzero terms appear on one side of the inequality sign. If the nonzero side of the inequality involves quotients, bring them to a common denominator.

2. **Factor.**   Factor the nonzero side of the inequality.

3. **Find the Intervals.**   Determine the values for which each factor is zero. These numbers will divide the real line into intervals. List the intervals that are determined by these numbers.

4. **Make a Table or Diagram.**   Use test values to make a table or diagram of the signs of each factor on each interval. In the last row of the table determine the sign of the product (or quotient) of these factors.

5. **Solve.**   Determine the solution of the inequality from the last row of the sign table. Be sure to check whether the inequality is satisfied by some or all of the endpoints of the intervals. (This may happen if the inequality involves ≤ or ≥.)

   The factoring technique that is described in these guidelines works only if all nonzero terms appear on one side of the inequality symbol. If the inequality is not written in this form, first rewrite it, as indicated in Step 1.

▶ **EXAMPLE 3** | Solving a Quadratic Inequality

Solve the inequality $x^2 \le 5x - 6$.

▼ **SOLUTION**   We will follow the guidelines above.

Move all terms to one side.   We move all the terms to the left-hand side.

$$x^2 \le 5x - 6 \qquad \text{Given inequality}$$
$$x^2 - 5x + 6 \le 0 \qquad \text{Subtract 5x, add 6}$$

Factor.   Factoring the left-hand side of the inequality, we get

$$(x - 2)(x - 3) \le 0 \qquad \text{Factor}$$

Find the intervals.   The factors of the left-hand side are $x - 2$ and $x - 3$. These factors are zero when x is 2 and 3, respectively. As shown in Figure 3, the numbers 2 and 3 divide the real line into the three intervals

$$(-\infty, 2), (2, 3), (3, \infty)$$

The factors $x - 2$ and $x - 3$ change sign only at 2 and 3, respectively. So these factors maintain their sign on each of these three intervals.

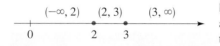

FIGURE 3

Make a table or diagram.   To determine the sign of each factor on each of the intervals that we found, we use **test values.** We choose a number inside each interval and check the sign of the factors $x - 2$ and $x - 3$ at the number we chose. For the interval $(-\infty, 2)$ let's choose the test value 1 (see Figure 4). Substituting 1 for x in the factors $x - 2$ and $x - 3$, we get

$$x - 2 = 1 - 2 = -1 < 0$$
$$x - 3 = 1 - 3 = -2 < 0$$

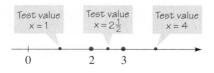

FIGURE 4

So both factors are negative on this interval. Notice that we need to check only one test value for each interval because the factors $x - 2$ and $x - 3$ do not change sign on any of the three intervals we found.

Using the test values $x = 2\frac{1}{2}$ and $x = 4$ for the intervals $(2, 3)$ and $(3, \infty)$ (see Figure 4), respectively, we construct the following sign table. The final row of the table is obtained from the fact that the expression in the last row is the product of the two factors.

Interval	$(-\infty, 2)$	$(2, 3)$	$(3, \infty)$
Sign of $x - 2$	−	+	+
Sign of $x - 3$	−	−	+
Sign of $(x - 2)(x - 3)$	+	−	+

If you prefer, you can represent this information on a real line, as in the following sign diagram. The vertical lines indicate the points at which the real line is divided into intervals:

		2		3	
Sign of $x - 2$	−		+		+
Sign of $x - 3$	−		−		+
Sign of $(x - 2)(x - 3)$	+		−		+

Solve. We read from the table or the diagram that $(x - 2)(x - 3)$ is negative on the interval $(2, 3)$. Thus, the solution of the inequality $(x - 2)(x - 3) \leq 0$ is

$$\{x \mid 2 \leq x \leq 3\} = [2, 3]$$

We have included the endpoints 2 and 3 because we seek values of x such that the product is either less than *or equal to* zero. The solution is illustrated in Figure 5. ▲

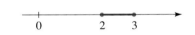

FIGURE 5

▶ **EXAMPLE 4** | Solving an Inequality

Solve the inequality $2x^2 - x > 1$.

▼ **SOLUTION** We will follow the guidelines on page 84.

Move all terms to one side. We move all the terms to the left-hand side.

$$2x^2 - x > 1 \qquad \textit{Given inequality}$$
$$2x^2 - x - 1 > 0 \qquad \textit{Subtract 1}$$

Factor. Factoring the left-hand side of the inequality, we get

$$(2x + 1)(x - 1) > 0 \qquad \textit{Factor}$$

Find the intervals. The factors of the left-hand side are $2x + 1$ and $x - 1$. These factors are zero when x is $-\frac{1}{2}$ and 1. These numbers divide the real line into the intervals

$$\left(-\infty, -\tfrac{1}{2}\right), \left(-\tfrac{1}{2}, 1\right), \left(1, \infty\right)$$

Make a diagram. We make the following diagram, using test points to determine the sign of each factor in each interval.

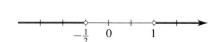

The solution set is graphed in Figure 6.

		$-\dfrac{1}{2}$		1	
Sign of $2x + 1$	$-$		$+$		$+$
Sign of $x - 1$	$-$		$-$		$+$
Sign of $(2x + 1)(x - 1)$	$+$		$-$		$+$

Solve. From the diagram we see that $(2x + 1)(x - 1) > 0$ for x in the interval $\left(-\infty, -\frac{1}{2}\right)$ or for x in $(1, \infty)$. So the solution set is the union of these two intervals:

$$\left(-\infty, -\tfrac{1}{2}\right) \cup (1, \infty)$$

The solution set is graphed in Figure 6.   ▲

FIGURE 6

▶ **EXAMPLE 5** | Solving an Inequality with Repeated Factors

Solve the inequality $x(x - 1)^2(x - 3) < 0$.

▼ **SOLUTION**   All nonzero terms are already on one side of the inequality, and the nonzero side of the inequality is already factored. So we begin by finding the intervals for this inequality.

Find the intervals.   The factors of the left-hand side are x, $(x - 1)^2$, and $x - 3$. These are zero when $x = 0, 1, 3$. These numbers divide the real line into the intervals

$$(-\infty, 0), (0, 1), (1, 3), (3, \infty)$$

Make a diagram.   We make the following diagram, using test points to determine the sign of each factor in each interval.

		0		1		3	
Sign of x	$-$		$+$		$+$		$+$
Sign of $(x - 1)^2$	$+$		$+$		$+$		$+$
Sign of $(x - 3)$	$-$		$-$		$-$		$+$
Sign of $x(x - 1)^2(x - 3)$	$+$		$-$		$-$		$+$

Solve.   From the diagram we see that $x(x - 1)^2(x - 3) < 0$ for x in the interval $(0, 1)$ or for x in $(1, 3)$. So the solution set is the union of these two intervals:

$$(0, 1) \cup (1, 3)$$

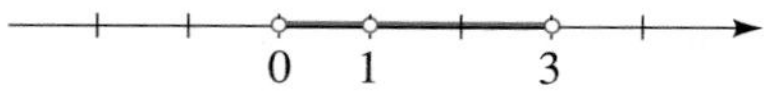

The solution set is graphed in Figure 7.   ▲

FIGURE 7

▶ **EXAMPLE 6** | Solving an Inequality Involving a Quotient

Solve the inequality $\dfrac{1 + x}{1 - x} \geq 1$.

▼ **SOLUTION**

Move all terms to one side.   We move the terms to the left-hand side and simplify using a common denominator.

⊘ It is tempting to multiply both sides of the inequality by $1 - x$ (as you would if this were an *equation*). But this doesn't work because we don't know whether $1 - x$ is positive or negative, so we can't tell whether the inequality needs to be reversed.

$$\frac{1 + x}{1 - x} \geq 1 \qquad \text{Given inequality}$$

$$\frac{1 + x}{1 - x} - 1 \geq 0 \qquad \text{Subtract 1}$$

$$\frac{1 + x}{1 - x} - \frac{1 - x}{1 - x} \geq 0 \qquad \text{Common denominator } 1 - x$$

$$\frac{1 + x - 1 + x}{1 - x} \geq 0 \qquad \text{Combine the fractions}$$

$$\frac{2x}{1 - x} \geq 0 \qquad \text{Simplify}$$

Find the intervals. The factors of the left-hand side are $2x$ and $1 - x$. These are zero when x is 0 and 1. These numbers divide the real line into the intervals

$$(-\infty, 0), (0, 1), (1, \infty)$$

Make a diagram. We make the following diagram using test points to determine the sign of each factor in each interval.

	0		1	
Sign of $2x$	$-$	$+$		$+$
Sign of $1 - x$	$+$	$+$		$-$
Sign of $\dfrac{2x}{1 - x}$	$-$	$+$		$-$

Solve. From the diagram we see that $\dfrac{2x}{1 - x} \geq 0$ for x in the interval $[0, 1)$. We include the endpoint 0 because the original inequality requires that the quotient be greater than *or equal to* 1. However, we do not include the other endpoint 1 because the quotient in the inequality is not defined at 1. So the solution set is the interval

$$[0, 1)$$

The solution set is graphed in Figure 8.   ▲

FIGURE 8

⊘   Example 6 shows that we should always check the endpoints of the solution set to see whether they satisfy the original inequality.

■ Modeling with Inequalities

Modeling real-life problems frequently leads to inequalities because we are often interested in determining when one quantity is more (or less) than another.

▶ **EXAMPLE 7** | Carnival Tickets

A carnival has two plans for tickets.

> Plan A: $5 entrance fee and 25¢ each ride
>
> Plan B: $2 entrance fee and 50¢ each ride

How many rides would you have to take for Plan A to be less expensive than Plan B?

▼ **SOLUTION** We are asked for the number of rides for which Plan A is less expensive than Plan B. So let

Identify the variable ▶

$$x = \text{number of rides}$$

The information in the problem may be organized as follows.

In Words	In Algebra
Number of rides	x
Cost with Plan A	$5 + 0.25x$
Cost with Plan B	$2 + 0.50x$

Translate from words to algebra ▶

Now we set up the model.

Set up the model ▶

$$\underset{\text{Plan A}}{\text{cost with}} < \underset{\text{Plan B}}{\text{cost with}}$$

$$5 + 0.25x < 2 + 0.50x$$

Solve ▶

$$3 + 0.25x < 0.50x \qquad \text{Subtract 2}$$

$$3 < 0.25x \qquad \text{Subtract } 0.25x$$

$$12 < x \qquad \text{Divide by } 0.25$$

So if you plan to take *more than* 12 rides, Plan A is less expensive. ▲

▶ **EXAMPLE 8** │ Relationship Between Fahrenheit and Celsius Scales

The instructions on a bottle of medicine indicate that the bottle should be stored at a temperature between 5 °C and 30 °C. What range of temperatures does this correspond to on the Fahrenheit scale?

▼ **SOLUTION** The relationship between degrees Celsius (C) and degrees Fahrenheit (F) is given by the equation $C = \frac{5}{9}(F - 32)$. Expressing the statement on the bottle in terms of inequalities, we have

$$5 < C < 30$$

So the corresponding Fahrenheit temperatures satisfy the inequalities

$$5 < \tfrac{5}{9}(F - 32) < 30 \qquad \text{Substitute } C = \tfrac{5}{9}(F-32)$$

$$\tfrac{9}{5} \cdot 5 < F - 32 < \tfrac{9}{5} \cdot 30 \qquad \text{Multiply by } \tfrac{9}{5}$$

$$9 < F - 32 < 54 \qquad \text{Simplify}$$

$$9 + 32 < F < 54 + 32 \qquad \text{Add 32}$$

$$41 < F < 86 \qquad \text{Simplify}$$

The medicine should be stored at a temperature between 41°F and 86°F. ▲

1.7 | Absolute Value Equations and Inequalities

LEARNING OBJECTIVES

After completing this section, you will be able to:

▪ Solve absolute value equations
▪ Solve absolute value inequalities

Recall from Section P.3 that the absolute value of a number a is given by

$$|a| = \begin{cases} a & \text{if } a \geq 0 \\ -a & \text{if } a < 0 \end{cases}$$

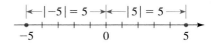

−5   0   5

FIGURE 1

and that it represents the distance from a to the origin on the real number line (see Figure 1). More generally, $|x - a|$ is the distance between x and a on the real number line. Figure 2 illustrates the fact that the distance between 2 and 5 is 3.

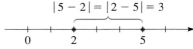

$|5 - 2| = |2 - 5| = 3$

0   2   5

FIGURE 2

▪ Absolute Value Equations

We use the following property to solve equations that involve absolute value.

$$|x| = C \quad \text{is equivalent to} \quad x = \pm C$$

This property says that to solve an absolute value equation, we must solve *two* separate equations. For example, the equation $|x| = 5$ is equivalent to the two equations $x = 5$ and $x = -5$.

▶ **EXAMPLE 1** | Solving an Absolute Value Equation

Solve the equation $|2x - 5| = 3$.

▼ **SOLUTION**   The equation $|2x - 5| = 3$ is equivalent to two equations:

$$2x - 5 = 3 \quad \text{or} \quad 2x - 5 = -3$$
$$2x = 8 \qquad\qquad 2x = 2 \qquad \text{Add 5}$$
$$x = 4 \qquad\qquad x = 1 \qquad \text{Divide by 2}$$

The solutions are 1 and 4.   ▲

Check Your Answers

$x = 1$:

$$\begin{aligned} \text{LHS} &= |2 \cdot 1 - 5| \\ &= |-3| = 3 = \text{RHS} \quad ✔ \end{aligned}$$

$x = 4$:

$$\begin{aligned} \text{LHS} &= |2 \cdot 4 - 5| \\ &= |3| = 3 = \text{RHS} \quad ✔ \end{aligned}$$

▶ **EXAMPLE 2** | Solving an Absolute Value Equation

Solve the equation $3|x - 7| + 5 = 14$.

▼ **SOLUTION**   We first isolate the absolute value on one side of the equal sign.

$$3|x - 7| + 5 = 14$$
$$3|x - 7| = 9 \qquad \text{Subtract 5}$$
$$|x - 7| = 3 \qquad \text{Divide by 3}$$

$$x - 7 = 3 \qquad \text{or} \qquad x - 7 = -3 \qquad \text{Take cases}$$

$$x = 10 \qquad\qquad\qquad x = 4 \qquad \text{Add 7}$$

The solutions are 4 and 10.   ▲

■ Absolute Value Inequalities

We use the following properties to solve inequalities that involve absolute value.

These properties hold when x is replaced by any algebraic expression.

PROPERTIES OF ABSOLUTE VALUE INEQUALITIES

Inequality	Equivalent form	Graph
1. $\lvert x \rvert < c$	$-c < x < c$	
2. $\lvert x \rvert \le c$	$-c \le x \le c$	
3. $\lvert x \rvert > c$	$x < -c \ $ or $\ c < x$	
4. $\lvert x \rvert \ge c$	$x \le -c \ $ or $\ c \le x$	

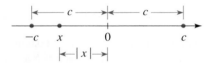

FIGURE 3

These properties can be proved by using the definition of absolute value. To prove Property 1, for example, note that the inequality $\lvert x \rvert < c$ says that the distance from x to 0 is less than c, and from Figure 3 you can see that this is true if and only if x is between c and $-c$.

▶ **EXAMPLE 3** | Solving an Absolute Value Inequality

Solve the inequality $\lvert x - 5 \rvert < 2$.

▼ **SOLUTION 1**   The inequality $\lvert x - 5 \rvert < 2$ is equivalent to

$$-2 < x - 5 < 2 \qquad \text{Property 1}$$

$$3 < x < 7 \qquad \text{Add 5}$$

The solution set is the open interval $(3, 7)$.

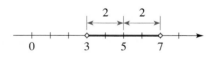

FIGURE 4

▼ **SOLUTION 2**   Geometrically, the solution set consists of all numbers x whose distance from 5 is less than 2. From Figure 4 we see that this is the interval $(3, 7)$.   ▲

▶ **EXAMPLE 4** | Solving an Absolute Value Inequality

Solve the inequality $\lvert 3x + 2 \rvert \ge 4$.

▼ **SOLUTION**   By Property 4 the inequality $\lvert 3x + 2 \rvert \ge 4$ is equivalent to

$$3x + 2 \ge 4 \qquad \text{or} \qquad 3x + 2 \le -4$$

$$3x \ge 2 \qquad\qquad\qquad 3x \le -6 \qquad \text{Subtract 2}$$

$$x \ge \tfrac{2}{3} \qquad\qquad\qquad x \le -2 \qquad \text{Divide by 3}$$

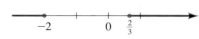

FIGURE 5

So the solution set is

$$\{x \mid x \le -2 \quad \text{or} \quad x \ge \tfrac{2}{3}\} = (-\infty, -2] \cup [\tfrac{2}{3}, \infty)$$

The solution set is graphed in Figure 5.

▶ **EXAMPLE 5** | Piston Tolerances

The specifications for a car engine indicate that the pistons have diameter 3.8745 in. with a tolerance of 0.0015 in. This means that the diameters can vary from the indicated specification by as much as 0.0015 in. and still be acceptable.

(a) Find an inequality involving absolute values that describes the range of possible diameters for the pistons.

(b) Solve the inequality.

▼ **SOLUTION**

(a) Let d represent the actual diameter of a piston. Since the difference between the actual diameter (d) and the specified diameter (3.8745) is less than 0.0015, we have

$$|d - 3.8745| \le 0.0015$$

(b) The inequality is equivalent to

$$-0.0015 \le d - 3.8745 \le 0.0015 \qquad \text{Property 1}$$
$$3.8730 \le d \le 3.8760 \qquad \text{Add 3.8745}$$

Acceptable piston diameters may vary between 3.8730 in. and 3.8760 in.

▶ **CHAPTER 1** | REVIEW

▼ **PROPERTIES AND FORMULAS**

Properties of Equality (p. 50)

$A = B \iff A + C = B + C$

$A = B \iff CA = CB \quad (C \ne 0)$

Linear Equations (p. 50)

A **linear equation** is an equation of the form $ax + b = 0$

Power Equations (p. 53)

A **power equation** is an equation of the form $X^n = a$. Its solutions are

$$X = \sqrt[n]{a} \quad \text{if } n \text{ is odd}$$
$$X = \pm\sqrt[n]{a} \quad \text{if } n \text{ is even}$$

If n is even and $a < 0$, the equation has no real solution.

Simple Interest (p. 57)

If a principal P is invested at simple annual interest rate r for t years, then the interest I is given by

$$I = Prt$$

Zero-Product Property (p. 64)

If $AB = 0$ then $A = 0$ or $B = 0$.

Completing the Square (p. 65)

To make $x^2 + bx$ a perfect square, add $\left(\dfrac{b}{2}\right)^2$. This gives the perfect square

$$x^2 + bx + \left(\frac{b}{2}\right)^2 = \left(x + \frac{b}{2}\right)^2$$

Quadratic Formula (pp. 66–68)

A **quadratic equation** is an equation of the form

$$ax^2 + bx + c = 0$$

Its solutions are given by the **Quadratic Formula**:

$$x = \frac{-b \pm \sqrt{b^2 - 4ac}}{2a}$$

The **discriminant** is $D = b^2 - 4ac$.

If $D > 0$, the equation has two real solutions.

If $D = 0$, the equation has one solution.

If $D < 0$, the equation has two complex solutions.

Complex Numbers (pp. 71–74)

A **complex number** is a number of the form $a + bi$, where $i = \sqrt{-1}$.

The **complex conjugate** of $a + bi$ is

$$\overline{a + bi} = a - bi$$

To **multiply** complex numbers, treat them as binomials and use $i^2 = -1$ to simplify the result.

To **divide** complex numbers, multiply numerator and denominator by the complex conjugate of the denominator:

$$\frac{a + bi}{c + di} = \left(\frac{a + bi}{c + di}\right) \cdot \left(\frac{c - di}{c - di}\right) = \frac{(a + bi)(c - di)}{c^2 + d^2}$$

Inequalities (p. 82)

Adding the same quantity to each side of an inequality gives an equivalent inequality:

$$A < B \quad \Leftrightarrow \quad A + C < B + C$$

Multiplying each side of an inequality by the same *positive* quantity gives an equivalent inequality. Multiplying each side by the same *negative* quantity reverses the direction of the inequality:

$$A < B \quad \Leftrightarrow \quad CA < CB \quad \text{if } C > 0$$
$$A < B \quad \Leftrightarrow \quad CA > CB \quad \text{if } C < 0$$

Absolute Value Equations (p. 89)

To solve an absolute value equation, we use

$$|x| = C \quad \Leftrightarrow \quad x = C \quad \text{or} \quad x = -C$$

Absolute Value Inequalities (p. 90)

To solve an absolute value inequality, we use

$$|x| < C \quad \Leftrightarrow \quad -C < x < C$$
$$|x| > C \quad \Leftrightarrow \quad x < -C \quad \text{or} \quad x > C$$

▼ CONCEPT SUMMARY

Section 1.1	Review Exercises
▪ Solve linear equations	1–12
▪ Solve power equations	13–18, 23–28
▪ Solve for one variable in terms of another	19–22

Section 1.2	Review Exercises
▪ Make equations that model real-world situations	49–56
▪ Use equations to answer questions about real-world situations	49–56

Section 1.3	Review Exercises
▪ Solve quadratic equations by factoring, completing the square, or using the quadratic formula	31–38
▪ Model with quadratic equations	52, 53, 55, 56

Section 1.4	Review Exercises
▪ Add and subtract complex numbers	57–58
▪ Multiply and divide complex numbers	59–64
▪ Simplify expressions with roots of negative numbers	65–66
▪ Find complex roots of quadratic equations	67–70

Section 1.5	Review Exercises
▪ Solve basic polynomial equations	69–74
▪ Solve equations involving radicals or fractional powers	25–28, 42–44
▪ Solve equations of quadratic type	41–44
▪ Model with equations	51, 53, 54

Section 1.6	Review Exercises
▪ Solve linear inequalities	75–78
▪ Solve nonlinear inequalities	79–84, 89
▪ Model with inequalities	90

Section 1.7	Review Exercises
▪ Solve absolute value equations	45–48
▪ Solve absolute value inequalities	85–88

▼ EXERCISES

1–18 ■ Find all real solutions of the equation.

1. $5x + 11 = 36$

2. $3 - x = 5 + 3x$

3. $3x + 12 = 24$

4. $5x - 7 = 42$

5. $7x - 6 = 4x + 9$

6. $8 - 2x = 14 + x$

7. $\frac{1}{3}x - \frac{1}{2} = 2$

8. $\frac{2}{3}x + \frac{3}{5} = \frac{1}{5} - 2x$

9. $2(x + 3) - 4(x - 5) = 8 - 5x$

10. $\dfrac{x - 5}{2} - \dfrac{2x + 5}{3} = \dfrac{5}{6}$

11. $\dfrac{x + 1}{x - 1} = \dfrac{2x - 1}{2x + 1}$

12. $\dfrac{x}{x + 2} - 3 = \dfrac{1}{x + 2}$

13. $x^2 = 144$

14. $4x^2 = 49$

15. $5x^4 - 16 = 0$

16. $x^3 - 27 = 0$

17. $5x^3 - 15 = 0$

18. $6x^4 + 15 = 0$

19–22 ■ Solve the equation for the indicated variable.

19. $A = \dfrac{x + y}{2}$; solve for x

20. $V = xy + yz + xz$; solve for y

21. $J = \dfrac{1}{t} + \dfrac{1}{2t} + \dfrac{1}{3t}$; solve for t

22. $F = k\dfrac{q_1 q_2}{r^2}$; solve for r

23–48 ■ Find all real solutions of the equation.

23. $(x + 1)^3 = -64$

24. $(x + 2)^2 - 2 = 0$

25. $\sqrt[3]{x} = -3$

26. $x^{2/3} - 4 = 0$

27. $4x^{3/4} - 500 = 0$

28. $(x - 2)^{1/5} = 2$

29. $\dfrac{x + 1}{x - 1} = \dfrac{3x}{3x - 6}$

30. $(x + 2)^2 = (x - 4)^2$

31. $x^2 - 9x + 14 = 0$

32. $x^2 + 24x + 144 = 0$

33. $2x^2 + x = 1$

34. $3x^2 + 5x - 2 = 0$

35. $4x^3 - 25x = 0$

36. $x^3 - 2x^2 - 5x + 10 = 0$

37. $3x^2 + 4x - 1 = 0$

38. $x^2 - 3x + 9 = 0$

39. $\dfrac{1}{x} + \dfrac{2}{x - 1} = 3$

40. $\dfrac{x}{x - 2} + \dfrac{1}{x + 2} = \dfrac{8}{x^2 - 4}$

41. $x^4 - 8x^2 - 9 = 0$

42. $x - 4\sqrt{x} = 32$

43. $x^{-1/2} - 2x^{1/2} + x^{3/2} = 0$

44. $(1 + \sqrt{x})^2 - 2(1 + \sqrt{x}) - 15 = 0$

45. $|x - 7| = 4$

46. $|3x| = 18$

47. $|2x - 5| = 9$

48. $4|3 - x| + 3 = 15$

49. A shopkeeper sells raisins for \$3.20 per pound and nuts for \$2.40 per pound. She decides to mix the raisins and nuts and sell 50 lb of the mixture for \$2.72 per pound. What quantities of raisins and nuts should she use?

50. Anthony leaves Kingstown at 2:00 P.M. and drives to Queensville, 160 mi distant, at 45 mi/h. At 2:15 P.M. Helen leaves Queensville and drives to Kingstown at 40 mi/h. At what time do they pass each other on the road?

51. A woman cycles 8 mi/h faster than she runs. Every morning she cycles 4 mi and runs $2\frac{1}{2}$ mi, for a total of 1 hour of exercise. How fast does she run?

52. The approximate distance d (in feet) that drivers travel after noticing that they must come to a sudden stop is given by the following formula, where x is the speed of the car (in mi/h):

$$d = x + \frac{x^2}{20}$$

If a car travels 75 ft before stopping, what was its speed before the brakes were applied?

53. The hypotenuse of a right triangle has length 20 cm. The sum of the lengths of the other two sides is 28 cm. Find the lengths of the other two sides of the triangle.

54. Abbie paints twice as fast as Beth and three times as fast as Cathie. If it takes them 60 min to paint a living room with all three working together, how long would it take Abbie if she works alone?

55. A rectangular swimming pool is 8 ft deep everywhere and twice as long as it is wide. If the pool holds 8464 ft³ of water, what are its dimensions?

56. A gardening enthusiast wishes to fence in three adjoining garden plots, one for each of his children, as shown in the figure. If each plot is to be 80 ft² in area and he has 88 ft of fencing material at hand, what dimensions should each plot have?

57–66 ■ Evaluate the expression and write the result in the form $a + bi$.

57. $(3 - 5i) - (6 + 4i)$

58. $(-2 + 3i) + (\frac{1}{2} - i)$

59. $(2 + 7i)(6 - i)$

60. $3(5 - 2i)\dfrac{i}{5}$

61. $\dfrac{2 - 3i}{2 + 3i}$

62. $\dfrac{2 + i}{4 - 3i}$

63. i^{45}

64. $(3 - i)^3$

65. $(1 - \sqrt{-3})(2 + \sqrt{-4})$

66. $\sqrt{-5} \cdot \sqrt{-20}$

67–74 ▪ Find all real and imaginary solutions of the equation.

67. $x^2 + 16 = 0$

68. $x^2 = -12$

69. $x^2 + 6x + 10 = 0$

70. $2x^2 - 3x + 2 = 0$

71. $x^4 - 256 = 0$

72. $x^3 - 2x^2 + 4x - 8 = 0$

73. $x^2 + 4x = (2x + 1)^2$

74. $x^3 = 125$

75–88 ▪ Solve the inequality. Express the solution using interval notation, and graph the solution set on the real number line.

75. $3x - 2 > -11$

76. $12 - x \geq 7x$

77. $-1 < 2x + 5 \leq 3$

78. $3 - x \leq 2x - 7$

79. $x^2 + 4x - 12 > 0$

80. $x^2 \leq 1$

81. $\dfrac{2x + 5}{x + 1} \leq 1$

82. $2x^2 \geq x + 3$

83. $\dfrac{x - 4}{x^2 - 4} \leq 0$

84. $\dfrac{5}{x^3 - x^2 - 4x + 4} < 0$

85. $|x - 5| \leq 3$

86. $|x - 4| < 0.02$

87. $|2x + 1| \geq 1$

88. $|x - 1| < |x - 3|$
[*Hint*: Interpret the quantities as distances.]

89. For what values of x is the algebraic expression defined as a real number?

(a) $\sqrt{24 - x - 3x^2}$

(b) $\dfrac{1}{\sqrt[4]{x - x^4}}$

90. The volume of a sphere is given by $V = \frac{4}{3}\pi r^3$, where r is the radius. Find the interval of values of the radius so that the volume is between 8 ft^3 and 12 ft^3, inclusive.

1. Find all real solutions of each equation.

   (a) $4x - 3 = 2x + 7$

   (b) $8x^3 = -125$

   (c) $x^{2/3} - 64 = 0$

   (d) $\dfrac{x}{2x - 5} = \dfrac{x + 3}{2x - 1}$

2. Einstein's famous equation $E = mc^2$ gives the relationship between energy E and mass m. In this equation c represents the speed of light. Solve the equation to express c in terms of E and m.

3. Natasha drove from Bedingfield to Portsmouth at an average speed of 100 km/h to attend a job interview. On the way back she decided to slow down to enjoy the scenery, so she drove at just 75 km/h. Her trip involved a total of 3.5 hours of driving time. What is the distance between Bedingfield and Portsmouth?

4. Calculate, and write the result in the standard form for complex numbers: $a + bi$.

   (a) $(3 - 5i) + (2 + i)$

   (b) $(3 - 5i) - (2 + i)$

   (c) $(1 + i)(5 - 2i)$

   (d) $\dfrac{1 + 2i}{3 - 4i}$

   (e) i^{25}

   (f) $(2 - \sqrt{-2})(\sqrt{8} - \sqrt{-4})$

5. Find all solutions, real and complex, of each equation.

   (a) $x^2 - x - 12 = 0$

   (b) $2x^2 + 4x + 3 = 0$

   (c) $\sqrt{3 - \sqrt{x + 5}} = 2$

   (d) $x^{1/2} - 3x^{1/4} + 2 = 0$

   (e) $x^4 - 16x^2 = 0$

   (f) $3\,|x - 4| - 10 = 0$

6. A rectangular parcel of land is 70 ft longer than it is wide. Each diagonal between opposite corners is 130 ft. What are the dimensions of the parcel?

7. Solve each inequality. Sketch the solution on a real number line, and write the answer using interval notation.

   (a) $-1 \le 5 - 2x < 10$

   (b) $x(x - 1)(x - 2) > 0$

   (c) $|x - 3| < 2$

   (d) $\dfrac{2x + 5}{x + 1} \le 1$

8. A bottle of medicine must be stored at a temperature between 5°C and 10°C. What range does this correspond to on the Fahrenheit scale?   [*Note:* The Fahrenheit (F) and Celsius (C) scales satisfy the relation $C = \frac{5}{9}(F - 32)$.]

9. For what values of x is the expression $\sqrt{4x - x^2}$ defined as a real number?

Coordinates and Graphs

In this chapter we study the coordinate plane. In the coordinate plane, we can draw a graph of an equation, thus allowing us to "see" the relationship of the variables in the equation. Thus, the coordinate plane is the link between algebra and geometry. We'll see how the graph of an equation can help us solve the equation.

The following Warm-Up Exercises will help you review the basic algebra skills that you will need for this chapter.

WARM-UP EXERCISES

1. Graph the following sets on a number line:

   (a) $\{x \mid x \geq 0\}$ **(b)** $\{x \mid |x| = 2\}$ **(c)** $\{x \mid |x| \leq 2\}$ **(d)** $\{x \mid x > 1\}$

2. Solve the following equations by using the quadratic formula:

   (a) $3x^2 + 5x - 2 = 0$ **(b)** $x^2 + 2x + 5 = 0$

3. Solve the following equation by completing the square: $2x^2 - 4x - 16 = 0$.

4. Simplify the following expression: $3(-x)^2 (-y)^4 + 6(-x) + (-x)^2 (-y)^3$.

5. Solve the following inequality: $3x - 2 > 0$.

6. Solve the following inequality: $x^2 + x - 2 \leq 0$.

7. Solve the following equations for b:

   (a) $5b = -1$ **(b)** $ab = -1$

8. For what value of a is $\dfrac{3 - b}{2 + a}$ undefined?

2.1 | The Coordinate Plane

LEARNING OBJECTIVES

After completing this section, you will be able to:

■ Graph points and regions in the coordinate plane
■ Use the Distance Formula
■ Use the Midpoint Formula

The Cartesian plane is named in honor of the French mathematician René Descartes (1596–1650), although another Frenchman, Pierre Fermat (1601–1665), also invented the principles of coordinate geometry at the same time.

The *coordinate plane* is the link between algebra and geometry. In the coordinate plane we can draw graphs of algebraic equations. The graphs, in turn, allow us to "see" the relationship between the variables in the equation.

Just as points on a line can be identified with real numbers to form the coordinate line, points in a plane can be identified with ordered pairs of numbers to form the **coordinate plane** or **Cartesian plane**. To do this, we draw two perpendicular real lines that intersect at 0 on each line. Usually, one line is horizontal with positive direction to the right and is called the **x-axis**; the other line is vertical with positive direction upward and is called the **y-axis**. The point of intersection of the x-axis and the y-axis is the **origin O**, and the two axes divide the plane into four **quadrants**, labeled I, II, III, and IV in Figure 1. (The points *on* the coordinate axes are not assigned to any quadrant.)

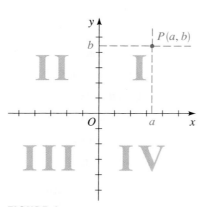

FIGURE 1   **FIGURE 2**

Although the notation for a point (a, b) is the same as the notation for an open interval (a, b), the context should make clear which meaning is intended.

Any point P in the coordinate plane can be located by a unique **ordered pair** of numbers (a, b), as shown in Figure 1. The first number a is called the **x-coordinate** of P; the second number b is called the **y-coordinate** of P. We can think of the coordinates of P as its "address," because they specify its location in the plane. Several points are labeled with their coordinates in Figure 2.

▶ **EXAMPLE 1** | Graphing Regions in the Coordinate Plane

Describe and sketch the regions given by each set.

(a) $\{(x, y) \mid x \geq 0\}$ **(b)** $\{(x, y) \mid y = 1\}$ **(c)** $\{(x, y) \mid |y| < 1\}$

▼ **SOLUTION**

(a) The points whose x-coordinates are 0 or positive lie on the y-axis or to the right of it, as shown in Figure 3(a).

(b) The set of all points with y-coordinate 1 is a horizontal line one unit above the x-axis, as in Figure 3(b).

MIDPOINT FORMULA

The midpoint of the line segment from $A(x_1, y_1)$ to $B(x_2, y_2)$ is

$$\left(\frac{x_1 + x_2}{2}, \frac{y_1 + y_2}{2} \right)$$

▶ **EXAMPLE 4** | Finding the Midpoint

The midpoint of the line segment that joins the points $(-2, 5)$ and $(4, 9)$ is

$$\left(\frac{-2 + 4}{2}, \frac{5 + 9}{2} \right) = (1, 7)$$

See Figure 8.

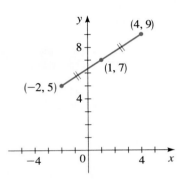

FIGURE 8

▶ **EXAMPLE 5** | Applying the Midpoint Formula

Show that the quadrilateral with vertices $P(1, 2)$, $Q(4, 4)$, $R(5, 9)$, and $S(2, 7)$ is a parallelogram by proving that its two diagonals bisect each other.

▼ **SOLUTION** If the two diagonals have the same midpoint, then they must bisect each other. The midpoint of the diagonal PR is

$$\left(\frac{1 + 5}{2}, \frac{2 + 9}{2} \right) = \left(3, \frac{11}{2} \right)$$

and the midpoint of the diagonal QS is

$$\left(\frac{4 + 2}{2}, \frac{4 + 7}{2} \right) = \left(3, \frac{11}{2} \right)$$

so each diagonal bisects the other, as shown in Figure 9. (A theorem from elementary geometry states that the quadrilateral is therefore a parallelogram.)

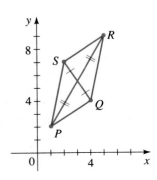

FIGURE 9

| **2.2** | Graphs of Equations in Two Variables |

LEARNING OBJECTIVES
After completing this section, you will be able to:

- Graph equations by plotting points
- Find intercepts of the graph of an equation
- Identify the equation of a circle
- Graph circles in a coordinate plane
- Determine symmetry properties of an equation

Fundamental Principle of Analytic Geometry
A point (x, y) lies on the graph of an equation if and only if its coordinates satisfy the equation.

An **equation in two variables**, such as $y = x^2 + 1$, expresses a relationship between two quantities. A point (x, y) **satisfies** the equation if it makes the equation true when the values for x and y are substituted into the equation. For example, the point $(3, 10)$ satisfies the equation $y = x^2 + 1$ because $10 = 3^2 + 1$, but the point $(1, 3)$ does not, because $3 \neq 1^2 + 1$.

THE GRAPH OF AN EQUATION

The **graph** of an equation in x and y is the set of all points (x, y) in the coordinate plane that satisfy the equation.

■ Graphing Equations by Plotting Points

The graph of an equation is a curve, so to graph an equation, we plot as many points as we can, then connect them by a smooth curve.

▶ **EXAMPLE 1** | Sketching a Graph by Plotting Points

Sketch the graph of the equation $2x - y = 3$.

▼ **SOLUTION**   We first solve the given equation for y to get

$$y = 2x - 3$$

This helps us to calculate the y-coordinates in the following table.

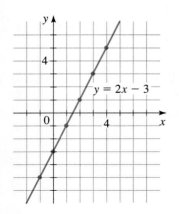

FIGURE 1

x	$y = 2x - 3$	(x, y)
-1	-5	$(-1, -5)$
0	-3	$(0, -3)$
1	-1	$(1, -1)$
2	1	$(2, 1)$
3	3	$(3, 3)$
4	5	$(4, 5)$

Of course, there are infinitely many points on the graph, and it is impossible to plot all of them. But the more points we plot, the better we can imagine what the graph represented by the equation looks like. We plot the points that we found in Figure 1; they appear to lie on a line. So we complete the graph by joining the points by a line. (In Section 2.4 we verify that the graph of this equation is indeed a line.)   ▲

▶ **EXAMPLE 2** | Sketching a Graph by Plotting Points

Sketch the graph of the equation $y = x^2 - 2$.

A detailed discussion of parabolas and their geometric properties is presented in Chapter 8.

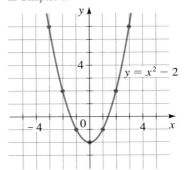

FIGURE 2

▼ **SOLUTION**   We find some of the points that satisfy the equation in the table below. In Figure 2 we plot these points and then connect them by a smooth curve. A curve with this shape is called a *parabola*.

x	$y = x^2 - 2$	(x, y)
-3	7	$(-3, 7)$
-2	2	$(-2, 2)$
-1	-1	$(-1, -1)$
0	-2	$(0, -2)$
1	-1	$(1, -1)$
2	2	$(2, 2)$
3	7	$(3, 7)$

▲

▶ **EXAMPLE 3** | Graphing an Absolute Value Equation

Sketch the graph of the equation $y = |x|$.

▼ **SOLUTION**   We make a table of values:

x	$y = |x|$	(x, y)
-3	3	$(-3, 3)$
-2	2	$(-2, 2)$
-1	1	$(-1, 1)$
0	0	$(0, 0)$
1	1	$(1, 1)$
2	2	$(2, 2)$
3	3	$(3, 3)$

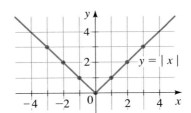

FIGURE 3

In Figure 3 we plot these points and use them to sketch the graph of the equation.   ▲

◼ Intercepts

The *x*-coordinates of the points where a graph intersects the *x*-axis are called the **x-intercepts** of the graph and are obtained by setting $y = 0$ in the equation of the graph. The *y*-coordinates of the points where a graph intersects the *y*-axis are called the **y-intercepts** of the graph and are obtained by setting $x = 0$ in the equation of the graph.

DEFINITION OF INTERCEPTS

Intercepts	How to find them	Where they are on the graph
x-intercepts: The *x*-coordinates of points where the graph of an equation intersects the *x*-axis	Set $y = 0$ and solve for x	
y-intercepts: The *y*-coordinates of points where the graph of an equation intersects the *y*-axis	Set $x = 0$ and solve for y	

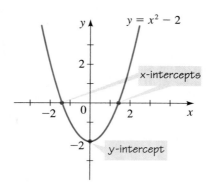

FIGURE 4

▶ **EXAMPLE 4** | Finding Intercepts

Find the x- and y-intercepts of the graph of the equation $y = x^2 - 2$.

▼ **SOLUTION** To find the x-intercepts, we set $y = 0$ and solve for x. Thus,

$$0 = x^2 - 2 \qquad \text{Set } y = 0$$
$$x^2 = 2 \qquad \text{Add 2 to each side}$$
$$x = \pm\sqrt{2} \qquad \text{Take the square root}$$

The x-intercepts are $\sqrt{2}$ and $-\sqrt{2}$.

To find the y-intercepts, we set $x = 0$ and solve for y. Thus,

$$y = 0^2 - 2 \qquad \text{Set } x = 0$$
$$y = -2$$

The y-intercept is -2.

The graph of this equation was sketched in Example 2. It is repeated in Figure 4 with the x- and y-intercepts labeled. ▲

▨ Circles

So far, we have discussed how to find the graph of an equation in x and y. The converse problem is to find an equation of a graph, that is, an equation that represents a given curve in the xy-plane. Such an equation is satisfied by the coordinates of the points on the curve and by no other point. This is the other half of the fundamental principle of analytic geometry as formulated by Descartes and Fermat. The idea is that if a geometric curve can be represented by an algebraic equation, then the rules of algebra can be used to analyze the curve.

As an example of this type of problem, let's find the equation of a circle with radius r and center (h, k). By definition, the circle is the set of all points $P(x, y)$ whose distance from the center $C(h, k)$ is r (see Figure 5). Thus, P is on the circle if and only if $d(P, C) = r$. From the Distance Formula we have

$$\sqrt{(x - h)^2 + (y - k)^2} = r$$
$$(x - h)^2 + (y - k)^2 = r^2 \qquad \text{Square each side}$$

This is the desired equation.

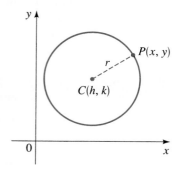

FIGURE 5

EQUATION OF A CIRCLE

An equation of the circle with center (h, k) and radius r is

$$(x - h)^2 + (y - k)^2 = r^2$$

This is called the **standard form** for the equation of the circle. If the center of the circle is the origin $(0, 0)$, then the equation is

$$x^2 + y^2 = r^2$$

▶ **EXAMPLE 5** | Graphing a Circle

Graph each equation.

(a) $x^2 + y^2 = 25$ **(b)** $(x - 2)^2 + (y + 1)^2 = 25$

▼ **SOLUTION**

(a) Rewriting the equation as $x^2 + y^2 = 5^2$, we see that this is an equation of the circle of radius 5 centered at the origin. Its graph is shown in Figure 6.

(b) Rewriting the equation as $(x - 2)^2 + (y + 1)^2 = 5^2$, we see that this is an equation of the circle of radius 5 centered at $(2, -1)$. Its graph is shown in Figure 7.

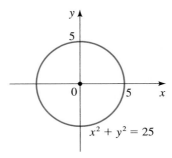

FIGURE 6   **FIGURE 7**

▶ **EXAMPLE 6** | Finding an Equation of a Circle

(a) Find an equation of the circle with radius 3 and center $(2, -5)$.

(b) Find an equation of the circle that has the points $P(1, 8)$ and $Q(5, -6)$ as the endpoints of a diameter.

▼ **SOLUTION**

(a) Using the equation of a circle with $r = 3$, $h = 2$, and $k = -5$, we obtain

$$(x - 2)^2 + (y + 5)^2 = 9$$

The graph is shown in Figure 8.

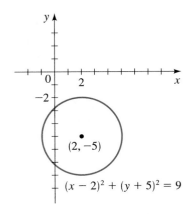

FIGURE 8

(b) We first observe that the center is the midpoint of the diameter PQ, so by the Midpoint Formula the center is

$$\left(\frac{1 + 5}{2}, \frac{8 - 6}{2} \right) = (3, 1)$$

The radius r is the distance from P to the center, so by the Distance Formula

$$r^2 = (3 - 1)^2 + (1 - 8)^2 = 2^2 + (-7)^2 = 53$$

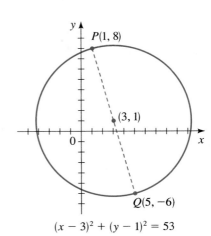

$(x - 3)^2 + (y - 1)^2 = 53$

FIGURE 9

Therefore, the equation of the circle is

$$(x - 3)^2 + (y - 1)^2 = 53$$

The graph is shown in Figure 9 on the previous page.   ▲

Completing the square is used in many contexts in algebra. In Section 1.3 we used completing the square to solve quadratic equations.

Let's expand the equation of the circle in the preceding example.

$(x - 3)^2 + (y - 1)^2 = 53$	Standard form
$x^2 - 6x + 9 + y^2 - 2y + 1 = 53$	Expand the squares
$x^2 - 6x + y^2 - 2y = 43$	Subtract 10 to get expanded form

Suppose we are given the equation of a circle in expanded form. Then to find its center and radius, we must put the equation back in standard form. That means that we must reverse the steps in the preceding calculation, and to do that, we need to know what to add to an expression such as $x^2 - 6x$ to make it a perfect square—that is, we need to complete the square, as in the next example.

▶ **EXAMPLE 7** | Identifying an Equation of a Circle

Show that the equation $x^2 + y^2 + 2x - 6y + 7 = 0$ represents a circle, and find the center and radius of the circle.

▼ **SOLUTION**   We first group the x-terms and y-terms. Then we complete the square within each grouping. That is, we complete the square for $x^2 + 2x$ by adding $\left(\frac{1}{2} \cdot 2\right)^2 = 1$, and we complete the square for $y^2 - 6y$ by adding $\left[\frac{1}{2} \cdot (-6)\right]^2 = 9$.

$(x^2 + 2x \quad\;\;) + (y^2 - 6y \quad\;\;) = -7$	Group terms
$(x^2 + 2x + 1) + (y^2 - 6y + 9) = -7 + 1 + 9$	Complete the square by adding 1 and 9 to each side
$(x + 1)^2 + (y - 3)^2 = 3$	Factor and simplify

⊘ We must add the same numbers to *each side* to maintain equality.

Comparing this equation with the standard equation of a circle, we see that $h = -1$, $k = 3$, and $r = \sqrt{3}$, so the given equation represents a circle with center $(-1, 3)$ and radius $\sqrt{3}$.   ▲

■ Symmetry

Figure 10 shows the graph of $y = x^2$. Notice that the part of the graph to the left of the y-axis is the mirror image of the part to the right of the y-axis. The reason is that if the point (x, y) is on the graph, then so is $(-x, y)$, and these points are reflections of each other about the y-axis. In this situation we say that the graph is **symmetric with respect to the y-axis**. Similarly, we say that a graph is **symmetric with respect to the x-axis** if whenever the point (x, y) is on the graph, then so is $(x, -y)$. A graph is **symmetric with respect to the origin** if whenever (x, y) is on the graph, so is $(-x, -y)$.

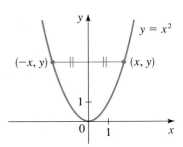

FIGURE 10

DEFINITION OF SYMMETRY

Type of symmetry	How to test for symmetry	What the graph looks like (figures in this section)	Geometric meaning
Symmetry with respect to the _x_-axis	The equation is unchanged when y is replaced by $-y$	 (Figures 6, 11, 12)	Graph is unchanged when reflected in the _x_-axis
Symmetry with respect to the _y_-axis	The equation is unchanged when x is replaced by $-x$	 (Figures 2, 3, 4, 6, 10, 12)	Graph is unchanged when reflected in the _y_-axis
Symmetry with respect to the origin	The equation is unchanged when x is replaced by $-x$ and y by $-y$	 (Figures 6, 12)	Graph is unchanged when rotated 180° about the origin

The remaining examples in this section show how symmetry helps us to sketch the graphs of equations.

▶ **EXAMPLE 8** | Using Symmetry to Sketch a Graph

Test the equation $x = y^2$ for symmetry, and sketch the graph.

▼ **SOLUTION**   If y is replaced by $-y$ in the equation $x = y^2$, we get

$$x = (-y)^2 \qquad \text{Replace } y \text{ by } -y$$

$$x = y^2 \qquad \text{Simplify}$$

so the equation is unchanged. Therefore, the graph is symmetric about the _x_-axis. But changing x to $-x$ gives the equation $-x = y^2$, which is not the same as the original equation, so the graph is not symmetric about the _y_-axis.

We use the symmetry about the _x_-axis to sketch the graph by first plotting points just for $y > 0$ and then reflecting the graph in the _x_-axis, as shown in Figure 11 on the next page.

y	$x = y^2$	(x, y)
0	0	$(0, 0)$
1	1	$(1, 1)$
2	4	$(4, 2)$
3	9	$(9, 3)$

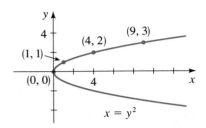

FIGURE 11

▶ **EXAMPLE 9** | Testing an Equation for Symmetry

Test the equation $y = x^3 - 9x$ for symmetry.

▼ **SOLUTION**   If we replace x by $-x$ and y by $-y$ in the equation, we get

$$-y = (-x)^3 - 9(-x) \qquad \text{Replace x by } -x \text{ and y by } -y$$

$$-y = -x^3 + 9x \qquad \text{Simplify}$$

$$y = x^3 - 9x \qquad \text{Multiply by } -1$$

so the equation is unchanged. This means that the graph is symmetric with respect to the origin. ▲

▶ **EXAMPLE 10** | A Circle That Has All Three Types of Symmetry

Test the equation of the circle $x^2 + y^2 = 4$ for symmetry.

▼ **SOLUTION**   The equation $x^2 + y^2 = 4$ remains unchanged when x is replaced by $-x$ and y is replaced by $-y$, since $(-x)^2 = x^2$ and $(-y)^2 = y^2$, so the circle exhibits all three types of symmetry. It is symmetric with respect to the x-axis, the y-axis, and the origin, as shown in Figure 12. ▲

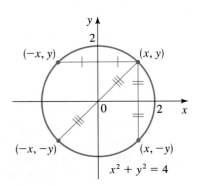

FIGURE 12

Graphing Calculators: Solving Equations and Inequalities Graphically

2.3

LEARNING OBJECTIVES

After completing this section, you will be able to:

- Use a graphing calculator to graph equations
- Solve equations graphically
- Solve inequalities graphically

In Chapter 1 we solved equations and inequalities algebraically. In the preceding section we learned how to sketch the graph of an equation in a coordinate plane. In this section we use graphs to solve equations and inequalities. To do this, we must first draw a graph using a graphing device. So we begin by giving a few guidelines to help us use graphing devices effectively.

■ Using a Graphing Calculator

A graphing calculator or computer displays a rectangular portion of the graph of an equation in a display window or viewing screen, which we call a **viewing rectangle**. The default screen often gives an incomplete or misleading picture, so it is important to choose the viewing rectangle with care. If we choose the x-values to range from a minimum value of $\mathtt{Xmin} = a$ to a maximum value of $\mathtt{Xmax} = b$ and the y-values to range from a minimum value of $\mathtt{Ymin} = c$ to a maximum value of $\mathtt{Ymax} = d$, then the displayed portion of the graph lies in the rectangle

$$[a, b] \times [c, d] = \{(x, y) \mid a \le x \le b, c \le y \le d\}$$

as shown in Figure 1. We refer to this as the $[a, b]$ by $[c, d]$ viewing rectangle.

The graphing device draws the graph of an equation much as you would. It plots points of the form (x, y) for a certain number of values of x, equally spaced between a and b. If the equation is not defined for an x-value or if the corresponding y-value lies outside the viewing rectangle, the device ignores this value and moves on to the next x-value. The machine connects each point to the preceding plotted point to form a representation of the graph of the equation.

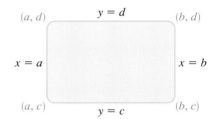

FIGURE 1 The viewing rectangle $[a, b]$ by $[c, d]$

▶ **EXAMPLE 1** | Choosing an Appropriate Viewing Rectangle

Graph the equation $y = x^2 + 3$ in an appropriate viewing rectangle.

▼ **SOLUTION**   Let's experiment with different viewing rectangles. We start with the viewing rectangle $[-2, 2]$ by $[-2, 2]$, so we set

$$\mathtt{Xmin} = -2 \qquad \mathtt{Ymin} = -2$$
$$\mathtt{Xmax} = 2 \qquad \mathtt{Ymax} = 2$$

The resulting graph in Figure 2(a) is blank! This is because $x^2 \ge 0$, so $x^2 + 3 \ge 3$ for all x. Thus, the graph lies entirely above the viewing rectangle, so this viewing rectangle is not appropriate. If we enlarge the viewing rectangle to $[-4, 4]$ by $[-4, 4]$, as in Figure 2(b), we begin to see a portion of the graph.

Now let's try the viewing rectangle $[-10, 10]$ by $[-5, 30]$. The graph in Figure 2(c) seems to give a more complete view of the graph. If we enlarge the viewing rectangle even further, as in Figure 2(d), the graph doesn't show clearly that the y-intercept is 3.

So the viewing rectangle $[-10, 10]$ by $[-5, 30]$ gives an appropriate representation of the graph.

FIGURE 2 Graphs of $y = x^2 + 3$

▶ **EXAMPLE 2** | Two Graphs on the Same Screen

Graph the equations $y = 3x^2 - 6x + 1$ and $y = 0.23x - 2.25$ together in the viewing rectangle $[-1, 3]$ by $[-2.5, 1.5]$. Do the graphs intersect in this viewing rectangle?

▼ **SOLUTION** Figure 3(a) shows the essential features of both graphs. One is a parabola, and the other is a line. It looks as if the graphs intersect near the point $(1, -2)$. However, if we zoom in on the area around this point as shown in Figure 3(b), we see that although the graphs almost touch, they do not actually intersect.

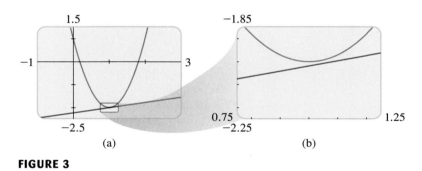

(a)   (b)

FIGURE 3

You can see from Examples 1 and 2 that the choice of a viewing rectangle makes a big difference in the appearance of a graph. If you want an overview of the essential features of a graph, you must choose a relatively large viewing rectangle to obtain a global view of the graph. If you want to investigate the details of a graph, you must zoom in to a small viewing rectangle that shows just the feature of interest.

Most graphing calculators can only graph equations in which y is isolated on one side of the equal sign. The next example shows how to graph equations that don't have this property.

▶ **EXAMPLE 3** | Graphing a Circle

Graph the circle $x^2 + y^2 = 1$.

▼ **SOLUTION** We first solve for y, to isolate it on one side of the equal sign.

$$y^2 = 1 - x^2 \qquad \text{Subtract } x^2$$

$$y = \pm\sqrt{1 - x^2} \qquad \text{Take square roots}$$

The graph in Figure 4(c) looks somewhat flattened. Most graphing calculators allow you to set the scales on the axes so that circles really look like circles. On the TI-82 and TI-83, from the $\boxed{\text{ZOO}}$ menu, choose ZSquare to set the scales appropriately. (On the TI-86 the command is Zsq.)

Therefore, the circle is described by the graphs of *two* equations:

$$y = \sqrt{1 - x^2} \quad \text{and} \quad y = -\sqrt{1 - x^2}$$

The first equation represents the top half of the circle (because $y \geq 0$), and the second represents the bottom half of the circle (because $y \leq 0$). If we graph the first equation in the viewing rectangle $[-2, 2]$ by $[-2, 2]$, we get the semicircle shown in Figure 4(a). The graph of the second equation is the semicircle in Figure 4(b). Graphing these semicircles together on the same viewing screen, we get the full circle in Figure 4(c).

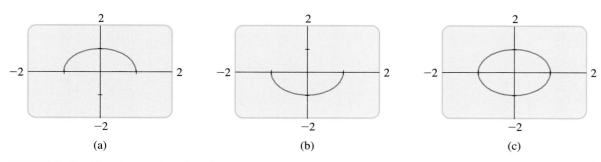

(a)   (b)   (c)

FIGURE 4 Graphing the equation $x^2 + y^2 = 1$ ▲

■ Solving Equations Graphically

In Chapter 1 we learned how to solve equations. To solve an equation such as

$$3x - 5 = 0$$

we used the **algebraic method**. This means that we used the rules of algebra to isolate x on one side of the equation. We view x as an *unknown,* and we use the rules of algebra to hunt it down. Here are the steps in the solution:

$$
\begin{aligned}
3x - 5 &= 0 \\
3x &= 5 && \text{Add 5} \\
x &= \tfrac{5}{3} && \text{Divide by 3}
\end{aligned}
$$

So the solution is $x = \frac{5}{3}$.

We can also solve this equation by the **graphical method**. In this method we view x as a *variable* and sketch the graph of the equation

$$y = 3x - 5$$

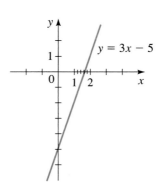

FIGURE 5

Different values for x give different values for y. Our goal is to find the value of x for which $y = 0$. From the graph in Figure 5 we see that $y = 0$ when $x \approx 1.7$. Thus, the solution is $x \approx 1.7$. Note that from the graph we obtain an approximate solution. We summarize these methods in the box on the following page.

The advantage of the algebraic method is that it gives exact answers. Also, the process of unraveling the equation to arrive at the answer helps us to understand the algebraic structure of the equation. On the other hand, for many equations it is difficult or impossible to isolate x.

The graphical method gives a numerical approximation to the answer. This is an advantage when a numerical answer is desired. (For example, an engineer might find an answer expressed as $x \approx 2.6$ more immediately useful than $x = \sqrt{7}$.) Also, graphing an equation helps us to visualize how the solution is related to other values of the variable.

SOLVING AN EQUATION

Algebraic method

Use the rules of algebra to isolate the unknown x on one side of the equation.

Example: $2x = 6 - x$

$$3x = 6 \qquad \text{Add } x$$
$$x = 2 \qquad \text{Divide by 3}$$

The solution is $x = 2$.

Graphical method

Move all terms to one side and set equal to y. Sketch the graph to find the value of x where $y = 0$.

Example: $2x = 6 - x$

$$0 = 6 - 3x$$

Set $y = 6 - 3x$ and graph.

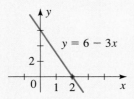

From the graph the solution is $x \approx 2$.

▶ **EXAMPLE 4** | Solving a Quadratic Equation Algebraically and Graphically

Solve the quadratic equations algebraically and graphically.

(a) $x^2 - 4x + 2 = 0$

(b) $x^2 - 4x + 4 = 0$

(c) $x^2 - 4x + 6 = 0$

▼ **SOLUTION 1:** Algebraic

We use the Quadratic Formula to solve each equation.

The Quadratic Formula is discussed on page 66.

(a) $x = \dfrac{-(-4) \pm \sqrt{(-4)^2 - 4 \cdot 1 \cdot 2}}{2} = \dfrac{4 \pm \sqrt{8}}{2} = 2 \pm \sqrt{2}$

There are two solutions, $x = 2 + \sqrt{2}$ and $x = 2 - \sqrt{2}$.

(b) $x = \dfrac{-(-4) \pm \sqrt{(-4)^2 - 4 \cdot 1 \cdot 4}}{2} = \dfrac{4 \pm \sqrt{0}}{2} = 2$

There is just one solution, $x = 2$.

(c) $x = \dfrac{-(-4) \pm \sqrt{(-4)^2 - 4 \cdot 1 \cdot 6}}{2} = \dfrac{4 \pm \sqrt{-8}}{2}$

There is no real solution.

▼ **SOLUTION 2:** Graphical

We graph the equations $y = x^2 - 4x + 2$, $y = x^2 - 4x + 4$, and $y = x^2 - 4x + 6$ in Figure 6. By determining the x-intercepts of the graphs, we find the following solutions.

(a) $x \approx 0.6$ and $x \approx 3.4$

(b) $x = 2$

(c) There is no x-intercept, so the equation has no solution.

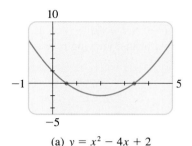
(a) $y = x^2 - 4x + 2$

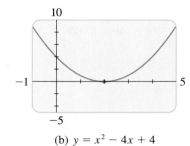
(b) $y = x^2 - 4x + 4$

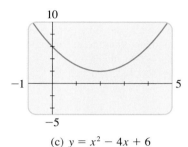
(c) $y = x^2 - 4x + 6$

FIGURE 6

The graphs in Figure 6 show visually why a quadratic equation may have two solutions, one solution, or no real solution. We proved this fact algebraically in Section 1.3 when we studied the discriminant.

▶ **EXAMPLE 5** | Another Graphical Method

Solve the equation algebraically and graphically: $5 - 3x = 8x - 20$

▼ **SOLUTION 1:** Algebraic

$$5 - 3x = 8x - 20$$
$$-3x = 8x - 25 \qquad \text{Subtract 5}$$
$$-11x = -25 \qquad \text{Subtract 8x}$$
$$x = \frac{-25}{-11} = 2\tfrac{3}{11} \qquad \text{Divide by −11 and simplify}$$

▼ **SOLUTION 2:** Graphical

We could move all terms to one side of the equal sign, set the result equal to y, and graph the resulting equation. But to avoid all this algebra, we graph two equations instead:

$$y_1 = 5 - 3x \qquad \text{and} \qquad y_2 = 8x - 20$$

The solution of the original equation will be the value of x that makes y_1 equal to y_2; that is, the solution is the x-coordinate of the intersection point of the two graphs. Using the $\boxed{\text{TRACE}}$ feature or the `intersect` command on a graphing calculator, we see from Figure 7 that the solution is $x \approx 2.27$.

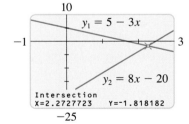

FIGURE 7

In the next example we use the graphical method to solve an equation that is extremely difficult to solve algebraically.

▶ **EXAMPLE 6** | Solving an Equation in an Interval

Solve the equation

$$x^3 - 6x^2 + 9x = \sqrt{x}$$

in the interval $[1, 6]$.

▼ **SOLUTION** We are asked to find all solutions x that satisfy $1 \le x \le 6$, so we will graph the equation in a viewing rectangle for which the x-values are restricted to this interval.

$$x^3 - 6x^2 + 9x = \sqrt{x}$$

$$x^3 - 6x^2 + 9x - \sqrt{x} = 0 \qquad \text{Subtract } \sqrt{x}$$

We can also use the z e r o command to find the solutions, as shown in Figures 8(a) and 8(b).

Figure 8 shows the graph of the equation $y = x^3 - 6x^2 + 9x - \sqrt{x}$ in the viewing rectangle $[1, 6]$ by $[-5, 5]$. There are two x-intercepts in this viewing rectangle; zooming in, we see that the solutions are $x \approx 2.18$ and $x \approx 3.72$.

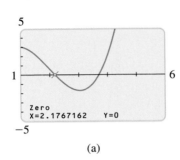

(a)

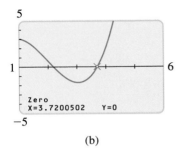

(b)

FIGURE 8

The equation in Example 6 actually has four solutions. You are asked to find the other two in the exercise set.

▶ **EXAMPLE 7** | Intensity of Light

Two light sources are 10 m apart. One is three times as intense as the other. The light intensity L (in lux) at a point x meters from the weaker source is given by

$$L = \frac{10}{x^2} + \frac{30}{(10 - x)^2}$$

(See Figure 9.) Find the points at which the light intensity is 4 lux.

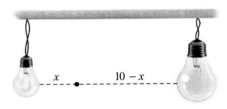

FIGURE 9

▼ **SOLUTION** We need to solve the equation

$$4 = \frac{10}{x^2} + \frac{30}{(10 - x)^2}$$

The graphs of

$$y_1 = 4 \qquad \text{and} \qquad y_2 = \frac{10}{x^2} + \frac{30}{(10 - x)^2}$$

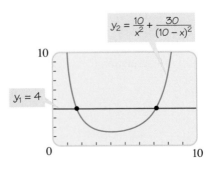

FIGURE 10

are shown in Figure 10. Zooming in (or using the i n t e r s e c t command) we find two solutions, $x \approx 1.67431$ and $x \approx 7.1927193$. So the light intensity is 4 lux at the points that are 1.67 m and 7.19 m from the weaker source. ▲

■ Solving Inequalities Graphically

Inequalities can be solved graphically. To describe the method, we solve

$$x^2 - 5x + 6 \le 0$$

This inequality was solved algebraically in Section 1.6, Example 3. To solve the inequality graphically, we draw the graph of

$$y = x^2 - 5x + 6$$

Our goal is to find those values of x for which $y \le 0$. These are simply the x-values for which the graph lies below the x-axis. From Figure 11 we see that the solution of the inequality is the interval $[2, 3]$.

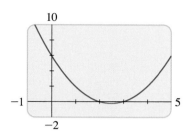

FIGURE 11
$x^2 - 5x + 6 \le 0$

▶ **EXAMPLE 8** | Solving an Inequality Graphically

Solve the inequality $3.7x^2 + 1.3x - 1.9 \le 2.0 - 1.4x$.

▼ **SOLUTION**   We graph the equations

$$y_1 = 3.7x^2 + 1.3x - 1.9 \qquad \text{and} \qquad y_2 = 2.0 - 1.4x$$

in the same viewing rectangle in Figure 12. We are interested in those values of x for which $y_1 \le y_2$; these are points for which the graph of y_2 lies on or above the graph of y_1. To determine the appropriate interval, we look for the x-coordinates of points where the graphs intersect. We conclude that the solution is (approximately) the interval $[-1.45, 0.72]$. ▲

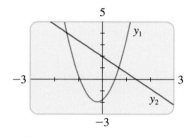

FIGURE 12
$y_1 = 3.7x^2 + 1.3x - 1.9$
$y_2 = 2.0 - 1.4x$

▶ **EXAMPLE 9** | Solving an Inequality Graphically

Solve the inequality $x^3 - 5x^2 \ge -8$.

▼ **SOLUTION**   We write the inequality as

$$x^3 - 5x^2 + 8 \ge 0$$

and then graph the equation

$$y = x^3 - 5x^2 + 8$$

in the viewing rectangle $[-6, 6]$ by $[-15, 15]$, as shown in Figure 13. The solution of the inequality consists of those intervals on which the graph lies on or above the x-axis. By moving the cursor to the x-intercepts, we find that, correct to one decimal place, the solution is $[-1.1, 1.5] \cup [4.6, \infty)$.

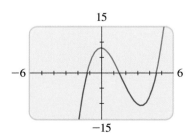

FIGURE 13 $x^3 - 5x^2 + 8 \ge 0$

2.4 | Lines

LEARNING OBJECTIVES

After completing this section, you will be able to:

■ Find the slope of a line
■ Find the point-slope form of the equation of a line
■ Find the slope-intercept form of the equation of a line
■ Find equations of horizontal and vertical lines
■ Find the general equation of a line
■ Find equations for parallel and perpendicular lines
■ Model with linear equations: interpret slope as rate of change

In this section we find equations for straight lines lying in a coordinate plane. The equations will depend on how the line is inclined, so we begin by discussing the concept of slope.

■ The Slope of a Line

We first need a way to measure the "steepness" of a line, or how quickly it rises (or falls) as we move from left to right. We define *run* to be the distance we move to the right and *rise* to be the corresponding distance that the line rises (or falls). The *slope* of a line is the ratio of rise to run:

$$\text{slope} = \frac{\text{rise}}{\text{run}}$$

Figure 1 shows situations in which slope is important. Carpenters use the term *pitch* for the slope of a roof or a staircase; the term *grade* is used for the slope of a road.

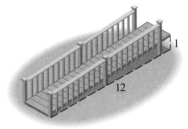

Slope of a ramp
Slope $= \frac{1}{12}$

Pitch of a roof
Slope $= \frac{1}{3}$

Grade of a road
Slope $= \frac{8}{100}$

FIGURE 1

If a line lies in a coordinate plane, then the **run** is the change in the *x*-coordinate and the **rise** is the corresponding change in the *y*-coordinate between any two points on the line (see Figure 2). This gives us the following definition of slope.

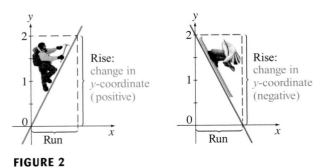

FIGURE 2

SLOPE OF A LINE

The **slope** m of a nonvertical line that passes through the points $A(x_1, y_1)$ and $B(x_2, y_2)$ is

$$m = \frac{\text{rise}}{\text{run}} = \frac{y_2 - y_1}{x_2 - x_1}$$

The slope of a vertical line is not defined.

The slope is independent of which two points are chosen on the line. We can see that this is true from the similar triangles in Figure 3:

$$\frac{y_2 - y_1}{x_2 - x_1} = \frac{y_2' - y_1'}{x_2' - x_1'}$$

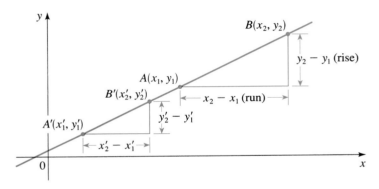

FIGURE 3

Figure 4 shows several lines labeled with their slopes. Notice that lines with positive slope slant upward to the right, whereas lines with negative slope slant downward to the right. The steepest lines are those for which the absolute value of the slope is the largest; a horizontal line has slope zero.

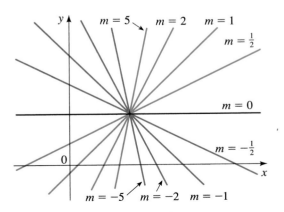

FIGURE 4 Lines with various slopes

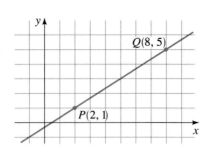

FIGURE 5

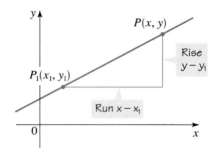

FIGURE 6

▶ **EXAMPLE 1** | Finding the Slope of a Line Through Two Points

Find the slope of the line that passes through the points $P(2, 1)$ and $Q(8, 5)$.

▼ **SOLUTION** Since any two different points determine a line, only one line passes through these two points. From the definition the slope is

$$m = \frac{y_2 - y_1}{x_2 - x_1} = \frac{5 - 1}{8 - 2} = \frac{4}{6} = \frac{2}{3}$$

This says that for every 3 units we move to the right, the line rises 2 units. The line is drawn in Figure 5. ▲

■ Point-Slope Form of the Equation of a Line

Now let's find the equation of the line that passes through a given point $P(x_1, y_1)$ and has slope m. A point $P(x, y)$ with $x \neq x_1$ lies on this line if and only if the slope of the line through P_1 and P is equal to m (see Figure 6), that is,

$$\frac{y - y_1}{x - x_1} = m$$

This equation can be rewritten in the form $y - y_1 = m(x - x_1)$; note that the equation is also satisfied when $x = x_1$ and $y = y_1$. Therefore, it is an equation of the given line.

> **POINT-SLOPE FORM OF THE EQUATION OF A LINE**
>
> An equation of the line that passes through the point (x_1, y_1) and has slope m is
>
> $$y - y_1 = m(x - x_1)$$

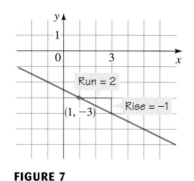

FIGURE 7

▶ **EXAMPLE 2** | Finding the Equation of a Line with Given Point and Slope

(a) Find an equation of the line through $(1, -3)$ with slope $-\frac{1}{2}$.

(b) Sketch the line.

▼ **SOLUTION**

(a) Using the point-slope form with $m = -\frac{1}{2}$, $x_1 = 1$, and $y_1 = -3$, we obtain an equation of the line as

$$y + 3 = -\tfrac{1}{2}(x - 1) \qquad \text{From point-slope equation}$$

$$2y + 6 = -x + 1 \qquad \text{Multiply by 2}$$

$$x + 2y + 5 = 0 \qquad \text{Rearrange}$$

(b) The fact that the slope is $-\frac{1}{2}$ tells us that when we move to the right 2 units, the line drops 1 unit. This enables us to sketch the line in Figure 7. ▲

▶ **EXAMPLE 3** | Finding the Equation of a Line Through Two Given Points

Find an equation of the line through the points $(-1, 2)$ and $(3, -4)$.

We can use *either* point, $(-1, 2)$ *or* $(3, -4)$, in the point-slope equation. We will end up with the same final answer.

▼ **SOLUTION** The slope of the line is

$$m = \frac{-4 - 2}{3 - (-1)} = -\frac{6}{4} = -\frac{3}{2}$$

Using the point-slope form with $x_1 = -1$ and $y_1 = 2$, we obtain

$$y - 2 = -\tfrac{3}{2}(x + 1) \qquad \text{From point-slope equation}$$

$$2y - 4 = -3x - 3 \qquad \text{Multiply by 2}$$

$$3x + 2y - 1 = 0 \qquad \text{Rearrange}$$

Slope-Intercept Form of the Equation of a Line

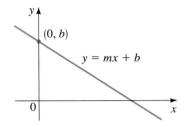

FIGURE 8

Suppose a nonvertical line has slope m and y-intercept b (see Figure 8). This means that the line intersects the y-axis at the point $(0, b)$, so the point-slope form of the equation of the line, with $x = 0$ and $y = b$, becomes

$$y - b = m(x - 0)$$

This simplifies to $y = mx + b$, which is called the **slope-intercept form** of the equation of a line.

SLOPE-INTERCEPT FORM OF THE EQUATION OF A LINE

An equation of the line that has slope m and y-intercept b is

$$y = mx + b$$

▶ **EXAMPLE 4** | Lines in Slope-Intercept Form

(a) Find the equation of the line with slope 3 and y-intercept -2.

(b) Find the slope and y-intercept of the line $3y - 2x = 1$.

▼ **SOLUTION**

(a) Since $m = 3$ and $b = -2$, from the slope-intercept form of the equation of a line we get

$$y = 3x - 2$$

(b) We first write the equation in the form $y = mx + b$:

Slope   y-intercept

$y = \tfrac{2}{3}x + \tfrac{1}{3}$

$$3y - 2x = 1$$

$$3y = 2x + 1 \qquad \text{Add } 2x$$

$$y = \tfrac{2}{3}x + \tfrac{1}{3} \qquad \text{Divide by 3}$$

From the slope-intercept form of the equation of a line, we see that the slope is $m = \tfrac{2}{3}$ and the y-intercept is $b = \tfrac{1}{3}$.

Vertical and Horizontal Lines

If a line is horizontal, its slope is $m = 0$, so its equation is $y = b$, where b is the y-intercept (see Figure 9). A vertical line does not have a slope, but we can write its equation as $x = a$, where a is the x-intercept, because the x-coordinate of every point on the line is a.

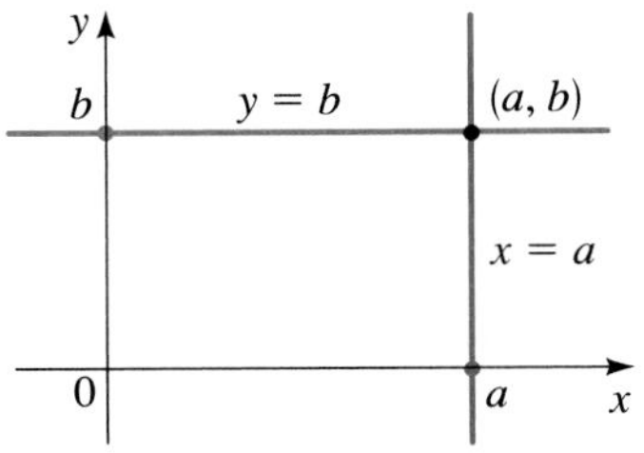

FIGURE 9

VERTICAL AND HORIZONTAL LINES

An equation of the vertical line through (a, b) is $x = a$.

An equation of the horizontal line through (a, b) is $y = b$.

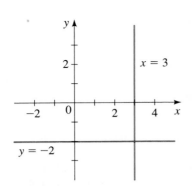

FIGURE 10

EXAMPLE 5 | Vertical and Horizontal Lines

(a) An equation for the vertical line through $(3, 5)$ is $x = 3$.

(b) The graph of the equation $x = 3$ is a vertical line with x-intercept 3.

(c) An equation for the horizontal line through $(8, -2)$ is $y = -2$.

(d) The graph of the equation $y = -2$ is a horizontal line with y-intercept -2.

The lines are graphed in Figure 10. ▲

▪ General Equation of a Line

A **linear equation** is an equation of the form

$$Ax + By + C = 0$$

where A, B, and C are constants and A and B are not both 0. The equation of a line is a linear equation:

- A nonvertical line has the equation $y = mx + b$ or $-mx + y - b = 0$, which is a linear equation with $A = -m$, $B = 1$, and $C = -b$.
- A vertical line has the equation $x = a$ or $x - a = 0$, which is a linear equation with $A = 1$, $B = 0$, and $C = -a$.

Conversely, the graph of a linear equation is a line:

- If $B \neq 0$, the equation becomes

$$y = -\frac{A}{B}x - \frac{C}{B} \qquad \text{Divide by } B$$

and this is the slope-intercept form of the equation of a line (with $m = -A/B$ and $b = -C/B$).

- If $B = 0$, the equation becomes

$$Ax + C = 0 \qquad \text{Set } B = 0$$

or $x = -C/A$, which represents a vertical line.

We have proved the following.

GENERAL EQUATION OF A LINE

The graph of every **linear equation**

$$Ax + By + C = 0 \qquad (A, B \text{ not both zero})$$

is a line. Conversely, every line is the graph of a linear equation.

▶ **EXAMPLE 6** | Graphing a Linear Equation

Sketch the graph of the equation $2x - 3y - 12 = 0$.

▼ **SOLUTION 1** Since the equation is linear, its graph is a line. To draw the graph, it is enough to find any two points on the line. The intercepts are the easiest points to find.

x-intercept: Substitute $y = 0$, to get $2x - 12 = 0$, so $x = 6$

y-intercept: Substitute $x = 0$, to get $-3y - 12 = 0$, so $y = -4$

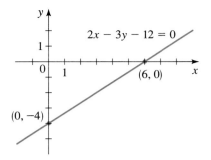

FIGURE 11

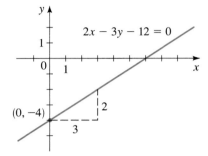

FIGURE 12

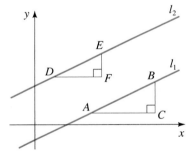

FIGURE 13

With these points we can sketch the graph in Figure 11.

▼ **SOLUTION 2**   We write the equation in slope-intercept form:

$$2x - 3y - 12 = 0$$
$$2x - 3y = 12 \qquad \text{Add 12}$$
$$-3y = -2x + 12 \qquad \text{Subtract 2x}$$
$$y = \tfrac{2}{3}x - 4 \qquad \text{Divide by } -3$$

This equation is in the form $y = mx + b$, so the slope is $m = \tfrac{2}{3}$ and the y-intercept is $b = -4$. To sketch the graph, we plot the y-intercept and then move 3 units to the right and 2 units up as shown in Figure 12.   ▲

Parallel and Perpendicular Lines

Since slope measures the steepness of a line, it seems reasonable that parallel lines should have the same slope. In fact, we can prove this.

> **PARALLEL LINES**
>
> Two nonvertical lines are parallel if and only if they have the same slope.

▼ **PROOF**   Let the lines l_1 and l_2 in Figure 13 have slopes m_1 and m_2. If the lines are parallel, then the right triangles ABC and DEF are similar, so

$$m_1 = \frac{d(B, C)}{d(A, C)} = \frac{d(E, F)}{d(D, F)} = m_2$$

Conversely, if the slopes are equal, then the triangles will be similar, so $\angle BAC = \angle EDF$ and the lines are parallel.   ▲

▶ **EXAMPLE 7** | Finding the Equation of a Line Parallel to a Given Line

Find an equation of the line through the point $(5, 2)$ that is parallel to the line $4x + 6y + 5 = 0$.

▼ **SOLUTION**   First we write the equation of the given line in slope-intercept form.

$$4x + 6y + 5 = 0$$
$$6y = -4x - 5 \qquad \text{Subtract 4x + 5}$$
$$y = -\tfrac{2}{3}x - \tfrac{5}{6} \qquad \text{Divide by 6}$$

So the line has slope $m = -\tfrac{2}{3}$. Since the required line is parallel to the given line, it also has slope $m = -\tfrac{2}{3}$. From the point-slope form of the equation of a line, we get

$$y - 2 = -\tfrac{2}{3}(x - 5) \qquad \text{Slope } m = -\tfrac{2}{3}, \text{ point } (5, 2)$$
$$3y - 6 = -2x + 10 \qquad \text{Multiply by 3}$$
$$2x + 3y - 16 = 0 \qquad \text{Rearrange}$$

Thus, the equation of the required line is $2x + 3y - 16 = 0$.   ▲

The condition for perpendicular lines is not as obvious as that for parallel lines.

PERPENDICULAR LINES

Two lines with slopes m_1 and m_2 are perpendicular if and only if $m_1 m_2 = -1$, that is, their slopes are negative reciprocals:

$$m_2 = -\frac{1}{m_1}$$

Also, a horizontal line (slope 0) is perpendicular to a vertical line (no slope).

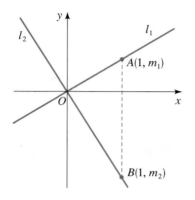

FIGURE 14

▼ **PROOF**   In Figure 14 we show two lines intersecting at the origin. (If the lines intersect at some other point, we consider lines parallel to these that intersect at the origin. These lines have the same slopes as the original lines.)

If the lines l_1 and l_2 have slopes m_1 and m_2, then their equations are $y = m_1 x$ and $y = m_2 x$. Notice that $A(1, m_1)$ lies on l_1 and $B(1, m_2)$ lies on l_2. By the Pythagorean Theorem and its converse $OA \perp OB$ if and only if

$$[d(O, A)]^2 + [d(O, B)]^2 = [d(A, B)]^2$$

By the Distance Formula this becomes

$$(1^2 + m_1^2) + (1^2 + m_2^2) = (1 - 1)^2 + (m_2 - m_1)^2$$

$$2 + m_1^2 + m_2^2 = m_2^2 - 2m_1 m_2 + m_1^2$$

$$2 = -2m_1 m_2$$

$$m_1 m_2 = -1 \qquad \blacktriangle$$

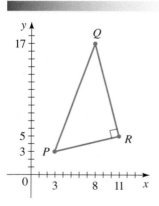

FIGURE 15

▶ **EXAMPLE 8** | Perpendicular Lines

Show that the points $P(3, 3)$, $Q(8, 17)$, and $R(11, 5)$ are the vertices of a right triangle.

▼ **SOLUTION**   The slopes of the lines containing PR and QR are, respectively,

$$m_1 = \frac{5 - 3}{11 - 3} = \frac{1}{4} \qquad \text{and} \qquad m_2 = \frac{5 - 17}{11 - 8} = -4$$

Since $m_1 m_2 = -1$, these lines are perpendicular, so PQR is a right triangle. It is sketched in Figure 15. $\qquad \blacktriangle$

▶ **EXAMPLE 9** | Finding an Equation of a Line Perpendicular to a Given Line

Find an equation of the line that is perpendicular to the line $4x + 6y + 5 = 0$ and passes through the origin.

▼ **SOLUTION**   In Example 7 we found that the slope of the line $4x + 6y + 5 = 0$ is $-\frac{2}{3}$. Thus, the slope of a perpendicular line is the negative reciprocal, that is, $\frac{3}{2}$. Since the required line passes through $(0, 0)$, the point-slope form gives

$$y - 0 = \tfrac{3}{2}(x - 0)$$

$$y = \tfrac{3}{2}x \qquad \blacktriangle$$

▶ **EXAMPLE 10** | Graphing a Family of Lines

Use a graphing calculator to graph the family of lines

$$y = 0.5x + b$$

for $b = -2, -1, 0, 1, 2$. What property do the lines share?

▼ **SOLUTION**   The lines are graphed in Figure 16 in the viewing rectangle $[-6, 6]$ by $[-6, 6]$. The lines all have the same slope, so they are parallel.

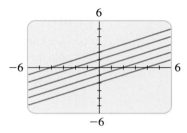

FIGURE 16 $y = 0.5x + b$ ▲

▓ Modeling with Linear Equations: Slope as Rate of Change

When a line is used to model the relationship between two quantities, the slope of the line is the **rate of change** of one quantity with respect to the other. For example, the graph in Figure 17(a) gives the amount of gas in a tank that is being filled. The slope between the indicated points is

$$m = \frac{6 \text{ gallons}}{3 \text{ minutes}} = 2 \text{ gal/min}$$

The slope is the *rate* at which the tank is being filled, 2 gallons per minute. In Figure 17(b) the tank is being drained at the *rate* of 0.03 gallon per minute, and the slope is -0.03.

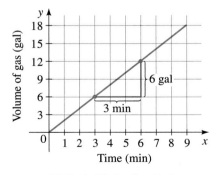

(a) Tank filled at 2 gal/min
Slope of line is 2

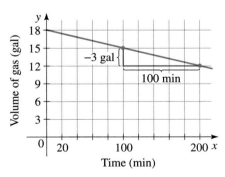

(b) Tank drained at 0.03 gal/min
Slope of line is −0.03

FIGURE 17

   The next two examples give other situations in which the slope of a line is a rate of change.

▶ **EXAMPLE 11** | Slope as Rate of Change

A dam is built on a river to create a reservoir. The water level w in the reservoir is given by the equation

$$w = 4.5t + 28$$

where t is the number of years since the dam was constructed and w is measured in feet.

(a) Sketch a graph of this equation.

(b) What do the slope and w-intercept of this graph represent?

▼ **SOLUTION**

(a) This equation is linear, so its graph is a line. Since two points determine a line, we plot two points that lie on the graph and draw a line through them.

When $t = 0$, then $w = 4.5(0) + 28 = 28$, so $(0, 28)$ is on the line.

When $t = 2$, then $w = 4.5(2) + 28 = 37$, so $(2, 37)$ is on the line.

The line that is determined by these points is shown in Figure 18.

(b) The slope is $m = 4.5$; it represents the rate of change of water level with respect to time. This means that the water level *increases* 4.5 ft per year. The w-intercept is 28 and occurs when $t = 0$, so it represents the water level when the dam was constructed. ▲

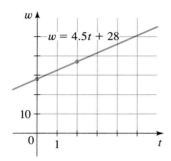

FIGURE 18

▶ **EXAMPLE 12** | Linear Relationship Between Temperature and Elevation

(a) As dry air moves upward, it expands and cools. If the ground temperature is 20°C and the temperature at a height of 1 km is 10°C, express the temperature T (in °C) in terms of the height h (in kilometers). (Assume that the relationship between T and h is linear.)

(b) Draw the graph of the linear equation. What does its slope represent?

(c) What is the temperature at a height of 2.5 km?

▼ **SOLUTION**

(a) Because we are assuming a linear relationship between T and h, the equation must be of the form

$$T = mh + b$$

where m and b are constants. When $h = 0$, we are given that $T = 20$, so

$$20 = m(0) + b$$

$$b = 20$$

Thus, we have

$$T = mh + 20$$

When $h = 1$, we have $T = 10$, so

$$10 = m(1) + 20$$

$$m = 10 - 20 = -10$$

The required expression is

$$T = -10h + 20$$

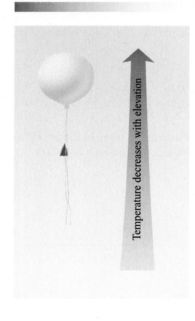

Temperature decreases with elevation

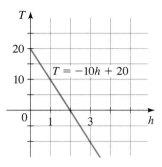

FIGURE 19

(b) The graph is sketched in Figure 19. The slope is $m = -10°C/km$, and this represents the rate of change of temperature with respect to distance above the ground. So the temperature *decreases* 10°C per kilometer of height.

(c) At a height of $h = 2.5$ km the temperature is

$$T = -10(2.5) + 20 = -25 + 20 = -5°C$$

▲

2.5 | Making Models Using Variation

LEARNING OBJECTIVES

After completing this section, you will be able to:

■ Find equations for direct variation
■ Find equations for inverse variation
■ Find equations for joint variation

When scientists talk about a mathematical model for a real-world phenomenon, they often mean an equation that describes the relationship between two quantities. For instance, the model might describe how the population of an animal species varies with time or how the pressure of a gas varies as its temperature changes. In this section we study a kind of modeling called *variation*.

■ Direct Variation

Two types of mathematical models occur so often that they are given special names. The first is called *direct variation* and occurs when one quantity is a constant multiple of the other, so we use an equation of the form $y = kx$ to model this dependence.

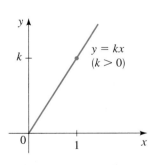

FIGURE 1

> **DIRECT VARIATION**
>
> If the quantities x and y are related by an equation
>
> $$y = kx$$
>
> for some constant $k \neq 0$, we say that y **varies directly as** x, or y is **directly proportional to** x, or simply y **is proportional to** x. The constant k is called the **constant of proportionality**.

Recall that the graph of an equation of the form $y = mx + b$ is a line with slope m and y-intercept b. So the graph of an equation $y = kx$ that describes direct variation is a line with slope k and y-intercept 0 (see Figure 1).

▶ **EXAMPLE 1** | Direct Variation

During a thunderstorm you see the lightning before you hear the thunder because light travels much faster than sound. The distance between you and the storm varies directly as the time interval between the lightning and the thunder.

(a) Suppose that the thunder from a storm 5400 ft away takes 5 s to reach you. Determine the constant of proportionality, and write the equation for the variation.

(b) Sketch the graph of this equation. What does the constant of proportionality represent?

(c) If the time interval between the lightning and thunder is now 8 s, how far away is the storm?

▼ **SOLUTION**

(a) Let d be the distance from you to the storm, and let t be the length of the time interval. We are given that d varies directly as t, so

$$d = kt$$

where k is a constant. To find k, we use the fact that $t = 5$ when $d = 5400$. Substituting these values in the equation, we get

$$5400 = k(5) \qquad \text{Substitute}$$

$$k = \frac{5400}{5} = 1080 \qquad \text{Solve for } k$$

Substituting this value of k in the equation for d, we obtain

$$d = 1080t$$

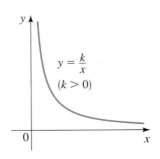

FIGURE 2

as the equation for d as a function of t.

(b) The graph of the equation $d = 1080t$ is a line through the origin with slope 1080 and is shown in Figure 2. The constant $k = 1080$ is the approximate speed of sound (in ft/s).

(c) When $t = 8$, we have

$$d = 1080 \cdot 8 = 8640$$

So the storm is 8640 ft $\approx$ 1.6 mi away. ▲

▦ Inverse Variation

Another equation that is frequently used in mathematical modeling is $y = k/x$, where k is a constant.

INVERSE VARIATION

If the quantities x and y are related by the equation

$$y = \frac{k}{x}$$

for some constant $k \neq 0$, we say that y **is inversely proportional to** x or y **varies inversely as** x.

FIGURE 3 Inverse variation

The graph of $y = k/x$ for $x > 0$ is shown in Figure 3 for the case $k > 0$. It gives a picture of what happens when y is inversely proportional to x.

▶ **EXAMPLE 2** | Inverse Variation

Boyle's Law states that when a sample of gas is compressed at a constant temperature, the pressure of the gas is inversely proportional to the volume of the gas.

(a) Suppose the pressure of a sample of air that occupies 0.106 m³ at 25 °C is 50 kPa. Find the constant of proportionality, and write the equation that expresses the inverse proportionality.

(b) If the sample expands to a volume of 0.3 m³, find the new pressure.

▼ **SOLUTION**

(a) Let P be the pressure of the sample of gas and let V be its volume. Then, by the definition of inverse proportionality, we have

$$P = \frac{k}{V}$$

where k is a constant. To find k, we use the fact that $P = 50$ when $V = 0.106$. Substituting these values in the equation, we get

$$50 = \frac{k}{0.106} \qquad \text{Substitute}$$

$$k = (50)(0.106) = 5.3 \qquad \text{Solve for } k$$

Putting this value of k in the equation for P, we have

$$P = \frac{5.3}{V}$$

(b) When $V = 0.3$, we have

$$P = \frac{5.3}{0.3} \approx 17.7$$

So the new pressure is about 17.7 kPa. ▲

■ Joint Variation

A physical quantity often depends on more than one other quantity. If one quantity is proportional to two or more other quantities, we call this relationship *joint variation*.

JOINT VARIATION

If the quantities x, y, and z are related by the equation

$$z = kxy$$

where k is a nonzero constant, we say that z **varies jointly as** x and y or z **is jointly proportional to** x and y.

In the sciences, relationships between three or more variables are common, and any combination of the different types of proportionality that we have discussed is possible. For example, if

$$z = k\frac{x}{y}$$

we say that z **is proportional to** x and **inversely proportional to** y.

▶ **EXAMPLE 3** | Newton's Law of Gravitation

Newton's Law of Gravitation says that two objects with masses m_1 and m_2 attract each other with a force F that is jointly proportional to their masses and inversely proportional to the square of the distance r between the objects. Express Newton's Law of Gravitation as an equation.

▼ **SOLUTION**   Using the definitions of joint and inverse variation and the traditional notation G for the gravitational constant of proportionality, we have

$$F = G\frac{m_1 m_2}{r^2}$$

If m_1 and m_2 are fixed masses, then the gravitational force between them is $F = C/r^2$ (where $C = Gm_1 m_2$ is a constant). Figure 4 shows the graph of this equation for $r > 0$ with $C = 1$. Observe how the gravitational attraction decreases with increasing distance.

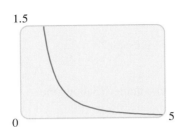

FIGURE 4 Graph of $F = \dfrac{1}{r^2}$

▶ CHAPTER 2 | REVIEW

▼ PROPERTIES AND FORMULAS

The Distance Formula (p. 99)

The distance between the points $A(x_1, y_1)$ and $B(x_2, y_2)$ is

$$d(A, B) = \sqrt{(x_1 - x_2)^2 + (y_1 - y_2)^2}$$

The Midpoint Formula (p. 101)

The midpoint of the line segment from $A(x_1, y_1)$ to $B(x_2, y_2)$ is

$$\left(\frac{x_1 + x_2}{2}, \frac{y_1 + y_2}{2}\right)$$

Intercepts (p. 103)

To find the **x-intercepts** of the graph of an equation, set $y = 0$ and solve for x.

To find the **y-intercepts** of the graph of an equation, set $x = 0$ and solve for y.

Circles (p. 104)

The circle with center $(0, 0)$ and radius r has equation

$$x^2 + y^2 = r^2$$

The circle with center (h, k) and radius r has equation

$$(x - h)^2 + (y - k)^2 = r^2$$

Symmetry (p. 107)

The graph of an equation is **symmetric with respect to the x-axis** if the equation remains unchanged when you replace y by $-y$.

The graph of an equation is **symmetric with respect to the y-axis** if the equation remains unchanged when you replace x by $-x$.

The graph of an equation is **symmetric with respect to the origin** if the equation remains unchanged when you replace x by $-x$ and y by $-y$.

Slope of a Line (p. 117)

The slope of the nonvertical line that contains the points $A(x_1, y_1)$ and $B(x_2, y_2)$ is

$$m = \frac{\text{rise}}{\text{run}} = \frac{y_2 - y_1}{x_2 - x_1}$$

Equations of Lines (pp. 118–120)

If a line has slope m, has y-intercept b, and contains the point (x_1, y_1), then:

the **point-slope form** of its equation is

$$y - y_1 = m(x - x_1)$$

the **slope-intercept form** of its equation is

$$y = mx + b$$

The equation of any line can be expressed in the **general form**

$$Ax + By + C = 0$$

(where A and B can't both be 0).

Vertical and Horizontal Lines (p. 119)

The **vertical** line containing the point (a, b) has the equation $x = a$.

The **horizontal** line containing the point (a, b) has the equation $y = b$.

Parallel and Perpendicular Lines (pp. 121–122)

Two lines with slopes m_1 and m_2 are

parallel if and only if $m_1 = m_2$

perpendicular if and only if $m_1 m_2 = -1$

Variation (pp. 125–127)

If y is **directly proportional** to x, then

$$y = kx$$

If y is **inversely proportional** to x, then

$$y = \frac{k}{x}$$

If z is **jointly proportional** to x and y, then

$$z = kxy$$

In each case, k is the **constant of proportionality**.

▼ CONCEPT SUMMARY

Section 2.1

	Review Exercises
▪ Graph points and regions in the coordinate plane	1(a)–4(a), 5, 6
▪ Use the Distance Formula	1(b)–4(b), 7
▪ Use the Midpoint Formula	1(c)–4(c)

Section 2.2

	Review Exercises
▪ Graph equations by plotting points	15–24
▪ Find intercepts of the graph of an equation	25–30
▪ Identify the equation of a circle	1(e)–4(e), 8–14, 63, 64
▪ Graph circles in a coordinate plane	1(e)–4(e), 11–14
▪ Determine symmetry properties of an equation	25–30

Section 2.3

	Review Exercises
▪ Use a graphing calculator to graph equations	31–34
▪ Solve equations graphically	35–38
▪ Solve inequalities graphically	39–42

Section 2.4

	Review Exercises
▪ Find the slope of a line	1(d)–4(d)
▪ Find the point-slope form of the equation of a line	1(d)–4(d)
▪ Find the slope-intercept form of the equation of a line	1(d)–4(d), 43–52, 63, 64
▪ Find equations for horizontal and vertical lines	47–48
▪ Find the general form for the equation of a line	43–52
▪ Find equations for parallel or perpendicular lines	49–52
▪ Model with linear equations: interpret slope as rate of change	53–54

Section 2.5

	Review Exercises
▪ Find equations for direct variation	55, 59, 60
▪ Find equations for inverse variation	56–58
▪ Find equations for joint variation	61–62

▼ EXERCISES

1–4 ▪ Two points P and Q are given.
(a) Plot P and Q on a coordinate plane.
(b) Find the distance from P to Q.
(c) Find the midpoint of the segment PQ.
(d) Find the slope of the line determined by P and Q, and find equations for the line in point-slope form and in slope-intercept form. Then sketch a graph of the line.
(e) Sketch the circle that passes through Q and has center P, and find the equation of this circle.

1. $P(0, 3), \quad Q(3, 7)$ **2.** $P(2, 0), \quad Q(-4, 8)$

3. $P(-6, 2), \quad Q(4, -14)$ **4.** $P(5, -2), \quad Q(-3, -6)$

5–6 ▪ Sketch the region given by the set.

5. $\{(x, y) \mid -4 < x < 4 \text{ and } -2 < y < 2\}$

6. $\{(x, y) \mid x \geq 4 \text{ or } y \geq 2\}$

7. Which of the points $A(4, 4)$ or $B(5, 3)$ is closer to the point $C(-1, -3)$?

8. Find an equation of the circle that has center $(2, -5)$ and radius $\sqrt{2}$.

9. Find an equation of the circle that has center $(-5, -1)$ and passes through the origin.

10. Find an equation of the circle that contains the points $P(2, 3)$ and $Q(-1, 8)$ and has the midpoint of the segment PQ as its center.

11–14 ▪ (a) Complete the square to determine whether the equation represents a circle or a point or has no graph. (b) If the equation is that of a circle, find its center and radius, and sketch its graph.

11. $x^2 + y^2 + 2x - 6y + 9 = 0$

12. $2x^2 + 2y^2 - 2x + 8y = \frac{1}{2}$

13. $x^2 + y^2 + 72 = 12x$

14. $x^2 + y^2 - 6x - 10y + 34 = 0$

15–24 ▪ Sketch the graph of the equation by making a table and plotting points.

15. $y = 2 - 3x$ **16.** $2x - y + 1 = 0$

17. $x + 3y = 21$ **18.** $x = 2y + 12$

19. $\frac{x}{2} - \frac{y}{7} = 1$ **20.** $\frac{x}{4} + \frac{y}{5} = 0$

21. $y = 16 - x^2$ **22.** $8x + y^2 = 0$

23. $x = \sqrt{y}$ **24.** $y = -\sqrt{1 - x^2}$

25–30 ▪ (a) Test the equation for symmetry with respect to the x-axis, the y-axis, and the origin. (b) Find the x- and y-intercepts of the graph of the equation.

25. $y = 9 - x^2$ **26.** $6x + y^2 = 36$

27. $x^2 + (y - 1)^2 = 1$ **28.** $x^4 = 16 + y$

29. $9x^2 - 16y^2 = 144$ **30.** $y = \dfrac{4}{x}$

31–34 ▪ Use a graphing device to graph the equation in an appropriate viewing rectangle.

31. $y = x^2 - 6x$ **32.** $y = \sqrt{5 - x}$

33. $y = x^3 - 4x^2 - 5x$ **34.** $\dfrac{x^2}{4} + y^2 = 1$

35–38 ▪ Solve the equation graphically.

35. $x^2 - 4x = 2x + 7$ **36.** $\sqrt{x + 4} = x^2 - 5$

37. $x^4 - 9x^2 = x - 9$ **38.** $||x + 3| - 5| = 2$

39–42 ▪ Solve the inequality graphically.

39. $4x - 3 \geq x^2$ **40.** $x^3 - 4x^2 - 5x > 2$

41. $x^4 - 4x^2 < \frac{1}{2}x - 1$ **42.** $|x^2 - 16| - 10 \geq 0$

43–52 ▪ A description of a line is given. Find an equation for the line in (a) slope-intercept form and (b) general form.

43. The line that has slope 2 and y-intercept 6

44. The line that has slope $-\frac{1}{2}$ and passes through the point $(6, -3)$

45. The line that passes through the points $(-1, -6)$ and $(2, -4)$

46. The line that has x-intercept 4 and y-intercept 12

47. The vertical line that passes through the point $(3, -2)$

48. The horizontal line with y-intercept 5

49. The line that passes through the point $(1, 1)$ and is parallel to the line $2x - 5y = 10$

50. The line that passes through the origin and is parallel to the line containing $(2, 4)$ and $(4, -4)$

51. The line that passes through the origin and is perpendicular to the line $y = \frac{1}{2}x - 10$

52. The line that passes through the point $(1, 7)$ and is perpendicular to the line $x - 3y + 16 = 0$

53. Hooke's Law states that if a weight w is attached to a hanging spring, then the stretched length s of the spring is linearly related to w. For a particular spring we have

$$s = 0.3w + 2.5$$

where s is measured in inches and w in pounds.
(a) What do the slope and s-intercept in this equation represent?
(b) How long is the spring when a 5-lb weight is attached?

54. Margarita is hired by an accounting firm at a salary of \$60,000 per year. Three years later her annual salary has increased to \$70,500. Assume that her salary increases linearly.
(a) Find an equation that relates her annual salary S and the number of years t that she has worked for the firm.
(b) What do the slope and S-intercept of her salary equation represent?
(c) What will her salary be after 12 years with the firm?

55. Suppose that M varies directly as z and that $M = 120$ when $z = 15$. Write an equation that expresses this variation.

56. Suppose that z is inversely proportional to y and that $z = 12$ when $y = 16$. Write an equation that expresses z in terms of y.

57. The intensity of illumination I from a light varies inversely as the square of the distance d from the light.
  (a) Write this statement as an equation.
  (b) Determine the constant of proportionality if it is known that a lamp has an intensity of 1000 candles at a distance of 8 m.
  (c) What is the intensity of this lamp at a distance of 20 m?

58. The frequency of a vibrating string under constant tension is inversely proportional to its length. If a violin string 12 inches long vibrates 440 times per second, to what length must it be shortened to vibrate 660 times per second?

59. The terminal velocity of a parachutist is directly proportional to the square root of his weight. A 160-lb parachutist attains a terminal velocity of 9 mi/h. What is the terminal velocity for a parachutist who weighs 240 lb?

60. The maximum range of a projectile is directly proportional to the square of its velocity. A baseball pitcher throws a ball at 60 mi/h, with a maximum range of 242 ft. What is his maximum range if he throws the ball at 70 mi/h?

61. Suppose that F is jointly proportional to q_1 and q_2 and that $F = 0.006$ when $q_1 = 4$ and $q_2 = 12$. Find an equation that expresses F in terms of q_1 and q_2.

62. The kinetic energy E of a moving object is jointly proportional to the object's mass m and the square of its speed v. A rock with mass 10 kg that is moving at 6 m/s has a kinetic energy of 180 J (joules). What is the kinetic energy of a car with mass 1700 kg that is moving at 30 m/s?

63–64 ▪ Find equations for the circle and the line in the figure.

63.

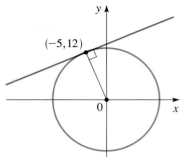

64.

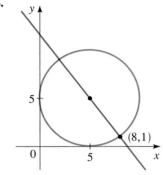

1. Let $P(1, -3)$ and $Q(7, 5)$ be two points in the coordinate plane.

   (a) Plot P and Q in the coordinate plane.

   (b) Find the distance between P and Q.

   (c) Find the midpoint of the segment PQ.

   (d) Find the slope of the line that contains P and Q.

   (e) Find the perpendicular bisector of the line that contains P and Q.

   (f) Find an equation for the circle for which the segment PQ is a diameter.

2. Find the center and radius of each circle, and sketch its graph.

   (a) $x^2 + y^2 = \frac{25}{4}$ (b) $(x - 3)^2 + y^2 = 9$ (c) $x^2 + 6x + y^2 - 2y + 6 = 0$

3. Test each equation for symmetry. Find the x- and y-intercepts, and sketch a graph of the equation.

   (a) $x = 4 - y^2$ (b) $y = |x - 2|$

4. A line has the general linear equation $3x - 5y = 15$.

   (a) Find the x- and y-intercepts of the graph of this line.

   (b) Graph the line. Use the intercepts that you found in part (a) to help you.

   (c) Write the equation of the line in slope-intercept form.

   (d) What is the slope of the line?

   (e) What is the slope of any line perpendicular to the given line?

5. Find an equation for the line with the given property.

   (a) It passes through the point $(3, -6)$ and is parallel to the line $3x + y - 10 = 0$.

   (b) It has x-intercept 6 and y-intercept 4.

6. A geologist uses a probe to measure the temperature T (in °C) of the soil at various depths below the surface, and finds that at a depth of x cm, the temperature is given by the linear equation $T = 0.08x - 4$.

   (a) What is the temperature at a depth of one meter (100 cm)?

   (b) Sketch a graph of the linear equation.

   (c) What do the slope, the x-intercept, and the T-intercept of the graph of this equation represent?

7. Solve the equation or inequality graphically, correct to two decimals.

   (a) $x^3 - 9x - 1 = 0$

   (b) $\frac{1}{2}x + 2 \geq \sqrt{x^2 + 1}$

8. The maximum weight M that can be supported by a beam is jointly proportional to its width w and the square of its height h and inversely proportional to its length L.

   (a) Write an equation that expresses this proportionality.

   (b) Determine the constant of proportionality if a beam 4 in. wide, 6 in. high, and 12 ft long can support a weight of 4800 lb.

   (c) If a 10-ft beam made of the same material is 3 in. wide and 10 in. high, what is the maximum weight it can support?

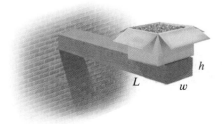

1. Johanna earns 4.75% simple interest on her bank account annually. If she earns $380 interest in a given year, what amount of money did she have in her account at the beginning of the year?

2. Calculate each complex number and write the result in the form $a + bi$.

   (a) $\left(3 + \frac{5}{2}i\right) - \left(\frac{10}{3} - \frac{1}{2}i\right)$ (b) $\left(\sqrt{3} + \sqrt{-2}\right)\left(\sqrt{12} - \sqrt{-8}\right)$

   (c) $\dfrac{-15i}{4 + 3i}$ (d) The complex conjugate of $\dfrac{4 - \sqrt{-3}}{2}$

3. Find all solutions, real and complex, of each equation.

   (a) $x + 7 = 2 - \frac{3}{2}x$ (b) $\dfrac{4x - 1}{2x + 6} = \dfrac{2x - 3}{x + 2}$

   (c) $x^2 - 5x = 6$ (d) $3x^2 - 4x + 2 = 0$

   (e) $x^4 = 16$ (f) $|x - 0.5| = 2.25$

4. Solve each inequality. Graph the solution on a real number line, and express the solution in interval notation.

   (a) $x - 5 \geq 4 - 2x$ (b) $7 - 2|x + 1| < 1$

   (c) $2x^2 - x - 1 \leq 0$ (d) $\dfrac{x}{x - 3} > 1$

 5. (a) Use a graphing calculator to graph the equation $y = x^4 - 2x^3 - 7x^2 + 8x + 12$.

   (b) Use your graph to solve the equation $x^4 - 2x^3 - 7x^2 + 8x + 12 = 0$.

   (c) Use your graph to solve the inequality $x^4 - 2x^3 - 7x^2 + 8x + 12 \leq 0$.

6. Let $P(-3, 6)$ and $Q(10, 1)$ be two points in the coordinate plane.

   (a) Find the distance between P and Q.

   (b) Find the slope-intercept form of the equation of the line that contains P and Q.

   (c) Find an equation of the circle that contains P and Q and whose center is the midpoint of the segment PQ.

   (d) Find an equation for the line that contains P and that is perpendicular to the segment PQ.

   (e) Find an equation for the line that passes through the origin and that is parallel to the segment PQ.

7. Sketch a graph of each equation. Label all x- and y-intercepts. If the graph is a circle, find its center and radius.

   (a) $2x - 5y = 20$ (b) $x^2 - 6x + y^2 = 0$

8. Find an equation for each graph.

   (a)   (b)

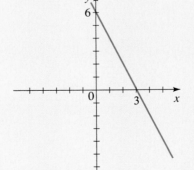

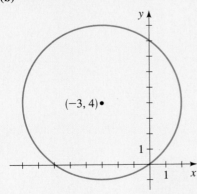

9. A sailboat departed from an island and traveled north for 2 hours, then east for $1\frac{1}{2}$ hours, all at the same speed. At the end of the trip it was 20 miles from where it started. At what speed did the boat travel?

10. A survey finds that the average starting salary for young people in their first full-time job is proportional to the square of the number of years of education they have completed. College graduates with 16 years of education have an average starting salary of $48,000.

    (a) Write an equation that expresses the relationship between years of education x and average starting salary S.

    (b) What is the average starting salary of a person who drops out of high school after completing the tenth grade?

    (c) A person with a master's degree has an average starting salary of $60,750. How many years of education does this represent?

Functions

One of the most basic and important ideas in all of mathematics is the concept of a function. In this chapter we study functions, their graphs, and some of their applications.

The following Warm-Up Exercises will help you review the basic algebra skills that you will need for this chapter.

WARM-UP EXERCISES

1. Simplify the following, if possible:

   (a) $(x^2 + 3x + 2) - (x^2 - 2x + 3)$ **(b)** $(x + 4)(x^2 + 2)$

2. Simplify the following, if possible:

   (a) $(x - 2)^2 + (x - 2) + 4$ **(b)** $(x^2 + x - 3) + 4$

3. Simplify the following: $\dfrac{(x + h)^2 - x^2}{h}$.

4. For what values of x are the following expressions defined?

   (a) $\dfrac{x + 3}{\sqrt{x - 2}}$ **(b)** $\dfrac{x + 3}{\sqrt[3]{x - 2}}$

5. Find the slope of the line that goes through the points $(3, 2)$ and $(5, -3)$.

6. Simplify the following: $\dfrac{[(4x - 5)^{1/7}]^7 + 5}{4}$.

7. Solve for y: $x = \dfrac{6y - 2}{3}$.

8. Graph the equations $y = \dfrac{2}{x - 3}$ and $y = \dfrac{2}{x} + 3$ on your calculator on the same screen.

3.1 | What Is a Function?

LEARNING OBJECTIVES

After completing this section, you will be able to:

▣ Recognize functions in the real world
▣ Work with function notation
▣ Find domains of functions
▣ Represent functions verbally, algebraically, graphically, and numerically

Perhaps the most useful mathematical idea for modeling the real world is the concept of function, which we study in this chapter. In this section we explore the idea of a function and then give the mathematical definition of function.

▣ Functions All Around Us

In nearly every physical phenomenon we observe that one quantity depends on another. For example, your height depends on your age, the temperature depends on the date, the cost of mailing a package depends on its weight (see Figure 1). We use the term *function* to describe this dependence of one quantity on another. That is, we say the following:

- Height is a function of age.
- Temperature is a function of date.
- Cost of mailing a package is a function of weight.

The U.S. Post Office uses a simple rule to determine the cost of mailing a first-class parcel on the basis of its weight. But it's not so easy to describe the rule that relates height to age or the rule that relates temperature to date.

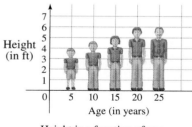

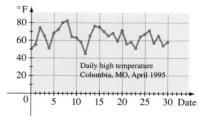

w (ounces)	Postage (dollars)
$0 < w \le 1$	1.13
$1 < w \le 2$	1.30
$2 < w \le 3$	1.47
$3 < w \le 4$	1.64
$4 < w \le 5$	1.81
$5 < w \le 6$	1.98

FIGURE 1   Height is a function of age.   Temperature is a function of date.   Postage is a function of weight.

Can you think of other functions? Here are some more examples:

- The area of a circle is a function of its radius.
- The number of bacteria in a culture is a function of time.
- The weight of an astronaut is a function of her elevation.
- The price of a commodity is a function of the demand for that commodity.

The rule that describes how the area A of a circle depends on its radius r is given by the formula $A = \pi r^2$. Even when a precise rule or formula describing a function is not available, we can still describe the function by a graph. For example, when you turn on a hot water faucet, the temperature of the water depends on how long the water has been running. So we can say

- Temperature of water from the faucet is a function of time.

Figure 2 shows a rough graph of the temperature T of the water as a function of the time t that has elapsed since the faucet was turned on. The graph shows that the initial temperature of the water is close to room temperature. When the water from the hot water tank reaches the faucet, the water's temperature T increases quickly. In the next phase, T is constant at the temperature of the water in the tank. When the tank is drained, T decreases to the temperature of the cold water supply.

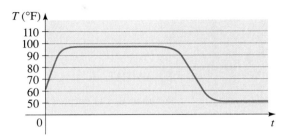

FIGURE 2 Graph of water temperature T as a function of time t

■ Definition of Function

We have previously used letters to stand for numbers. Here we do something quite different. We use letters to represent *rules*.

A function is a rule. To talk about a function, we need to give it a name. We will use letters such as f, g, h, $\ldots$ to represent functions. For example, we can use the letter f to represent a rule as follows:

"f"   is the rule   "square the number"

When we write $f(2)$, we mean "apply the rule f to the number 2." Applying the rule gives $f(2) = 2^2 = 4$. Similarly, $f(3) = 3^2 = 9$, $f(4) = 4^2 = 16$, and in general $f(x) = x^2$.

> **DEFINITION OF A FUNCTION**
>
> A **function** f is a rule that assigns to each element x in a set A exactly one element, called $f(x)$, in a set B.

We usually consider functions for which the sets A and B are sets of real numbers. The symbol $f(x)$ is read "f of x" or "f at x" and is called the **value of f at x**, or the **image of x under f**. The set A is called the **domain** of the function. The **range** of f is the set of all possible values of $f(x)$ as x varies throughout the domain, that is,

$$\text{range of } f = \{f(x) \mid x \in A\}$$

The symbol that represents an arbitrary number in the domain of a function f is called an **independent variable**. The symbol that represents a number in the range of f is called a **dependent variable**. So if we write $y = f(x)$, then x is the independent variable and y is the dependent variable.

It is helpful to think of a function as a **machine** (see Figure 3). If x is in the domain of the function f, then when x enters the machine, it is accepted as an **input** and the machine produces an **output** $f(x)$ according to the rule of the function. Thus, we can think of the domain as the set of all possible inputs and the range as the set of all possible outputs.

The $\boxed{\sqrt{}}$ key on your calculator is a good example of a function as a machine. First you input x into the display. Then you press the key labeled $\boxed{\sqrt{}}$. (On most *graphing* calculators the order of these operations is reversed.) If $x < 0$, then x is not in the domain of this function; that is, x is not an acceptable input, and the calculator will indicate an error. If $x \geq 0$, then an approximation to $\sqrt{x}$ appears in the display, correct to a certain number of decimal places. (Thus, the $\boxed{\sqrt{}}$ key on your calculator is not quite the same as the exact mathematical function f defined by $f(x) = \sqrt{x}$.)

FIGURE 3 Machine diagram of f

x input → f → $f(x)$ output

Another way to picture a function is by an **arrow diagram** as in Figure 4 on the next page. Each arrow connects an element of A to an element of B. The arrow indicates that $f(x)$ is associated with x, $f(a)$ is associated with a, and so on.

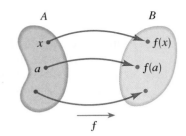

FIGURE 4 Arrow diagram of f

▶ **EXAMPLE 1** | Analyzing a Function

A function f is defined by the formula

$$f(x) = x^2 + 4$$

(a) Express in words how f acts on the input x to produce the output $f(x)$.

(b) Evaluate $f(3)$, $f(-2)$, and $f(\sqrt{5})$.

(c) Find the domain and range of f.

(d) Draw a machine diagram for f.

▼ **SOLUTION**

(a) The formula tells us that f first squares the input x and then adds 4 to the result. So f is the function

"square, then add 4"

(b) The values of f are found by substituting for x in the formula $f(x) = x^2 + 4$.

$$f(3) = 3^2 + 4 = 13 \qquad \text{Replace } x \text{ by } 3$$

$$f(-2) = (-2)^2 + 4 = 8 \qquad \text{Replace } x \text{ by } -2$$

$$f(\sqrt{5}) = (\sqrt{5})^2 + 4 = 9 \qquad \text{Replace } x \text{ by } \sqrt{5}$$

(c) The domain of f consists of all possible inputs for f. Since we can evaluate the formula $f(x) = x^2 + 4$ for every real number x, the domain of f is the set $\mathbb{R}$ of all real numbers.

The range of f consists of all possible outputs of f. Because $x^2 \geq 0$ for all real numbers x, we have $x^2 + 4 \geq 4$, so for every output of f we have $f(x) \geq 4$. Thus, the range of f is $\{y \mid y \geq 4\} = [4, \infty)$.

(d) A machine diagram for f is shown in Figure 5. ▲

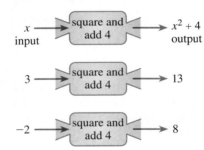

FIGURE 5 Machine diagram

Evaluating a Function

In the definition of a function the independent variable x plays the role of a "placeholder." For example, the function $f(x) = 3x^2 + x - 5$ can be thought of as

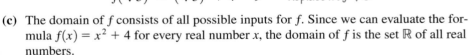

To evaluate f at a number, we substitute the number for the placeholder.

▶ **EXAMPLE 2** | Evaluating a Function

Let $f(x) = 3x^2 + x - 5$. Evaluate each function value.

(a) $f(-2)$ **(b)** $f(0)$

(c) $f(4)$ **(d)** $f\left(\tfrac{1}{2}\right)$

▼ **SOLUTION** To evaluate f at a number, we substitute the number for x in the definition of f.

(a) $f(-2) = 3 \cdot (-2)^2 + (-2) - 5 = 5$

(b) $f(0) = 3 \cdot 0^2 + 0 - 5 = -5$

(c) $f(4) = 3 \cdot (4)^2 + 4 - 5 = 47$

(d) $f\left(\frac{1}{2}\right) = 3 \cdot \left(\frac{1}{2}\right)^2 + \frac{1}{2} - 5 = -\frac{15}{4}$ ▲

A piecewise-defined function is defined by different formulas on different parts of its domain. The function C of Example 3 is piecewise defined.

► **EXAMPLE 3** | A Piecewise Defined Function

A cell phone plan costs \$39 a month. The plan includes 400 free minutes and charges 20¢ for each additional minute of usage. The monthly charges are a function of the number of minutes used, given by

$$C(x) = \begin{cases} 39 & \text{if } 0 \le x \le 400 \\ 39 + 0.20(x - 400) & \text{if } x > 400 \end{cases}$$

Find $C(100)$, $C(400)$, and $C(480)$.

▼ **SOLUTION** Remember that a function is a rule. Here is how we apply the rule for this function. First we look at the value of the input x. If $0 \le x \le 400$, then the value of $C(x)$ is 39. On the other hand, if $x > 400$, then the value of $C(x)$ is $39 + 0.20(x - 400)$.

Since $100 \le 400$, we have $C(100) = 39$.

Since $400 \le 400$, we have $C(400) = 39$.

Since $480 > 400$, we have $C(480) = 39 + 0.20(480 - 400) = 55$.

Thus, the plan charges \$39 for 100 minutes, \$39 for 400 minutes, and \$55 for 480 minutes.

▲

Expressions like the one in part (d) of Example 4 occur frequently in calculus; they are called *difference quotients*, and they represent the average change in the value of f between $x = a$ and $x = a + h$.

► **EXAMPLE 4** | Evaluating a Function

If $f(x) = 2x^2 + 3x - 1$, evaluate the following.

(a) $f(a)$

(b) $f(-a)$

(c) $f(a + h)$

(d) $\dfrac{f(a + h) - f(a)}{h}, \quad h \neq 0$

▼ **SOLUTION**

(a) $f(a) = 2a^2 + 3a - 1$

(b) $f(-a) = 2(-a)^2 + 3(-a) - 1 = 2a^2 - 3a - 1$

(c) $f(a + h) = 2(a + h)^2 + 3(a + h) - 1$

$$= 2(a^2 + 2ah + h^2) + 3(a + h) - 1$$

$$= 2a^2 + 4ah + 2h^2 + 3a + 3h - 1$$

(d) Using the results from parts (c) and (a), we have

$$\frac{f(a + h) - f(a)}{h} = \frac{(2a^2 + 4ah + 2h^2 + 3a + 3h - 1) - (2a^2 + 3a - 1)}{h}$$

$$= \frac{4ah + 2h^2 + 3h}{h} = 4a + 2h + 3$$

▲

The weight of an object on or near the earth is the gravitational force that the earth exerts on it. When in orbit around the earth, an astronaut experiences the sensation of "weightlessness" because the centripetal force that keeps her in orbit is exactly the same as the gravitational pull of the earth.

▶ **EXAMPLE 5** | The Weight of an Astronaut

If an astronaut weighs 130 pounds on the surface of the earth, then her weight when she is h miles above the earth is given by the function

$$w(h) = 130\left(\frac{3960}{3960 + h}\right)^2$$

(a) What is her weight when she is 100 mi above the earth?

(b) Construct a table of values for the function w that gives her weight at heights from 0 to 500 mi. What do you conclude from the table?

▼ **SOLUTION**

(a) We want the value of the function w when $h = 100$; that is, we must calculate $w(100)$.

$$w(100) = 130\left(\frac{3960}{3960 + 100}\right)^2 \approx 123.67$$

So at a height of 100 mi, she weighs about 124 lb.

(b) The table gives the astronaut's weight, rounded to the nearest pound, at 100-mile increments. The values in the table are calculated as in part (a).

h	$w(h)$
0	130
100	124
200	118
300	112
400	107
500	102

The table indicates that the higher the astronaut travels, the less she weighs. ▲

■ The Domain of a Function

Recall that the *domain* of a function is the set of all inputs for the function. The domain of a function may be stated explicitly. For example, if we write

$$f(x) = x^2 \qquad 0 \le x \le 5$$

then the domain is the set of all real numbers x for which $0 \le x \le 5$. If the function is given by an algebraic expression and the domain is not stated explicitly, then by convention *the domain of the function is the domain of the algebraic expression—that is, the set of all real numbers for which the expression is defined as a real number.* For example, consider the functions

Domains of algebraic expressions are discussed on page 32.

$$f(x) = \frac{1}{x - 4} \qquad g(x) = \sqrt{x}$$

The function f is not defined at $x = 4$, so its domain is $\{x \mid x \ne 4\}$. The function g is not defined for negative x, so its domain is $\{x \mid x \ge 0\}$.

▶ **EXAMPLE 6** | Finding Domains of Functions

Find the domain of each function.

(a) $f(x) = \dfrac{1}{x^2 - x}$ **(b)** $g(x) = \sqrt{9 - x^2}$ **(c)** $h(t) = \dfrac{t}{\sqrt{t + 1}}$

▼ **SOLUTION**

(a) The function is not defined when the denominator is 0. Since

$$f(x) = \frac{1}{x^2 - x} = \frac{1}{x(x-1)}$$

we see that $f(x)$ is not defined when $x = 0$ or $x = 1$. Thus, the domain of f is

$$\{x \mid x \neq 0, x \neq 1\}$$

The domain may also be written in interval notation as

$$(\infty, 0) \cup (0, 1) \cup (1, \infty)$$

(b) We can't take the square root of a negative number, so we must have $9 - x^2 \geq 0$. Using the methods of Section 1.6, we can solve this inequality to find that $-3 \leq x \leq 3$. Thus, the domain of g is

$$\{x \mid -3 \leq x \leq 3\} = [-3, 3]$$

(c) We can't take the square root of a negative number, and we can't divide by 0, so we must have $t + 1 > 0$, that is, $t > -1$. So the domain of h is

$$\{t \mid t > -1\} = (-1, \infty)$$
▲

■ Four Ways to Represent a Function

To help us understand what a function is, we have used machine and arrow diagrams. We can describe a specific function in the following four ways:

- verbally (by a description in words)
- algebraically (by an explicit formula)
- visually (by a graph)
- numerically (by a table of values)

A single function may be represented in all four ways, and it is often useful to go from one representation to another to gain insight into the function. However, certain functions are described more naturally by one method than by the others. An example of a verbal description is the following rule for converting between temperature scales:

"To find the Fahrenheit equivalent of a Celsius temperature, multiply the Celsius temperature by $\frac{9}{5}$, then add 32."

In Example 7 we see how to describe this verbal rule or function algebraically, graphically, and numerically. A useful representation of the area of a circle as a function of its radius is the algebraic formula

$$A(r) = \pi r^2$$

The graph produced by a seismograph (see the box on the next page) is a visual representation of the vertical acceleration function $a(t)$ of the ground during an earthquake. As a final example, consider the function $C(w)$, which is described verbally as "the cost of mailing a first-class letter with weight w." The most convenient way of describing this function is numerically—that is, using a table of values.

We will be using all four representations of functions throughout this book. We summarize them in the following box.

FOUR WAYS TO REPRESENT A FUNCTION

Verbal

Using words:

"To convert from Celsius to Fahrenheit, multiply the Celsius temperature by $\frac{9}{5}$, then add 32."

Relation between Celsius and Fahrenheit temperature scales

Algebraic

Using a formula:

$$A(r) = \pi r^2$$

Area of a circle

Visual

Using a graph:

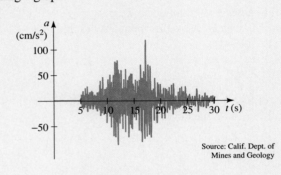

Source: Calif. Dept. of Mines and Geology

Vertical acceleration during an earthquake

Numerical

Using a table of values:

w (ounces)	$C(w)$ (dollars)
$0 < w \leq 1$	1.13
$1 < w \leq 2$	1.30
$2 < w \leq 3$	1.47
$3 < w \leq 4$	1.64
$4 < w \leq 5$	1.81
$\vdots$	$\vdots$

Cost of mailing a first-class parcel

▶ **EXAMPLE 7** | Representing a Function Verbally, Algebraically, Numerically, and Graphically

Let $F(C)$ be the Fahrenheit temperature corresponding to the Celsius temperature C. (Thus, F is the function that converts Celsius inputs to Fahrenheit outputs.) The box above gives a verbal description of this function. Find ways to represent this function

(a) Algebraically (using a formula)

(b) Numerically (using a table of values)

(c) Visually (using a graph)

▼ **SOLUTION**

(a) The verbal description tells us that we should first multiply the input C by $\frac{9}{5}$ and then add 32 to the result. So we get

$$F(C) = \tfrac{9}{5}C + 32$$

(b) We use the algebraic formula for F that we found in part (a) to construct a table of values:

C (Celsius)	F (Fahrenheit)
-10	14
0	32
10	50
20	68
30	86
40	104

(c) We use the points tabulated in part (b) to help us draw the graph of this function in Figure 6.

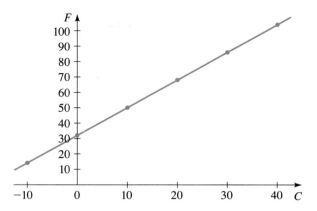

FIGURE 6 Celsius and Fahrenheit

3.2 | Graphs of Functions

LEARNING OBJECTIVES
After completing this section, you will be able to:

- Graph a function by plotting points
- Graph a function using a graphing calculator
- Graph piecewise defined functions
- Use the Vertical Line Test
- Determine whether an equation defines a function

The most important way to visualize a function is through its graph. In this section we investigate in more detail the concept of graphing functions.

■ Graphing Functions by Plotting Points

To graph a function f, we plot the points $(x, f(x))$ in a coordinate plane. In other words, we plot the points (x, y) whose x-coordinate is an input and whose y-coordinate is the corresponding output of the function.

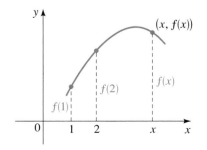

FIGURE 1 The height of the graph above the point x is the value of $f(x)$.

> **THE GRAPH OF A FUNCTION**
>
> If f is a function with domain A, then the **graph** of f is the set of ordered pairs
>
> $$\{(x, f(x)) \mid x \in A\}$$
>
> In other words, the graph of f is the set of all points (x, y) such that $y = f(x)$; that is, the graph of f is the graph of the equation $y = f(x)$.

The graph of a function f gives a picture of the behavior or "life history" of the function. We can read the value of $f(x)$ from the graph as being the height of the graph above the point x (see Figure 1).

A function f of the form $f(x) = mx + b$ is called a **linear function** because its graph is the graph of the equation $y = mx + b$, which represents a line with slope m and y-intercept b. A special case of a linear function occurs when the slope is $m = 0$. The function $f(x) = b$, where b is a given number, is called a **constant function** because all its values are the same number, namely, b. Its graph is the horizontal line $y = b$. Figure 2 shows the graphs of the constant function $f(x) = 3$ and the linear function $f(x) = 2x + 1$.

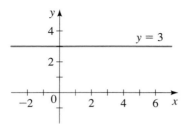

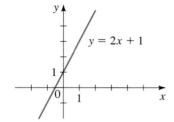

FIGURE 2   The constant function $f(x) = 3$ The linear function $f(x) = 2x + 1$

▶ **EXAMPLE 1** | Graphing Functions by Plotting Points

Sketch the graphs of the following functions.

(a) $f(x) = x^2$ **(b)** $g(x) = x^3$ **(c)** $h(x) = \sqrt{x}$

▼ **SOLUTION**   We first make a table of values. Then we plot the points given by the table and join them by a smooth curve to obtain the graph. The graphs are sketched in Figure 3.

x	$f(x) = x^2$
0	0
$\pm\frac{1}{2}$	$\frac{1}{4}$
± 1	1
± 2	4
± 3	9

x	$g(x) = x^3$
0	0
$\frac{1}{2}$	$\frac{1}{8}$
1	1
2	8
$-\frac{1}{2}$	$-\frac{1}{8}$
-1	-1
-2	-8

x	$h(x) = \sqrt{x}$
0	0
1	1
2	$\sqrt{2}$
3	$\sqrt{3}$
4	2
5	$\sqrt{5}$

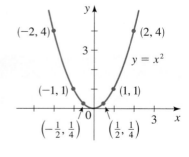

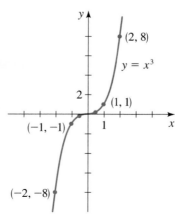

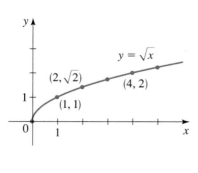

FIGURE 3   (a) $f(x) = x^2$ (b) $g(x) = x^3$ (c) $h(x) = \sqrt{x}$

■ Graphing Functions with a Graphing Calculator

A convenient way to graph a function is to use a graphing calculator. Because the graph of a function f is the graph of the equation $y = f(x)$, we can use the methods of Section 2.3 to graph functions on a graphing calculator.

▶ **EXAMPLE 2** | Graphing a Function with a Graphing Calculator

Use a graphing calculator to graph the function $f(x) = x^3 - 8x^2$ in an appropriate viewing rectangle.

▼ **SOLUTION** To graph the function $f(x) = x^3 - 8x^2$, we graph the equation $y = x^3 - 8x^2$. On the TI-83 graphing calculator the default viewing rectangle gives the graph in Figure 4(a). But this graph appears to spill over the top and bottom of the screen. We need to expand the vertical axis to get a better representation of the graph. The viewing rectangle $[-4, 10]$ by $[-100, 100]$ gives a more complete picture of the graph, as shown in Figure 4(b).

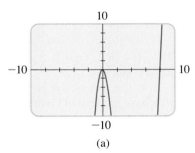
(a)

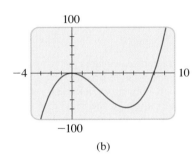
(b)

FIGURE 4 Graphing the function $f(x) = x^3 - 8x^2$

▶ **EXAMPLE 3** | A Family of Power Functions

(a) Graph the functions $f(x) = x^n$ for $n = 2, 4,$ and 6 in the viewing rectangle $[-2, 2]$ by $[-1, 3]$.

(b) Graph the functions $f(x) = x^n$ for $n = 1, 3,$ and 5 in the viewing rectangle $[-2, 2]$ by $[-2, 2]$.

(c) What conclusions can you draw from these graphs?

▼ **SOLUTION** To graph the function $f(x) = x^n$, we graph the equation $y = x^n$. The graphs for parts (a) and (b) are shown in Figure 5.

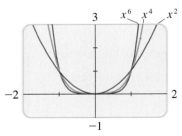
(a) Even powers of x

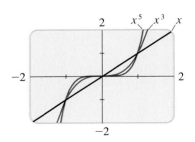
(b) Odd powers of x

FIGURE 5 A family of power functions $f(x) = x^n$

(c) We see that the general shape of the graph of $f(x) = x^n$ depends on whether n is even or odd.

If n is even, the graph of $f(x) = x^n$ is similar to the parabola $y = x^2$.

If n is odd, the graph of $f(x) = x^n$ is similar to that of $y = x^3$. ▲

Notice from Figure 5 that as n increases, the graph of $y = x^n$ becomes flatter near 0 and steeper when $x > 1$. When $0 < x < 1$, the lower powers of x are the "bigger" functions. But when $x > 1$, the higher powers of x are the dominant functions.

■ Graphing Piecewise Defined Functions

A piecewise defined function is defined by different formulas on different parts of its domain. As you might expect, the graph of such a function consists of separate pieces.

▶ **EXAMPLE 4** | Graph of a Piecewise Defined Function

Sketch the graph of the function.

$$f(x) = \begin{cases} x^2 & \text{if } x \le 1 \\ 2x + 1 & \text{if } x > 1 \end{cases}$$

▼ **SOLUTION**   If $x \le 1$, then $f(x) = x^2$, so the part of the graph to the left of $x = 1$ coincides with the graph of $y = x^2$, which we sketched in Figure 3. If $x > 1$, then $f(x) = 2x + 1$, so the part of the graph to the right of $x = 1$ coincides with the line $y = 2x + 1$, which we graphed in Figure 2. This enables us to sketch the graph in Figure 6.

The solid dot at $(1, 1)$ indicates that this point is included in the graph; the open dot at $(1, 3)$ indicates that this point is excluded from the graph.

On many graphing calculators the graph in Figure 6 can be produced by using the logical functions in the calculator. For example, on the TI-83 the following equation gives the required graph:

$Y_1 = (X \le 1)X^2 + (X > 1)(2X + 1)$

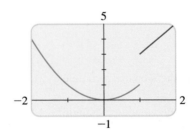

(To avoid the extraneous vertical line between the two parts of the graph, put the calculator in **Dot** mode.)

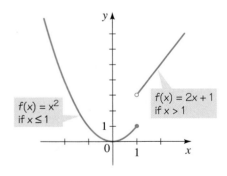

$f(x) = x^2$
if $x \le 1$

$f(x) = 2x + 1$
if $x > 1$

FIGURE 6

$$f(x) = \begin{cases} x^2 & \text{if } x \le 1 \\ 2x + 1 & \text{if } x > 1 \end{cases}$$

▲

▶ **EXAMPLE 5** | Graph of the Absolute Value Function

Sketch the graph of the absolute value function $f(x) = |x|$.

▼ **SOLUTION**   Recall that

$$|x| = \begin{cases} x & \text{if } x \ge 0 \\ -x & \text{if } x < 0 \end{cases}$$

Using the same method as in Example 4, we note that the graph of f coincides with the line $y = x$ to the right of the y-axis and coincides with the line $y = -x$ to the left of the y-axis (see Figure 7).

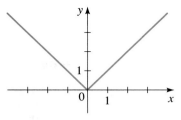

FIGURE 7 Graph of $f(x) = |x|$

The **greatest integer function** is defined by

$$[\![x]\!] = \text{greatest integer less than or equal to } x$$

For example, $[\![2]\!] = 2$, $[\![2.3]\!] = 2$, $[\![1.999]\!] = 1$, $[\![0.002]\!] = 0$, $[\![-3.5]\!] = -4$, and $[\![-0.5]\!] = -1$.

▶ **EXAMPLE 6** | Graph of the Greatest Integer Function

Sketch the graph of $f(x) = [\![x]\!]$.

▼ **SOLUTION**   The table shows the values of f for some values of x. Note that $f(x)$ is constant between consecutive integers, so the graph between integers is a horizontal line segment, as shown in Figure 8.

x	$[\![x]\!]$
⋮	⋮
$-2 \le x < -1$	-2
$-1 \le x < 0$	-1
$0 \le x < 1$	0
$1 \le x < 2$	1
$2 \le x < 3$	2
⋮	⋮

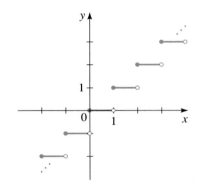

FIGURE 8 The greatest integer function, $y = [\![x]\!]$ ▲

The greatest integer function is an example of a **step function**. The next example gives a real-world example of a step function.

▶ **EXAMPLE 7** | The Cost Function for Long-Distance Phone Calls

The cost of a long-distance daytime phone call from Toronto to Mumbai, India, is 69 cents for the first minute and 58 cents for each additional minute (or part of a minute). Draw the graph of the cost C (in dollars) of the phone call as a function of time t (in minutes).

▼ **SOLUTION**   Let $C(t)$ be the cost for t minutes. Since $t > 0$, the domain of the function is $(0, \infty)$. From the given information we have

$$
\begin{aligned}
C(t) &= 0.69 && \text{if } 0 < t \le 1 \\
C(t) &= 0.69 + 0.58 = 1.27 && \text{if } 1 < t \le 2 \\
C(t) &= 0.69 + 2(0.58) = 1.85 && \text{if } 2 < t \le 3 \\
C(t) &= 0.69 + 3(0.58) = 2.43 && \text{if } 3 < t \le 4
\end{aligned}
$$

FIGURE 9 Cost of a long-distance call

and so on. The graph is shown in Figure 9. ▲

■ The Vertical Line Test

The graph of a function is a curve in the xy-plane. But the question arises: Which curves in the xy-plane are graphs of functions? This is answered by the following test.

> **THE VERTICAL LINE TEST**
>
> A curve in the coordinate plane is the graph of a function if and only if no vertical line intersects the curve more than once.

We can see from Figure 10 why the Vertical Line Test is true. If each vertical line $x = a$ intersects a curve only once at (a, b), then exactly one functional value is defined by $f(a) = b$. But if a line $x = a$ intersects the curve twice, at (a, b) and at (a, c), then the curve cannot represent a function because a function cannot assign two different values to a.

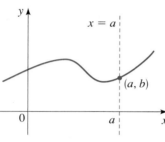

FIGURE 10 Vertical Line Test

Graph of a function   Not a graph of a function

▶ **EXAMPLE 8** | Using the Vertical Line Test

Using the Vertical Line Test, we see that the curves in parts (b) and (c) of Figure 11 represent functions, whereas those in parts (a) and (d) do not.

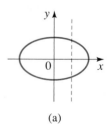
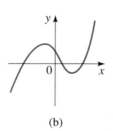
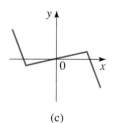
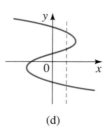

(a)   (b)   (c)   (d)

FIGURE 11   ▲

■ Equations That Define Functions

Any equation in the variables x and y defines a relationship between these variables. For example, the equation

$$y - x^2 = 0$$

defines a relationship between y and x. Does this equation define y as a *function* of x? To find out, we solve for y and get

$$y = x^2$$

We see that the equation defines a rule, or function, that gives one value of y for each value of x. We can express this rule in function notation as

$$f(x) = x^2$$

But not every equation defines y as a function of x, as the following example shows.

▶ **EXAMPLE 9** | Equations That Define Functions

Does the equation define y as a function of x?
(a) $y - x^2 = 2$
(b) $x^2 + y^2 = 4$

▼ **SOLUTION**

(a) Solving for y in terms of x gives

$$y - x^2 = 2$$
$$y = x^2 + 2 \qquad \text{Add } x^2$$

The last equation is a rule that gives one value of y for each value of x, so it defines y as a function of x. We can write the function as $f(x) = x^2 + 2$.

(b) We try to solve for y in terms of x:

$$x^2 + y^2 = 4$$
$$y^2 = 4 - x^2 \qquad \text{Subtract } x^2$$
$$y = \pm\sqrt{4 - x^2} \qquad \text{Take square roots}$$

The last equation gives two values of y for a given value of x. Thus, the equation does not define y as a function of x. ▲

The graphs of the equations in Example 9 are shown in Figure 12. The Vertical Line Test shows graphically that the equation in Example 9(a) defines a function but the equation in Example 9(b) does not.

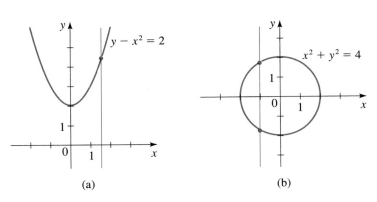

(a)   (b)

FIGURE 12

The following table shows the graphs of some functions that you will see frequently in this book.

SOME FUNCTIONS AND THEIR GRAPHS

Linear functions
$f(x) = mx + b$

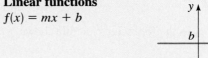

$f(x) = b$ $f(x) = mx + b$

Power functions
$f(x) = x^n$

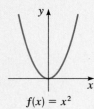 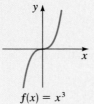 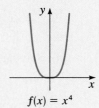 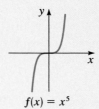

$f(x) = x^2$ $f(x) = x^3$ $f(x) = x^4$ $f(x) = x^5$

Root functions
$f(x) = \sqrt[n]{x}$

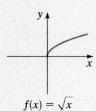 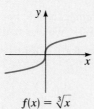 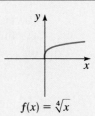 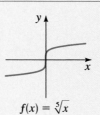

$f(x) = \sqrt{x}$ $f(x) = \sqrt[3]{x}$ $f(x) = \sqrt[4]{x}$ $f(x) = \sqrt[5]{x}$

Reciprocal functions
$f(x) = 1/x^n$

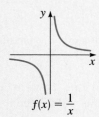 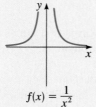

$f(x) = \dfrac{1}{x}$ $f(x) = \dfrac{1}{x^2}$

Absolute value function
$f(x) = |x|$

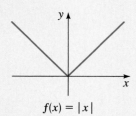

$f(x) = |x|$

Greatest integer function
$f(x) = [\![x]\!]$

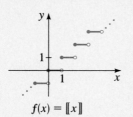

$f(x) = [\![x]\!]$

3.3 | Getting Information from the Graph of a Function

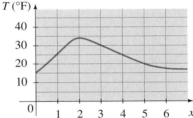

LEARNING OBJECTIVES

After completing this section, you will be able to:

■ Find function values from a graph
■ Find the domain and range of a function from a graph
■ Find where a function is increasing or decreasing from a graph
■ Find local maxima and minima of functions from a graph

Many properties of a function are more easily obtained from a graph than from the rule that describes the function. We will see in this section how a graph tells us whether the values of a function are increasing or decreasing and also where the maximum and minimum values of a function are.

■ Values of a Function; Domain and Range

A complete graph of a function contains all the information about a function, because the graph tells us which input values correspond to which output values. To analyze the graph of a function, we must keep in mind that *the height of the graph is the value of the function*. So we can read off the values of a function from its graph.

▶ **EXAMPLE 1** | Finding the Values of a Function from a Graph

The function T graphed in Figure 1 gives the temperature between noon and 6:00 P.M. at a certain weather station.

(a) Find $T(1)$, $T(3)$, and $T(5)$.   **(b)** Which is larger, $T(2)$ or $T(4)$?

(c) Find the value(s) of x for which $T(x) = 25$.

(d) Find the value(s) of x for which $T(x) \geq 25$.

▼ **SOLUTION**

(a) $T(1)$ is the temperature at 1:00 P.M. It is represented by the height of the graph above the x-axis at $x = 1$. Thus, $T(1) = 25$. Similarly, $T(3) = 30$ and $T(5) = 20$.

(b) Since the graph is higher at $x = 2$ than at $x = 4$, it follows that $T(2)$ is larger than $T(4)$.

(c) The height of the graph is 25 when x is 1 and when x is 4. In other words, the temperature is 25 at 1:00 P.M. and 4:00 P.M.

(d) The graph is higher than 25 for x between 1 and 4. In other words, the temperature was 25 or greater between 1:00 P.M. and 4:00 P.M.   ▲

The graph of a function helps us to picture the domain and range of the function on the x-axis and y-axis, as shown in Figure 2.

FIGURE 1 Temperature function

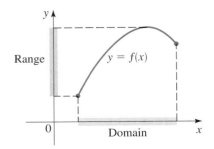

FIGURE 2 Domain and range of f

▶ **EXAMPLE 2** | Finding the Domain and Range from a Graph

(a) Use a graphing calculator to draw the graph of $f(x) = \sqrt{4 - x^2}$.

(b) Find the domain and range of f.

▼ **SOLUTION**

(a) The graph is shown in Figure 3.

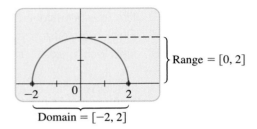

Range $= [0, 2]$

Domain $= [-2, 2]$

FIGURE 3 Graph of $f(x) = \sqrt{4 - x^2}$

(b) From the graph in Figure 3 we see that the domain is $[-2, 2]$ and the range is $[0, 2]$.

▲

■ Increasing and Decreasing Functions

It is very useful to know where the graph of a function rises and where it falls. The graph shown in Figure 4 rises, falls, then rises again as we move from left to right: It rises from A to B, falls from B to C, and rises again from C to D. The function f is said to be *increasing* when its graph rises and *decreasing* when its graph falls.

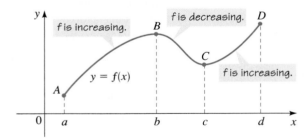

FIGURE 4 f is increasing on $[a, b]$ and $[c, d]$. f is decreasing on $[b, c]$.

We have the following definition.

DEFINITION OF INCREASING AND DECREASING FUNCTIONS

f is **increasing** on an interval I if $f(x_1) < f(x_2)$ whenever $x_1 < x_2$ in I.

f is **decreasing** on an interval I if $f(x_1) > f(x_2)$ whenever $x_1 < x_2$ in I.

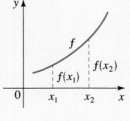

f is increasing

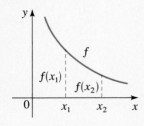

f is decreasing

▶ **EXAMPLE 3** | Intervals on Which a Function Increases and Decreases

The graph in Figure 5 gives the weight W of a person at age x. Determine the intervals on which the function W is increasing and on which it is decreasing.

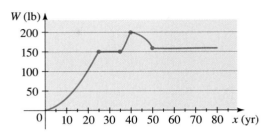

FIGURE 5 Weight as a function of age

▼ **SOLUTION**   The function W is increasing on $[0, 25]$ and $[35, 40]$. It is decreasing on $[40, 50]$. The function W is constant (neither increasing nor decreasing) on $[25, 30]$ and $[50, 80]$. This means that the person gained weight until age 25, then gained weight again between ages 35 and 40. He lost weight between ages 40 and 50.   ▲

▶ **EXAMPLE 4** | Finding Intervals Where a Function Increases and Decreases

  (a) Sketch a graph of the function $f(x) = 12x^2 + 4x^3 - 3x^4$.

  (b) Find the domain and range of f.

  (c) Find the intervals on which f increases and decreases.

▼ **SOLUTION**

  (a) We use a graphing calculator to sketch the graph in Figure 6.

  (b) The domain of f is $\mathbb{R}$ because f is defined for all real numbers. Using the [TRACE] feature on the calculator, we find that the highest value is $f(2) = 32$. So the range of f is $(-\infty, 32]$.

  (c) From the graph we see that f is increasing on the intervals $(-\infty, -1]$ and $[0, 2]$ and is decreasing on $[-1, 0]$ and $[2, \infty)$.

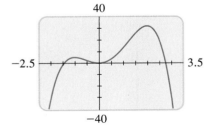

FIGURE 6 Graph of
$f(x) = 12x^2 + 4x^3 - 3x^4$

▶ **EXAMPLE 5** | Finding Intervals Where a Function Increases and Decreases

  (a) Sketch the graph of the function $f(x) = x^{2/3}$.

  (b) Find the domain and range of the function.

  (c) Find the intervals on which f increases and decreases.

▼ **SOLUTION**

(a) We use a graphing calculator to sketch the graph in Figure 7.

(b) From the graph we observe that the domain of f is $\mathbb{R}$ and the range is $[0, \infty)$.

(c) From the graph we see that f is decreasing on $(-\infty, 0]$ and increasing on $[0, \infty)$.

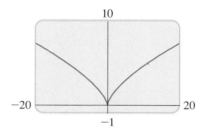

FIGURE 7 Graph of $f(x) = x^{2/3}$

▲

Local Maximum and Minimum Values of a Function

Finding the largest or smallest values of a function is important in many applications. For example, if a function represents revenue or profit, then we are interested in its maximum value. For a function that represents cost, we would want to find its minimum value. We can easily find these values from the graph of a function. We first define what we mean by a local maximum or minimum.

LOCAL MAXIMUMS AND MINIMUMS OF A FUNCTION

1. The function value $f(a)$ is a **local maximum value** of f if

$$f(a) \geq f(x) \quad \text{when } x \text{ is near } a$$

(This means that $f(a) \geq f(x)$ for all x in some open interval containing a.) In this case we say that f has a **local maximum** at $x = a$.

2. The function value $f(a)$ is a **local minimum** of f if

$$f(a) \leq f(x) \quad \text{when } x \text{ is near } a$$

(This means that $f(a) \leq f(x)$ for all x in some open interval containing a.) In this case we say that f has a **local minimum** at $x = a$.

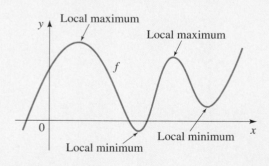

We can find the local maximum and minimum values of a function using a graphing calculator.

If there is a viewing rectangle such that the point $(a, f(a))$ is the highest point on the graph of f *within* the viewing rectangle (not on the edge), then the number $f(a)$ is a local maximum value of f (see Figure 8 on the next page). Notice that $f(a) \geq f(x)$ for all numbers x that are close to a.

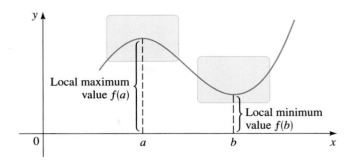

FIGURE 8

Similarly, if there is a viewing rectangle such that the point $(b, f(b))$ is the lowest point on the graph of f within the viewing rectangle, then the number $f(b)$ is a local minimum value of f. In this case, $f(b) \leq f(x)$ for all numbers x that are close to b.

▶ **EXAMPLE 6** | Finding Local Maxima and Minima from a Graph

Find the local maximum and minimum values of the function $f(x) = x^3 - 8x + 1$, correct to three decimals.

▼ **SOLUTION**   The graph of f is shown in Figure 9. There appears to be one local maximum between $x = -2$ and $x = -1$, and one local minimum between $x = 1$ and $x = 2$.

Let's find the coordinates of the local maximum point first. We zoom in to enlarge the area near this point, as shown in Figure 10. Using the $\boxed{\texttt{TRACE}}$ feature on the graphing device, we move the cursor along the curve and observe how the y-coordinates change. The local maximum value of y is 9.709, and this value occurs when x is -1.633, correct to three decimals.

We locate the minimum value in a similar fashion. By zooming in to the viewing rectangle shown in Figure 11, we find that the local minimum value is about -7.709, and this value occurs when $x \approx 1.633$.

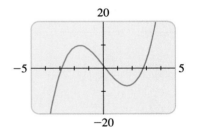

FIGURE 9 Graph of
$f(x) = x^3 - 8x + 1$

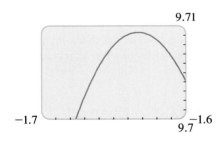

FIGURE 10

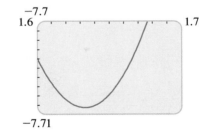

FIGURE 11   ▲

The `maximum` and `minimum` commands on a TI-82 or TI-83 calculator provide another method for finding extreme values of functions. We use this method in the next example.

▶ **EXAMPLE 7** | A Model for the Food Price Index

A model for the food price index (the price of a representative "basket" of foods) between 1990 and 2000 is given by the function

$$I(t) = -0.0113t^3 + 0.0681t^2 + 0.198t + 99.1$$

where t is measured in years since midyear 1990, so $0 \leq t \leq 10$, and $I(t)$ is scaled so that $I(3) = 100$. Estimate the time when food was most expensive during the period 1990–2000.

▼ **SOLUTION**   The graph of I as a function of t is shown in Figure 12(a). There appears to be a maximum between $t = 4$ and $t = 7$. Using the `maximum` command, as shown in Figure 12(b), we see that the maximum value of I is about 100.38, and it occurs when $t \approx 5.15$, which corresponds to August 1995.

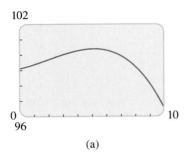

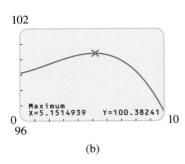

FIGURE 12
(a)   (b)   ▲

| **3.4** | Average Rate of Change of a Function |

LEARNING OBJECTIVES
After completing this section, you will be able to:

■ Find the average rate of change of a function
■ Interpret average rate of change in real-world situations
■ Recognize that a function with constant average rate of change is linear

Functions are often used to model changing quantities. In this section we learn how to find the rate at which the values of a function change as the input variable changes.

■ Average Rate of Change

We are all familiar with the concept of speed: If you drive a distance of 120 miles in 2 hours, then your average speed, or rate of travel, is $\frac{120 \text{ mi}}{2 \text{ h}} = 60$ mi/h.

Now suppose you take a car trip and record the distance that you travel every few minutes. The distance s you have traveled is a function of the time t:

$$s(t) = \text{total distance traveled at time } t$$

We graph the function s as shown in Figure 1. The graph shows that you have traveled a total of 50 miles after 1 hour, 75 miles after 2 hours, 140 miles after 3 hours, and so on. To find your *average* speed between any two points on the trip, we divide the distance traveled by the time elapsed.

Let's calculate your average speed between 1:00 P.M. and 4:00 P.M. The time elapsed is $4 - 1 = 3$ hours. To find the distance you traveled, we subtract the distance at 1:00 P.M. from the distance at 4:00 P.M., that is, $200 - 50 = 150$ mi. Thus, your average speed is

$$\text{average speed} = \frac{\text{distance traveled}}{\text{time elapsed}} = \frac{150 \text{ mi}}{3 \text{ h}} = 50 \text{ mi/h}$$

FIGURE 1 Average speed

The average speed that we have just calculated can be expressed by using function notation:

$$\text{average speed} = \frac{s(4) - s(1)}{4 - 1} = \frac{200 - 50}{3} = 50 \text{ mi/h}$$

Note that the average speed is different over different time intervals. For example, between 2:00 P.M. and 3:00 P.M. we find that

$$\text{average speed} = \frac{s(3) - s(2)}{3 - 2} = \frac{140 - 75}{1} = 65 \text{ mi/h}$$

Finding average rates of change is important in many contexts. For instance, we might be interested in knowing how quickly the air temperature is dropping as a storm approaches or how fast revenues are increasing from the sale of a new product. So we need to know how to determine the average rate of change of the functions that model these quantities. In fact, the concept of average rate of change can be defined for any function.

AVERAGE RATE OF CHANGE

The **average rate of change** of the function $y = f(x)$ between $x = a$ and $x = b$ is

$$\text{average rate of change} = \frac{\text{change in } y}{\text{change in } x} = \frac{f(b) - f(a)}{b - a}$$

The average rate of change is the slope of the **secant line** between $x = a$ and $x = b$ on the graph of f, that is, the line that passes through $(a, f(a))$ and $(b, f(b))$.

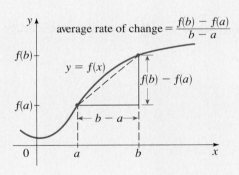

▶ **EXAMPLE 1** | Calculating the Average Rate of Change

For the function $f(x) = (x - 3)^2$, whose graph is shown in Figure 2, find the average rate of change between the following points:

(a) $x = 1$ and $x = 3$ **(b)** $x = 4$ and $x = 7$

▼ **SOLUTION**

(a) Average rate of change $= \dfrac{f(3) - f(1)}{3 - 1}$ Definition

$$= \frac{(3 - 3)^2 - (1 - 3)^2}{3 - 1} \qquad \text{Use } f(x) = (x - 3)^2$$

$$= \frac{0 - 4}{2} = -2$$

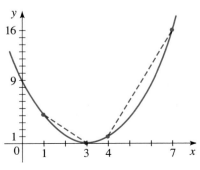

FIGURE 2 $f(x) = (x - 3)^2$

(b) Average rate of change $= \dfrac{f(7) - f(4)}{7 - 4}$ Definition

$= \dfrac{(7 - 3)^2 - (4 - 3)^2}{7 - 4}$ Use $f(x) = (x - 3)^2$

$= \dfrac{16 - 1}{3} = 5$ ▲

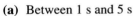

▶ **EXAMPLE 2** | Average Speed of a Falling Object

If an object is dropped from a high cliff or a tall building, then the distance it has fallen after t seconds is given by the function $d(t) = 16t^2$. Find its average speed (average rate of change) over the following intervals:

(a) Between 1 s and 5 s   **(b)** Between $t = a$ and $t = a + h$

▼ **SOLUTION**

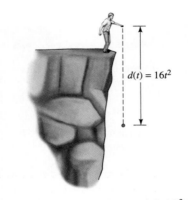

$d(t) = 16t^2$

Function: In t seconds, the stone falls $16t^2$ ft.

(a) Average rate of change $= \dfrac{d(5) - d(1)}{5 - 1}$ Definition

$= \dfrac{16(5)^2 - 16(1)^2}{5 - 1}$ Use $d(t) = 16t^2$

$= \dfrac{400 - 16}{4}$

$= 96$ ft/s

(b) Average rate of change $= \dfrac{d(a + h) - d(a)}{(a + h) - a}$ Definition

$= \dfrac{16(a + h)^2 - 16(a)^2}{(a + h) - a}$ Use $d(t) = 16t^2$

$= \dfrac{16(a^2 + 2ah + h^2 - a^2)}{h}$ Expand and factor 16

$= \dfrac{16(2ah + h^2)}{h}$ Simplify numerator

$= \dfrac{16h(2a + h)}{h}$ Factor h

$= 16(2a + h)$ Simplify   ▲

The average rate of change calculated in Example 2(b) is known as a *difference quotient*. In calculus we use difference quotients to calculate *instantaneous* rates of change. An example of an instantaneous rate of change is the speed shown on the speedometer of your car. This changes from one instant to the next as your car's speed changes.

The graphs in Figure 3 show that if a function is increasing on an interval, then the average rate of change between any two points is positive, whereas if a function is decreasing on an interval, then the average rate of change between any two points is negative.

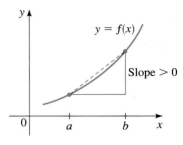
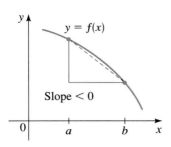

f increasing
Average rate of change positive

f decreasing
Average rate of change negative

FIGURE 3

Time	Temperature (°F)
8:00 A.M.	38
9:00 A.M.	40
10:00 A.M.	44
11:00 A.M.	50
12:00 NOON	56
1:00 P.M.	62
2:00 P.M.	66
3:00 P.M.	67
4:00 P.M.	64
5:00 P.M.	58
6:00 P.M.	55
7:00 P.M.	51

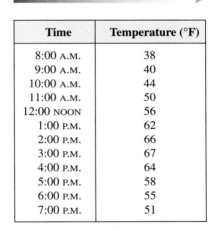

FIGURE 4

▶ **EXAMPLE 3** | Average Rate of Temperature Change

The table gives the outdoor temperatures observed by a science student on a spring day. Draw a graph of the data, and find the average rate of change of temperature between the following times:

(a) 8:00 A.M. and 9:00 A.M.

(b) 1:00 P.M. and 3:00 P.M.

(c) 4:00 P.M. and 7:00 P.M.

▼ **SOLUTION**   A graph of the temperature data is shown in Figure 4. Let t represent time, measured in hours since midnight (so, for example, 2:00 P.M. corresponds to $t = 14$). Define the function F by

$$F(t) = \text{temperature at time } t$$

(a) Average rate of change $= \dfrac{\text{temperature at 9 A.M.} - \text{temperature at 8 A.M.}}{9 - 8}$

$$= \frac{F(9) - F(8)}{9 - 8} = \frac{40 - 38}{9 - 8} = 2$$

The average rate of change was 2°F per hour.

(b) Average rate of change $= \dfrac{\text{temperature at 3 P.M.} - \text{temperature at 1 P.M.}}{15 - 13}$

$$= \frac{F(15) - F(13)}{15 - 13} = \frac{67 - 62}{2} = 2.5$$

The average rate of change was 2.5°F per hour.

(c) Average rate of change $= \dfrac{\text{temperature at 7 P.M.} - \text{temperature at 4 P.M.}}{19 - 16}$

$$= \frac{F(19) - F(16)}{19 - 16} = \frac{51 - 64}{3} \approx -4.3$$

The average rate of change was about −4.3°F per hour during this time interval. The negative sign indicates that the temperature was dropping. ▲

■ Linear Functions Have Constant Rate of Change

For a linear function $f(x) = mx + b$ the average rate of change between any two points is the same constant m. This agrees with what we learned in Section 2.4: that the slope of a line $y = mx + b$ is the average rate of change of y with respect to x. On the other hand, if a function f has constant average rate of change, then it must be a linear function. You are asked to prove this fact in the exercise set. In the next example we find the average rate of change for a particular linear function.

▶ **EXAMPLE 4** | Linear Functions Have Constant Rate of Change

Let $f(x) = 3x - 5$. Find the average rate of change of f between the following points.

(a) $x = 0$ and $x = 1$

(b) $x = 3$ and $x = 7$

(c) $x = a$ and $x = a + h$

What conclusion can you draw from your answers?

▼ **SOLUTION**

(a) Average rate of change $= \dfrac{f(1) - f(0)}{1 - 0} = \dfrac{(3 \cdot 1 - 5) - (3 \cdot 0 - 5)}{1}$

$= \dfrac{(-2) - (-5)}{1} = 3$

(b) Average rate of change $= \dfrac{f(7) - f(3)}{7 - 3} = \dfrac{(3 \cdot 7 - 5) - (3 \cdot 3 - 5)}{4}$

$= \dfrac{16 - 4}{4} = 3$

(c) Average rate of change $= \dfrac{f(a + h) - f(a)}{(a + h) - a} = \dfrac{[3(a + h) - 5] - [3a - 5]}{h}$

$= \dfrac{3a + 3h - 5 - 3a + 5}{h} = \dfrac{3h}{h} = 3$

It appears that the average rate of change is always 3 for this function. In fact, part (c) proves that the rate of change between any two arbitrary points $x = a$ and $x = a + h$ is 3.

▲

3.5 | Transformations of Functions

LEARNING OBJECTIVES

After completing this section, you will be able to:

■ Shift graphs vertically
■ Shift graphs horizontally
■ Stretch or shrink graphs vertically
■ Stretch or shrink graphs horizontally
■ Determine whether a function is odd or even

In this section we study how certain transformations of a function affect its graph. This will give us a better understanding of how to graph functions. The transformations that we study are shifting, reflecting, and stretching.

▇ Vertical Shifting

Adding a constant to a function shifts its graph vertically: upward if the constant is positive and downward if it is negative.

In general, suppose we know the graph of $y = f(x)$. How do we obtain from it the graphs of

$$y = f(x) + c \qquad \text{and} \qquad y = f(x) - c \qquad (c > 0)$$

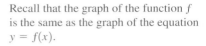

Recall that the graph of the function f is the same as the graph of the equation $y = f(x)$.

The y-coordinate of each point on the graph of $y = f(x) + c$ is c units above the y-coordinate of the corresponding point on the graph of $y = f(x)$. So we obtain the graph of $y = f(x) + c$ simply by shifting the graph of $y = f(x)$ upward c units. Similarly, we obtain the graph of $y = f(x) - c$ by shifting the graph of $y = f(x)$ downward c units.

VERTICAL SHIFTS OF GRAPHS

Suppose $c > 0$.

To graph $y = f(x) + c$, shift the graph of $y = f(x)$ upward c units.
To graph $y = f(x) - c$, shift the graph of $y = f(x)$ downward c units.

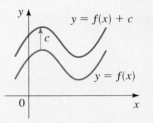 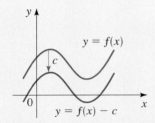

▶ **EXAMPLE 1** | Vertical Shifts of Graphs

Use the graph of $f(x) = x^2$ to sketch the graph of each function.

(a) $g(x) = x^2 + 3$ **(b)** $h(x) = x^2 - 2$

▼ **SOLUTION**   The function $f(x) = x^2$ was graphed in Example 1(a), Section 3.2. It is sketched again in Figure 1.

(a) Observe that

$$g(x) = x^2 + 3 = f(x) + 3$$

So the y-coordinate of each point on the graph of g is 3 units above the corresponding point on the graph of f. This means that to graph g we shift the graph of f upward 3 units, as in Figure 1.

(b) Similarly, to graph h, we shift the graph of f downward 2 units, as shown.

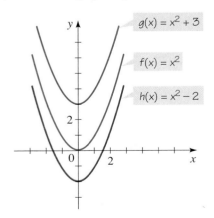

FIGURE 1

■ Horizontal Shifting

Suppose that we know the graph of $y = f(x)$. How do we use it to obtain the graphs of

$$y = f(x + c) \qquad \text{and} \qquad y = f(x - c) \qquad (c > 0)$$

The value of $f(x - c)$ at x is the same as the value of $f(x)$ at $x - c$. Since $x - c$ is c units to the left of x, it follows that the graph of $y = f(x - c)$ is just the graph of $y = f(x)$ shifted to the right c units. Similar reasoning shows that the graph of $y = f(x + c)$ is the graph of $y = f(x)$ shifted to the left c units. The following box summarizes these facts.

HORIZONTAL SHIFTS OF GRAPHS

Suppose $c > 0$.
    To graph $y = f(x - c)$, shift the graph of $y = f(x)$ to the right c units.
    To graph $y = f(x + c)$, shift the graph of $y = f(x)$ to the left c units.

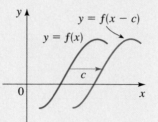 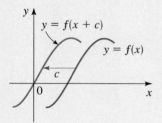

▶ **EXAMPLE 2** | Horizontal Shifts of Graphs

Use the graph of $f(x) = x^2$ to sketch the graph of each function.
(a) $g(x) = (x + 4)^2$ **(b)** $h(x) = (x - 2)^2$

▼ **SOLUTION**

(a) To graph g, we shift the graph of f to the left 4 units.

(b) To graph h, we shift the graph of f to the right 2 units.

The graphs of g and h are sketched in Figure 2.

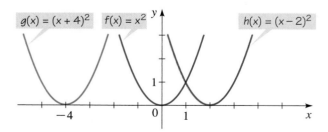

FIGURE 2

▶ **EXAMPLE 3** | Combining Horizontal and Vertical Shifts

Sketch the graph of $f(x) = \sqrt{x - 3} + 4$.

▼ **SOLUTION**   We start with the graph of $y = \sqrt{x}$ (Example 1(c), Section 3.2) and shift it to the right 3 units to obtain the graph of $y = \sqrt{x - 3}$. Then we shift the resulting graph upward 4 units to obtain the graph of $f(x) = \sqrt{x - 3} + 4$ shown in Figure 3.

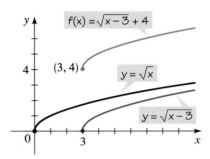

FIGURE 3

■ Reflecting Graphs

Suppose we know the graph of $y = f(x)$. How do we use it to obtain the graphs of $y = -f(x)$ and $y = f(-x)$? The y-coordinate of each point on the graph of $y = -f(x)$ is simply the negative of the y-coordinate of the corresponding point on the graph of $y = f(x)$. So the desired graph is the reflection of the graph of $y = f(x)$ in the x-axis. On the other hand, the value of $y = f(-x)$ at x is the same as the value of $y = f(x)$ at $-x$, so the desired graph here is the reflection of the graph of $y = f(x)$ in the y-axis. The following box summarizes these observations.

REFLECTING GRAPHS

To graph $y = -f(x)$, reflect the graph of $y = f(x)$ in the x-axis.

To graph $y = f(-x)$, reflect the graph of $y = f(x)$ in the y-axis.

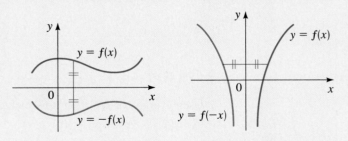

▶ **EXAMPLE 4** | Reflecting Graphs

Sketch the graph of each function.

(a) $f(x) = -x^2$ **(b)** $g(x) = \sqrt{-x}$

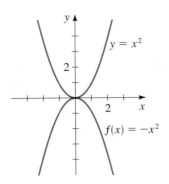

FIGURE 4

▼ **SOLUTION**

(a) We start with the graph of $y = x^2$. The graph of $f(x) = -x^2$ is the graph of $y = x^2$ reflected in the x-axis (see Figure 4).

(b) We start with the graph of $y = \sqrt{x}$ (Example 1(c) in Section 3.2.) The graph of $g(x) = \sqrt{-x}$ is the graph of $y = \sqrt{x}$ reflected in the y-axis (see Figure 5). Note that the domain of the function $g(x) = \sqrt{-x}$ is $\{x \mid x \le 0\}$.

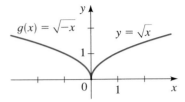

FIGURE 5

▲

Vertical Stretching and Shrinking

Suppose we know the graph of $y = f(x)$. How do we use it to obtain the graph of $y = cf(x)$? The y-coordinate of $y = cf(x)$ at x is the same as the corresponding y-coordinate of $y = f(x)$ multiplied by c. Multiplying the y-coordinates by c has the effect of vertically stretching or shrinking the graph by a factor of c.

VERTICAL STRETCHING AND SHRINKING OF GRAPHS

To graph $y = cf(x)$:

If $c > 1$, stretch the graph of $y = f(x)$ vertically by a factor of c.

If $0 < c < 1$, shrink the graph of $y = f(x)$ vertically by a factor of c.

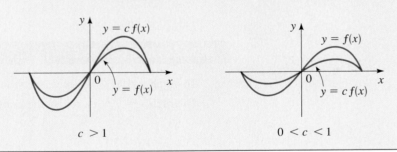

▶ **EXAMPLE 5** | Vertical Stretching and Shrinking of Graphs

Use the graph of $f(x) = x^2$ to sketch the graph of each function.

(a) $g(x) = 3x^2$ **(b)** $h(x) = \frac{1}{3}x^2$

▼ **SOLUTION**

(a) The graph of g is obtained by multiplying the y-coordinate of each point on the graph of f by 3. That is, to obtain the graph of g, we stretch the graph of f vertically by a factor of 3. The result is the narrower parabola in Figure 6.

(b) The graph of h is obtained by multiplying the y-coordinate of each point on the graph of f by $\frac{1}{3}$. That is, to obtain the graph of h, we shrink the graph of f vertically by a factor of $\frac{1}{3}$. The result is the wider parabola in Figure 6.

▲

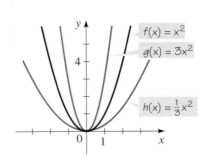

FIGURE 6

We illustrate the effect of combining shifts, reflections, and stretching in the following example.

▶ **EXAMPLE 6** | Combining Shifting, Stretching, and Reflecting

Sketch the graph of the function $f(x) = 1 - 2(x - 3)^2$.

▼ **SOLUTION**   Starting with the graph of $y = x^2$, we first shift to the right 3 units to get the graph of $y = (x - 3)^2$. Then we reflect in the x-axis and stretch by a factor of 2 to get the graph of $y = -2(x - 3)^2$. Finally, we shift upward 1 unit to get the graph of $f(x) = 1 - 2(x - 3)^2$ shown in Figure 7.

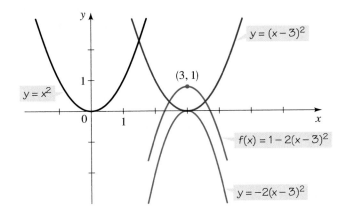

FIGURE 7

Horizontal Stretching and Shrinking

Now we consider horizontal shrinking and stretching of graphs. If we know the graph of $y = f(x)$, then how is the graph of $y = f(cx)$ related to it? The y-coordinate of $y = f(cx)$ at x is the same as the y-coordinate of $y = f(x)$ at cx. Thus, the x-coordinates in the graph of $y = f(x)$ correspond to the x-coordinates in the graph of $y = f(cx)$ multiplied by c. Looking at this the other way around, we see that the x-coordinates in the graph of $y = f(cx)$ are the x-coordinates in the graph of $y = f(x)$ multiplied by $1/c$. In other words, to change the graph of $y = f(x)$ to the graph of $y = f(cx)$, we must shrink (or stretch) the graph horizontally by a factor of $1/c$, as summarized in the following box.

HORIZONTAL SHRINKING AND STRETCHING OF GRAPHS

To graph $y = f(cx)$:

    If $c > 1$, shrink the graph of $y = f(x)$ horizontally by a factor of $1/c$.

    If $0 < c < 1$, stretch the graph of $y = f(x)$ horizontally by a factor of $1/c$.

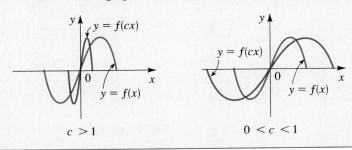

▶ **EXAMPLE 7** | Horizontal Stretching and Shrinking of Graphs

The graph of $y = f(x)$ is shown in Figure 8. Sketch the graph of each function.

(a) $y = f(2x)$ **(b)** $y = f\left(\frac{1}{2}x\right)$

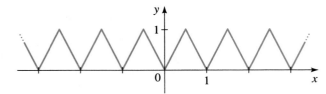

FIGURE 8
$y = f(x)$

▼ **SOLUTION**   Using the principles described in the preceding box, we obtain the graphs shown in Figures 9 and 10.

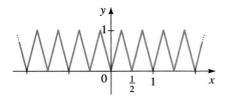

FIGURE 9 $y = f(2x)$

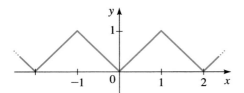

FIGURE 10 $y = f\left(\frac{1}{2}x\right)$ ▲

■ Even and Odd Functions

If a function f satisfies $f(-x) = f(x)$ for every number x in its domain, then f is called an **even function**. For instance, the function $f(x) = x^2$ is even because

$$f(-x) = (-x)^2 = (-1)^2 x^2 = x^2 = f(x)$$

The graph of an even function is symmetric with respect to the y-axis (see Figure 11). This means that if we have plotted the graph of f for $x \geq 0$, then we can obtain the entire graph simply by reflecting this portion in the y-axis.

If f satisfies $f(-x) = -f(x)$ for every number x in its domain, then f is called an **odd function**. For example, the function $f(x) = x^3$ is odd because

$$f(-x) = (-x)^3 = (-1)^3 x^3 = -x^3 = -f(x)$$

The graph of an odd function is symmetric about the origin (see Figure 12). If we have plotted the graph of f for $x \geq 0$, then we can obtain the entire graph by rotating this portion through 180° about the origin. (This is equivalent to reflecting first in the x-axis and then in the y-axis.)

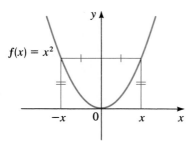

FIGURE 11 $f(x) = x^2$ is an even function.

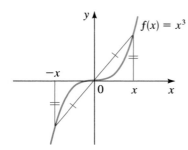

FIGURE 12 $f(x) = x^3$ is an odd function.

EVEN AND ODD FUNCTIONS

Let f be a function.

f is **even** if $f(-x) = f(x)$ for all x in the domain of f.

f is **odd** if $f(-x) = -f(x)$ for all x in the domain of f.

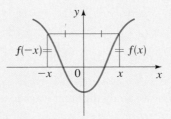 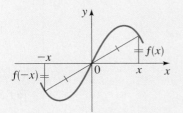

The graph of an even function is symmetric with respect to the y-axis.

The graph of an odd function is symmetric with respect to the origin.

▶ **EXAMPLE 8** | Even and Odd Functions

Determine whether the functions are even, odd, or neither even nor odd.

(a) $f(x) = x^5 + x$

(b) $g(x) = 1 - x^4$

(c) $h(x) = 2x - x^2$

▼ **SOLUTION**

(a)
$$f(-x) = (-x)^5 + (-x)$$
$$= -x^5 - x = -(x^5 + x)$$
$$= -f(x)$$

Therefore, f is an odd function.

(b) $g(-x) = 1 - (-x)^4 = 1 - x^4 = g(x)$

So g is even.

(c) $h(-x) = 2(-x) - (-x)^2 = -2x - x^2$

Since $h(-x) \neq h(x)$ and $h(-x) \neq -h(x)$, we conclude that h is neither even nor odd.   ▲

The graphs of the functions in Example 8 are shown in Figure 13. The graph of f is symmetric about the origin, and the graph of g is symmetric about the y-axis. The graph of h is not symmetric either about the y-axis or the origin.

FIGURE 13

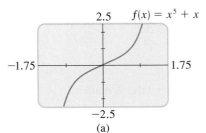
(a)

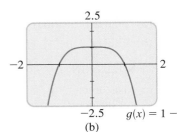
(b)

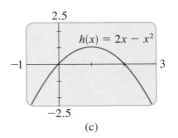
(c)

3.6 | Combining Functions

LEARNING OBJECTIVES

After completing this section, you will be able to:

- ■ Find sums, differences, products, and quotients of functions
- ■ Add functions graphically
- ■ Find the composition of two functions
- ■ Express a given function as a composite function

In this section we study different ways to combine functions to make new functions.

■ Sums, Differences, Products, and Quotients

The sum of f and g is defined by

$$(f + g)(x) = f(x) + g(x)$$

The name of the new function is "$f + g$." So this $+$ sign stands for the operation of addition of *functions*. The $+$ sign on the right side, however, stands for addition of the *numbers* $f(x)$ and $g(x)$.

Two functions f and g can be combined to form new functions $f + g$, $f - g$, fg, and f/g in a manner similar to the way we add, subtract, multiply, and divide real numbers. For example, we define the function $f + g$ by

$$(f + g)(x) = f(x) + g(x)$$

The new function $f + g$ is called the **sum** of the functions f and g; its value at x is $f(x) + g(x)$. Of course, the sum on the right-hand side makes sense only if both $f(x)$ and $g(x)$ are defined, that is, if x belongs to the domain of f and also to the domain of g. So if the domain of f is A and the domain of g is B, then the domain of $f + g$ is the intersection of these domains, that is, $A \cap B$. Similarly, we can define the **difference** $f - g$, the **product** fg, and the **quotient** f/g of the functions f and g. Their domains are $A \cap B$, but in the case of the quotient we must remember not to divide by 0.

ALGEBRA OF FUNCTIONS

Let f and g be functions with domains A and B. Then the functions $f + g$, $f - g$, fg, and f/g are defined as follows.

$$(f + g)(x) = f(x) + g(x) \qquad \text{Domain } A \cap B$$

$$(f - g)(x) = f(x) - g(x) \qquad \text{Domain } A \cap B$$

$$(fg)(x) = f(x)g(x) \qquad \text{Domain } A \cap B$$

$$\left(\frac{f}{g}\right)(x) = \frac{f(x)}{g(x)} \qquad \text{Domain } \{x \in A \cap B \mid g(x) \neq 0\}$$

▶ **EXAMPLE 1** | Combinations of Functions and Their Domains

Let $f(x) = \dfrac{1}{x - 2}$ and $g(x) = \sqrt{x}$.

(a) Find the functions $f + g$, $f - g$, fg, and f/g and their domains.

(b) Find $(f + g)(4)$, $(f - g)(4)$, $(fg)(4)$, and $(f/g)(4)$.

▼ **SOLUTION**

(a) The domain of f is $\{x \mid x \neq 2\}$, and the domain of g is $\{x \mid x \geq 0\}$. The intersection of the domains of f and g is

$$\{x \mid x \geq 0 \text{ and } x \neq 2\} = [0, 2) \cup (2, \infty)$$

Thus, we have

To divide fractions, invert the denominator and multiply:

$$\frac{1/(x-2)}{\sqrt{x}} = \frac{1/(x-2)}{\sqrt{x}/1}$$

$$= \frac{1}{x-2} \cdot \frac{1}{\sqrt{x}}$$

$$= \frac{1}{(x-2)\sqrt{x}}$$

$$(f+g)(x) = f(x) + g(x) = \frac{1}{x-2} + \sqrt{x} \qquad \text{Domain } \{x \mid x \geq 0 \text{ and } x \neq 2\}$$

$$(f-g)(x) = f(x) - g(x) = \frac{1}{x-2} - \sqrt{x} \qquad \text{Domain } \{x \mid x \geq 0 \text{ and } x \neq 2\}$$

$$(fg)(x) = f(x)g(x) = \frac{\sqrt{x}}{x-2} \qquad \text{Domain } \{x \mid x \geq 0 \text{ and } x \neq 2\}$$

$$\left(\frac{f}{g}\right)(x) = \frac{f(x)}{g(x)} = \frac{1}{(x-2)\sqrt{x}} \qquad \text{Domain } \{x \mid x > 0 \text{ and } x \neq 2\}$$

Note that in the domain of f/g we exclude 0 because $g(0) = 0$.

(b) Each of these values exist because $x = 4$ is in the domain of each function.

$$(f+g)(4) = f(4) + g(4) = \frac{1}{4-2} + \sqrt{4} = \frac{5}{2}$$

$$(f-g)(4) = f(4) - g(4) = \frac{1}{4-2} - \sqrt{4} = -\frac{3}{2}$$

$$(fg)(4) = f(4)g(4) = \left(\frac{1}{4-2}\right)\sqrt{4} = 1$$

$$\left(\frac{f}{g}\right)(4) = \frac{f(4)}{g(4)} = \frac{1}{(4-2)\sqrt{4}} = \frac{1}{4}$$

▲

The graph of the function $f + g$ can be obtained from the graphs of f and g by **graphical addition**. This means that we add corresponding y-coordinates, as illustrated in the next example.

FIGURE 1

► **EXAMPLE 2** | Using Graphical Addition

The graphs of f and g are shown in Figure 1. Use graphical addition to graph the function $f + g$.

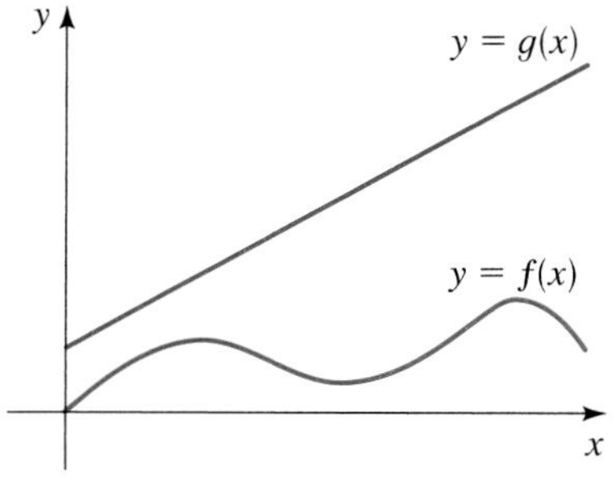

▼ **SOLUTION**   We obtain the graph of $f + g$ by "graphically adding" the value of $f(x)$ to $g(x)$ as shown in Figure 2. This is implemented by copying the line segment PQ on top of PR to obtain the point S on the graph of $f + g$.

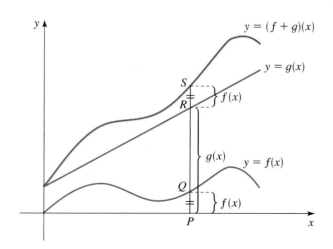

FIGURE 2 Graphical addition

▲

■ Composition of Functions

Now let's consider a very important way of combining two functions to get a new function. Suppose $f(x) = \sqrt{x}$ and $g(x) = x^2 + 1$. We may define a function h as

$$h(x) = f(g(x)) = f(x^2 + 1) = \sqrt{x^2 + 1}$$

The function h is made up of the functions f and g in an interesting way: Given a number x, we first apply to it the function g, then apply f to the result. In this case, f is the rule "take the square root," g is the rule "square, then add 1," and h is the rule "square, then add 1, then take the square root." In other words, we get the rule h by applying the rule g and then the rule f. Figure 3 shows a machine diagram for h.

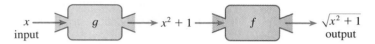

FIGURE 3 The h machine is composed of the g machine (first) and then the f machine.

In general, given any two functions f and g, we start with a number x in the domain of g and find its image $g(x)$. If this number $g(x)$ is in the domain of f, we can then calculate the value of $f(g(x))$. The result is a new function $h(x) = f(g(x))$ that is obtained by substituting g into f. It is called the *composition* (or *composite*) of f and g and is denoted by $f \circ g$ ("f composed with g").

COMPOSITION OF FUNCTIONS

Given two functions f and g, the **composite function** $f \circ g$ (also called the **composition** of f and g) is defined by

$$(f \circ g)(x) = f(g(x))$$

The domain of $f \circ g$ is the set of all x in the domain of g such that $g(x)$ is in the domain of f. In other words, $(f \circ g)(x)$ is defined whenever both $g(x)$ and $f(g(x))$ are defined. We can picture $f \circ g$ using an arrow diagram (Figure 4).

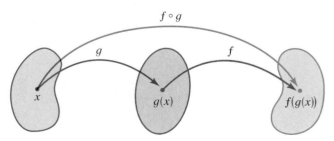

FIGURE 4 Arrow diagram for $f \circ g$

▶ **EXAMPLE 3** | Finding the Composition of Functions

Let $f(x) = x^2$ and $g(x) = x - 3$.

(a) Find the functions $f \circ g$ and $g \circ f$ and their domains.

(b) Find $(f \circ g)(5)$ and $(g \circ f)(7)$.

▼ **SOLUTION**

In Example 3, f is the rule "square" and g is the rule "subtract 3." The function $f \circ g$ first subtracts 3 and *then* squares; the function $g \circ f$ first squares and *then* subtracts 3.

(a) We have

$$(f \circ g)(x) = f(g(x)) \qquad \text{Definition of } f \circ g$$
$$= f(x - 3) \qquad \text{Definition of } g$$
$$= (x - 3)^2 \qquad \text{Definition of } f$$

and

$$(g \circ f)(x) = g(f(x)) \qquad \text{Definition of } g \circ f$$
$$= g(x^2) \qquad \text{Definition of } f$$
$$= x^2 - 3 \qquad \text{Definition of } g$$

The domains of both $f \circ g$ and $g \circ f$ are $\mathbb{R}$.

The graphs of f and g of Example 4, as well as $f \circ g$, $g \circ f$, $f \circ f$, and $g \circ g$, are shown below. These graphs indicate that the operation of composition can produce functions that are quite different from the original functions.

(b) We have

$$(f \circ g)(5) = f(g(5)) = f(2) = 2^2 = 4$$
$$(g \circ f)(7) = g(f(7)) = g(49) = 49 - 3 = 46$$

▲

You can see from Example 3 that, in general, $f \circ g \neq g \circ f$. Remember that the notation $f \circ g$ means that the function g is applied first and then f is applied second.

▶ **EXAMPLE 4** | Finding the Composition of Functions

If $f(x) = \sqrt{x}$ and $g(x) = \sqrt{2 - x}$, find the following functions and their domains.

(a) $f \circ g$ **(b)** $g \circ f$ **(c)** $f \circ f$ **(d)** $g \circ g$

▼ **SOLUTION**

(a) $(f \circ g)(x) = f(g(x))$ Definition of $f \circ g$
$= f(\sqrt{2 - x})$ Definition of g
$= \sqrt{\sqrt{2 - x}}$ Definition of f
$= \sqrt[4]{2 - x}$

The domain of $f \circ g$ is $\{x \mid 2 - x \geq 0\} = \{x \mid x \leq 2\} = (-\infty, 2]$.

(b) $(g \circ f)(x) = g(f(x))$ Definition of $g \circ f$
$= g(\sqrt{x})$ Definition of f
$= \sqrt{2 - \sqrt{x}}$ Definition of g

For $\sqrt{x}$ to be defined, we must have $x \geq 0$. For $\sqrt{2 - \sqrt{x}}$ to be defined, we must have $2 - \sqrt{x} \geq 0$, that is, $\sqrt{x} \leq 2$, or $x \leq 4$. Thus, we have $0 \leq x \leq 4$, so the domain of $g \circ f$ is the closed interval $[0, 4]$.

(c) $(f \circ f)(x) = f(f(x))$ Definition of $f \circ f$
$= f(\sqrt{x})$ Definition of f
$= \sqrt{\sqrt{x}}$ Definition of f
$= \sqrt[4]{x}$

The domain of $f \circ f$ is $[0, \infty)$.

(d) $(g \circ g)(x) = g(g(x))$ Definition of $g \circ g$
$= g(\sqrt{2 - x})$ Definition of g
$= \sqrt{2 - \sqrt{2 - x}}$ Definition of g

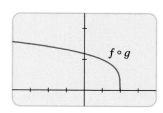

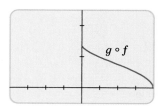

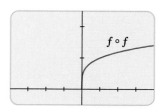

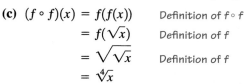

This expression is defined when both $2 - x \geq 0$ and $2 - \sqrt{2 - x} \geq 0$. The first inequality means $x \leq 2$, and the second is equivalent to $\sqrt{2 - x} \leq 2$, or $2 - x \leq 4$, or $x \geq -2$. Thus, $-2 \leq x \leq 2$, so the domain of $g \circ g$ is $[-2, 2]$. ▲

It is possible to take the composition of three or more functions. For instance, the composite function $f \circ g \circ h$ is found by first applying h, then g, and then f as follows:

$$(f \circ g \circ h)(x) = f(g(h(x)))$$

▶ **EXAMPLE 5** | A Composition of Three Functions

Find $f \circ g \circ h$ if $f(x) = x/(x + 1)$, $g(x) = x^{10}$, and $h(x) = x + 3$.

▼ **SOLUTION**

$$
\begin{aligned}
(f \circ g \circ h)(x) &= f(g(h(x))) && \text{Definition of } f \circ g \circ h \\
&= f(g(x + 3)) && \text{Definition of } h \\
&= f((x + 3)^{10}) && \text{Definition of } g \\
&= \frac{(x + 3)^{10}}{(x + 3)^{10} + 1} && \text{Definition of } f
\end{aligned}
$$

▲

So far, we have used composition to build complicated functions from simpler ones. But in calculus it is useful to be able to "decompose" a complicated function into simpler ones, as shown in the following example.

▶ **EXAMPLE 6** | Recognizing a Composition of Functions

Given $F(x) = \sqrt[4]{x + 9}$, find functions f and g such that $F = f \circ g$.

▼ **SOLUTION**   Since the formula for F says to first add 9 and then take the fourth root, we let

$$g(x) = x + 9 \qquad \text{and} \qquad f(x) = \sqrt[4]{x}$$

Then

$$
\begin{aligned}
(f \circ g)(x) &= f(g(x)) && \text{Definition of } f \circ g \\
&= f(x + 9) && \text{Definition of } g \\
&= \sqrt[4]{x + 9} && \text{Definition of } f \\
&= F(x)
\end{aligned}
$$

▲

▶ **EXAMPLE 7** | An Application of Composition of Functions

A ship is traveling at 20 mi/h parallel to a straight shoreline. The ship is 5 mi from shore. It passes a lighthouse at noon.

(a) Express the distance s between the lighthouse and the ship as a function of d, the distance the ship has traveled since noon; that is, find f so that $s = f(d)$.

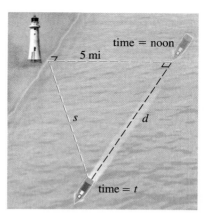

FIGURE 5

distance = rate × time

(b) Express d as a function of t, the time elapsed since noon; that is, find g so that $d = g(t)$.

(c) Find $f \circ g$. What does this function represent?

▼ **SOLUTION**   We first draw a diagram as in Figure 5.

(a) We can relate the distances s and d by the Pythagorean Theorem. Thus, s can be expressed as a function of d by

$$s = f(d) = \sqrt{25 + d^2}$$

(b) Since the ship is traveling at 20 mi/h, the distance d it has traveled is a function of t as follows:

$$d = g(t) = 20t$$

(c) We have

$$(f \circ g)(t) = f(g(t)) \qquad \text{Definition of } f \circ g$$
$$= f(20t) \qquad \text{Definition of } g$$
$$= \sqrt{25 + (20t)^2} \qquad \text{Definition of } f$$

The function $f \circ g$ gives the distance of the ship from the lighthouse as a function of time.   ▲

3.7 | One-to-One Functions and Their Inverses

LEARNING OBJECTIVES

After completing this section, you will be able to:

■ Determine whether a function is one-to-one
■ Find the inverse function of a one-to-one function
■ Draw the graph of an inverse function

The *inverse* of a function is a rule that acts on the output of the function and produces the corresponding input. So the inverse "undoes" or reverses what the function has done. Not all functions have inverses; those that do are called *one-to-one*.

■ One-to-One Functions

Let's compare the functions f and g whose arrow diagrams are shown in Figure 1 on the next page. Note that f never takes on the same value twice (any two numbers in A have different images), whereas g does take on the same value twice (both 2 and 3 have the same image, 4). In symbols, $g(2) = g(3)$ but $f(x_1) \neq f(x_2)$ whenever $x_1 \neq x_2$. Functions that have this latter property are called *one-to-one*.

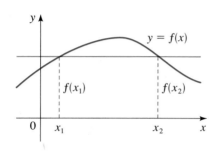

FIGURE 1

f is one-to-one

g is not one-to-one

DEFINITION OF A ONE-TO-ONE FUNCTION

A function with domain A is called a **one-to-one function** if no two elements of A have the same image, that is,

$$f(x_1) \neq f(x_2) \quad \text{whenever } x_1 \neq x_2$$

An equivalent way of writing the condition for a one-to-one function is this:

$$\text{If } f(x_1) = f(x_2), \text{ then } x_1 = x_2.$$

If a horizontal line intersects the graph of f at more than one point, then we see from Figure 2 that there are numbers $x_1 \neq x_2$ such that $f(x_1) = f(x_2)$. This means that f is not one-to-one. Therefore, we have the following geometric method for determining whether a function is one-to-one.

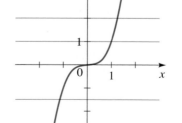

FIGURE 2 This function is not one-to-one because $f(x_1) = f(x_2)$.

HORIZONTAL LINE TEST

A function is one-to-one if and only if no horizontal line intersects its graph more than once.

▶ **EXAMPLE 1** | Deciding Whether a Function Is One-to-One

Is the function $f(x) = x^3$ one-to-one?

▼ **SOLUTION 1** If $x_1 \neq x_2$, then $x_1^3 \neq x_2^3$ (two different numbers cannot have the same cube). Therefore, $f(x) = x^3$ is one-to-one.

▼ **SOLUTION 2** From Figure 3 we see that no horizontal line intersects the graph of $f(x) = x^3$ more than once. Therefore, by the Horizontal Line Test, f is one-to-one. ▲

Notice that the function f of Example 1 is increasing and is also one-to-one. In fact, it can be proved that *every increasing function and every decreasing function is one-to-one*.

FIGURE 3 $f(x) = x^3$ is one-to-one.

▶ **EXAMPLE 2** | Deciding Whether a Function Is One-to-One

Is the function $g(x) = x^2$ one-to-one?

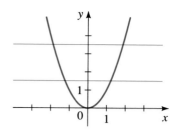

FIGURE 4 $f(x) = x^2$ is not one-to-one.

▼ **SOLUTION 1**   This function is not one-to-one because, for instance,

$$g(1) = 1 \quad \text{and} \quad g(-1) = 1$$

so 1 and -1 have the same image.

▼ **SOLUTION 2**   From Figure 4 we see that there are horizontal lines that intersect the graph of g more than once. Therefore, by the Horizontal Line Test, g is not one-to-one.

▲

Although the function g in Example 2 is not one-to-one, it is possible to restrict its domain so that the resulting function is one-to-one. In fact, if we define

$$h(x) = x^2 \quad x \geq 0$$

then h is one-to-one, as you can see from Figure 5 and the Horizontal Line Test.

▶   **EXAMPLE 3** | Showing That a Function Is One-to-One

Show that the function $f(x) = 3x + 4$ is one-to-one.

▼ **SOLUTION**   Suppose there are numbers x_1 and x_2 such that $f(x_1) = f(x_2)$. Then

$$3x_1 + 4 = 3x_2 + 4 \qquad \text{Suppose } f(x_1) = f(x_2)$$
$$3x_1 = 3x_2 \qquad \text{Subtract 4}$$
$$x_1 = x_2 \qquad \text{Divide by 3}$$

Therefore, f is one-to-one.

▲

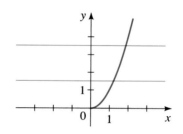

FIGURE 5 $f(x) = x^2 \ (x \geq 0)$ is one-to-one.

■ The Inverse of a Function

One-to-one functions are important because they are precisely the functions that possess inverse functions according to the following definition.

DEFINITION OF THE INVERSE OF A FUNCTION

Let f be a one-to-one function with domain A and range B. Then its **inverse function** f^{-1} has domain B and range A and is defined by

$$f^{-1}(y) = x \quad \Leftrightarrow \quad f(x) = y$$

for any y in B.

This definition says that if f takes x to y, then f^{-1} takes y back to x. (If f were not one-to-one, then f^{-1} would not be defined uniquely.) The arrow diagram in Figure 6 indicates that f^{-1} reverses the effect of f. From the definition we have

$$\text{domain of } f^{-1} = \text{range of } f$$
$$\text{range of } f^{-1} = \text{domain of } f$$

FIGURE 6

▷ **EXAMPLE 4** | Finding f^{-1} for Specific Values

Don't mistake the -1 in f^{-1} for an exponent.

$$f^{-1} \quad \textit{does not mean} \quad \frac{1}{f(x)}$$

The reciprocal $1/f(x)$ is written as $(f(x))^{-1}$.

If $f(1) = 5$, $f(3) = 7$, and $f(8) = -10$, find $f^{-1}(5)$, $f^{-1}(7)$, and $f^{-1}(-10)$.

▼ **SOLUTION** From the definition of f^{-1} we have

$$f^{-1}(5) = 1 \quad \text{because} \quad f(1) = 5$$
$$f^{-1}(7) = 3 \quad \text{because} \quad f(3) = 7$$
$$f^{-1}(-10) = 8 \quad \text{because} \quad f(8) = -10$$

Figure 7 shows how f^{-1} reverses the effect of f in this case.

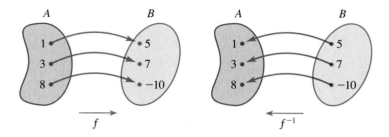

FIGURE 7

▲

By definition the inverse function f^{-1} undoes what f does: If we start with x, apply f, and then apply f^{-1}, we arrive back at x, where we started. Similarly, f undoes what f^{-1} does. In general, any function that reverses the effect of f in this way must be the inverse of f. These observations are expressed precisely as follows.

INVERSE FUNCTION PROPERTY

Let f be a one-to-one function with domain A and range B. The inverse function f^{-1} satisfies the following cancellation properties.

$$f^{-1}(f(x)) = x \quad \text{for every } x \text{ in } A$$
$$f(f^{-1}(x)) = x \quad \text{for every } x \text{ in } B$$

Conversely, any function f^{-1} satisfying these equations is the inverse of f.

These properties indicate that f is the inverse function of f^{-1}, so we say that f and f^{-1} are *inverses of each other*.

▷ **EXAMPLE 5** | Verifying That Two Functions Are Inverses

Show that $f(x) = x^3$ and $g(x) = x^{1/3}$ are inverses of each other.

▼ **SOLUTION** Note that the domain and range of both f and g is $\mathbb{R}$. We have

$$g(f(x)) = g(x^3) = (x^3)^{1/3} = x$$
$$f(g(x)) = f(x^{1/3}) = (x^{1/3})^3 = x$$

So by the Property of Inverse Functions, f and g are inverses of each other. These equations simply say that the cube function and the cube root function, when composed, cancel each other. ▲

Now let's examine how we compute inverse functions. We first observe from the definition of f^{-1} that

$$y = f(x) \quad \Leftrightarrow \quad f^{-1}(y) = x$$

So if $y = f(x)$ and if we are able to solve this equation for x in terms of y, then we must have $x = f^{-1}(y)$. If we then interchange x and y, we have $y = f^{-1}(x)$, which is the desired equation.

In Example 6 note how f^{-1} reverses the effect of f. The function f is the rule "multiply by 3, then subtract 2," whereas f^{-1} is the rule "add 2, then divide by 3."

Check Your Answer

We use the Inverse Function Property.

$$f^{-1}(f(x)) = f^{-1}(3x - 2)$$
$$= \frac{(3x - 2) + 2}{3}$$
$$= \frac{3x}{3} = x$$

$$f(f^{-1}(x)) = f\left(\frac{x + 2}{3}\right)$$
$$= 3\left(\frac{x + 2}{3}\right) - 2$$
$$= x + 2 - 2 = x \quad ✔$$

In Example 7 note how f^{-1} reverses the effect of f. The function f is the rule "Take the fifth power, subtract 3, then divide by 2," whereas f^{-1} is the rule "Multiply by 2, add 3, then take the fifth root."

Check Your Answer

We use the Inverse Function Property.

$$f^{-1}(f(x)) = f^{-1}\left(\frac{x^5 - 3}{2}\right)$$
$$= \left[2\left(\frac{x^5 - 3}{2}\right) + 3\right]^{1/5}$$
$$= (x^5 - 3 + 3)^{1/5}$$
$$= (x^5)^{1/5} = x$$

$$f(f^{-1}(x)) = f((2x + 3)^{1/5})$$
$$= \frac{[(2x + 3)^{1/5}]^5 - 3}{2}$$
$$= \frac{2x + 3 - 3}{2}$$
$$= \frac{2x}{2} = x \quad ✔$$

> ## HOW TO FIND THE INVERSE OF A ONE-TO-ONE FUNCTION
>
> **1.** Write $y = f(x)$.
>
> **2.** Solve this equation for x in terms of y (if possible).
>
> **3.** Interchange x and y. The resulting equation is $y = f^{-1}(x)$.

Note that Steps 2 and 3 can be reversed. In other words, we can interchange x and y first and then solve for y in terms of x.

▶ **EXAMPLE 6** | Finding the Inverse of a Function

Find the inverse of the function $f(x) = 3x - 2$.

▼ **SOLUTION**   First we write $y = f(x)$.

$$y = 3x - 2$$

Then we solve this equation for x.

$$3x = y + 2 \qquad \text{Add 2}$$
$$x = \frac{y + 2}{3} \qquad \text{Divide by 3}$$

Finally, we interchange x and y.

$$y = \frac{x + 2}{3}$$

Therefore, the inverse function is $f^{-1}(x) = \dfrac{x + 2}{3}$.   ▲

▶ **EXAMPLE 7** | Finding the Inverse of a Function

Find the inverse of the function $f(x) = \dfrac{x^5 - 3}{2}$.

▼ **SOLUTION**   We first write $y = (x^5 - 3)/2$ and solve for x.

$$y = \frac{x^5 - 3}{2} \qquad \text{Equation defining function}$$
$$2y = x^5 - 3 \qquad \text{Multiply by 2}$$
$$x^5 = 2y + 3 \qquad \text{Add 3 (and switch sides)}$$
$$x = (2y + 3)^{1/5} \qquad \text{Take fifth root of each side}$$

Then we interchange x and y to get $y = (2x + 3)^{1/5}$. Therefore, the inverse function is $f^{-1}(x) = (2x + 3)^{1/5}$.   ▲

■ Graphing the Inverse of a Function

The principle of interchanging x and y to find the inverse function also gives us a method for obtaining the graph of f^{-1} from the graph of f. If $f(a) = b$, then $f^{-1}(b) = a$. Thus, the point (a, b) is on the graph of f if and only if the point (b, a) is on the graph of f^{-1}. But we get the point (b, a) from the point (a, b) by reflecting in the line $y = x$ (see Figure 8). Therefore, as Figure 9 illustrates, the following is true.

The graph of f^{-1} is obtained by reflecting the graph of f in the line $y = x$.

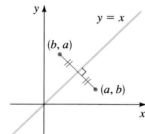

FIGURE 8   **FIGURE 9**

▶ **EXAMPLE 8** | Graphing the Inverse of a Function

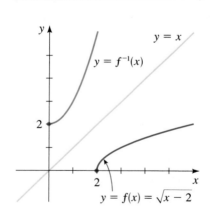

FIGURE 10

(a) Sketch the graph of $f(x) = \sqrt{x - 2}$.

(b) Use the graph of f to sketch the graph of f^{-1}.

(c) Find an equation for f^{-1}.

▼ **SOLUTION**

(a) Using the transformations from Section 3.5, we sketch the graph of $y = \sqrt{x - 2}$ by plotting the graph of the function $y = \sqrt{x}$ (Example 1(c) in Section 3.2) and moving it to the right 2 units.

(b) The graph of f^{-1} is obtained from the graph of f in part (a) by reflecting it in the line $y = x$, as shown in Figure 10.

(c) Solve $y = \sqrt{x - 2}$ for x, noting that $y \geq 0$.

$$\sqrt{x - 2} = y$$

$$x - 2 = y^2 \qquad \text{Square each side}$$

$$x = y^2 + 2 \qquad y \geq 0 \qquad \text{Add 2}$$

Interchange x and y:

$$y = x^2 + 2 \qquad x \geq 0$$

Thus, $$f^{-1}(x) = x^2 + 2 \qquad x \geq 0$$

This expression shows that the graph of f^{-1} is the right half of the parabola $y = x^2 + 2$, and from the graph shown in Figure 10, this seems reasonable. ▲

In Example 8 note how f^{-1} reverses the effect of f. The function f is the rule "Subtract 2, then take the square root," whereas f^{-1} is the rule "Square, then add 2."

▶ CHAPTER 3 | REVIEW

▼ PROPERTIES AND FORMULAS

Function Notation (p. 137)

If a function is given by the formula $y = f(x)$, then x is the independent variable and denotes the **input**; y is the dependent variable and denotes the **output**; the **domain** is the set of all possible inputs x; the **range** is the set of all possible outputs y.

The Graph of a Function (p. 143)

The graph of a function f is the graph of the equation $y = f(x)$ that defines f.

The Vertical Line Test (p. 148)

A curve in the coordinate plane is the graph of a function if and only if no vertical line intersects the graph more than once.

Increasing and Decreasing Functions (p. 152)

A function f is **increasing** on an interval if $f(x_1) < f(x_2)$ whenever $x_1 < x_2$ in the interval.

A function f is **decreasing** on an interval if $f(x_1) > f(x_2)$ whenever $x_1 < x_2$ in the interval.

Local Maximum and Minimum Values (p. 154)

The function value $f(a)$ is a **local maximum value** of the function f if $f(a) \geq f(x)$ for all x near a. In this case we also say that f has a **local maximum** at $x = a$.

The function value $f(b)$ is a **local minimum value** of the function f if $f(b) \leq f(x)$ for all x near b. In this case we also say that f has a **local minimum** at $x = b$.

Average Rate of Change (p. 157)

The **average rate of change** of the function f between $x = a$ and $x = b$ is the slope of the **secant** line between $(a, f(a))$ and $(b, f(b))$:

$$\text{average rate of change} = \frac{f(b) - f(a)}{b - a}$$

Vertical and Horizontal Shifts of Graphs (pp. 161–162)

Let c be a positive constant.

To graph $y = f(x) + c$, shift the graph of $y = f(x)$ **upward** by c units.

To graph $y = f(x) - c$, shift the graph of $y = f(x)$ **downward** by c units.

To graph $y = f(x - c)$, shift the graph of $y = f(x)$ **to the right** by c units.

To graph $y = f(x + c)$, shift the graph of $y = f(x)$ **to the left** by c units.

Reflecting Graphs (p. 163)

To graph $y = -f(x)$, **reflect** the graph of $y = f(x)$ in the **x-axis**.

To graph $y = f(-x)$, **reflect** the graph of $y = f(x)$ in the **y-axis**.

Vertical and Horizontal Stretching and Shrinking of Graphs (pp. 164–165)

If $c > 1$, then to graph $y = cf(x)$, **stretch** the graph of $y = f(x)$ **vertically** by a factor of c.

If $0 < c < 1$, then to graph $y = cf(x)$, **shrink** the graph of $y = f(x)$ **vertically** by a factor of c.

If $c > 1$, then to graph $y = f(cx)$, **shrink** the graph of $y = f(x)$ **horizontally** by a factor of $1/c$.

If $0 < c < 1$, then to graph $y = f(cx)$, **stretch** the graph of $y = f(x)$ **horizontally** by a factor of $1/c$.

Even and Odd Functions (p. 167)

A function f is

  even if $f(-x) = f(x)$

  odd if $f(-x) = -f(x)$

for every x in the domain of f.

Composition of Functions (p. 170)

Given two functions f and g, the **composition** of f and g is the function $f \circ g$ defined by

$$(f \circ g)(x) = f(g(x))$$

The **domain** of $f \circ g$ is the set of all x for which both $g(x)$ and $f(g(x))$ are defined.

One-to-One Functions (p. 174)

A function f is **one-to-one** if $f(x_1) \neq f(x_2)$ whenever x_1 and x_2 are *different* elements of the domain of f.

To see from its graph whether a function is one-to-one, use the **Horizontal Line Test**: A function is one-to-one if and only if no horizontal line intersects its graph more than once.

Inverse of a Function (pp. 175–176)

Let f be a one-to-one function with domain A and range B.

The **inverse** of f is the function f^{-1} defined by

$$f^{-1}(y) = x \quad \Leftrightarrow \quad f(x) = y$$

The inverse function f^{-1} has domain B and range A.

The functions f and f^{-1} satisfy the following **cancellation properties**:

$$f^{-1}(f(x)) = x \quad \text{for every } x \text{ in } A$$
$$f(f^{-1}(x)) = x \quad \text{for every } x \text{ in } B$$

▼ CONCEPT SUMMARY _____

Section 3.1	**Review Exercises**
▪ Recognize functions in the real world	7–8
▪ Work with function notation	9–10
▪ Find domains of functions	13–22
▪ Represent functions verbally, algebraically, graphically, and numerically	1–6, 23–40

Section 3.2	**Review Exercises**
▪ Graph a function by plotting points	23–40
▪ Graph a function with a graphing calculator	45–50
▪ Graph piecewise defined functions	37–40
▪ Use the Vertical Line Test	11
▪ Determine whether an equation defines a function	41–44

Section 3.3	**Review Exercises**
▪ Find function values from a graph	12(a)
▪ Find the domain and range of a function from a graph	12(b), 12(c)
▪ Find where a function is increasing or decreasing from a graph	2(d), 53–54
▪ Find local maxima and minima of functions from a graph	12(e), 67–72

Section 3.4	**Review Exercises**
▪ Find the average rate of change of a function	55–58
▪ Interpret average rate of change in real-world situations	59–60
▪ Recognize that a function with constant average rate of change is linear	61–62

Section 3.5	**Review Exercises**
▪ Shift graphs vertically	63–64
▪ Shift graphs horizontally	63–64
▪ Stretch or shrink graphs vertically	63–64
▪ Stretch or shrink graphs horizontally	63–64
▪ Determine whether a function is odd or even	65–66

Section 3.6	**Review Exercises**
▪ Find sums, difference, products, and quotients of functions	75
▪ Add functions graphically	73–74
▪ Find the composition of two functions	75–78
▪ Express a given function as a composite function	79–80

Section 3.7	**Review Exercises**
▪ Determine whether a function is one-to-one	11, 12(f), 81–86
▪ Find the inverse function of a one-to-one function	87–90
▪ Draw the graph of an inverse function	91–92

▼ EXERCISES _____

1–2 ▪ A verbal description of a function f is given. Find a formula that expresses f in function notation.

1. "Square, then subtract 5."

2. "Divide by 2, then add 9."

3–4 ▪ A formula for a function f is given. Give a verbal description of the function.

3. $f(x) = 3(x + 10)$

4. $f(x) = \sqrt{6x - 10}$

5–6 ▪ Complete the table of values for the given function.

5. $g(x) = x^2 - 4x$

x	$g(x)$
−1	
0	
1	
2	
3	

6. $h(x) = 3x^2 + 2x - 5$

x	$h(x)$
−2	
−1	
0	
1	
2	

7. A publisher estimates that the cost $C(x)$ of printing a run of x copies of a certain mathematics textbook is given by the function $C(x) = 5000 + 30x - 0.001x^2$.
  (a) Find $C(1000)$ and $C(10,000)$.
  (b) What do your answers in part (a) represent?
  (c) Find $C(0)$. What does this number represent?

8. Reynalda works as a salesperson in the electronics division of a department store. She earns a base weekly salary plus a commission based on the retail price of the goods she has sold. If she sells x dollars worth of goods in a week, her earnings for that week are given by the function $E(x) = 400 + 0.03x$.
  (a) Find $E(2000)$ and $E(15,000)$.
  (b) What do your answers in part (a) represent?
  (c) Find $E(0)$. What does this number represent?
  (d) From the formula for E, determine what percentage Reynalda earns on the goods that she sells.

9. If $f(x) = x^2 - 4x + 6$, find $f(0)$, $f(2)$, $f(-2)$, $f(a)$, $f(-a)$, $f(x + 1)$, $f(2x)$, and $2f(x) - 2$.

10. If $f(x) = 4 - \sqrt{3x - 6}$, find $f(5)$, $f(9)$, $f(a + 2)$, $f(-x)$, $f(x^2)$, and $[f(x)]^2$.

11. Which of the following figures are graphs of functions? Which of the functions are one-to-one?

(a)

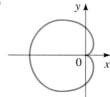

(b)
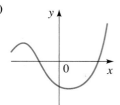

(c)
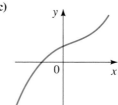

(d)
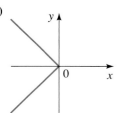

12. The graph of a function f is given.
  (a) Find $f(-2)$ and $f(2)$.
  (b) Find the domain of f.
  (c) Find the range of f.
  (d) On what intervals is f increasing? On what intervals is f decreasing?
  (e) What are the local maximum values of f?
  (f) Is f one-to-one?

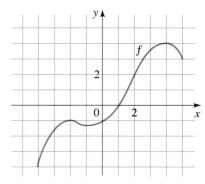

13–14 ▪ Find the domain and range of the function.

13. $f(x) = \sqrt{x + 3}$

14. $F(t) = t^2 + 2t + 5$

15–22 ▪ Find the domain of the function.

15. $f(x) = 7x + 15$

16. $f(x) = \dfrac{2x + 1}{2x - 1}$

17. $f(x) = \sqrt{x + 4}$

18. $f(x) = 3x - \dfrac{2}{\sqrt{x + 1}}$

19. $f(x) = \dfrac{1}{x} + \dfrac{1}{x + 1} + \dfrac{1}{x + 2}$

20. $g(x) = \dfrac{2x^2 + 5x + 3}{2x^2 - 5x - 3}$

21. $h(x) = \sqrt{4 - x} + \sqrt{x^2 - 1}$

22. $f(x) = \dfrac{\sqrt[3]{2x + 1}}{\sqrt[3]{2x + 2}}$

23–40 ▪ Sketch the graph of the function.

23. $f(x) = 1 - 2x$

24. $f(x) = \frac{1}{3}(x - 5)$, $2 \le x \le 8$

25. $f(t) = 1 - \frac{1}{2}t^2$

26. $g(t) = t^2 - 2t$

27. $f(x) = x^2 - 6x + 6$

28. $f(x) = 3 - 8x - 2x^2$

29. $g(x) = 1 - \sqrt{x}$

30. $g(x) = -|x|$

31. $h(x) = \frac{1}{2}x^3$

32. $h(x) = \sqrt{x + 3}$

33. $h(x) = \sqrt[3]{x}$

34. $H(x) = x^3 - 3x^2$

35. $g(x) = \dfrac{1}{x^2}$

36. $G(x) = \dfrac{1}{(x - 3)^2}$

37. $f(x) = \begin{cases} 1 - x & \text{if } x < 0 \\ 1 & \text{if } x \ge 0 \end{cases}$

38. $f(x) = \begin{cases} 1 - 2x & \text{if } x \le 0 \\ 2x - 1 & \text{if } x > 0 \end{cases}$

39. $f(x) = \begin{cases} x + 6 & \text{if } x < -2 \\ x^2 & \text{if } x \ge -2 \end{cases}$

40. $f(x) = \begin{cases} -x & \text{if } x < 0 \\ x^2 & \text{if } 0 \le x < 2 \\ 1 & \text{if } x \ge 2 \end{cases}$

41–44 ▪ Determine whether the equation defines y as a function of x.

41. $x + y^2 = 14$

42. $3x - \sqrt{y} = 8$

43. $x^3 - y^3 = 27$

44. $2x = y^4 - 16$

45. Determine which viewing rectangle produces the most appropriate graph of the function
$$f(x) = 6x^3 - 15x^2 + 4x - 1.$$
  (i) $[-2, 2]$ by $[-2, 2]$ (ii) $[-8, 8]$ by $[-8, 8]$
  (iii) $[-4, 4]$ by $[-12, 12]$ (iv) $[-100, 100]$ by $[-100, 100]$

46. Determine which viewing rectangle produces the most appropriate graph of the function $f(x) = \sqrt{100 - x^3}$.
  (i) $[-4, 4]$ by $[-4, 4]$
  (ii) $[-10, 10]$ by $[-10, 10]$
  (iii) $[-10, 10]$ by $[-10, 40]$
  (iv) $[-100, 100]$ by $[-100, 100]$

47–50 ▪ Draw the graph of the function in an appropriate viewing rectangle.

47. $f(x) = x^2 + 25x + 173$

48. $f(x) = 1.1x^3 - 9.6x^2 - 1.4x + 3.2$

49. $f(x) = \dfrac{x}{\sqrt{x^2 + 16}}$

50. $f(x) = |x(x + 2)(x + 4)|$

51. Find, approximately, the domain of the function

$$f(x) = \sqrt{x^3 - 4x + 1}.$$

52. Find, approximately, the range of the function

$$f(x) = x^4 - x^3 + x^2 + 3x - 6.$$

53–54 ▪ Draw a graph of the function f, and determine the intervals on which f is increasing and on which f is decreasing.

53. $f(x) = x^3 - 4x^2$ **54.** $f(x) = |x^4 - 16|$

55–58 ▪ Find the average rate of change of the function between the given points.

55. $f(x) = x^2 + 3x$; $x = 0, x = 2$

56. $f(x) = \dfrac{1}{x - 2}$; $x = 4, x = 8$

57. $f(x) = \dfrac{1}{x}$; $x = 3, x = 3 + h$

58. $f(x) = (x + 1)^2$; $x = a, x = a + h$

59. The population of a planned seaside community in Florida is given by the function $P(t) = 3000 + 200t + 0.1t^2$, where t represents the number of years since the community was incorporated in 1985.
    (a) Find $P(10)$ and $P(20)$. What do these values represent?
    (b) Find the average rate of change of P between $t = 10$ and $t = 20$. What does this number represent?

60. Ella is saving for her retirement by making regular deposits into a 401(k) plan. As her salary rises, she finds that she can deposit increasing amounts each year. Between 1995 and 2008, the annual amount (in dollars) that she deposited was given by the function $D(t) = 3500 + 15t^2$, where t represents the year of the deposit measured from the start of the plan (so 1995 corresponds to $t = 0$ and 1996 corresponds to $t = 1$, and so on).
    (a) Find $D(0)$ and $D(15)$. What do these values represent?
    (b) Assuming that her deposits continue to be modeled by the function D, in what year will she deposit $17,000?
    (c) Find the average rate of change of D between $t = 0$ and $t = 15$. What does this number represent?

61–62 ▪ A function f is given. **(a)** Find the average rate of change of f between $x = 0$ and $x = 2$, and the average rate of change of f between $x = 15$ and $x = 50$. **(b)** Were the two average rates of change that you found in part (a) the same? Explain why or why not.

61. $f(x) = \frac{1}{2}x - 6$ **62.** $f(x) = 8 - 3x$

63. Suppose the graph of f is given. Describe how the graphs of the following functions can be obtained from the graph of f.
    (a) $y = f(x) + 8$ **(b)** $y = f(x + 8)$
    (c) $y = 1 + 2f(x)$ **(d)** $y = f(x - 2) - 2$
    (e) $y = f(-x)$ **(f)** $y = -f(-x)$
    (g) $y = -f(x)$ **(h)** $y = f^{-1}(x)$

64. The graph of f is given. Draw the graphs of the following functions.
    (a) $y = f(x - 2)$ **(b)** $y = -f(x)$
    (c) $y = 3 - f(x)$ **(d)** $y = \frac{1}{2}f(x) - 1$
    (e) $y = f^{-1}(x)$ **(f)** $y = f(-x)$

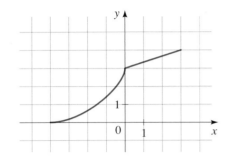

65. Determine whether f is even, odd, or neither.
    (a) $f(x) = 2x^5 - 3x^2 + 2$ **(b)** $f(x) = x^3 - x^7$
    (c) $f(x) = \dfrac{1 - x^2}{1 + x^2}$ **(d)** $f(x) = \dfrac{1}{x + 2}$

66. Determine whether the function in the figure is even, odd, or neither.

(a) 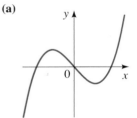   **(b)**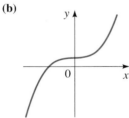

(c) 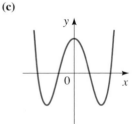   **(d)**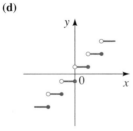

67. Find the minimum value of the function $g(x) = 2x^2 + 4x - 5$.

68. Find the maximum value of the function $f(x) = 1 - x - x^2$.

69. A stone is thrown upward from the top of a building. Its height (in feet) above the ground after t seconds is given by $h(t) = -16t^2 + 48t + 32$. What maximum height does it reach?

70. The profit P (in dollars) generated by selling x units of a certain commodity is given by

$$P(x) = -1500 + 12x - 0.0004x^2$$

What is the maximum profit, and how many units must be sold to generate it?

71–72 ▪ Find the local maximum and minimum values of the function and the values of x at which they occur. State each answer correct to two decimal places.

71. $f(x) = 3.3 + 1.6x - 2.5x^3$ **72.** $f(x) = x^{2/3}(6 - x)^{1/3}$

73–74 ▪ Two functions, f and g, are given. Draw graphs of f, g, and $f + g$ on the same graphing calculator screen to illustrate the concept of graphical addition.

73. $f(x) = x + 2$, $g(x) = x^2$

74. $f(x) = x^2 + 1$, $g(x) = 3 - x^2$

75. If $f(x) = x^2 - 3x + 2$ and $g(x) = 4 - 3x$, find the following functions.
    (a) $f + g$ **(b)** $f - g$ **(c)** fg
    (d) f/g **(e)** $f \circ g$ **(f)** $g \circ f$

76. If $f(x) = 1 + x^2$ and $g(x) = \sqrt{x - 1}$, find the following.
    (a) $f \circ g$ **(b)** $g \circ f$ **(c)** $(f \circ g)(2)$
    (d) $(f \circ f)(2)$ **(e)** $f \circ g \circ f$ **(f)** $g \circ f \circ g$

77–78 ▪ Find the functions $f \circ g$, $g \circ f$, $f \circ f$, and $g \circ g$ and their domains.

77. $f(x) = 3x - 1$, $g(x) = 2x - x^2$

78. $f(x) = \sqrt{x}$, $g(x) = \dfrac{2}{x - 4}$

79. Find $f \circ g \circ h$, where $f(x) = \sqrt{1 - x}$, $g(x) = 1 - x^2$, and $h(x) = 1 + \sqrt{x}$.

80. If $T(x) = \dfrac{1}{\sqrt{1 + \sqrt{x}}}$, find functions f, g, and h such that $f \circ g \circ h = T$.

81–86 ▪ Determine whether the function is one-to-one.

81. $f(x) = 3 + x^3$

82. $g(x) = 2 - 2x + x^2$

83. $h(x) = \dfrac{1}{x^4}$

84. $r(x) = 2 + \sqrt{x + 3}$

85. $p(x) = 3.3 + 1.6x - 2.5x^3$

86. $q(x) = 3.3 + 1.6x + 2.5x^3$

87–90 ▪ Find the inverse of the function.

87. $f(x) = 3x - 2$ **88.** $f(x) = \dfrac{2x + 1}{3}$

89. $f(x) = (x + 1)^3$ **90.** $f(x) = 1 + \sqrt[5]{x - 2}$

91. (a) Sketch the graph of the function
$$f(x) = x^2 - 4 \qquad x \geq 0$$
    (b) Use part (a) to sketch the graph of f^{-1}.
    (c) Find an equation for f^{-1}.

92. (a) Show that the function $f(x) = 1 + \sqrt[4]{x}$ is one-to-one.
    (b) Sketch the graph of f.
    (c) Use part (b) to sketch the graph of f^{-1}.
    (d) Find an equation for f^{-1}.

1. Which of the following are graphs of functions? If the graph is that of a function, is it one-to-one?

(a)

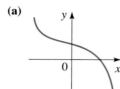

(b)

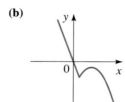

(c)

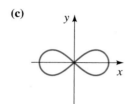

(d)

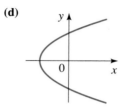

2. Let $f(x) = \dfrac{\sqrt{x+1}}{x}$.

   (a) Evaluate $f(3)$, $f(5)$, and $f(a-1)$.

   (b) Find the domain of f.

3. A function f has the following verbal description: "Subtract 2, then cube the result."

   (a) Find a formula that expresses f algebraically.

   (b) Make a table of values of f, for the inputs $-1, 0, 1, 2, 3$, and 4.

   (c) Sketch a graph of f, using the table of values from part (b) to help you.

   (d) How do we know that f has an inverse? Give a verbal description for f^{-1}.

   (e) Find a formula that expresses f^{-1} algebraically.

4. A school fund-raising group sells chocolate bars to help finance a swimming pool for their physical education program. The group finds that when they set their price at x dollars per bar (where $0 < x \le 5$), their total sales revenue (in dollars) is given by the function $R(x) = -500x^2 + 3000x$.

   (a) Evaluate $R(2)$ and $R(4)$. What do these values represent?

   (b) Use a graphing calculator to draw a graph of R. What does the graph tell us about what happens to revenue as the price increases from 0 to 5 dollars?

   (c) What is the maximum revenue, and at what price is it achieved?

5. Determine the average rate of change for the function $f(t) = t^2 - 2t$ between $t = 2$ and $t = 5$.

6. (a) Sketch the graph of the function $f(x) = x^3$.

   (b) Use part (a) to graph the function $g(x) = (x-1)^3 - 2$.

7. (a) How is the graph of $y = f(x-3) + 2$ obtained from the graph of f?

   (b) How is the graph of $y = f(-x)$ obtained from the graph of f?

8. Let $f(x) = \begin{cases} 1 - x & \text{if } x \le 1 \\ 2x + 1 & \text{if } x > 1 \end{cases}$

   (a) Evaluate $f(-2)$ and $f(1)$.

   (b) Sketch the graph of f.

9. If $f(x) = x^2 + 1$ and $g(x) = x - 3$, find the following.

   (a) $f \circ g$ (b) $g \circ f$

   (c) $f(g(2))$ (d) $g(f(2))$

   (e) $g \circ g \circ g$

10. (a) If $f(x) = \sqrt{3 - x}$, find the inverse function f^{-1}.

(b) Sketch the graphs of f and f^{-1} on the same coordinate axes.

11. The graph of a function f is given.

(a) Find the domain and range of f.

(b) Sketch the graph of f^{-1}.

(c) Find the average rate of change of f between $x = 2$ and $x = 6$.

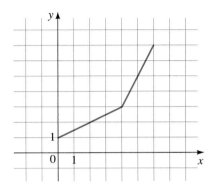

12. Let $f(x) = 3x^4 - 14x^2 + 5x - 3$.

(a) Draw the graph of f in an appropriate viewing rectangle.

(b) Is f one-to-one?

(c) Find the local maximum and minimum values of f and the values of x at which they occur. State each answer correct to two decimal places.

(d) Use the graph to determine the range of f.

(e) Find the intervals on which f is increasing and on which f is decreasing.

Polynomial and Rational Functions

In this chapter we study functions defined by polynomial expressions. Polynomials are constructed using just addition, subtraction, multiplication, and taking powers, so their values are easy to calculate. We will use polynomials to model real-world problems, including ones that involve areas and volumes. We will also study rational functions, which are quotients of polynomial functions.

The following Warm-Up Exercises will help you review the basic algebra skills that you will need for this chapter.

WARM-UP EXERCISES

1. Solve the following quadratic equation by completing the square: $2x^2 - x - 1 = 0$.

2. Let $f(x) = (x - 30)(x + 1)^2(x - 2)$. Compute the following without using a calculator:

$$f(30), f(-1), f(2), f(0).$$

3. Consider the arithmetic problem $21{,}198 \div 15$. Use long division to compute the quotient and remainder.

4. Find all the factors of 90.

5. Factor the following polynomial: $x^3 - x^2 - 6x$.

6. Let $f(x) = \dfrac{2x^3 - 5}{x^3 - 8}$.

    (a) Compute $f(0)$.

    (b) Compute $f(10)$, $f(100)$, and $f(1000)$. What is happening to $f(x)$ as x gets very large?

    (c) Compute $f(1.999)$, $f(2.00001)$, and $f(1.999999)$. What is happening to $f(x)$ as x gets very close to 2?

<table>
<tr><td>**4.1**</td><td># Quadratic Functions and Models</td></tr>
</table>

LEARNING OBJECTIVES

After completing this section, you will be able to:

▨ Express a quadratic function in standard form
▨ Graph a quadratic function using its standard form
▨ Find maximum and minimum values of quadratic functions
▨ Model with quadratic functions

A polynomial function is a function that is defined by a polynomial expression. So a **polynomial function of degree n** is a function of the form

$$P(x) = a_n x^n + a_{n-1} x^{n-1} + \cdots + a_1 x + a_0$$

Polynomial expressions are defined in Section P.5.

We have already studied polynomial functions of degree 0 and 1. These are functions of the form $P(x) = a_0$ and $P(x) = a_1 x + a_0$, respectively, whose graphs are lines. In this section we study polynomial functions of degree 2. These are called quadratic functions.

QUADRATIC FUNCTIONS

A **quadratic function** is a polynomial function of degree 2. So a quadratic function is a function of the form

$$f(x) = ax^2 + bx + c \qquad (a \neq 0)$$

We see in this section how quadratic functions model many real-world phenomena. We begin by analyzing the graphs of quadratic functions.

▨ Graphing Quadratic Functions Using the Standard Form

If we take $a = 1$ and $b = c = 0$ in the quadratic function

$$f(x) = ax^2 + bx + c$$

we get the quadratic function

$$f(x) = x^2$$

whose graph is the parabola graphed in Example 1 of Section 3.2. In fact, the graph of any quadratic function is a **parabola**; it can be obtained from the graph of $f(x) = x^2$ by the transformations given in Section 3.5.

STANDARD FORM OF A QUADRATIC FUNCTION

A quadratic function $f(x) = ax^2 + bx + c$ can be expressed in the **standard form**

$$f(x) = a(x - h)^2 + k$$

by completing the square. The graph of f is a parabola with **vertex** (h, k); the parabola opens upward if $a > 0$ or downward if $a < 0$.

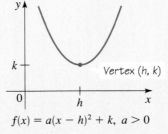

$$f(x) = a(x - h)^2 + k, \ a > 0$$

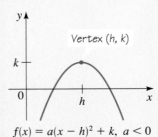

$$f(x) = a(x - h)^2 + k, \ a < 0$$

▶ **EXAMPLE 1** | Standard Form of a Quadratic Function

Let $f(x) = 2x^2 - 12x + 23$.

(a) Express f in standard form.   **(b)** Sketch the graph of f.

▼ **SOLUTION**

Completing the square is discussed in Section 1.3.

$f(x) = 2(x - 3)^2 + 5$

Vertex is $(3, 5)$

(a) Since the coefficient of x^2 is not 1, we must factor this coefficient from the terms involving x before we complete the square.

$$f(x) = 2x^2 - 12x + 23$$
$$= 2(x^2 - 6x) + 23 \qquad \text{Factor 2 from the x-terms}$$
$$= 2(x^2 - 6x + 9) + 23 - 2 \cdot 9 \qquad \begin{array}{l}\text{Complete the square: Add 9 inside} \\ \text{parentheses, subtract } 2 \cdot 9 \text{ outside}\end{array}$$
$$= 2(x - 3)^2 + 5 \qquad \text{Factor and simplify}$$

The standard form is $f(x) = 2(x - 3)^2 + 5$.

(b) The standard form tells us that we get the graph of f by taking the parabola $y = x^2$, shifting it to the right 3 units, stretching it by a factor of 2, and moving it upward 5 units. The vertex of the parabola is at $(3, 5)$, and the parabola opens upward. We sketch the graph in Figure 1 after noting that the y-intercept is $f(0) = 23$.

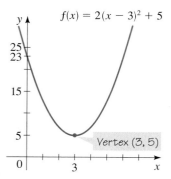

FIGURE 1

Maximum and Minimum Values of Quadratic Functions

If a quadratic function has vertex (h, k), then the function has a minimum value at the vertex if its graph opens upward and a maximum value at the vertex if its graph opens downward. For example, the function graphed in Figure 1 has minimum value 5 when $x = 3$, since the vertex $(3, 5)$ is the lowest point on the graph.

MAXIMUM OR MINIMUM VALUE OF A QUADRATIC FUNCTION

Let f be a quadratic function with standard form $f(x) = a(x - h)^2 + k$. The maximum or minimum value of f occurs at $x = h$.

If $a > 0$, then the **minimum value** of f is $f(h) = k$.

If $a < 0$, then the **maximum value** of f is $f(h) = k$.

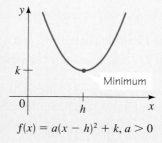

$f(x) = a(x - h)^2 + k, a > 0$

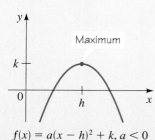

$f(x) = a(x - h)^2 + k, a < 0$

▶ **EXAMPLE 2** | Minimum Value of a Quadratic Function

Consider the quadratic function $f(x) = 5x^2 - 30x + 49$.

(a) Express f in standard form.

(b) Sketch the graph of f.

(c) Find the minimum value of f.

▼ **SOLUTION**

(a) To express this quadratic function in standard form, we complete the square.

$$f(x) = 5x^2 - 30x + 49$$

$$= 5(x^2 - 6x) + 49 \qquad \textit{Factor 5 from the x-terms}$$

$$= 5(x^2 - 6x + 9) + 49 - 5 \cdot 9 \qquad \begin{array}{l}\textit{Complete the square: Add 9 inside}\\ \textit{parentheses, subtract 5 · 9 outside}\end{array}$$

$$= 5(x - 3)^2 + 4 \qquad \textit{Factor and simplify}$$

(b) The graph is a parabola that has its vertex at $(3, 4)$ and opens upward, as sketched in Figure 2.

(c) Since the coefficient of x^2 is positive, f has a minimum value. The minimum value is $f(3) = 4$. ▲

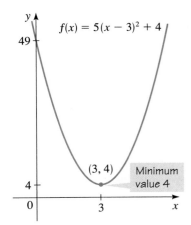

$f(x) = 5(x - 3)^2 + 4$

$(3, 4)$ Minimum value 4

FIGURE 2

▶ **EXAMPLE 3** | Maximum Value of a Quadratic Function

Consider the quadratic function $f(x) = -x^2 + x + 2$.

(a) Express f in standard form.

(b) Sketch the graph of f.

(c) Find the maximum value of f.

▼ **SOLUTION**

(a) To express this quadratic function in standard form, we complete the square.

$$y = -x^2 + x + 2$$

$$= -(x^2 - x) + 2 \qquad \textit{Factor −1 from the x-terms}$$

$$= -\left(x^2 - x + \tfrac{1}{4}\right) + 2 - (-1)\tfrac{1}{4} \qquad \begin{array}{l}\textit{Complete the square: Add }\tfrac{1}{4}\textit{ inside}\\ \textit{parentheses, subtract }(-1)\tfrac{1}{4}\textit{ outside}\end{array}$$

$$= -\left(x - \tfrac{1}{2}\right)^2 + \tfrac{9}{4} \qquad \textit{Factor and simplify}$$

(b) From the standard form we see that the graph is a parabola that opens downward and has vertex $\left(\tfrac{1}{2}, \tfrac{9}{4}\right)$. As an aid to sketching the graph, we find the intercepts. The y-intercept is $f(0) = 2$. To find the x-intercepts, we set $f(x) = 0$ and factor the resulting equation.

$$-x^2 + x + 2 = 0 \qquad \textit{Set y = 0}$$

$$x^2 - x - 2 = 0 \qquad \textit{Multiply by −1}$$

$$(x - 2)(x + 1) = 0 \qquad \textit{Factor}$$

Thus, the x-intercepts are $x = 2$ and $x = -1$. The graph of f is sketched in Figure 3.

(c) Since the coefficient of x^2 is negative, f has a maximum value, which is $f\left(\tfrac{1}{2}\right) = \tfrac{9}{4}$. ▲

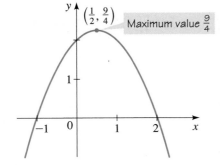

$\left(\tfrac{1}{2}, \tfrac{9}{4}\right)$ Maximum value $\tfrac{9}{4}$

FIGURE 3 Graph of $f(x) = -x^2 + x + 2$

Expressing a quadratic function in standard form helps us to sketch its graph as well as find its maximum or minimum value. If we are interested only in finding the maximum or minimum value, then a formula is available for doing so. This formula is obtained by completing the square for the general quadratic function as follows:

$$f(x) = ax^2 + bx + c$$

$$= a\left(x^2 + \frac{b}{a}x\right) + c \qquad \text{Factor } a \text{ from the x-terms}$$

$$= a\left(x^2 + \frac{b}{a}x + \frac{b^2}{4a^2}\right) + c - a\left(\frac{b^2}{4a^2}\right) \qquad \begin{array}{l}\text{Complete the square:} \\ \text{Add } \frac{b^2}{4a^2} \text{ inside parentheses,} \\ \text{subtract } a\left(\frac{b^2}{4a^2}\right) \text{ outside}\end{array}$$

$$= a\left(x + \frac{b}{2a}\right)^2 + c - \frac{b^2}{4a} \qquad \text{Factor}$$

This equation is in standard form with $h = -b/(2a)$ and $k = c - b^2/(4a)$. Since the maximum or minimum value occurs at $x = h$, we have the following result.

MAXIMUM OR MINIMUM VALUE OF A QUADRATIC FUNCTION

The maximum or minimum value of a quadratic function $f(x) = ax^2 + bx + c$ occurs at

$$x = -\frac{b}{2a}$$

If $a > 0$, then the **minimum value** is $f\left(-\dfrac{b}{2a}\right)$.

If $a < 0$, then the **maximum value** is $f\left(-\dfrac{b}{2a}\right)$.

▶ **EXAMPLE 4** | Finding Maximum and Minimum Values of Quadratic Functions

Find the maximum or minimum value of each quadratic function.

(a) $f(x) = x^2 + 4x$ **(b)** $g(x) = -2x^2 + 4x - 5$

▼ **SOLUTION**

(a) This is a quadratic function with $a = 1$ and $b = 4$. Thus, the maximum or minimum value occurs at

$$x = -\frac{b}{2a} = -\frac{4}{2 \cdot 1} = -2$$

Since $a > 0$, the function has the *minimum* value

$$f(-2) = (-2)^2 + 4(-2) = -4$$

(b) This is a quadratic function with $a = -2$ and $b = 4$. Thus, the maximum or minimum value occurs at

$$x = -\frac{b}{2a} = -\frac{4}{2 \cdot (-2)} = 1$$

Since $a < 0$, the function has the *maximum* value

$$f(1) = -2(1)^2 + 4(1) - 5 = -3$$

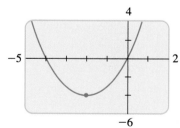

The minimum value occurs at $x = -2$.

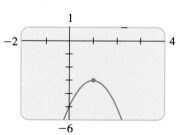

The maximum value occurs at $x = 1$.

▲

■ Modeling with Quadratic Functions

We study some examples of real-world phenomena that are modeled by quadratic functions. These examples and the *Application* exercises for this section show some of the variety of situations that are naturally modeled by quadratic functions.

▶ **EXAMPLE 5** | Maximum Gas Mileage for a Car

Most cars get their best gas mileage when traveling at a relatively modest speed. The gas mileage M for a certain new car is modeled by the function

$$M(s) = -\frac{1}{28}s^2 + 3s - 31, \qquad 15 \le s \le 70$$

where s is the speed in mi/h and M is measured in mi/gal. What is the car's best gas mileage, and at what speed is it attained?

▼ **SOLUTION**   The function M is a quadratic function with $a = -\frac{1}{28}$ and $b = 3$. Thus, its maximum value occurs when

$$s = -\frac{b}{2a} = -\frac{3}{2(-\frac{1}{28})} = 42$$

The maximum is $M(42) = -\frac{1}{28}(42)^2 + 3(42) - 31 = 32$. So the car's best gas mileage is 32 mi/gal, when it is traveling at 42 mi/h.   ▲

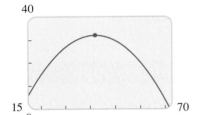

The maximum gas mileage occurs at 42 mi/h.

▶ **EXAMPLE 6** | Maximizing Revenue from Ticket Sales

A hockey team plays in an arena that has a seating capacity of 15,000 spectators. With the ticket price set at $14, average attendance at recent games has been 9500. A market survey indicates that for each dollar the ticket price is lowered, the average attendance increases by 1000.

(a) Find a function that models the revenue in terms of ticket price.

(b) Find the price that maximizes revenue from ticket sales.

(c) What ticket price is so high that no one attends and so no revenue is generated?

▼ **SOLUTION**

(a) The model that we want is a function that gives the revenue for any ticket price.

Express the model in words ▶   revenue = ticket price × attendance

There are two varying quantities: ticket price and attendance. Since the function we want depends on price, we let

Choose the variable ▶ x = ticket price

Next, we express attendance in terms of x.

In Words	In Algebra
Ticket price	x
Amount ticket price is lowered	$14 - x$
Increase in attendance	$1000(14 - x)$
Attendance	$9500 + 1000(14 - x)$

The model that we want is the function R that gives the revenue for a given ticket price x.

Set up the model ▶

$$\text{revenue} = \boxed{\text{ticket price}} \times \boxed{\text{attendance}}$$

$$R(x) = x \times [9500 + 1000(14 - x)]$$
$$R(x) = x(23{,}500 - 1000x)$$
$$R(x) = 23{,}500x - 1000x^2$$

Use the model ▶ **(b)** Since R is a quadratic function with $a = -1000$ and $b = 23{,}500$, the maximum occurs at

$$x = -\frac{b}{2a} = -\frac{23{,}500}{2(-1000)} = 11.75$$

So a ticket price of $11.75 gives the maximum revenue.

Use the model ▶ **(c)** We want to find the ticket price for which $R(x) = 0$:

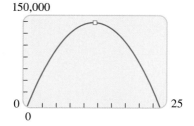

150,000

0
0   25

Maximum attendance occurs when ticket price is $11.75.

$$23{,}500x - 1000x^2 = 0 \qquad \text{Set } R(x) = 0$$
$$23.5x - x^2 = 0 \qquad \text{Divide by 1000}$$
$$x(23.5 - x) = 0 \qquad \text{Factor}$$
$$x = 0 \quad \text{or} \quad x = 23.5 \qquad \text{Solve for } x$$

So according to this model, a ticket price of $23.50 is just too high; at that price, no one attends to watch this team play. (Of course, revenue is also zero if the ticket price is zero.)   ▲

4.2 | Polynomial Functions and Their Graphs

LEARNING OBJECTIVES
After completing this section, you will be able to:

▨ Graph basic polynomial functions
▨ Use end behavior of a polynomial function to help sketch its graph
▨ Use the zeros of a polynomial function to sketch its graph
▨ Use the multiplicity of a zero to help sketch the graph of a polynomial function
▨ Find local maxima and minima of polynomial functions

In this section we study polynomial functions of any degree. But before we work with polynomial functions, we must agree on some terminology.

POLYNOMIAL FUNCTIONS

A **polynomial function of degree n** is a function of the form

$$P(x) = a_n x^n + a_{n-1} x^{n-1} + \cdots + a_1 x + a_0$$

where n is a nonnegative integer and $a_n \neq 0$.

The numbers $a_0, a_1, a_2, \ldots, a_n$ are called the **coefficients** of the polynomial.

The number a_0 is the **constant coefficient** or **constant term**.

The number a_n, the coefficient of the highest power, is the **leading coefficient**, and the term $a_n x^n$ is the **leading term**.

We often refer to polynomial functions simply as *polynomials*. The following polynomial has degree 5, leading coefficient 3, and constant term -6.

Leading coefficient 3   Degree 5   Constant term -6

$$3x^5 + 6x^4 - 2x^3 + x^2 + 7x - 6$$

Leading term $3x^5$

Coefficients 3, 6, -2, 1, 7, and -6

Here are some more examples of polynomials.

$$P(x) = 3 \qquad \text{Degree 0}$$

$$Q(x) = 4x - 7 \qquad \text{Degree 1}$$

$$R(x) = x^2 + x \qquad \text{Degree 2}$$

$$S(x) = 2x^3 - 6x^2 - 10 \qquad \text{Degree 3}$$

If a polynomial consists of just a single term, then it is called a **monomial**. For example, $P(x) = x^3$ and $Q(x) = -6x^5$ are monomials.

■ Graphing Basic Polynomial Functions

The graphs of polynomials of degree 0 or 1 are lines (Section 2.4), and the graphs of polynomials of degree 2 are parabolas (Section 4.1). The greater the degree of the polynomial, the more complicated its graph can be. However, the graph of a polynomial function is always a smooth curve; that is, it has no breaks or corners (see Figure 1). The proof of this fact requires calculus.

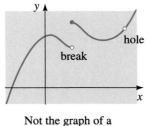

Not the graph of a polynomial function

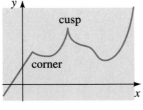

Not the graph of a polynomial function

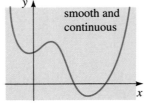

Graph of a polynomial function

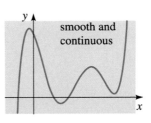

Graph of a polynomial function

FIGURE 1

The simplest polynomial functions are the monomials $P(x) = x^n$, whose graphs are shown in Figure 2. As the figure suggests, the graph of $P(x) = x^n$ has the same general shape as $y = x^2$ when n is even and the same general shape as $y = x^3$ when n is odd. However, as the degree n becomes larger, the graphs become flatter around the origin and steeper elsewhere.

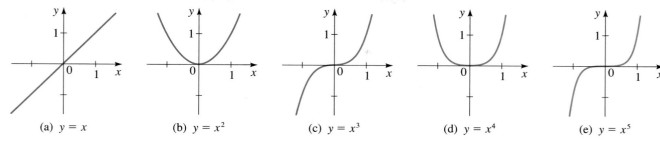

(a) $y = x$ (b) $y = x^2$ (c) $y = x^3$ (d) $y = x^4$ (e) $y = x^5$

FIGURE 2 Graphs of monomials

▶ **EXAMPLE 1** | Transformations of Monomials

Sketch the graphs of the following functions.

(a) $P(x) = -x^3$

(b) $Q(x) = (x - 2)^4$

(c) $R(x) = -2x^5 + 4$

▼ **SOLUTION**   We use the graphs in Figure 2 and transform them using the techniques of Section 3.5.

(a) The graph of $P(x) = -x^3$ is the reflection of the graph of $y = x^3$ in the x-axis, as shown in Figure 3(a) below.

(b) The graph of $Q(x) = (x - 2)^4$ is the graph of $y = x^4$ shifted to the right 2 units, as shown in Figure 3(b).

(c) We begin with the graph of $y = x^5$. The graph of $y = -2x^5$ is obtained by stretching the graph vertically and reflecting it in the x-axis (see the dashed blue graph in Figure 3(c)). Finally, the graph of $R(x) = -2x^5 + 4$ is obtained by shifting upward 4 units (see the red graph in Figure 3(c)).

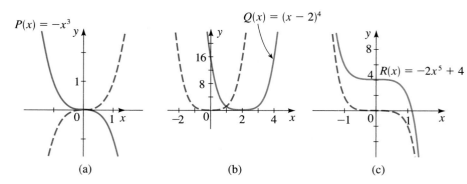

(a)   (b)   (c)

FIGURE 3

◼ End Behavior and the Leading Term

The **end behavior** of a polynomial is a description of what happens as x becomes large in the positive or negative direction. To describe end behavior, we use the following notation:

| $x \to \infty$ | means | "x becomes large in the positive direction" |

| $x \to -\infty$ | means | "x becomes large in the negative direction" |

For example, the monomial $y = x^2$ in Figure 2(b) has the following end behavior:

$$y \to \infty \quad \text{as} \quad x \to \infty \quad \text{and} \quad y \to \infty \quad \text{as} \quad x \to -\infty$$

The monomial $y = x^3$ in Figure 2(c) has the following end behavior:

$$y \to \infty \quad \text{as} \quad x \to \infty \quad \text{and} \quad y \to -\infty \quad \text{as} \quad x \to -\infty$$

For any polynomial, *the end behavior is determined by the term that contains the highest power of x*, because when x is large, the other terms are relatively insignificant in size. The following box shows the four possible types of end behavior, based on the highest power and the sign of its coefficient.

END BEHAVIOR OF POLYNOMIALS

The end behavior of the polynomial $P(x) = a_n x^n + a_{n-1} x^{n-1} + \cdots + a_1 x + a_0$ is determined by the degree n and the sign of the leading coefficient a_n, as indicated in the following graphs.

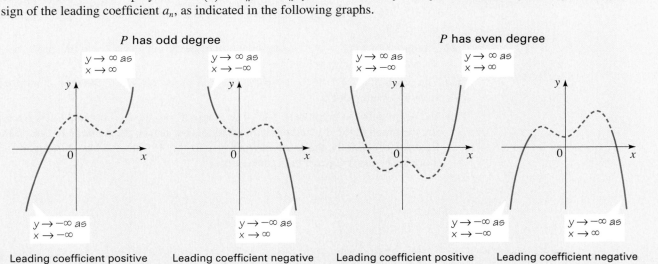

P has odd degree

$y \to \infty$ as $x \to \infty$

$y \to \infty$ as $x \to -\infty$

$y \to -\infty$ as $x \to -\infty$

$y \to -\infty$ as $x \to \infty$

Leading coefficient positive   Leading coefficient negative

P has even degree

$y \to \infty$ as $x \to -\infty$

$y \to \infty$ as $x \to \infty$

$y \to -\infty$ as $x \to -\infty$

$y \to -\infty$ as $x \to \infty$

Leading coefficient positive   Leading coefficient negative

▶ **EXAMPLE 2** | End Behavior of a Polynomial

Determine the end behavior of the polynomial

$$P(x) = -2x^4 + 5x^3 + 4x - 7$$

▼ **SOLUTION**   The polynomial P has degree 4 and leading coefficient -2. Thus, P has *even* degree and *negative* leading coefficient, so it has the following end behavior:

$$y \to -\infty \quad \text{as} \quad x \to \infty \quad \text{and} \quad y \to -\infty \quad \text{as} \quad x \to -\infty$$

The graph in Figure 4, on the next page, illustrates the end behavior of P.

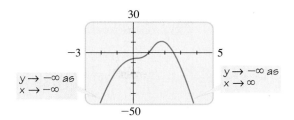

FIGURE 4

$P(x) = -2x^4 + 5x^3 + 4x - 7$

▶ **EXAMPLE 3** | End Behavior of a Polynomial

(a) Determine the end behavior of the polynomial $P(x) = 3x^5 - 5x^3 + 2x$.

(b) Confirm that P and its leading term $Q(x) = 3x^5$ have the same end behavior by graphing them together.

▼ **SOLUTION**

(a) Since P has odd degree and positive leading coefficient, it has the following end behavior:

$$y \to \infty \quad \text{as} \quad x \to \infty \qquad \text{and} \qquad y \to -\infty \quad \text{as} \quad x \to -\infty$$

(b) Figure 5 shows the graphs of P and Q in progressively larger viewing rectangles. The larger the viewing rectangle, the more the graphs look alike. This confirms that they have the same end behavior.

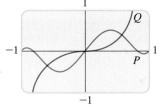 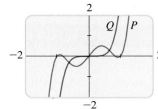 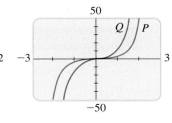 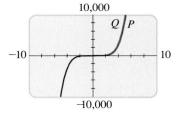

FIGURE 5

$P(x) = 3x^5 - 5x^3 + 2x$

$Q(x) = 3x^5$

To see algebraically why P and Q in Example 3 have the same end behavior, factor P as follows and compare with Q.

$$P(x) = 3x^5\left(1 - \frac{5}{3x^2} + \frac{2}{3x^4}\right) \qquad Q(x) = 3x^5$$

When x is large, the terms $5/3x^2$ and $2/3x^4$ are close to 0. So for large x, we have

$$P(x) \approx 3x^5(1 - 0 - 0) = 3x^5 = Q(x)$$

x	$P(x)$	$Q(x)$
15	2,261,280	2,278,125
30	72,765,060	72,900,000
50	936,875,100	937,500,000

So when x is large, P and Q have approximately the same values. We can also see this numerically by making a table like the one in the margin.

By the same reasoning we can show that the end behavior of *any* polynomial is determined by its leading term.

■ Using Zeros to Graph Polynomials

If P is a polynomial function, then c is called a **zero** of P if $P(c) = 0$. In other words, the zeros of P are the solutions of the polynomial equation $P(x) = 0$. Note that if $P(c) = 0$, then the graph of P has an x-intercept at $x = c$, so the x-intercepts of the graph are the zeros of the function.

REAL ZEROS OF POLYNOMIALS

If P is a polynomial and c is a real number, then the following are equivalent:

1. c is a zero of P.
2. $x = c$ is a solution of the equation $P(x) = 0$.
3. $x - c$ is a factor of $P(x)$.
4. c is an x-intercept of the graph of P.

To find the zeros of a polynomial P, we factor and then use the Zero-Product Property (see page 64). For example, to find the zeros of $P(x) = x^2 + x - 6$, we factor P to get

$$P(x) = (x - 2)(x + 3)$$

From this factored form we easily see that

1. 2 is a zero of P.
2. $x = 2$ is a solution of the equation $x^2 + x - 6 = 0$.
3. $x - 2$ is a factor of $x^2 + x - 6$.
4. 2 is an x-intercept of the graph of P.

The same facts are true for the other zero, -3.

The following theorem has many important consequences. Here we use it to help us graph polynomial functions.

INTERMEDIATE VALUE THEOREM FOR POLYNOMIALS

If P is a polynomial function and $P(a)$ and $P(b)$ have opposite signs, then there exists at least one value c between a and b for which $P(c) = 0$.

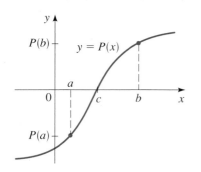

FIGURE 6

We will not prove this theorem, but Figure 6 shows why it is intuitively plausible.

One important consequence of this theorem is that between any two successive zeros the values of a polynomial are either all positive or all negative. That is, between two successive zeros the graph of a polynomial lies *entirely above* or *entirely below* the x-axis. To see why, suppose c_1 and c_2 are successive zeros of P. If P has both positive and negative values between c_1 and c_2, then by the Intermediate Value Theorem P must have another zero between c_1 and c_2. But that's not possible because c_1 and c_2 are successive zeros. This observation allows us to use the following guidelines to graph polynomial functions.

GUIDELINES FOR GRAPHING POLYNOMIAL FUNCTIONS

1. Zeros. Factor the polynomial to find all its real zeros; these are the x-intercepts of the graph.

2. Test Points. Make a table of values for the polynomial. Include test points to determine whether the graph of the polynomial lies above or below the x-axis on the intervals determined by the zeros. Include the y-intercept in the table.

3. End Behavior. Determine the end behavior of the polynomial.

4. Graph. Plot the intercepts and other points you found in the table. Sketch a smooth curve that passes through these points and exhibits the required end behavior.

▶ **EXAMPLE 4** | Using Zeros to Graph a Polynomial Function

Sketch the graph of the polynomial function $P(x) = (x + 2)(x - 1)(x - 3)$.

▼ **SOLUTION**   The zeros are $x = -2, 1$, and 3. These determine the intervals $(-\infty, -2)$, $(-2, 1)$, $(1, 3)$, and $(3, \infty)$. Using test points in these intervals, we get the information in the following sign diagram (see Section 1.6).

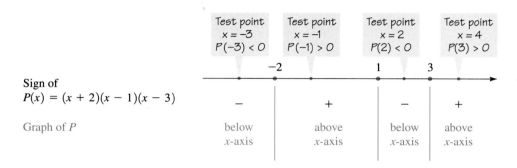

Sign of
$P(x) = (x + 2)(x - 1)(x - 3)$

Graph of P

Plotting a few additional points and connecting them with a smooth curve helps us to complete the graph in Figure 7.

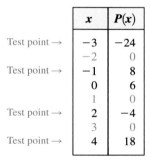

x	$P(x)$
-3	-24
-2	0
-1	8
0	6
1	0
2	-4
3	0
4	18

Test point → -3
Test point → -1
Test point → 2
Test point → 4

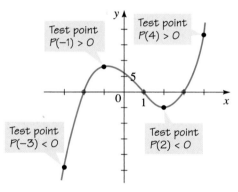

FIGURE 7 $P(x) = (x + 2)(x - 1)(x - 3)$ ▲

▶ **EXAMPLE 5** | Finding Zeros and Graphing a Polynomial Function

Let $P(x) = x^3 - 2x^2 - 3x$.

(a) Find the zeros of P.   **(b)** Sketch the graph of P.

▼ **SOLUTION**

(a) To find the zeros, we factor completely.

$$P(x) = x^3 - 2x^2 - 3x$$
$$= x(x^2 - 2x - 3) \quad \text{Factor } x$$
$$= x(x - 3)(x + 1) \quad \text{Factor quadratic}$$

Thus, the zeros are $x = 0$, $x = 3$, and $x = -1$.

(b) The x-intercepts are $x = 0$, $x = 3$, and $x = -1$. The y-intercept is $P(0) = 0$. We make a table of values of $P(x)$, making sure that we choose test points between (and to the right and left of) successive zeros.

Since P is of odd degree and its leading coefficient is positive, it has the following end behavior:

$$y \to \infty \quad \text{as} \quad x \to \infty \quad \text{and} \quad y \to -\infty \quad \text{as} \quad x \to -\infty$$

We plot the points in the table and connect them by a smooth curve to complete the graph, as shown in Figure 8.

x	$P(x)$
Test point → -2	-10
-1	0
Test point → $-\frac{1}{2}$	$\frac{7}{8}$
0	0
Test point → 1	-4
2	-6
3	0
Test point → 4	20

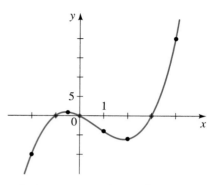

FIGURE 8 $P(x) = x^3 - 2x^2 - 3x$

▶ **EXAMPLE 6** | Finding Zeros and Graphing a Polynomial Function

Let $P(x) = -2x^4 - x^3 + 3x^2$.

(a) Find the zeros of P.   **(b)** Sketch the graph of P.

▼ **SOLUTION**

(a) To find the zeros, we factor completely.

$$P(x) = -2x^4 - x^3 + 3x^2$$
$$= -x^2(2x^2 + x - 3) \qquad \text{Factor } -x^2$$
$$= -x^2(2x + 3)(x - 1) \qquad \text{Factor quadratic}$$

Thus, the zeros are $x = 0$, $x = -\frac{3}{2}$, and $x = 1$.

(b) The x-intercepts are $x = 0$, $x = -\frac{3}{2}$, and $x = 1$. The y-intercept is $P(0) = 0$. We make a table of values of $P(x)$, making sure that we choose test points between (and to the right and left of) successive zeros.

Since P is of even degree and its leading coefficient is negative, it has the following end behavior:

$$y \to -\infty \quad \text{as} \quad x \to \infty \quad \text{and} \quad y \to -\infty \quad \text{as} \quad x \to -\infty$$

We plot the points from the table and connect the points by a smooth curve to complete the graph in Figure 9.

A table of values is most easily calculated by using a programmable calculator or a graphing calculator.

x	$P(x)$
-2	-12
-1.5	0
-1	2
-0.5	0.75
0	0
0.5	0.5
1	0
1.5	-6.75

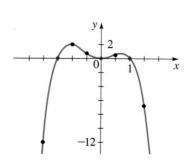

FIGURE 9 $P(x) = -2x^4 - x^3 + 3x^2$

▶ **EXAMPLE 7** | Finding Zeros and Graphing a Polynomial Function

Let $P(x) = x^3 - 2x^2 - 4x + 8$.

(a) Find the zeros of P.

(b) Sketch the graph of P.

▼ SOLUTION

(a) To find the zeros, we factor completely.

$$\begin{aligned}
P(x) &= x^3 - 2x^2 - 4x + 8 \\
&= x^2(x - 2) - 4(x - 2) && \text{Group and factor} \\
&= (x^2 - 4)(x - 2) && \text{Factor } x - 2 \\
&= (x + 2)(x - 2)(x - 2) && \text{Difference of squares} \\
&= (x + 2)(x - 2)^2 && \text{Simplify}
\end{aligned}$$

Thus, the zeros are $x = -2$ and $x = 2$.

(b) The x-intercepts are $x = -2$ and $x = 2$. The y-intercept is $P(0) = 8$. The table gives additional values of $P(x)$.

Since P is of odd degree and its leading coefficient is positive, it has the following end behavior:

$$y \to \infty \quad \text{as} \quad x \to \infty \qquad \text{and} \qquad y \to -\infty \quad \text{as} \quad x \to -\infty$$

We connect the points by a smooth curve to complete the graph in Figure 10.

x	$P(x)$
-3	-25
-2	0
-1	9
0	8
1	3
2	0
3	5

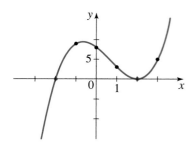

FIGURE 10 $P(x) = x^3 - 2x^2 - 4x + 8$ ▲

■ Shape of the Graph Near a Zero

Although $x = 2$ is a zero of the polynomial in Example 7, the graph does not cross the x-axis at the x-intercept 2. This is because the factor $(x - 2)^2$ corresponding to that zero is raised to an even power, so it doesn't change sign as we test points on either side of 2. In the same way the graph does not cross the x-axis at $x = 0$ in Example 6.

In general, if c is a zero of P and the corresponding factor $x - c$ occurs exactly m times in the factorization of P then we say that c is a **zero of multiplicity** m. By considering test points on either side of the x-intercept c, we conclude that the graph crosses the x-axis at c if the multiplicity m is odd and does not cross the x-axis if m is even. Moreover, it can be shown by using calculus that near $x = c$ the graph has the same general shape as $A(x - c)^m$.

SHAPE OF THE GRAPH NEAR A ZERO OF MULTIPLICITY m

If c is a zero of P of multiplicity m, then the shape of the graph of P near c is as follows.

Multiplicity of c	Shape of the graph of P near the x-intercept c
m odd, $m > 1$	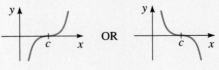
m even, $m > 1$	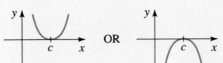

▶ **EXAMPLE 8** | Graphing a Polynomial Function Using Its Zeros

Graph the polynomial $P(x) = x^4(x-2)^3(x+1)^2$.

▼ **SOLUTION**   The zeros of P are -1, 0, and 2 with multiplicities 2, 4, and 3, respectively.

> 0 is a zero of multiplicity 4.   2 is a zero of multiplicity 3.   −1 is a zero of multiplicity 2.

$$P(x) = x^4(x-2)^3(x+1)^2$$

The zero 2 has *odd* multiplicity, so the graph crosses the x-axis at the x-intercept 2. But the zeros 0 and -1 have *even* multiplicity, so the graph does not cross the x-axis at the x-intercepts 0 and -1.

Since P is a polynomial of degree 9 and has positive leading coefficient, it has the following end behavior:

$$y \to \infty \quad \text{as} \quad x \to \infty \quad \text{and} \quad y \to -\infty \quad \text{as} \quad x \to -\infty$$

With this information and a table of values, we sketch the graph in Figure 11.

x	$P(x)$
-1.3	-9.2
-1	0
-0.5	-3.9
0	0
1	-4
2	0
2.3	8.2

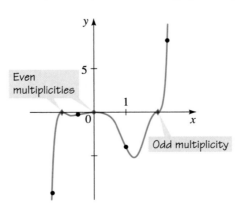

FIGURE 11 $P(x) = x^4(x-2)^3(x+1)^2$

■ Local Maxima and Minima of Polynomials

Recall from Section 3.3 that if the point $(a, f(a))$ is the highest point on the graph of f within some viewing rectangle, then $f(a)$ is a local maximum value of f, and if $(b, f(b))$ is the lowest point on the graph of f within a viewing rectangle, then $f(b)$ is a local minimum value (see Figure 12). We say that such a point $(a, f(a))$ is a **local maximum point** on the graph and that $(b, f(b))$ is a **local minimum point**. The set of all local maximum and minimum points on the graph of a function is called its **local extrema**.

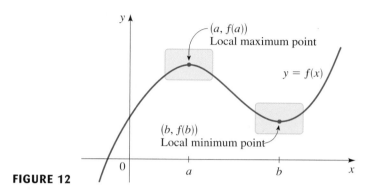

FIGURE 12

For a polynomial function the number of local extrema must be less than the degree, as the following principle indicates. (A proof of this principle requires calculus.)

LOCAL EXTREMA OF POLYNOMIALS

If $P(x) = a_n x^n + a_{n-1} x^{n-1} + \cdots + a_1 x + a_0$ is a polynomial of degree n, then the graph of P has at most $n - 1$ local extrema.

A polynomial of degree n may in fact have less than $n - 1$ local extrema. For example, $P(x) = x^5$ (graphed in Figure 2) has *no* local extrema, even though it is of degree 5. The preceding principle tells us only that a polynomial of degree n can have no more than $n - 1$ local extrema.

▶ **EXAMPLE 9** | The Number of Local Extrema

Determine how many local extrema each polynomial has.

(a) $P_1(x) = x^4 + x^3 - 16x^2 - 4x + 48$

(b) $P_2(x) = x^5 + 3x^4 - 5x^3 - 15x^2 + 4x - 15$

(c) $P_3(x) = 7x^4 + 3x^2 - 10x$

▼ **SOLUTION**   The graphs are shown in Figure 13 on the next page.

(a) P_1 has two local minimum points and one local maximum point, for a total of three local extrema.

(b) P_2 has two local minimum points and two local maximum points, for a total of four local extrema.

(c) P_3 has just one local extremum, a local minimum.

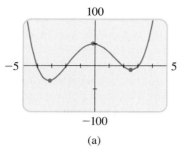

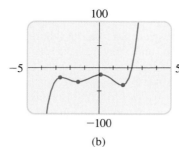

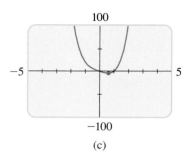

(a)   (b)   (c)

FIGURE 13 $P_1(x) = x^4 + x^3 - 16x^2 - 4x + 48$ $P_2(x) = x^5 + 3x^4 - 5x^3 - 15x^2 + 4x - 15$ $P_3(x) = 7x^4 + 3x^2 - 10x$ ▲

With a graphing calculator we can quickly draw the graphs of many functions at once, on the same viewing screen. This allows us to see how changing a value in the definition of the functions affects the shape of its graph. In the next example we apply this principle to a family of third-degree polynomials.

▶ **EXAMPLE 10** | A Family of Polynomials

Sketch the family of polynomials $P(x) = x^3 - cx^2$ for $c = 0, 1, 2,$ and 3. How does changing the value of c affect the graph?

▼ **SOLUTION**   The polynomials

$$P_0(x) = x^3 \qquad\qquad P_1(x) = x^3 - x^2$$
$$P_2(x) = x^3 - 2x^2 \qquad P_3(x) = x^3 - 3x^2$$

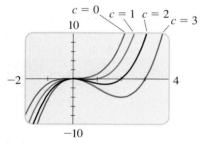

FIGURE 14 A family of polynomials $P(x) = x^3 - cx^2$

are graphed in Figure 14. We see that increasing the value of c causes the graph to develop an increasingly deep "valley" to the right of the y-axis, creating a local maximum at the origin and a local minimum at a point in Quadrant IV. This local minimum moves lower and farther to the right as c increases. To see why this happens, factor $P(x) = x^2(x - c)$. The polynomial P has zeros at 0 and c, and the larger c gets, the farther to the right the minimum between 0 and c will be. ▲

4.3 | Dividing Polynomials

LEARNING OBJECTIVES

After completing this section, you will be able to:

▨ Use long division to divide polynomials
▨ Use synthetic division to divide polynomials
▨ Use the Remainder Theorem to find values of a polynomial
▨ Use the Factor Theorem to factor a polynomial
▨ Find a polynomial with specified zeros

So far in this chapter we have been studying polynomial functions *graphically*. In this section we begin to study polynomials *algebraically*. Most of our work will be concerned with factoring polynomials, and to factor, we need to know how to divide polynomials.

■ Long Division of Polynomials

Dividing polynomials is much like the familiar process of dividing numbers. When we divide 38 by 7, the quotient is 5 and the remainder is 3. We write

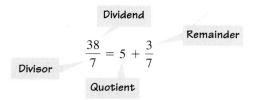

To divide polynomials, we use long division, as follows.

DIVISION ALGORITHM

If $P(x)$ and $D(x)$ are polynomials, with $D(x) \neq 0$, then there exist unique polynomials $Q(x)$ and $R(x)$, where $R(x)$ is either 0 or of degree less than the degree of $D(x)$, such that

$$P(x) = D(x) \cdot Q(x) + R(x)$$

Dividend   Divisor   Quotient   Remainder

The polynomials $P(x)$ and $D(x)$ are called the **dividend** and **divisor**, respectively, $Q(x)$ is the **quotient**, and $R(x)$ is the **remainder**.

To write the division algorithm another way, divide through by $D(x)$:

$$\frac{P(x)}{D(x)} = Q(x) + \frac{R(x)}{D(x)}$$

▶ **EXAMPLE 1** | Long Division of Polynomials

Divide $6x^2 - 26x + 12$ by $x - 4$.

▼ **SOLUTION**   The *dividend* is $6x^2 - 26x + 12$ and the *divisor* is $x - 4$. We begin by arranging them as follows:

$$x - 4 \overline{)6x^2 - 26x + 12}$$

Next we divide the leading term in the dividend by the leading term in the divisor to get the first term of the quotient: $6x^2/x = 6x$. Then we multiply the divisor by $6x$ and subtract the result from the dividend.

$$
\begin{array}{r}
6x \\
x - 4 \overline{)6x^2 - 26x + 12} \\
\underline{6x^2 - 24x} \\
-2x + 12
\end{array}
$$

Divide leading terms: $\dfrac{6x^2}{x} = 6x$

Multiply: $6x(x - 4) = 6x^2 - 24x$

Subtract and "bring down" 12

We repeat the process using the last line $-2x + 12$ as the dividend.

$$
\begin{array}{r}
6x^2 - 2 \\
x - 4 \overline{)6x^2 - 26x + 12} \\
\underline{6x^2 - 24x} \\
-2x + 12 \\
\underline{-2x + 8} \\
4
\end{array}
$$

Divide leading terms: $\dfrac{-2x}{x} = -2$

Multiply: $-2(x - 4) = -2x + 8$

Subtract

The division process ends when the last line is of lesser degree than the divisor. The last line then contains the *remainder*, and the top line contains the *quotient*. The result of the division can be interpreted in either of two ways.

$$\underbrace{\frac{6x^2 - 26x + 12}{x - 4}}_{\substack{\text{Dividend} \\ \text{Divisor}}} = 6x - 2 + \frac{4}{x - 4} \quad \text{Remainder}$$

Dividend · Quotient · Remainder

or

$$6x^2 - 26x + 12 = (x - 4)(6x - 2) + 4 \quad \text{Remainder}$$

Dividend · Divisor · Quotient

▶ **EXAMPLE 2** | Long Division of Polynomials

Let $P(x) = 8x^4 + 6x^2 - 3x + 1$ and $D(x) = 2x^2 - x + 2$. Find polynomials $Q(x)$ and $R(x)$ such that $P(x) = D(x) \cdot Q(x) + R(x)$.

▼ **SOLUTION**   We use long division after first inserting the term $0x^3$ into the dividend to ensure that the columns line up correctly.

$$
\begin{array}{r}
4x^2 + 2x \\
2x^2 - x + 2 \overline{)\, 8x^4 + 0x^3 + 6x^2 - 3x + 1} \\
\underline{8x^4 - 4x^3 + 8x^2} \\
4x^3 - 2x^2 - 3x \\
\underline{4x^3 - 2x^2 + 4x} \\
-7x + 1
\end{array}
$$

Multiply divisor by $4x^2$

Subtract

Multiply divisor by $2x$

Subtract

The process is complete at this point because $-7x + 1$ is of lesser degree than the divisor $2x^2 - x + 2$. From the above long division we see that $Q(x) = 4x^2 + 2x$ and $R(x) = -7x + 1$, so

$$8x^4 + 6x^2 - 3x + 1 = (2x^2 - x + 2)(4x^2 + 2x) + (-7x + 1)$$ ▲

■ Synthetic Division

Synthetic division is a quick method of dividing polynomials; it can be used when the divisor is of the form $x - c$. In synthetic division we write only the essential parts of the long division. Compare the following long and synthetic divisions, in which we divide $2x^3 - 7x^2 + 5$ by $x - 3$. (We'll explain how to perform the synthetic division in Example 3.)

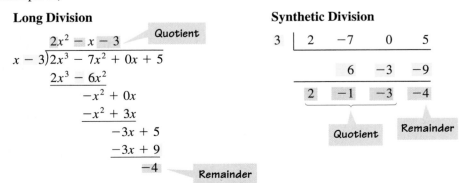

Long Division

$$
\begin{array}{r}
2x^2 - x - 3 \quad \text{Quotient}\\
x - 3 \overline{)\, 2x^3 - 7x^2 + 0x + 5} \\
\underline{2x^3 - 6x^2} \\
-x^2 + 0x \\
\underline{-x^2 + 3x} \\
-3x + 5 \\
\underline{-3x + 9} \\
-4 \quad \text{Remainder}
\end{array}
$$

Synthetic Division

$$
\begin{array}{r|rrrr}
3 & 2 & -7 & 0 & 5 \\
  &   & 6 & -3 & -9 \\
\hline
  & 2 & -1 & -3 & -4
\end{array}
$$

Quotient   Remainder

Note that in synthetic division we abbreviate $2x^3 - 7x^2 + 5$ by writing only the coefficients: 2   −7   0   5, and instead of $x - 3$, we simply write 3. (Writing 3 instead of −3 allows us to add instead of subtract, but this changes the sign of all the numbers that appear in the gold boxes.)

The next example shows how synthetic division is performed.

▶ **EXAMPLE 3** | Synthetic Division

Use synthetic division to divide $2x^3 - 7x^2 + 5$ by $x - 3$.

▼ **SOLUTION**   We begin by writing the appropriate coefficients to represent the divisor and the dividend.

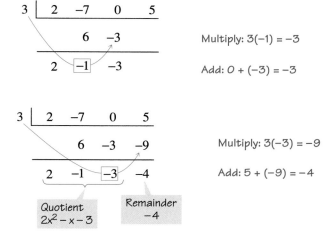

We bring down the 2, multiply $3 \cdot 2 = 6$, and write the result in the middle row. Then we add.

$$
\begin{array}{c|cccc}
3 & 2 & -7 & 0 & 5 \\
 & & 6 & & \\
\hline
 & \boxed{2} & -1 & &
\end{array}
$$

Multiply: $3 \cdot 2 = 6$

Add: $-7 + 6 = -1$

We repeat this process of multiplying and then adding until the table is complete.

$$
\begin{array}{c|cccc}
3 & 2 & -7 & 0 & 5 \\
 & & 6 & -3 & \\
\hline
 & 2 & \boxed{-1} & -3 &
\end{array}
$$

Multiply: $3(-1) = -3$

Add: $0 + (-3) = -3$

$$
\begin{array}{c|cccc}
3 & 2 & -7 & 0 & 5 \\
 & & 6 & -3 & -9 \\
\hline
 & 2 & -1 & \boxed{-3} & -4
\end{array}
$$

Multiply: $3(-3) = -9$

Add: $5 + (-9) = -4$

Quotient
$2x^2 - x - 3$

Remainder
-4

From the last line of the synthetic division we see that the quotient is $2x^2 - x - 3$ and the remainder is -4. Thus

$$2x^3 - 7x^2 + 5 = (x - 3)(2x^2 - x - 3) - 4 \qquad ▲$$

The Remainder and Factor Theorems

The next theorem shows how synthetic division can be used to evaluate polynomials easily.

REMAINDER THEOREM

If the polynomial $P(x)$ is divided by $x - c$, then the remainder is the value $P(c)$.

▼ **PROOF**   If the divisor in the Division Algorithm is of the form $x - c$ for some real number c, then the remainder must be a constant (since the degree of the remainder is less than the degree of the divisor). If we call this constant r, then

$$P(x) = (x - c) \cdot Q(x) + r$$

Replacing x by c in this equation, we get $P(c) = (c - c) \cdot Q(x) + r = 0 + r = r$, that is, $P(c)$ is the remainder r.   ▲

▶ **EXAMPLE 4** | Using the Remainder Theorem to Find the Value of a Polynomial

Let $P(x) = 3x^5 + 5x^4 - 4x^3 + 7x + 3$.

(a) Find the quotient and remainder when $P(x)$ is divided by $x + 2$.

(b) Use the Remainder Theorem to find $P(-2)$.

▼ **SOLUTION**

(a) Since $x + 2 = x - (-2)$, the synthetic division for this problem takes the following form.

$$
\begin{array}{r|rrrrrr}
-2 & 3 & 5 & -4 & 0 & 7 & 3 \\
   &   & -6 & 2 & 4 & -8 & 2 \\
\hline
   & 3 & -1 & -2 & 4 & -1 & 5
\end{array}
$$

Remainder is 5, so $P(-2) = 5$.

The quotient is $3x^4 - x^3 - 2x^2 + 4x - 1$, and the remainder is 5.

(b) By the Remainder Theorem, $P(-2)$ is the remainder when $P(x)$ is divided by $x - (-2) = x + 2$. From part (a) the remainder is 5, so $P(-2) = 5$.   ▲

The next theorem says that *zeros* of polynomials correspond to *factors*; we used this fact in Section 4.2 to graph polynomials.

FACTOR THEOREM

c is a zero of P if and only if $x - c$ is a factor of $P(x)$.

▼ **PROOF**   If $P(x)$ factors as $P(x) = (x - c) \cdot Q(x)$, then

$$P(c) = (c - c) \cdot Q(c) = 0 \cdot Q(c) = 0$$

Conversely, if $P(c) = 0$, then by the Remainder Theorem

$$P(x) = (x - c) \cdot Q(x) + 0 = (x - c) \cdot Q(x)$$

so $x - c$ is a factor of $P(x)$.   ▲

▶ **EXAMPLE 5** | Factoring a Polynomial Using the Factor Theorem

Let $P(x) = x^3 - 7x + 6$. Show that $P(1) = 0$, and use this fact to factor $P(x)$ completely.

$$
\begin{array}{r|rrrr}
1 & 1 & 0 & -7 & 6 \\
  &   & 1 & 1 & -6 \\
\hline
  & 1 & 1 & -6 & 0
\end{array}
$$

▼ **SOLUTION**   Substituting, we see that $P(1) = 1^3 - 7 \cdot 1 + 6 = 0$. By the Factor Theorem this means that $x - 1$ is a factor of $P(x)$. Using synthetic or long division (shown in the margin), we see that

$$\frac{x^2 + x - 6}{x - 1 \overline{)\smash{x^3 + 0x^2 - 7x + 6}}}$$
$$\underline{x^3 - x^2}$$
$$x^2 - 7x$$
$$\underline{x^2 - x}$$
$$-6x + 6$$
$$\underline{-6x + 6}$$
$$0$$

$$P(x) = x^3 - 7x + 6$$
$$= (x - 1)(x^2 + x - 6) \qquad \text{See margin}$$
$$= (x - 1)(x - 2)(x + 3) \qquad \text{Factor quadratic } x^2 + x - 6 \qquad \blacktriangle$$

▶ **EXAMPLE 6** | Finding a Polynomial with Specified Zeros

Find a polynomial of degree 4 that has zeros $-3, 0, 1$, and 5.

▼ **SOLUTION** By the Factor Theorem $x - (-3)$, $x - 0$, $x - 1$, and $x - 5$ must all be factors of the desired polynomial, so let

$$P(x) = (x + 3)(x - 0)(x - 1)(x - 5)$$
$$= x^4 - 3x^3 - 13x^2 + 15x$$

Since $P(x)$ is of degree 4, it is a solution of the problem. Any other solution of the problem must be a constant multiple of $P(x)$, since only multiplication by a constant does not change the degree.   ▲

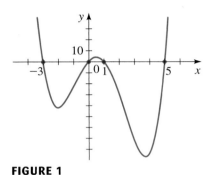

FIGURE 1
$P(x) = (x + 3)x(x - 1)(x - 5)$ has zeros $-3, 0, 1$, and 5.

The polynomial P of Example 6 is graphed in Figure 1. Note that the zeros of P correspond to the x-intercepts of the graph.

4.4 | Real Zeros of Polynomials

LEARNING OBJECTIVES
After completing this section, you will be able to:

▪ Use the Rational Zeros Theorem to find the rational zeros of a polynomial
▪ Use Descartes' Rule of Signs to determine the number of positive and negative zeros of a polynomial
▪ Use the Upper and Lower Bounds Theorem to find upper and lower bounds for the zeros of a polynomial
▪ Use algebra and graphing devices to solve polynomial equations

The Factor Theorem tells us that finding the zeros of a polynomial is really the same thing as factoring it into linear factors. In this section we study some algebraic methods that help us to find the real zeros of a polynomial and thereby factor the polynomial. We begin with the *rational* zeros of a polynomial.

■ Rational Zeros of Polynomials

To help us understand the next theorem, let's consider the polynomial

$$P(x) = (x - 2)(x - 3)(x + 4) \qquad \text{Factored form}$$
$$= x^3 - x^2 - 14x + 24 \qquad \text{Expanded form}$$

From the factored form we see that the zeros of P are 2, 3, and -4. When the polynomial is expanded, the constant 24 is obtained by multiplying $(-2) \times (-3) \times 4$. This means that the zeros of the polynomial are all factors of the constant term. The following generalizes this observation.

RATIONAL ZEROS THEOREM

If the polynomial $P(x) = a_n x^n + a_{n-1} x^{n-1} + \cdots + a_1 x + a_0$ has integer coefficients, then every rational zero of P is of the form

$$\frac{p}{q}$$

where p is a factor of the constant coefficient a_0
and q is a factor of the leading coefficient a_n.

▼ **PROOF**   If p/q is a rational zero, in lowest terms, of the polynomial P, then we have

$$a_n \left(\frac{p}{q} \right)^n + a_{n-1} \left(\frac{p}{q} \right)^{n-1} + \cdots + a_1 \left(\frac{p}{q} \right) + a_0 = 0$$

$$a_n p^n + a_{n-1} p^{n-1} q + \cdots + a_1 p q^{n-1} + a_0 q^n = 0 \qquad \text{Multiply by } q^n$$

$$p(a_n p^{n-1} + a_{n-1} p^{n-2} q + \cdots + a_1 q^{n-1}) = -a_0 q^n \qquad \begin{array}{l}\text{Subtract } a_0 q^n \\ \text{and factor LHS}\end{array}$$

Now p is a factor of the left side, so it must be a factor of the right side as well. Since p/q is in lowest terms, p and q have no factor in common, and so p must be a factor of a_0. A similar proof shows that q is a factor of a_n.   ▲

We see from the Rational Zeros Theorem that if the leading coefficient is 1 or -1, then the rational zeros must be factors of the constant term.

▶ **EXAMPLE 1** | Using the Rational Zeros Theorem

Find the rational zeros of $P(x) = x^3 - 3x + 2$.

▼ **SOLUTION**   Since the leading coefficient is 1, any rational zero must be a divisor of the constant term 2. So the possible rational zeros are ± 1 and ± 2. We test each of these possibilities.

$$P(1) = (1)^3 - 3(1) + 2 = 0$$

$$P(-1) = (-1)^3 - 3(-1) + 2 = 4$$

$$P(2) = (2)^3 - 3(2) + 2 = 4$$

$$P(-2) = (-2)^3 - 3(-2) + 2 = 0$$

The rational zeros of P are 1 and -2.   ▲

The following box explains how we use the Rational Zeros Theorem with synthetic division to factor a polynomial.

FINDING THE RATIONAL ZEROS OF A POLYNOMIAL

1. **List Possible Zeros.**   List all possible rational zeros, using the Rational Zeros Theorem.
2. **Divide.**   Use synthetic division to evaluate the polynomial at each of the candidates for the rational zeros that you found in Step 1. When the remainder is 0, note the quotient you have obtained.
3. **Repeat.**   Repeat Steps 1 and 2 for the quotient. Stop when you reach a quotient that is quadratic or factors easily, and use the quadratic formula or factor to find the remaining zeros.

▶ **EXAMPLE 2** | Finding Rational Zeros

Factor the polynomial $P(x) = 2x^3 + x^2 - 13x + 6$, and find all its zeros.

▼ **SOLUTION**   By the Rational Zeros Theorem the rational zeros of P are of the form

$$\text{possible rational zero of } P = \frac{\text{factor of constant term}}{\text{factor of leading coefficient}}$$

The constant term is 6 and the leading coefficient is 2, so

$$\text{possible rational zero of } P = \frac{\text{factor of 6}}{\text{factor of 2}}$$

The factors of 6 are $\pm 1, \pm 2, \pm 3, \pm 6$ and the factors of 2 are $\pm 1, \pm 2$. Thus, the possible rational zeros of P are

$$\pm\frac{1}{1}, \quad \pm\frac{2}{1}, \quad \pm\frac{3}{1}, \quad \pm\frac{6}{1}, \quad \pm\frac{1}{2}, \quad \pm\frac{2}{2}, \quad \pm\frac{3}{2}, \quad \pm\frac{6}{2}$$

Simplifying the fractions and eliminating duplicates, we get the following list of possible rational zeros:

$$\pm 1, \quad \pm 2, \quad \pm 3, \quad \pm 6, \quad \pm\frac{1}{2}, \quad \pm\frac{3}{2}$$

To check which of these *possible* zeros actually *are* zeros, we need to evaluate P at each of these numbers. An efficient way to do this is to use synthetic division.

Test whether 1 is a zero

```
1 | 2    1   -13    6
  |      2    3   -10
  +------------------
    2    3   -10   -4
```

Remainder is not 0, so 1 is not a zero.

Test whether 2 is a zero

```
2 | 2    1   -13    6
  |      4    10   -6
  +------------------
    2    5   -3     0
```

Remainder is 0, so 2 is a zero.

From the last synthetic division we see that 2 is a zero of P and that P factors as

$$
\begin{aligned}
P(x) &= 2x^3 + x^2 - 13x + 6 \\
&= (x - 2)(2x^2 + 5x - 3) && \text{From synthetic division} \\
&= (x - 2)(2x - 1)(x + 3) && \text{Factor } 2x^2 + 5x - 3
\end{aligned}
$$

From the factored form we see that the zeros of P are 2, $\frac{1}{2}$, and –3. ▲

▶ **EXAMPLE 3** | Using the Rational Zeros Theorem and the Quadratic Formula

Let $P(x) = x^4 - 5x^3 - 5x^2 + 23x + 10$.

(a) Find the zeros of P.

(b) Sketch the graph of P.

▼ **SOLUTION**

(a) The leading coefficient of P is 1, so all the rational zeros are integers: They are divisors of the constant term 10. Thus, the possible candidates are

$$\pm 1, \quad \pm 2, \quad \pm 5, \quad \pm 10$$

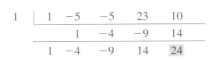

1	1	−5	−5	23	10
		1	−4	−9	14
	1	−4	−9	14	**24**

2	1	−5	−5	23	10
		2	−6	−22	2
	1	−3	−11	1	**12**

5	1	−5	−5	23	10
		5	0	−25	−10
	1	0	−5	−2	**0**

Using synthetic division (see the margin), we find that 1 and 2 are not zeros but that 5 is a zero and that P factors as

$$x^4 - 5x^3 - 5x^2 + 23x + 10 = (x - 5)(x^3 - 5x - 2)$$

We now try to factor the quotient $x^3 - 5x - 2$. Its possible zeros are the divisors of −2, namely,

$$\pm 1, \quad \pm 2$$

Since we already know that 1 and 2 are not zeros of the original polynomial P, we don't need to try them again. Checking the remaining candidates −1 and −2, we see that −2 is a zero (see the margin), and P factors as

$$
\begin{aligned}
x^4 - 5x^3 - 5x^2 + 23x + 10 &= (x - 5)(x^3 - 5x - 2) \\
&= (x - 5)(x + 2)(x^2 - 2x - 1)
\end{aligned}
$$

−2	1	0	−5	−2
		−2	4	2
	1	−2	−1	**0**

Now we use the quadratic formula to obtain the two remaining zeros of P:

$$x = \frac{2 \pm \sqrt{(-2)^2 - 4(1)(-1)}}{2} = 1 \pm \sqrt{2}$$

The zeros of P are 5, −2, $1 + \sqrt{2}$, and $1 - \sqrt{2}$.

(b) Now that we know the zeros of P, we can use the methods of Section 4.2 to sketch the graph. If we want to use a graphing calculator instead, knowing the zeros allows us to choose an appropriate viewing rectangle—one that is wide enough to contain all the x-intercepts of P. Numerical approximations to the zeros of P are

$$5, \quad -2, \quad 2.4, \quad \text{and} \quad -0.4$$

So in this case we choose the rectangle $[-3, 6]$ by $[-50, 50]$ and draw the graph shown in Figure 1. ▲

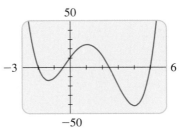

FIGURE 1

$P(x) = x^4 - 5x^3 - 5x^2 + 23x + 10$

Descartes' Rule of Signs and Upper and Lower Bounds for Roots

In some cases, the following rule—discovered by the French philosopher and mathematician René Descartes around 1637—is helpful in eliminating candidates from lengthy lists of possible rational roots. To describe this rule, we need the concept of *variation in sign*. If $P(x)$ is a polynomial with real coefficients, written with descending powers of x (and omitting powers with coefficient 0), then a **variation in sign** occurs whenever adjacent coefficients have opposite signs. For example,

$$P(x) = 5x^7 - 3x^5 - x^4 + 2x^2 + x - 3$$

has three variations in sign.

Polynomial	Variations in sign
$x^2 + 4x + 1$	0
$2x^3 + x - 6$	1
$x^4 - 3x^2 - x + 4$	2

DESCARTES' RULE OF SIGNS

Let P be a polynomial with real coefficients.

1. The number of positive real zeros of $P(x)$ either is equal to the number of variations in sign in $P(x)$ or is less than that by an even whole number.

2. The number of negative real zeros of $P(x)$ either is equal to the number of variations in sign in $P(-x)$ or is less than that by an even whole number.

▶ **EXAMPLE 4** | Using Descartes' Rule

Use Descartes' Rule of Signs to determine the possible number of positive and negative real zeros of the polynomial

$$P(x) = 3x^6 + 4x^5 + 3x^3 - x - 3$$

▼ **SOLUTION**   The polynomial has one variation in sign, so it has one positive zero. Now

$$P(-x) = 3(-x)^6 + 4(-x)^5 + 3(-x)^3 - (-x) - 3$$
$$= 3x^6 - 4x^5 - 3x^3 + x - 3$$

So $P(-x)$ has three variations in sign. Thus, $P(x)$ has either three or one negative zero(s), making a total of either two or four real zeros.   ▲

We say that a is a **lower bound** and b is an **upper bound** for the zeros of a polynomial if every real zero c of the polynomial satisfies $a \leq c \leq b$. The next theorem helps us to find such bounds for the zeros of a polynomial.

THE UPPER AND LOWER BOUNDS THEOREM

Let P be a polynomial with real coefficients.

1. If we divide $P(x)$ by $x - b$ (with $b > 0$) using synthetic division and if the row that contains the quotient and remainder has no negative entry, then b is an upper bound for the real zeros of P.

2. If we divide $P(x)$ by $x - a$ (with $a < 0$) using synthetic division and if the row that contains the quotient and remainder has entries that are alternately nonpositive and nonnegative, then a is a lower bound for the real zeros of P.

The phrase "alternately nonpositive and nonnegative" simply means that the signs of the numbers alternate, with 0 considered to be positive or negative as required.

▶ **EXAMPLE 5** | Upper and Lower Bounds for Zeros of a Polynomial

Show that all the real zeros of the polynomial $P(x) = x^4 - 3x^2 + 2x - 5$ lie between -3 and 2.

▼ **SOLUTION**   We divide $P(x)$ by $x - 2$ and $x + 3$ using synthetic division.

2	1	0	-3	2	-5
		2	4	2	8
	1	2	1	4	3

All entries positive

-3	1	0	-3	2	-5
		-3	9	-18	48
	1	-3	6	-16	43

Entries alternate in sign.

By the Upper and Lower Bounds Theorem, -3 is a lower bound and 2 is an upper bound for the zeros. Since neither -3 nor 2 is a zero (the remainders are not 0 in the division table), all the real zeros lie between these numbers.   ▲

▶ **EXAMPLE 6** | Factoring a Fifth-Degree Polynomial

Factor completely the polynomial

$$P(x) = 2x^5 + 5x^4 - 8x^3 - 14x^2 + 6x + 9$$

▼ **SOLUTION**   The possible rational zeros of P are $\pm\frac{1}{2}$, ±1, $\pm\frac{3}{2}$, ±3, $\pm\frac{9}{2}$, and ±9. We check the positive candidates first, beginning with the smallest.

$\frac{1}{2}$	2	5	-8	-14	6	9
		1	3	$-\frac{5}{2}$	$-\frac{33}{4}$	$-\frac{9}{8}$
	2	6	-5	$-\frac{33}{2}$	$-\frac{9}{4}$	$\frac{63}{8}$

$\frac{1}{2}$ is not a zero

1	2	5	-8	-14	6	9
		2	7	-1	-15	-9
	2	7	-1	-15	-9	0

$P(1) = 0$

So 1 is a zero, and $P(x) = (x - 1)(2x^4 + 7x^3 - x^2 - 15x - 9)$. We continue by factoring the quotient. We still have the same list of possible zeros except that $\frac{1}{2}$ has been eliminated.

1	2	7	-1	-15	-9
		2	9	8	-7
	2	9	8	-7	-16

1 is not a zero.

$\frac{3}{2}$	2	7	-1	-15	-9
		3	15	21	9
	2	10	14	6	0

$P\left(\frac{3}{2}\right) = 0$, all entries nonnegative

We see that $\frac{3}{2}$ is both a zero and an upper bound for the zeros of $P(x)$, so we do not need to check any further for positive zeros, because all the remaining candidates are greater than $\frac{3}{2}$.

$$P(x) = (x - 1)(x - \tfrac{3}{2})(2x^3 + 10x^2 + 14x + 6)$$ From synthetic division

$$= (x - 1)(2x - 3)(x^3 + 5x^2 + 7x + 3)$$ Factor 2 from last factor, multiply into second factor

By Descartes' Rule of Signs, $x^3 + 5x^2 + 7x + 3$ has no positive zero, so its only possible rational zeros are -1 and -3.

$$
\begin{array}{r|rrrr}
-1 & 1 & 5 & 7 & 3 \\
   &   & -1 & -4 & -3 \\
\hline
   & 1 & 4 & 3 & 0
\end{array}
$$

$P(-1) = 0$

Therefore

$$P(x) = (x - 1)(2x - 3)(x + 1)(x^2 + 4x + 3)$$ From synthetic division

$$= (x - 1)(2x - 3)(x + 1)^2(x + 3)$$ Factor quadratic

This means that the zeros of P are $1, \frac{3}{2}, -1$, and -3. The graph of the polynomial is shown in Figure 2. ▲

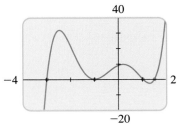

FIGURE 2
$P(x) = 2x^5 + 5x^4 - 8x^3 - 14x^2 + 6x + 9$
$= (x - 1)(2x - 3)(x + 1)^2(x + 3)$

■ Using Algebra and Graphing Devices to Solve Polynomial Equations

In Section 2.3 we used graphing devices to solve equations graphically. We can now use the algebraic techniques that we've learned to select an appropriate viewing rectangle when solving a polynomial equation graphically.

▶ **EXAMPLE 7** | Solving a Fourth-Degree Equation Graphically

Find all real solutions of the following equation, correct to the nearest tenth.

$$3x^4 + 4x^3 - 7x^2 - 2x - 3 = 0$$

▼ **SOLUTION** To solve the equation graphically, we graph

$$P(x) = 3x^4 + 4x^3 - 7x^2 - 2x - 3$$

We use the Upper and Lower Bounds Theorem to see where the solutions can be found.

First we use the Upper and Lower Bounds Theorem to find two numbers between which all the solutions must lie. This allows us to choose a viewing rectangle that is certain to contain all the x-intercepts of P. We use synthetic division and proceed by trial and error.

To find an upper bound, we try the whole numbers, $1, 2, 3, \ldots$, as potential candidates. We see that 2 is an upper bound for the solutions.

$$
\begin{array}{r|rrrrr}
2 & 3 & 4 & -7 & -2 & -3 \\
  &   & 6 & 20 & 26 & 48 \\
\hline
  & 3 & 10 & 13 & 24 & 45
\end{array}
$$

All positive

Now we look for a lower bound, trying the numbers -1, -2, and -3 as potential candidates. We see that -3 is a lower bound for the solutions.

$$
\begin{array}{r|rrrrr}
-3 & 3 & 4 & -7 & -2 & -3 \\
   &   & -9 & 15 & -24 & 78 \\
\hline
   & 3 & -5 & 8 & -26 & 75
\end{array}
$$

Entries alternate in sign.

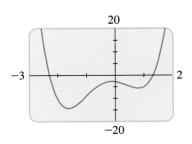

FIGURE 3
$y = 3x^4 + 4x^3 - 7x^2 - 2x - 3$

Thus, all the solutions lie between -3 and 2. So the viewing rectangle $[-3, 2]$ by $[-20, 20]$ contains all the x-intercepts of P. The graph in Figure 3 has two x-intercepts, one between -3 and -2 and the other between 1 and 2. Zooming in, we find that the solutions of the equation, to the nearest tenth, are -2.3 and 1.3. ▲

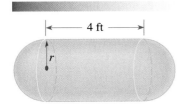

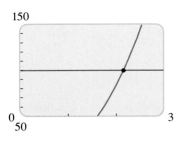

FIGURE 4

Volume of a cylinder: $V = \pi r^2 h$

Volume of a sphere: $V = \frac{4}{3}\pi r^3$

FIGURE 5

$y = \frac{4}{3}\pi x^3 + 4\pi x^2$ and $y = 100$

▶ **EXAMPLE 8** | Determining the Size of a Fuel Tank

A fuel tank consists of a cylindrical center section that is 4 ft long and two hemispherical end sections, as shown in Figure 4. If the tank has a volume of 100 ft³, what is the radius r shown in the figure, correct to the nearest hundredth of a foot?

▼ **SOLUTION**   Using the volume formula, we see that the volume of the cylindrical section of the tank is

$$\pi \cdot r^2 \cdot 4$$

The two hemispherical parts together form a complete sphere whose volume is

$$\frac{4}{3}\pi r^3$$

Because the total volume of the tank is 100 ft³, we get the following equation:

$$\frac{4}{3}\pi r^3 + 4\pi r^2 = 100$$

A negative solution for r would be meaningless in this physical situation, and by substitution we can verify that $r = 3$ leads to a tank that is over 226 ft³ in volume, much larger than the required 100 ft³. Thus, we know the correct radius lies somewhere between 0 and 3 ft, so we use a viewing rectangle of $[0, 3]$ by $[50, 150]$ to graph the function $y = \frac{4}{3}\pi x^3 + 4\pi x^2$, as shown in Figure 5. Since we want the value of this function to be 100, we also graph the horizontal line $y = 100$ in the same viewing rectangle. The correct radius will be the x-coordinate of the point of intersection of the curve and the line. Using the cursor and zooming in, we see that at the point of intersection $x \approx 2.15$, correct to two decimal places. Thus, the tank has a radius of about 2.15 ft.   ▲

Note that we also could have solved the equation in Example 8 by first writing it as

$$\frac{4}{3}\pi r^3 + 4\pi r^2 - 100 = 0$$

and then finding the x-intercept of the function $y = \frac{4}{3}\pi x^3 + 4\pi x^2 - 100$.

4.5 | Complex Zeros and the Fundamental Theorem of Algebra

LEARNING OBJECTIVES

After completing this section, you will be able to:

■ State the Fundamental Theorem of Algebra
■ Factor a polynomial completely (into linear factors) over the complex numbers
■ Determine the multiplicity of a zero of a polynomial
■ Use the Conjugate Roots Theorem to find polynomials with specified zeros
■ Factor a polynomial completely (into linear and quadratic factors) over the real numbers

We have already seen that an nth-degree polynomial can have at most n real zeros. In the complex number system an nth-degree polynomial has exactly n zeros and so can be factored into exactly n linear factors. This fact is a consequence of the Fundamental Theorem of Algebra, which was proved by the German mathematician C. F. Gauss in 1799.

The Fundamental Theorem of Algebra and Complete Factorization

The following theorem is the basis for much of our work in factoring polynomials and solving polynomial equations.

FUNDAMENTAL THEOREM OF ALGEBRA

Every polynomial

$$P(x) = a_n x^n + a_{n-1} x^{n-1} + \cdots + a_1 x + a_0 \qquad (n \geq 1, \ a_n \neq 0)$$

with complex coefficients has at least one complex zero.

Because any real number is also a complex number, the theorem applies to polynomials with real coefficients as well.

The Fundamental Theorem of Algebra and the Factor Theorem together show that a polynomial can be factored completely into linear factors, as we now prove.

COMPLETE FACTORIZATION THEOREM

If $P(x)$ is a polynomial of degree $n \geq 1$, then there exist complex numbers $a, c_1, c_2, \ldots, c_n$ (with $a \neq 0$) such that

$$P(x) = a(x - c_1)(x - c_2) \cdots (x - c_n)$$

▼ **PROOF** By the Fundamental Theorem of Algebra, P has at least one zero. Let's call it c_1. By the Factor Theorem (see page 208), $P(x)$ can be factored as

$$P(x) = (x - c_1) \cdot Q_1(x)$$

where $Q_1(x)$ is of degree $n - 1$. Applying the Fundamental Theorem to the quotient $Q_1(x)$ gives us the factorization

$$P(x) = (x - c_1) \cdot (x - c_2) \cdot Q_2(x)$$

where $Q_2(x)$ is of degree $n - 2$ and c_2 is a zero of $Q_1(x)$. Continuing this process for n steps, we get a final quotient $Q_n(x)$ of degree 0, a nonzero constant that we will call a. This means that P has been factored as

$$P(x) = a(x - c_1)(x - c_2) \cdots (x - c_n) \qquad \blacktriangle$$

To actually find the complex zeros of an nth-degree polynomial, we usually first factor as much as possible, then use the quadratic formula on parts that we can't factor further.

▶ **EXAMPLE 1** | Factoring a Polynomial Completely

Let $P(x) = x^3 - 3x^2 + x - 3$.

(a) Find all the zeros of P.

(b) Find the complete factorization of P.

▼ **SOLUTION**

(a) We first factor P as follows.

$$P(x) = x^3 - 3x^2 + x - 3 \qquad \text{Given}$$
$$= x^2(x - 3) + (x - 3) \qquad \text{Group terms}$$
$$= (x - 3)(x^2 + 1) \qquad \text{Factor } x - 3$$

We find the zeros of P by setting each factor equal to 0:

$$P(x) = (x - 3)(x^2 + 1)$$

This factor is 0 when x = 3.   This factor is 0 when x = i or −i.

Setting $x - 3 = 0$, we see that $x = 3$ is a zero. Setting $x^2 + 1 = 0$, we get $x^2 = -1$, so $x = \pm i$. So the zeros of P are 3, i, and $-i$.

(b) Since the zeros are 3, i, and $-i$, by the Complete Factorization Theorem P factors as

$$P(x) = (x - 3)(x - i)[x - (-i)]$$
$$= (x - 3)(x - i)(x + i)$$

▲

▶ **EXAMPLE 2** | Factoring a Polynomial Completely

Let $P(x) = x^3 - 2x + 4$.

(a) Find all the zeros of P.

(b) Find the complete factorization of P.

▼ **SOLUTION**

(a) The possible rational zeros are the factors of 4, which are ± 1, ± 2, ± 4. Using synthetic division (see the margin), we find that -2 is a zero, and the polynomial factors as

$$P(x) = (x + 2)(x^2 - 2x + 2)$$

This factor is 0 when x = −2.   Use the quadratic formula to find when this factor is 0.

To find the zeros, we set each factor equal to 0. Of course, $x + 2 = 0$ means that $x = -2$. We use the quadratic formula to find when the other factor is 0.

$$x^2 - 2x + 2 = 0 \qquad \text{Set factor equal to 0}$$
$$x = \frac{2 \pm \sqrt{4 - 8}}{2} \qquad \text{Quadratic Formula}$$
$$x = \frac{2 \pm 2i}{2} \qquad \text{Take square root}$$
$$x = 1 \pm i \qquad \text{Simplify}$$

So the zeros of P are -2, $1 + i$, and $1 - i$.

(b) Since the zeros are -2, $1 + i$, and $1 - i$, by the Complete Factorization Theorem P factors as

$$P(x) = [x - (-2)][x - (1 + i)][x - (1 - i)]$$
$$= (x + 2)(x - 1 - i)(x - 1 + i)$$

▲

Zeros and Their Multiplicities

In the Complete Factorization Theorem the numbers $c_1, c_2, \ldots, c_n$ are the zeros of P. These zeros need not all be different. If the factor $x - c$ appears k times in the complete factorization of $P(x)$, then we say that c is a zero of **multiplicity k** (see pages 201–202). For example, the polynomial

$$P(x) = (x - 1)^3(x + 2)^2(x + 3)^5$$

has the following zeros:

$$1 \,(\text{multiplicity } 3), \qquad -2 \,(\text{multiplicity } 2), \qquad -3 \,(\text{multiplicity } 5)$$

The polynomial P has the same number of zeros as its degree: It has degree 10 and has 10 zeros, provided we count multiplicities. This is true for all polynomials, as we prove in the following theorem.

ZEROS THEOREM

Every polynomial of degree $n \geq 1$ has exactly n zeros, provided that a zero of multiplicity k is counted k times.

▼ **PROOF**   Let P be a polynomial of degree n. By the Complete Factorization Theorem

$$P(x) = a(x - c_1)(x - c_2) \cdots (x - c_n)$$

Now suppose that c is a zero of P other than $c_1, c_2, \ldots, c_n$. Then

$$P(c) = a(c - c_1)(c - c_2) \cdots (c - c_n) = 0$$

Thus, by the Zero-Product Property one of the factors $c - c_i$ must be 0, so $c = c_i$ for some i. It follows that P has exactly the n zeros $c_1, c_2, \ldots, c_n$.   ▲

▶ **EXAMPLE 3** | Factoring a Polynomial with Complex Zeros

Find the complete factorization and all five zeros of the polynomial

$$P(x) = 3x^5 + 24x^3 + 48x$$

▼ **SOLUTION**   Since $3x$ is a common factor, we have

$$P(x) = 3x(x^4 + 8x^2 + 16)$$
$$= 3x(x^2 + 4)^2$$

> This factor is 0 when $x = 0$.   This factor is 0 when $x = 2i$ or $x = -2i$.

To factor $x^2 + 4$, note that $2i$ and $-2i$ are zeros of this polynomial. Thus, $x^2 + 4 = (x - 2i)(x + 2i)$, so

$$P(x) = 3x[(x - 2i)(x + 2i)]^2$$
$$= 3x(x - 2i)^2(x + 2i)^2$$

> 0 is a zero of multiplicity 1. $2i$ is a zero of multiplicity 2. $-2i$ is a zero of multiplicity 2.

The zeros of P are 0, $2i$, and $-2i$. Since the factors $x - 2i$ and $x + 2i$ each occur twice in the complete factorization of P, the zeros $2i$ and $-2i$ are of multiplicity 2 (or *double* zeros). Thus, we have found all five zeros. ▲

The following table gives further examples of polynomials with their complete factorizations and zeros.

Degree	Polynomial	Zero(s)	Number of zeros
1	$P(x) = x - 4$	4	1
2	$P(x) = x^2 - 10x + 25$ $= (x - 5)(x - 5)$	5 (multiplicity 2)	2
3	$P(x) = x^3 + x$ $= x(x - i)(x + i)$	$0, i, -i$	3
4	$P(x) = x^4 + 18x^2 + 81$ $= (x - 3i)^2(x + 3i)^2$	$3i$ (multiplicity 2), $-3i$ (multiplicity 2)	4
5	$P(x) = x^5 - 2x^4 + x^3$ $= x^3(x - 1)^2$	0 (multiplicity 3), 1 (multiplicity 2)	5

▶ **EXAMPLE 4** | Finding Polynomials with Specified Zeros

(a) Find a polynomial $P(x)$ of degree 4, with zeros i, $-i$, 2, and -2, and with $P(3) = 25$.

(b) Find a polynomial $Q(x)$ of degree 4, with zeros -2 and 0, where -2 is a zero of multiplicity 3.

▼ **SOLUTION**

(a) The required polynomial has the form

$$P(x) = a(x - i)(x - (-i))(x - 2)(x - (-2))$$

$$= a(x^2 + 1)(x^2 - 4) \qquad \text{Difference of squares}$$

$$= a(x^4 - 3x^2 - 4) \qquad \text{Multiply}$$

We know that $P(3) = a(3^4 - 3 \cdot 3^2 - 4) = 50a = 25$, so $a = \frac{1}{2}$. Thus,

$$P(x) = \tfrac{1}{2}x^4 - \tfrac{3}{2}x^2 - 2$$

(b) We require

$$Q(x) = a[x - (-2)]^3(x - 0)$$

$$= a(x + 2)^3 x$$

$$= a(x^3 + 6x^2 + 12x + 8)x \qquad \text{Special Product Formula 4 (Section P.5)}$$

$$= a(x^4 + 6x^3 + 12x^2 + 8x)$$

Since we are given no information about Q other than its zeros and their multiplicity, we can choose any number for a. If we use $a = 1$, we get

$$Q(x) = x^4 + 6x^3 + 12x^2 + 8x$$

▲

▶ **EXAMPLE 5** | Finding All the Zeros of a Polynomial

Find all four zeros of $P(x) = 3x^4 - 2x^3 - x^2 - 12x - 4$.

▼ **SOLUTION**   Using the Rational Zeros Theorem from Section 4.4, we obtain the following list of possible rational zeros: $\pm 1, \pm 2, \pm 4, \pm\frac{1}{3}, \pm\frac{2}{3}, \pm\frac{4}{3}$. Checking these using synthetic division, we find that 2 and $-\frac{1}{3}$ are zeros, and we get the following factorization.

$$P(x) = 3x^4 - 2x^3 - x^2 - 12x - 4$$
$$= (x - 2)(3x^3 + 4x^2 + 7x + 2) \qquad \text{Factor } x - 2$$
$$= (x - 2)\left(x + \tfrac{1}{3}\right)(3x^2 + 3x + 6) \qquad \text{Factor } x + \tfrac{1}{3}$$
$$= 3(x - 2)\left(x + \tfrac{1}{3}\right)(x^2 + x + 2) \qquad \text{Factor } 3$$

The zeros of the quadratic factor are

$$x = \frac{-1 \pm \sqrt{1 - 8}}{2} = -\frac{1}{2} \pm i\frac{\sqrt{7}}{2} \qquad \text{Quadratic formula}$$

so the zeros of $P(x)$ are

$$2, \quad -\frac{1}{3}, \quad -\frac{1}{2} + i\frac{\sqrt{7}}{2}, \quad \text{and} \quad -\frac{1}{2} - i\frac{\sqrt{7}}{2} \qquad ▲$$

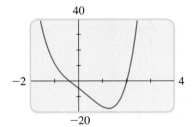

FIGURE 1
$P(x) = 3x^4 - 2x^3 - x^2 - 12x - 4$

Figure 1 shows the graph of the polynomial P in Example 5. The x-intercepts correspond to the real zeros of P. The imaginary zeros cannot be determined from the graph.

■ Complex Zeros Come in Conjugate Pairs

As you might have noticed from the examples so far, the complex zeros of polynomials with real coefficients come in pairs. Whenever $a + bi$ is a zero, its complex conjugate $a - bi$ is also a zero.

CONJUGATE ZEROS THEOREM

If the polynomial P has real coefficients and if the complex number z is a zero of P, then its complex conjugate $\bar{z}$ is also a zero of P.

▼ **PROOF**   Let

$$P(x) = a_n x^n + a_{n-1}x^{n-1} + \cdots + a_1 x + a_0$$

where each coefficient is real. Suppose that $P(z) = 0$. We must prove that $P(\bar{z}) = 0$. We use the facts that the complex conjugate of a sum of two complex numbers is the sum of the conjugates and that the conjugate of a product is the product of the conjugates.

$$P(\bar{z}) = a_n(\bar{z})^n + a_{n-1}(\bar{z})^{n-1} + \cdots + a_1\bar{z} + a_0$$
$$= \overline{a_n}\,\overline{z^n} + \overline{a_{n-1}}\,\overline{z^{n-1}} + \cdots + \overline{a_1}\,\overline{z} + \overline{a_0} \qquad \text{Because the coefficients are real}$$
$$= \overline{a_n z^n} + \overline{a_{n-1}z^{n-1}} + \cdots + \overline{a_1 z} + \overline{a_0}$$
$$= \overline{a_n z^n + a_{n-1}z^{n-1} + \cdots + a_1 z + a_0}$$
$$= \overline{P(z)} = \overline{0} = 0$$

This shows that $\bar{z}$ is also a zero of $P(x)$, which proves the theorem.   ▲

▶ **EXAMPLE 6** | A Polynomial with a Specified Complex Zero

Find a polynomial $P(x)$ of degree 3 that has integer coefficients and zeros $\frac{1}{2}$ and $3 - i$.

▼ **SOLUTION**   Since $3 - i$ is a zero, then so is $3 + i$ by the Conjugate Zeros Theorem. This means that $P(x)$ must have the following form.

$$
\begin{aligned}
P(x) &= a\left(x - \tfrac{1}{2}\right)[x - (3 - i)][x - (3 + i)] \\
&= a\left(x - \tfrac{1}{2}\right)[(x - 3) + i][(x - 3) - i] && \text{Regroup} \\
&= a\left(x - \tfrac{1}{2}\right)[(x - 3)^2 - i^2] && \text{Difference of Squares Formula} \\
&= a\left(x - \tfrac{1}{2}\right)(x^2 - 6x + 10) && \text{Expand} \\
&= a\left(x^3 - \tfrac{13}{2}x^2 + 13x - 5\right) && \text{Expand}
\end{aligned}
$$

To make all coefficients integers, we set $a = 2$ and get

$$
P(x) = 2x^3 - 13x^2 + 26x - 10
$$

Any other polynomial that satisfies the given requirements must be an integer multiple of this one. ▲

■ Linear and Quadratic Factors

We have seen that a polynomial factors completely into linear factors if we use complex numbers. If we don't use complex numbers, then a polynomial with real coefficients can always be factored into linear and quadratic factors. We use this property in Section 6.4 when we study partial fractions. A quadratic polynomial with no real zeros is called **irreducible** over the real numbers. Such a polynomial cannot be factored without using complex numbers.

LINEAR AND QUADRATIC FACTORS THEOREM

Every polynomial with real coefficients can be factored into a product of linear and irreducible quadratic factors with real coefficients.

▼ **PROOF**   We first observe that if $c = a + bi$ is a complex number, then

$$
\begin{aligned}
(x - c)(x - \bar{c}) &= [x - (a + bi)][x - (a - bi)] \\
&= [(x - a) - bi][(x - a) + bi] \\
&= (x - a)^2 - (bi)^2 \\
&= x^2 - 2ax + (a^2 + b^2)
\end{aligned}
$$

The last expression is a quadratic with *real* coefficients.

Now, if P is a polynomial with real coefficients, then by the Complete Factorization Theorem

$$
P(x) = a(x - c_1)(x - c_2) \cdots (x - c_n)
$$

Since the complex roots occur in conjugate pairs, we can multiply the factors corresponding to each such pair to get a quadratic factor with real coefficients. This results in P being factored into linear and irreducible quadratic factors. ▲

▶ **EXAMPLE 7** | Factoring a Polynomial into Linear and Quadratic Factors

Let $P(x) = x^4 + 2x^2 - 8$.

(a) Factor P into linear and irreducible quadratic factors with real coefficients.

(b) Factor P completely into linear factors with complex coefficients.

▼ **SOLUTION**

(a)
$$P(x) = x^4 + 2x^2 - 8$$
$$= (x^2 - 2)(x^2 + 4)$$
$$= (x - \sqrt{2})(x + \sqrt{2})(x^2 + 4)$$

The factor $x^2 + 4$ is irreducible, since it has no real zeros.

(b) To get the complete factorization, we factor the remaining quadratic factor.

$$P(x) = (x - \sqrt{2})(x + \sqrt{2})(x^2 + 4)$$
$$= (x - \sqrt{2})(x + \sqrt{2})(x - 2i)(x + 2i)$$

▲

4.6 | Rational Functions

LEARNING OBJECTIVES

After completing this section, you will be able to:

▪ Find the vertical asymptotes of a rational function
▪ Find the horizontal asymptote of a rational function
▪ Use asymptotes to graph a rational function
▪ Find the slant asymptote of a rational function

A rational function is a function of the form

$$r(x) = \frac{P(x)}{Q(x)}$$

where P and Q are polynomials. We assume that $P(x)$ and $Q(x)$ have no factor in common. Even though rational functions are constructed from polynomials, their graphs look quite different than the graphs of polynomial functions.

▪ Rational Functions and Asymptotes

Domains of rational expressions are discussed in Section P.7.

The *domain* of a rational function consists of all real numbers x except those for which the denominator is zero. When graphing a rational function, we must pay special attention to the behavior of the graph near those x-values. We begin by graphing a very simple rational function.

▶ **EXAMPLE 1** | A Simple Rational Function

Sketch a graph of the rational function $f(x) = \dfrac{1}{x}$.

▼ **SOLUTION**   The function f is not defined for $x = 0$. The following tables show that when x is close to zero, the value of $|f(x)|$ is large, and the closer x gets to zero, the larger $|f(x)|$ gets.

For positive real numbers,

$$\frac{1}{\text{BIG NUMBER}} = \text{small number}$$

$$\frac{1}{\text{small number}} = \text{BIG NUMBER}$$

x	$f(x)$
-0.1	-10
-0.01	-100
-0.00001	$-100{,}000$

Approaching 0^- Approaching $-\infty$

x	$f(x)$
0.1	10
0.01	100
0.00001	$100{,}000$

Approaching 0^+ Approaching ∞

We describe this behavior in words and in symbols as follows. The first table shows that as x approaches 0 from the left, the values of $y = f(x)$ decrease without bound. In symbols,

$$f(x) \to -\infty \quad \text{as} \quad x \to 0^-$$ "y approaches negative infinity as x approaches 0 from the left"

The second table shows that as x approaches 0 from the right, the values of $f(x)$ increase without bound. In symbols,

$$f(x) \to \infty \quad \text{as} \quad x \to 0^+$$ "y approaches infinity as x approaches 0 from the right"

The next two tables show how $f(x)$ changes as $|x|$ becomes large.

x	$f(x)$
-10	-0.1
-100	-0.01
$-100{,}000$	-0.00001

Approaching $-\infty$ Approaching 0

x	$f(x)$
10	0.1
100	0.01
$100{,}000$	0.00001

Approaching ∞ Approaching 0

These tables show that as $|x|$ becomes large, the value of $f(x)$ gets closer and closer to zero. We describe this situation in symbols by writing

$$f(x) \to 0 \quad \text{as} \quad x \to -\infty \qquad \text{and} \qquad f(x) \to 0 \quad \text{as} \quad x \to \infty$$

Using the information in these tables and plotting a few additional points, we obtain the graph shown in Figure 1.

x	$f(x) = \frac{1}{x}$
-2	$-\frac{1}{2}$
-1	-1
$-\frac{1}{2}$	-2
$\frac{1}{2}$	2
1	1
2	$\frac{1}{2}$

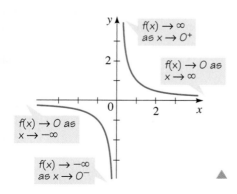

FIGURE 1
$f(x) = \frac{1}{x}$

In Example 1 we used the following arrow notation.

Symbol	Meaning
$x \to a^-$	x approaches a from the left
$x \to a^+$	x approaches a from the right
$x \to -\infty$	x goes to negative infinity; that is, x decreases without bound
$x \to \infty$	x goes to infinity; that is, x increases without bound

The line $x = 0$ is called a *vertical asymptote* of the graph in Figure 1, and the line $y = 0$ is a *horizontal asymptote*. Informally speaking, an asymptote of a function is a line to which the graph of the function gets closer and closer as one travels along that line.

DEFINITION OF VERTICAL AND HORIZONTAL ASYMPTOTES

1. The line $x = a$ is a **vertical asymptote** of the function $y = f(x)$ if y approaches $\pm\infty$ as x approaches a from the right or left.

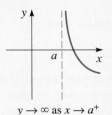

$y \to \infty$ as $x \to a^+$

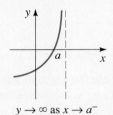

$y \to \infty$ as $x \to a^-$

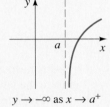

$y \to -\infty$ as $x \to a^+$

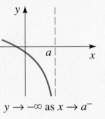

$y \to -\infty$ as $x \to a^-$

2. The line $y = b$ is a **horizontal asymptote** of the function $y = f(x)$ if y approaches b as x approaches $\pm\infty$.

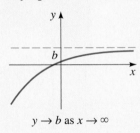

$y \to b$ as $x \to \infty$

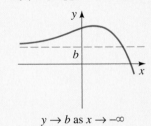

$y \to b$ as $x \to -\infty$

A rational function has vertical asymptotes where the function is undefined, that is, where the denominator is zero.

■ Transformations of $y = \dfrac{1}{x}$

A rational function of the form

$$r(x) = \frac{ax + b}{cx + d}$$

can be graphed by shifting, stretching, and/or reflecting the graph of $f(x) = \frac{1}{x}$ shown in Figure 1, using the transformations studied in Section 3.5. (Such functions are called *linear fractional transformations*.)

▶ **EXAMPLE 2** | Using Transformations to Graph Rational Functions

Sketch a graph of each rational function.

(a) $r(x) = \dfrac{2}{x - 3}$ **(b)** $s(x) = \dfrac{3x + 5}{x + 2}$

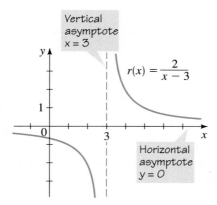

FIGURE 2

$$\begin{array}{r} 3 \\ x+2\overline{\smash{\big)}3x+5} \\ \underline{3x+6} \\ -1 \end{array}$$

▼ SOLUTION

(a) Let $f(x) = \frac{1}{x}$. Then we can express r in terms of f as follows:

$$r(x) = \frac{2}{x-3}$$

$$= 2\left(\frac{1}{x-3}\right) \qquad \text{Factor 2}$$

$$= 2(f(x-3)) \qquad \text{Since } f(x) = \frac{1}{x}$$

From this form we see that the graph of r is obtained from the graph of f by shifting 3 units to the right and stretching vertically by a factor of 2. Thus, r has vertical asymptote $x = 3$ and horizontal asymptote $y = 0$. The graph of r is shown in Figure 2.

(b) Using long division (see the margin), we get $s(x) = 3 - \frac{1}{x+2}$. Thus, we can express s in terms of f as follows:

$$s(x) = 3 - \frac{1}{x+2}$$

$$= -\frac{1}{x+2} + 3 \qquad \text{Rearrange terms}$$

$$= -f(x+2) + 3 \qquad \text{Since } f(x) = \frac{1}{x}$$

From this form we see that the graph of s is obtained from the graph of f by shifting 2 units to the left, reflecting in the x-axis, and shifting upward 3 units. Thus, s has vertical asymptote $x = -2$ and horizontal asymptote $y = 3$. The graph of s is shown in Figure 3.

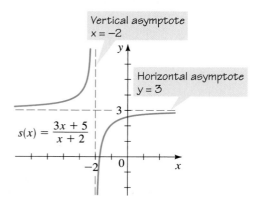

FIGURE 3

■ Asymptotes of Rational Functions

The methods of Example 2 work only for simple rational functions. To graph more complicated ones, we need to take a closer look at the behavior of a rational function near its vertical and horizontal asymptotes.

▶ **EXAMPLE 3** | Asymptotes of a Rational Function

Graph the rational function $r(x) = \dfrac{2x^2 - 4x + 5}{x^2 - 2x + 1}$.

▼ SOLUTION

Vertical asymptote: We first factor the denominator

$$r(x) = \frac{2x^2 - 4x + 5}{(x-1)^2}$$

The line $x = 1$ is a vertical asymptote because the denominator of r is zero when $x = 1$.

To see what the graph of r looks like near the vertical asymptote, we make tables of values for x-values to the left and to the right of 1. From the tables shown below we see that

$$y \to \infty \quad \text{as} \quad x \to 1^- \qquad \text{and} \qquad y \to \infty \quad \text{as} \quad x \to 1^+$$

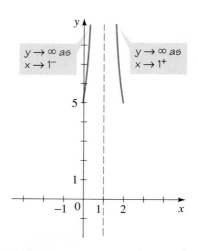

FIGURE 4

$x \to 1^-$

x	y
0	5
0.5	14
0.9	302
0.99	30,002

Approaching 1^- Approaching ∞

$x \to 1^+$

x	y
2	5
1.5	14
1.1	302
1.01	30,002

Approaching 1^+ Approaching ∞

Thus, near the vertical asymptote $x = 1$, the graph of r has the shape shown in Figure 4.

Horizontal asymptote: The horizontal asymptote is the value y approaches as $x \to \pm\infty$. To help us find this value, we divide both numerator and denominator by x^2, the highest power of x that appears in the expression:

$$y = \frac{2x^2 - 4x + 5}{x^2 - 2x + 1} \cdot \frac{\dfrac{1}{x^2}}{\dfrac{1}{x^2}} = \frac{2 - \dfrac{4}{x} + \dfrac{5}{x^2}}{1 - \dfrac{2}{x} + \dfrac{1}{x^2}}$$

The fractional expressions $\frac{4}{x}$, $\frac{5}{x^2}$, $\frac{2}{x}$, and $\frac{1}{x^2}$ all approach 0 as $x \to \pm\infty$. So as $x \to \pm\infty$, we have

These terms approach 0.

$$y = \frac{2 - \dfrac{4}{x} + \dfrac{5}{x^2}}{1 - \dfrac{2}{x} + \dfrac{1}{x^2}} \qquad \rightarrow \qquad \frac{2 - 0 + 0}{1 - 0 + 0} = 2$$

These terms approach 0.

Thus, the horizontal asymptote is the line $y = 2$.

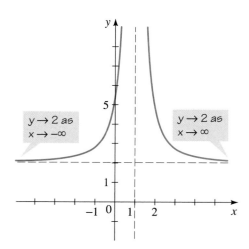

FIGURE 5

$$r(x) = \frac{2x^2 - 4x + 5}{x^2 - 2x + 1}$$

Since the graph must approach the horizontal asymptote, we can complete it as in Figure 5 on the previous page.   ▲

From Example 3 we see that the horizontal asymptote is determined by the leading coefficients of the numerator and denominator, since after dividing through by x^2 (the highest power of x) all other terms approach zero. In general, if $r(x) = P(x)/Q(x)$ and the degrees of P and Q are the same (both n, say), then dividing both numerator and denominator by x^n shows that the horizontal asymptote is

$$y = \frac{\text{leading coefficient of } P}{\text{leading coefficient of } Q}$$

The following box summarizes the procedure for finding asymptotes.

FINDING ASYMPTOTES OF RATIONAL FUNCTIONS

Let r be the rational function

$$r(x) = \frac{a_n x^n + a_{n-1} x^{n-1} + \cdots + a_1 x + a_0}{b_m x^m + b_{m-1} x^{m-1} + \cdots + b_1 x + b_0}$$

1. The vertical asymptotes of r are the lines $x = a$, where a is a zero of the denominator.

2. **(a)** If $n < m$, then r has horizontal asymptote $y = 0$.

    (b) If $n = m$, then r has horizontal asymptote $y = \dfrac{a_n}{b_m}$.

    (c) If $n > m$, then r has no horizontal asymptote.

▶ **EXAMPLE 4** | Asymptotes of a Rational Function

Find the vertical and horizontal asymptotes of $r(x) = \dfrac{3x^2 - 2x - 1}{2x^2 + 3x - 2}$.

▼ **SOLUTION**

Vertical asymptotes: We first factor

$$r(x) = \frac{3x^2 - 2x - 1}{(2x - 1)(x + 2)}$$

This factor is 0 when $x = \frac{1}{2}$.   This factor is 0 when $x = -2$.

The vertical asymptotes are the lines $x = \frac{1}{2}$ and $x = -2$.

Horizontal asymptote: The degrees of the numerator and denominator are the same, and

$$\frac{\text{leading coefficient of numerator}}{\text{leading coefficient of denominator}} = \frac{3}{2}$$

Thus, the horizontal asymptote is the line $y = \frac{3}{2}$.

To confirm our results, we graph r using a graphing calculator (see Figure 6).   ▲

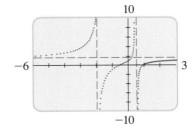

FIGURE 6
$$r(x) = \frac{3x^2 - 2x - 1}{2x^2 + 3x - 2}$$

Graph is drawn using dot mode to avoid extraneous lines.

Graphing Rational Functions

We have seen that asymptotes are important when graphing rational functions. In general, we use the following guidelines to graph rational functions.

SKETCHING GRAPHS OF RATIONAL FUNCTIONS

1. **Factor.** Factor the numerator and denominator.

2. **Intercepts.** Find the x-intercepts by determining the zeros of the numerator and the y-intercept from the value of the function at $x = 0$.

3. **Vertical Asymptotes.** Find the vertical asymptotes by determining the zeros of the denominator, and then see whether $y \to \infty$ or $y \to -\infty$ on each side of each vertical asymptote by using test values.

4. **Horizontal Asymptote.** Find the horizontal asymptote (if any), using the procedure described in the box on page 228.

5. **Sketch the Graph.** Graph the information provided by the first four steps. Then plot as many additional points as needed to fill in the rest of the graph of the function.

A fraction is 0 if and only if its numerator is 0.

▶ **EXAMPLE 5** | Graphing a Rational Function

Graph the rational function $r(x) = \dfrac{2x^2 + 7x - 4}{x^2 + x - 2}$.

▼ **SOLUTION** We factor the numerator and denominator, find the intercepts and asymptotes, and sketch the graph.

Factor: $\quad y = \dfrac{(2x - 1)(x + 4)}{(x - 1)(x + 2)}$

x-Intercepts: The x-intercepts are the zeros of the numerator, $x = \frac{1}{2}$ and $x = -4$.

y-Intercept: To find the y-intercept, we substitute $x = 0$ into the original form of the function.

$$r(0) = \frac{2(0)^2 + 7(0) - 4}{(0)2 + (0) - 2} = \frac{-4}{-2} = 2$$

The y-intercept is 2.

Vertical asymptotes: The vertical asymptotes occur where the denominator is 0, that is, where the function is undefined. From the factored form we see that the vertical asymptotes are the lines $x = 1$ and $x = -2$.

Behavior near vertical asymptotes: We need to know whether $y \to \infty$ or $y \to -\infty$ on each side of each vertical asymptote. To determine the sign of y for x-values near the vertical asymptotes, we use test values. For instance, as $x \to 1^-$, we use a test value close to and to the left of 1 ($x = 0.9$, say) to check whether y is positive or negative to the left of $x = 1$.

When choosing test values, we must make sure that there is no x-intercept between the test point and the vertical asymptote.

$$y = \frac{(2(0.9) - 1)((0.9) + 4)}{((0.9) - 1)((0.9) + 2)} \quad \text{whose sign is} \quad \frac{(+)(+)}{(-)(+)} \quad \text{(negative)}$$

So $y \to -\infty$ as $x \to 1^-$. On the other hand, as $x \to 1^+$, we use a test value close to and to the right of 1 ($x = 1.1$, say), to get

$$y = \frac{(2(1.1) - 1)((1.1) + 4)}{((1.1) - 1)((1.1) + 2)} \qquad \text{whose sign is} \qquad \frac{(+)(+)}{(+)(+)} \quad \text{(positive)}$$

So $y \to \infty$ as $x \to 1^+$. The other entries in the following table are calculated similarly.

As $x \to$	-2^-	-2^+	1^-	1^+
the sign of $y = \dfrac{(2x - 1)(x + 4)}{(x - 1)(x + 2)}$ is	$\dfrac{(-)(+)}{(-)(-)}$	$\dfrac{(-)(+)}{(-)(+)}$	$\dfrac{(+)(+)}{(-)(+)}$	$\dfrac{(+)(+)}{(+)(+)}$
so $y \to$	$-\infty$	∞	$-\infty$	∞

Horizontal asymptote: The degrees of the numerator and denominator are the same, and

$$\frac{\text{leading coefficient of numerator}}{\text{leading coefficient of denominator}} = \frac{2}{1} = 2$$

Thus, the horizontal asymptote is the line $y = 2$.

Graph: We use the information we have found, together with some additional values, to sketch the graph in Figure 7.

x	y
-6	0.93
-3	-1.75
-1	4.50
1.5	6.29
2	4.50
3	3.50

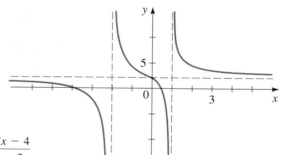

FIGURE 7

$$r(x) = \frac{2x^2 + 7x - 4}{x^2 + x - 2}$$

▶ **EXAMPLE 6** | Graphing a Rational Function

Graph the rational function $r(x) = \dfrac{5x + 21}{x^2 + 10x + 25}$.

▼ **SOLUTION**

Factor: $\quad y = \dfrac{5x + 21}{(x + 5)^2}$

x-Intercept: $\quad -\dfrac{21}{5}$, from $5x + 21 = 0$

y-Intercept: $\quad \dfrac{21}{25}$, because $r(0) = \dfrac{5 \cdot 0 + 21}{0^2 + 10 \cdot 0 + 25}$

$$= \frac{21}{25}$$

Vertical asymptote: $x = -5$, from the zeros of the denominator

Behavior near vertical asymptote:

As $x \to$	-5^-	-5^+
the sign of $y = \dfrac{5x + 21}{(x + 5)^2}$ is	$\dfrac{(-)}{(-)(-)}$	$\dfrac{(-)}{(+)(+)}$
so $y \to$	$-\infty$	$-\infty$

Horizontal asymptote: $y = 0$, because the degree of the numerator is less than the degree of the denominator

Graph: We use the information we have found, together with some additional values, to sketch the graph in Figure 8.

x	y
-15	-0.5
-10	-1.2
-3	1.5
-1	1.0
3	0.6
5	0.5
10	0.3

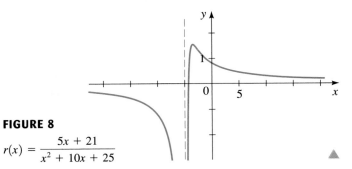

FIGURE 8

$$r(x) = \frac{5x + 21}{x^2 + 10x + 25}$$

    From the graph in Figure 8 we see that, contrary to the common misconception, a graph may cross a horizontal asymptote. The graph in Figure 8 crosses the x-axis (the horizontal asymptote) from below, reaches a maximum value near $x = -3$, and then approaches the x-axis from above as $x \to \infty$.

▶ **EXAMPLE 7** | Graphing a Rational Function

Graph the rational function $r(x) = \dfrac{x^2 - 3x - 4}{2x^2 + 4x}$.

▼ **SOLUTION**

Factor: $y = \dfrac{(x + 1)(x - 4)}{2x(x + 2)}$

x-Intercepts: -1 and 4, from $x + 1 = 0$ and $x - 4 = 0$

y-Intercept: None, because $r(0)$ is undefined

Vertical asymptotes: $x = 0$ and $x = -2$, from the zeros of the denominator

Behavior near vertical asymptotes:

As $x \to$	-2^-	-2^+	0^-	0^+
the sign of $y = \dfrac{(x+1)(x-4)}{2x(x+2)}$ is	$\dfrac{(-)(-)}{(-)(-)}$	$\dfrac{(-)(-)}{(-)(+)}$	$\dfrac{(+)(-)}{(-)(+)}$	$\dfrac{(+)(-)}{(+)(+)}$
so $y \to$	∞	$-\infty$	∞	$-\infty$

Horizontal asymptote: $y = \frac{1}{2}$, because the degree of the numerator and the degree of the denominator are the same and

$$\frac{\text{leading coefficient of numerator}}{\text{leading coefficient of denominator}} = \frac{1}{2}$$

Graph: We use the information we have found, together with some additional values, to sketch the graph in Figure 9.

x	y
-3	2.33
-2.5	3.90
-0.5	1.50
1	-1.00
3	-0.13
5	0.09

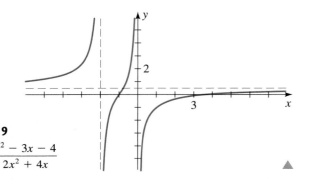

FIGURE 9

$$r(x) = \frac{x^2 - 3x - 4}{2x^2 + 4x}$$

■ Slant Asymptotes and End Behavior

If $r(x) = P(x)/Q(x)$ is a rational function in which the degree of the numerator is one more than the degree of the denominator, we can use the Division Algorithm to express the function in the form

$$r(x) = ax + b + \frac{R(x)}{Q(x)}$$

where the degree of R is less than the degree of Q and $a \neq 0$. This means that as $x \to \pm\infty$, $R(x)/Q(x) \to 0$, so for large values of $|x|$ the graph of $y = r(x)$ approaches the graph of the line $y = ax + b$. In this situation we say that $y = ax + b$ is a **slant asymptote**, or an **oblique asymptote**.

▶ **EXAMPLE 8** | A Rational Function with a Slant Asymptote

Graph the rational function $r(x) = \dfrac{x^2 - 4x - 5}{x - 3}$.

▼ **SOLUTION**

Factor: $y = \dfrac{(x+1)(x-5)}{x-3}$

x-Intercepts: −1 and 5, from $x + 1 = 0$ and $x − 5 = 0$

y-Intercepts: $\dfrac{5}{3}$, because $r(0) = \dfrac{0^2 − 4 \cdot 0 − 5}{0 − 3} = \dfrac{5}{3}$

Horizontal asymptote: None, because the degree of the numerator is greater than the degree of the denominator

Vertical asymptote: $x = 3$, from the zero of the denominator

Behavior near vertical asymptote: $y \to \infty$ as $x \to 3^-$ and $y \to -\infty$ as $x \to 3^+$

Slant asymptote: Since the degree of the numerator is one more than the degree of the denominator, the function has a slant asymptote. Dividing (see the margin), we obtain

$$r(x) = x − 1 − \frac{8}{x − 3}$$

Thus, $y = x − 1$ is the slant asymptote.

Graph: We use the information we have found, together with some additional values, to sketch the graph in Figure 10.

$$
\begin{array}{r}
x - 1 \\
x - 3 \overline{\smash{)}\ x^2 - 4x - 5} \\
\underline{x^2 - 3x} \\
-x - 5 \\
\underline{-x + 3} \\
-8
\end{array}
$$

x	*y*
−2	−1.4
1	4
2	9
4	−5
6	2.33

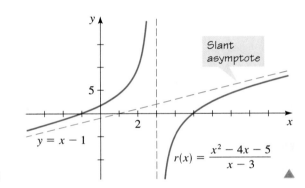

FIGURE 10

So far, we have considered only horizontal and slant asymptotes as end behaviors for rational functions. In the next example we graph a function whose end behavior is like that of a parabola.

▶ **EXAMPLE 9** | End Behavior of a Rational Function

Graph the rational function

$$r(x) = \frac{x^3 − 2x^2 + 3}{x − 2}$$

and describe its end behavior.

▼ **SOLUTION**

Factor: $y = \dfrac{(x + 1)(x^2 − 3x + 3)}{x − 2}$

x-Intercepts: −1, from $x + 1 = 0$ (The other factor in the numerator has no real zeros.)

y-Intercepts: $-\dfrac{3}{2}$, because $r(0) = \dfrac{0^3 − 2 \cdot 0^2 + 3}{0 − 2} = -\dfrac{3}{2}$

Vertical asymptote: $x = 2$, from the zero of the denominator

Behavior near vertical asymptote: $y \to -\infty$ as $x \to 2^-$ and $y \to \infty$ as $x \to 2^+$

Horizontal asymptote: None, because the degree of the numerator is greater than the degree of the denominator

End behavior: Dividing (see the margin), we get

$$x - 2 \overline{)x^3 - 2x^2 + 0x + 3} \quad \frac{x^2}{}$$
$$\underline{x^3 - 2x^2}$$
$$3$$

$$r(x) = x^2 + \frac{3}{x - 2}$$

This shows that the end behavior of r is like that of the parabola $y = x^2$ because $3/(x - 2)$ is small when $|x|$ is large. That is, $3/(x - 2) \to 0$ as $x \to \pm\infty$. This means that the graph of r will be close to the graph of $y = x^2$ for large $|x|$.

Graph: In Figure 11(a) we graph r in a small viewing rectangle; we can see the intercepts, the vertical asymptotes, and the local minimum. In Figure 11(b) we graph r in a larger viewing rectangle; here the graph looks almost like the graph of a parabola. In Figure 11(c) we graph both $y = r(x)$ and $y = x^2$; these graphs are very close to each other except near the vertical asymptote.

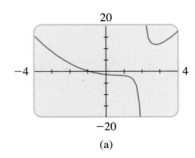

(a)

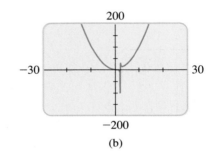

(b)

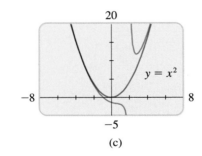

(c)

FIGURE 11
$r(x) = \dfrac{x^3 - 2x^2 + 3}{x - 2}$

■ Applications

Rational functions occur frequently in scientific applications of algebra. In the next example we analyze the graph of a function from the theory of electricity.

▶ **EXAMPLE 10** | Electrical Resistance

When two resistors with resistances R_1 and R_2 are connected in parallel, their combined resistance R is given by the formula

$$R = \frac{R_1 R_2}{R_1 + R_2}$$

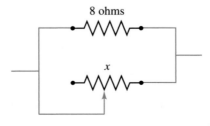

FIGURE 12

Suppose that a fixed 8-ohm resistor is connected in parallel with a variable resistor, as shown in Figure 12. If the resistance of the variable resistor is denoted by x, then the combined resistance R is a function of x. Graph R, and give a physical interpretation of the graph.

▼ **SOLUTION**   Substituting $R_1 = 8$ and $R_2 = x$ into the formula gives the function

$$R(x) = \frac{8x}{8 + x}$$

Since resistance cannot be negative, this function has physical meaning only when $x > 0$. The function is graphed in Figure 13(a) using the viewing rectangle $[0, 20]$ by $[0, 10]$. The function has no vertical asymptote when x is restricted to positive values. The combined resistance R increases as the variable resistance x increases. If we widen the viewing rectangle to $[0, 100]$ by $[0, 10]$, we obtain the graph in Figure 13(b). For large x the combined resistance R levels off, getting closer and closer to the horizontal asymptote $R = 8$. No matter how large the variable resistance x, the combined resistance is never greater than 8 ohms.

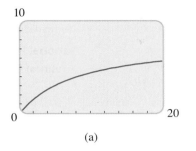

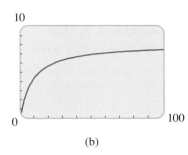

FIGURE 13

$R(x) = \dfrac{8x}{8 + x}$

(a)   (b)

▶ CHAPTER 4 | REVIEW

▼ PROPERTIES AND FORMULAS

Quadratic Functions (pp. 188–191)

A **quadratic function** is a function of the form

$$f(x) = ax^2 + bx + c$$

It can be expressed in the **standard form**

$$f(x) = a(x - h)^2 + k$$

by completing the square.

The graph of a quadratic function in standard form is a **parabola** with **vertex** (h, k).

If $a > 0$, then the quadratic function f has the **minimum value** k at $x = h$.

If $a < 0$, then the quadratic function f has the **maximum value** k at $x = h$.

Polynomial Functions (p. 194)

A **polynomial function** of **degree** n is a function P of the form

$$P(x) = a_n x^n + a_{n-1}x^{n-1} + \cdots + a_1 x + a_0$$

The numbers a_i are the **coefficients** of the polynomial; a_n is the **leading coefficient**, and a_0 is the **constant coefficient** (or **constant term**).

The graph of a polynomial function is a smooth, continuous curve.

Real Zeros of Polynomials (p. 198)

A **zero** of a polynomial P is a number c for which $P(c) = 0$. The following are equivalent ways of describing real zeros of polynomials:

1. c is a real zero of P.

2. $x = c$ is a solution of the equation $P(x) = 0$.

3. $x - c$ is a factor of $P(x)$.

4. c is an x-intercept of the graph of P.

Multiplicity of a Zero (p. 202)

A zero c of a polynomial P has multiplicity m if m is the highest power for which $(x - c)^m$ is a factor of $P(x)$.

Local Maxima and Minima (p. 203)

A polynomial function P of degree n has $n - 1$ or fewer **local extrema** (i.e., local maxima and minima).

Division of Polynomials (p. 205)

If P and D are any polynomials (with $D(x) \neq 0$), then we can divide P by D using either **long division** or (if D is linear) **synthetic division**. The result of the division can be expressed in one of the following equivalent forms:

$$P(x) = D(x) \cdot Q(x) + R(x)$$

$$\frac{P(x)}{D(x)} = Q(x) + \frac{R(x)}{D(x)}$$

In this division, P is the **dividend**, D is the **divisor**, Q is the **quotient**, and R is the **remainder**. When the division is continued to its completion, the degree of R will be less than the degree of D (or $R(x) = 0$).

Remainder Theorem (p. 207)

When $P(x)$ is divided by the linear divisor $D(x) = x - c$, the **remainder** is the constant $P(c)$. So one way to **evaluate** a polynomial function P at c is to use synthetic division to divide $P(x)$ by $x - c$ and observe the value of the remainder.

Rational Zeros of Polynomials (pp. 210–211)

If the polynomial P given by

$$P(x) = a_n x^n + a_{n-1}x^{n-1} + \cdots + a_1 x + a_0$$

has integer coefficients, then all the **rational zeros** of P have the form

$$x = \pm\frac{p}{q}$$

where p is a divisor of the constant term a_0 and q is a divisor of the leading coefficient a_n.

So to find all the rational zeros of a polynomial, we list all the *possible* rational zeros given by this principle and then check to see which *actually* are zeros by using synthetic division.

Descartes' Rule of Signs (p. 213)

Let P be a polynomial with real coefficients. Then:

The number of positive real zeros of P either is the number of **changes of sign** in the coefficients of $P(x)$ or is less than that by an even number.

The number of negative real zeros of P either is the number of **changes of sign** in the coefficients of $P(-x)$ or is less than that by an even number.

Upper and Lower Bounds Theorem (p. 213)

Suppose we divide the polynomial P by the linear expression $x - c$ and arrive at the result

$$P(x) = (x - c) \cdot Q(x) + r$$

If $c > 0$ and the coefficients of Q, followed by r, are all nonnegative, then c is an **upper bound** for the zeros of P.

If $c < 0$ and the coefficients of Q, followed by r (including zero coefficients), are alternately nonnegative and nonpositive, then c is a **lower bound** for the zeros of P.

The Fundamental Theorem of Algebra, Complete Factorization, and the Zeros Theorem (pp. 217–219)

Every polynomial P of degree n with complex coefficients has exactly n complex zeros, provided that each zero of multiplicity m is counted m times. P factors into n linear factors as follows:

$$P(x) = a(x - c_1)(x - c_2) \cdots (x - c_n)$$

where a is the leading coefficient of P and $c_1, c_2, \ldots, c_n$ are the zeros of P.

Conjugate Zeros Theorem (p. 221)

If the polynomial P has real coefficients and if $a + bi$ is a zero of P, then its complex conjugate $a - bi$ is also a zero of P.

Linear and Quadratic Factors Theorem (p. 222)

Every polynomial with real coefficients can be factored into linear and irreducible quadratic factors with real coefficients.

Rational Functions (p. 223)

A **rational function** r is a quotient of polynomial functions:

$$r(x) = \frac{P(x)}{Q(x)}$$

We generally assume that the polynomials P and Q have no factors in common.

Asymptotes (pp. 223–225)

The line $x = a$ is a **vertical asymptote** of the function $y = f(x)$ if

$$y \to \infty \quad \text{or} \quad y \to -\infty \quad \text{as} \quad x \to a^+ \quad \text{or} \quad x \to a^-$$

The line $y = b$ is a **horizontal asymptote** of the function $y = f(x)$ if

$$y \to b \quad \text{as} \quad x \to \infty \quad \text{or} \quad x \to -\infty$$

Asymptotes of Rational Functions (pp. 225–228)

Let $r(x) = \dfrac{P(x)}{Q(x)}$ be a rational function.

The vertical asymptotes of r are the lines $x = a$ where a is a zero of Q.

If the degree of P is less than the degree of Q, then the horizontal asymptote of r is the line $y = 0$.

If the degrees of P and Q are the same, then the horizontal asymptote of r is the line $y = b$, where

$$b = \frac{\text{leading coefficient of } P}{\text{leading coefficient of } Q}$$

If the degree of P is greater than the degree of Q, then r has no horizontal asymptote.

▼ CONCEPT SUMMARY

Section 4.1	Review Exercises
■ Express a quadratic function in standard form	1(a)–4(a)
■ Graph a quadratic function using its standard form	1(b)–4(b)
■ Find maximum and minimum values of quadratic functions	5–6
■ Model with quadratic functions	7–8

Section 4.2	Review Exercises
■ Graph basic polynomial functions	9–14
■ Use end behavior of a polynomial function to help sketch its graph	15–20, 39–46
■ Use the zeros of a polynomial function to sketch its graph	15–20, 39–46
■ Use the multiplicity of a zero to help sketch the graph of a polynomial function	15–16, 39–46
■ Find local maxima and minima of polynomial functions	17–20

Section 4.3	Review Exercises
■ Use long division to divide polynomials	29–30
■ Use synthetic division to divide polynomials	23–28
■ Use the remainder theorem to find values of a polynomial	31–32, 35–36
■ Use the Factor Theorem to factor a polynomial	33–34
■ Find a polynomial with specified zeros	47–50

Section 4.4   Review Exercises

- Use the Rational Zeros Theorem to find the rational zeros of a polynomial   37(a)–38(a)
- Use Descartes' Rule of Signs to determine the number of positive and negative zeros of a polynomial   37(b)–38(b)
- Use the Upper and Lower Bounds Theorem to find upper and lower bounds for the real zeros of   51–60
  a polynomial
- Use algebra and graphing devices to solve polynomial equations   61–64

Section 4.5   Review Exercises

- State the Fundamental Theorem of Algebra
- Factor a polynomial completely (into linear factors) over the complex numbers   51–60
- Determine the multiplicity of a zero of a polynomial   39–46, 51–60
- Use the Conjugate Roots Theorem to find polynomials with specified zeros   48–49
- Factor a polynomial completely (into linear and quadratic factors) over the real numbers   65–66

Section 4.6   Review Exercises

- Find the vertical asymptotes of a rational function   67–76
- Find the horizontal asymptote of a rational function   67–76
- Use asymptotes to graph a rational function   67–76
- Find the slant asymptote of a rational function   75–76

▼ EXERCISES

1–4 ▪ A quadratic function is given. **(a)** Express the function in standard form. **(b)** Graph the function.

1. $f(x) = x^2 + 4x + 1$ **2.** $f(x) = -2x^2 + 12x + 12$

3. $g(x) = 1 + 8x - x^2$ **4.** $g(x) = 6x - 3x^2$

5–6 ▪ Find the maximum or minimum value of the quadratic function.

5. $f(x) = 2x^2 + 4x - 5$ **6.** $g(x) = 1 - x - x^2$

7. A stone is thrown upward from the top of a building. Its height (in feet) above the ground after t seconds is given by the function $h(t) = -16t^2 + 48t + 32$. What maximum height does the stone reach?

8. The profit P (in dollars) generated by selling x units of a certain commodity is given by the function

$$P(x) = -1500 + 12x - 0.004x^2$$

What is the maximum profit, and how many units must be sold to generate it?

9–14 ▪ Graph the polynomial by transforming an appropriate graph of the form $y = x^n$. Show clearly all x- and y-intercepts.

9. $P(x) = -x^3 + 64$ **10.** $P(x) = 2x^3 - 16$

11. $P(x) = 2(x + 1)^4 - 32$ **12.** $P(x) = 81 - (x - 3)^4$

13. $P(x) = 32 + (x - 1)^5$ **14.** $P(x) = -3(x + 2)^5 + 96$

15–16 ▪ A polynomial function P is given. **(a)** Determine the multiplicity of each zero of P. **(b)** Sketch a graph of P.

15. $P(x) = x^3(x - 2)^2$ **16.** $P(x) = x(x + 1)^3(x - 1)^2$

17–20 ▪ Use a graphing device to graph the polynomial. Find the x- and y-intercepts and the coordinates of all local extrema, correct to the nearest decimal. Describe the end behavior of the polynomial.

17. $P(x) = x^3 - 4x + 1$ **18.** $P(x) = -2x^3 + 6x^2 - 2$

19. $P(x) = 3x^4 - 4x^3 - 10x - 1$

20. $P(x) = x^5 + x^4 - 7x^3 - x^2 + 6x + 3$

21. The strength S of a wooden beam of width x and depth y is given by the formula $S = 13.8xy^2$. A beam is to be cut from a log of diameter 10 in., as shown in the figure.
  (a) Express the strength S of this beam as a function of x only.
  (b) What is the domain of the function S?
  (c) Draw a graph of S.
  (d) What width will make the beam the strongest?

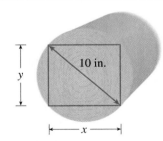

22. A small shelter for delicate plants is to be constructed of thin plastic material. It will have square ends and a rectangular top and back, with an open bottom and front, as shown in the figure. The total area of the four plastic sides is to be 1200 in².
  (a) Express the volume V of the shelter as a function of the depth x.
  (b) Draw a graph of V.
  (c) What dimensions will maximize the volume of the shelter?

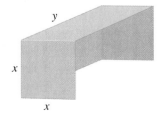

23–30 ▪ Find the quotient and remainder.

23. $\dfrac{x^2 - 3x + 5}{x - 2}$

24. $\dfrac{x^2 + x - 12}{x - 3}$

25. $\dfrac{x^3 - x^2 + 11x + 2}{x - 4}$

26. $\dfrac{x^3 + 2x^2 - 10}{x + 3}$

27. $\dfrac{x^4 - 8x^2 + 2x + 7}{x + 5}$

28. $\dfrac{2x^4 + 3x^3 - 12}{x + 4}$

29. $\dfrac{2x^3 + x^2 - 8x + 15}{x^2 + 2x - 1}$

30. $\dfrac{x^4 - 2x^2 + 7x}{x^2 - x + 3}$

31–32 ▪ Find the indicated value of the polynomial using the Remainder Theorem.

31. $P(x) = 2x^3 - 9x^2 - 7x + 13$;   find $P(5)$

32. $Q(x) = x^4 + 4x^3 + 7x^2 + 10x + 15$;   find $Q(-3)$

33. Show that $\frac{1}{2}$ is a zero of the polynomial
$$P(x) = 2x^4 + x^3 - 5x^2 + 10x - 4$$

34. Use the Factor Theorem to show that $x + 4$ is a factor of the polynomial
$$P(x) = x^5 + 4x^4 - 7x^3 - 23x^2 + 23x + 12$$

35. What is the remainder when the polynomial
$$P(x) = x^{500} + 6x^{201} - x^2 - 2x + 4$$
is divided by $x - 1$?

36. What is the remainder when $x^{101} - x^4 + 2$ is divided by $x + 1$?

37–38 ▪ A polynomial P is given.

(a) List all possible rational zeros (without testing to see whether they actually are zeros).

(b) Determine the possible number of positive and negative real zeros using Descartes' Rule of Signs.

37. $P(x) = x^5 - 6x^3 - x^2 + 2x + 18$

38. $P(x) = 6x^4 + 3x^3 + x^2 + 3x + 4$

39–46 ▪ A polynomial P is given.

(a) Find all real zeros of P, and state their multiplicities.

(b) Sketch the graph of P.

39. $P(x) = x^3 - 16x$

40. $P(x) = x^3 - 3x^2 - 4x$

41. $P(x) = x^4 + x^3 - 2x^2$

42. $P(x) = x^4 - 5x^2 + 4$

43. $P(x) = x^4 - 2x^3 - 7x^2 + 8x + 12$

44. $P(x) = x^4 - 2x^3 - 2x^2 + 8x - 8$

45. $P(x) = 2x^4 + x^3 + 2x^2 - 3x - 2$

46. $P(x) = 9x^5 - 21x^4 + 10x^3 + 6x^2 - 3x - 1$

47. Find a polynomial of degree 3 with constant coefficient 12 and zeros $-\frac{1}{2}$, 2, and 3.

48. Find a polynomial of degree 4 that has integer coefficients and zeros $3i$ and 4, with 4 a double zero.

49. Does there exist a polynomial of degree 4 with integer coefficients that has zeros i, $2i$, $3i$, and $4i$? If so, find it. If not, explain why.

50. Prove that the equation $3x^4 + 5x^2 + 2 = 0$ has no real root.

51–60 ▪ Find all rational, irrational, and complex zeros (and state their multiplicities). Use Descartes' Rule of Signs, the Upper and Lower Bounds Theorem, the Quadratic Formula, or other factoring techniques to help you whenever possible.

51. $P(x) = x^3 - 3x^2 - 13x + 15$

52. $P(x) = 2x^3 + 5x^2 - 6x - 9$

53. $P(x) = x^4 + 6x^3 + 17x^2 + 28x + 20$

54. $P(x) = x^4 + 7x^3 + 9x^2 - 17x - 20$

55. $P(x) = x^5 - 3x^4 - x^3 + 11x^2 - 12x + 4$

56. $P(x) = x^4 - 81$

57. $P(x) = x^6 - 64$

58. $P(x) = 18x^3 + 3x^2 - 4x - 1$

59. $P(x) = 6x^4 - 18x^3 + 6x^2 - 30x + 36$

60. $P(x) = x^4 + 15x^2 + 54$

61–64 ▪ Use a graphing device to find all real solutions of the equation.

61. $2x^2 = 5x + 3$

62. $x^3 + x^2 - 14x - 24 = 0$

63. $x^4 - 3x^3 - 3x^2 - 9x - 2 = 0$

64. $x^5 = x + 3$

65–66 ▪ A polynomial function P is given. Find all the real zeros of P, and factor P completely into linear and irreducible quadratic factors with real coefficients.

65. $P(x) = x^3 - 2x - 4$ **66.** $P(x) = x^4 + 3x^2 - 4$

67–72 ▪ Graph the rational function. Show clearly all x- and y-intercepts and asymptotes.

67. $r(x) = \dfrac{3x - 12}{x + 1}$

68. $r(x) = \dfrac{1}{(x + 2)^2}$

69. $r(x) = \dfrac{x - 2}{x^2 - 2x - 8}$

70. $r(x) = \dfrac{2x^2 - 6x - 7}{x - 4}$

71. $r(x) = \dfrac{x^2 - 9}{2x^2 + 1}$

72. $r(x) = \dfrac{x^3 + 27}{x + 4}$

73–76 ▪ Use a graphing device to analyze the graph of the rational function. Find all x- and y-intercepts and all vertical, horizontal, and slant asymptotes. If the function has no horizontal or slant asymptote, find a polynomial that has the same end behavior as the rational function.

73. $r(x) = \dfrac{x - 3}{2x + 6}$

74. $r(x) = \dfrac{2x - 7}{x^2 + 9}$

75. $r(x) = \dfrac{x^3 + 8}{x^2 - x - 2}$

76. $r(x) = \dfrac{2x^3 - x^2}{x + 1}$

77. Find the coordinates of all points of intersection of the graphs of
$$y = x^4 + x^2 + 24x \qquad \text{and} \qquad y = 6x^3 + 20$$

1. Express the quadratic function $f(x) = x^2 - x - 6$ in standard form, and sketch its graph.

2. Find the maximum or minimum value of the quadratic function $g(x) = 2x^2 + 6x + 3$.

3. A cannonball fired out to sea from a shore battery follows a parabolic trajectory given by the graph of the equation

$$h(x) = 10x - 0.01x^2$$

   where $h(x)$ is the height of the cannonball above the water when it has traveled a horizontal distance of x feet.

   (a) What is the maximum height that the cannonball reaches?

   (b) How far does the cannonball travel horizontally before splashing into the water?

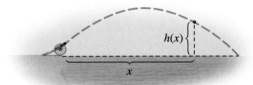

4. Graph the polynomial $P(x) = -(x + 2)^3 + 27$, showing clearly all x- and y-intercepts.

5. (a) Use synthetic division to find the quotient and remainder when $x^4 - 4x^2 + 2x + 5$ is divided by $x - 2$.

   (b) Use long division to find the quotient and remainder when $2x^5 + 4x^4 - x^3 - x^2 + 7$ is divided by $2x^2 - 1$.

6. Let $P(x) = 2x^3 - 5x^2 - 4x + 3$.

   (a) List all possible rational zeros of P.

   (b) Find the complete factorization of P.

   (c) Find the zeros of P.

   (d) Sketch the graph of P.

7. Find all real and complex zeros of $P(x) = x^3 - x^2 - 4x - 6$.

8. Find the complete factorization of $P(x) = x^4 - 2x^3 + 5x^2 - 8x + 4$.

9. Find a fourth-degree polynomial with integer coefficients that has zeros $3i$ and -1, with -1 a zero of multiplicity 2.

10. Let $P(x) = 2x^4 - 7x^3 + x^2 - 18x + 3$.

    (a) Use Descartes' Rule of Signs to determine how many positive and how many negative real zeros P can have.

    (b) Show that 4 is an upper bound and -1 is a lower bound for the real zeros of P.

    (c) Draw a graph of P, and use it to estimate the real zeros of P, correct to two decimal places.

    (d) Find the coordinates of all local extrema of P, correct to two decimals.

11. Consider the following rational functions:

$$r(x) = \frac{2x - 1}{x^2 - x - 2} \qquad s(x) = \frac{x^3 + 27}{x^2 + 4}$$

$$t(x) = \frac{x^3 - 9x}{x + 2} \qquad u(x) = \frac{x^2 + x - 6}{x^2 - 25}$$

    (a) Which of these rational functions has a horizontal asymptote?

    (b) Which of these functions has a slant asymptote?

    (c) Which of these functions has no vertical asymptote?

    (d) Graph $y = u(x)$, showing clearly any asymptotes and x- and y-intercepts the function may have.

    (e) Use long division to find a polynomial P that has the same end behavior as t. Graph both P and t on the same screen to verify that they have the same end behavior.

239

Exponential and Logarithmic Functions

So far we have studied relatively simple functions, such as linear, polynomial, and rational functions. We now turn our attention to two of the most important types of functions in mathematics: exponential functions and their inverses, and logarithmic functions. We use these functions to describe exponential growth in biology and economics and radioactive decay in physics and chemistry, as well as other real-world phenomena.

The following Warm-Up Exercises will help you review the basic algebra skills that you will need for this chapter.

WARM-UP EXERCISES

1. Compute the following without the use of a calculator:

    (a) 2^5 (b) $(-1)^8$ (c) $27^{-1/3}$ (d) $16^{3/2}$

2. Let $f(x)$ have the following graph:

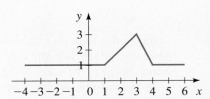

   Sketch graphs of the following:

    (a) $f(x + 4)$ (b) $f(x) + 2$ (c) $2f(x)$ (d) $f(-x)$

3. Consider the one-to-one function possessing the following table of values:

x	1	2	3	4	5	6	7	8
$f(x)$	1	3	4	7	8	11	14	17

   Compute each of the following, if possible:

   $f(1), f(4), f(5), f(14), f^{-1}(1), f^{-1}(4), f^{-1}(5), f^{-1}(14),$

4. Solve the following equation: $\dfrac{4x^{1/3} - 3}{5} = 1$.

5. Solve the following quadratic equation by factoring: $x^2 - 8x + 12 = 0$.

6. Let $f(x) = x^2 - \dfrac{11}{10}x + \dfrac{3}{10}$.

    (a) Use your calculator to graph this function and compute the x-intercepts of this graph.

    (b) Use your answer to part (a) to solve $f(x) = 0$.

7. Let $f(x) = 3x + 5 + a$. We know that $f(1) = 6$. Find a.

| **5.1** | # Exponential Functions |

LEARNING OBJECTIVES
After completing this section, you will be able to:

▨ Evaluate exponential functions
▨ Graph exponential functions
▨ Evaluate and graph the natural exponential function
▨ Find compound interest
▨ Find continuously compounded interest

In this chapter we study a new class of functions called *exponential functions*. For example,

$$f(x) = 2^x$$

is an exponential function (with base 2). Notice how quickly the values of this function increase:

$$f(3) = 2^3 = 8$$

$$f(10) = 2^{10} = 1024$$

$$f(30) = 2^{30} = 1{,}073{,}741{,}824$$

Compare this with the function $g(x) = x^2$, where $g(30) = 30^2 = 900$. The point is that when the variable is in the exponent, even a small change in the variable can cause a dramatic change in the value of the function.

In spite of this incomprehensibly huge growth, exponential functions are appropriate for modeling population growth for all living things, from bacteria to elephants. To understand how a population grows, consider the case of a single bacterium, which divides every hour. After one hour we would have 2 bacteria, after two hours 2^2 or 4 bacteria, after three hours 2^3 or 8 bacteria, and so on. After x hours we would have 2^x bacteria. This leads us to model the bacteria population by the function $f(x) = 2^x$.

| 0 | 1 | 2 | 3 | 4 | 5 | 6 |

The principle governing population growth is the following: The larger the population, the greater the number of offspring. This same principle is present in many other real-life situations. For example, the larger your bank account, the more interest you get. So we also use exponential functions to find compound interest.

To study exponential functions, we must first define what we mean by the exponential expression a^x when x is any real number.

▨ Exponential Functions

In Section P.5 we defined a^x for $a > 0$ and x a rational number, but we have not yet defined irrational powers. So what is meant by $5^{\sqrt{3}}$ or 2^π? To define a^x when x is irrational, we approximate x by rational numbers.

For example, since

$$\sqrt{3} \approx 1.73205\ldots$$

is an irrational number, we successively approximate $a^{\sqrt{3}}$ by the following rational powers:

$$a^{1.7}, a^{1.73}, a^{1.732}, a^{1.7320}, a^{1.73205}, \ldots$$

Intuitively, we can see that these rational powers of a are getting closer and closer to $a^{\sqrt{3}}$. It can be shown by using advanced mathematics that there is exactly one number that these powers approach. We define $a^{\sqrt{3}}$ to be this number.

For example, using a calculator, we find

$$5^{\sqrt{3}} \approx 5^{1.732}$$

$$\approx 16.2411\ldots$$

The more decimal places of $\sqrt{3}$ we use in our calculation, the better our approximation of $5^{\sqrt{3}}$.

It can be proved that the *Laws of Exponents are still true when the exponents are real numbers.*

The Laws of Exponents are listed on page 16.

EXPONENTIAL FUNCTIONS

The **exponential function with base a** is defined for all real numbers x by

$$f(x) = a^x$$

where $a > 0$ and $a \neq 1$.

We assume that $a \neq 1$ because the function $f(x) = 1^x = 1$ is just a constant function. Here are some examples of exponential functions:

$$f(x) = 2^x \qquad g(x) = 3^x \qquad h(x) = 10^x$$

Base 2   Base 3   Base 10

▶ **EXAMPLE 1** | Evaluating Exponential Functions

Let $f(x) = 3^x$, and evaluate the following:

(a) $f(2)$ (b) $f\left(-\frac{2}{3}\right)$ (c) $f(\pi)$ (d) $f(\sqrt{2})$

▼ **SOLUTION**   We use a calculator to obtain the values of f.

	Calculator keystrokes	Output
(a) $f(2) = 3^2 = 9$	3 ^ 2 ENTER	9
(b) $f\left(-\frac{2}{3}\right) = 3^{-2/3} \approx 0.4807$	3 ^ ((−) 2 ÷ 3) ENTER	0.4807498
(c) $f(\pi) = 3^{\pi} \approx 31.544$	3 ^ π ENTER	31.5442807
(d) $f(\sqrt{2}) = 3^{\sqrt{2}} \approx 4.7288$	3 ^ √ 2 ENTER	4.7288043

▲

■ Graphs of Exponential Functions

We first graph exponential functions by plotting points. We will see that the graphs of such functions have an easily recognizable shape.

▶ **EXAMPLE 2** | Graphing Exponential Functions by Plotting Points

Draw the graph of each function.

(a) $f(x) = 3^x$ **(b)** $g(x) = \left(\dfrac{1}{3}\right)^x$

▼ **SOLUTION** We calculate values of $f(x)$ and $g(x)$ and plot points to sketch the graphs in Figure 1.

x	$f(x) = 3^x$	$g(x) = \left(\frac{1}{3}\right)^x$
-3	$\frac{1}{27}$	27
-2	$\frac{1}{9}$	9
-1	$\frac{1}{3}$	3
0	1	1
1	3	$\frac{1}{3}$
2	9	$\frac{1}{9}$
3	27	$\frac{1}{27}$

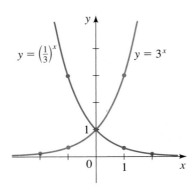

FIGURE 1

Notice that

$$g(x) = \left(\frac{1}{3}\right)^x = \frac{1}{3^x} = 3^{-x} = f(-x)$$

Reflecting graphs is explained in Section 3.5.

so we could have obtained the graph of g from the graph of f by reflecting in the y-axis. ▲

Figure 2 shows the graphs of the family of exponential functions $f(x) = a^x$ for various values of the base a. All of these graphs pass through the point $(0, 1)$ because $a^0 = 1$ for $a \neq 0$. You can see from Figure 2 that there are two kinds of exponential functions: If $0 < a < 1$, the exponential function decreases rapidly. If $a > 1$, the function increases rapidly (see the margin note).

To see just how quickly $f(x) = 2^x$ increases, let's perform the following thought experiment. Suppose we start with a piece of paper that is a thousandth of an inch thick, and we fold it in half 50 times. Each time we fold the paper, the thickness of the paper stack doubles, so the thickness of the resulting stack would be $2^{50}/1000$ inches. How thick do you think that is? It works out to be more than 17 million miles!

$y = \left(\frac{1}{3}\right)^x$ $\quad y = \left(\frac{1}{10}\right)^x$ $\quad y = 10^x$ $\quad y = 5^x$

$y = \left(\frac{1}{2}\right)^x$ $\quad y = \left(\frac{1}{5}\right)^x$ $\quad y = 3^x$

$y = \left(\frac{1}{2}\right)^x$ $\quad y = 2^x$

FIGURE 2 A family of exponential functions

See Section 4.6, page 225, where the "arrow notation" used here is explained.

The x-axis is a horizontal asymptote for the exponential function $f(x) = a^x$. This is because when $a > 1$, we have $a^x \to 0$ as $x \to -\infty$, and when $0 < a < 1$, we have $a^x \to 0$ as

$x \to \infty$ (see Figure 2). Also, $a^x > 0$ for all $x \in \mathbb{R}$, so the function $f(x) = a^x$ has domain $\mathbb{R}$ and range $(0, \infty)$. These observations are summarized in the following box.

GRAPHS OF EXPONENTIAL FUNCTIONS

The exponential function

$$f(x) = a^x \qquad (a > 0, a \neq 1)$$

has domain $\mathbb{R}$ and range $(0, \infty)$. The line $y = 0$ (the x-axis) is a horizontal asymptote of f. The graph of f has one of the following shapes.

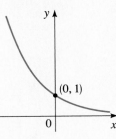

$f(x) = a^x$ for $a > 1$ $f(x) = a^x$ for $0 < a < 1$

▶ **EXAMPLE 3** | Identifying Graphs of Exponential Functions

Find the exponential function $f(x) = a^x$ whose graph is given.

(a)   **(b)**

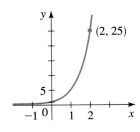   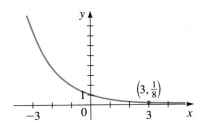

▼ **SOLUTION**

(a) Since $f(2) = a^2 = 25$, we see that the base is $a = 5$. So $f(x) = 5^x$.

(b) Since $f(3) = a^3 = \frac{1}{8}$, we see that the base is $a = \frac{1}{2}$. So $f(x) = \left(\frac{1}{2}\right)^x$.   ▲

In the next example we see how to graph certain functions, not by plotting points, but by taking the basic graphs of the exponential functions in Figure 2 and applying the shifting and reflecting transformations of Section 3.5.

▶ **EXAMPLE 4** | Transformations of Exponential Functions

Use the graph of $f(x) = 2^x$ to sketch the graph of each function.

(a) $g(x) = 1 + 2^x$

(b) $h(x) = -2^x$

(c) $k(x) = 2^{x-1}$

▼ **SOLUTION**

(a) To obtain the graph of $g(x) = 1 + 2^x$, we start with the graph of $f(x) = 2^x$ and shift it upward 1 unit. Notice from Figure 3(a) that the line $y = 1$ is now a horizontal asymptote.

Shifting and reflecting of graphs is explained in Section 3.5.

(b) Again we start with the graph of $f(x) = 2^x$, but here we reflect in the x-axis to get the graph of $h(x) = -2^x$ shown in Figure 3(b).

(c) This time we start with the graph of $f(x) = 2^x$ and shift it to the right by 1 unit to get the graph of $k(x) = 2^{x-1}$ shown in Figure 3(c).

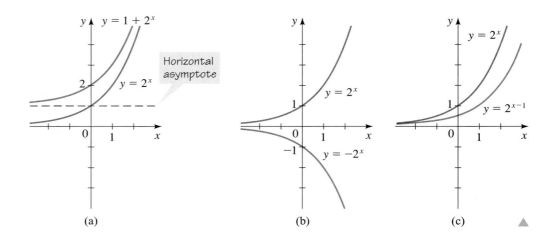

FIGURE 3

(a)   (b)   (c)

▶ **EXAMPLE 5** | Comparing Exponential and Power Functions

Compare the rates of growth of the exponential function $f(x) = 2^x$ and the power function $g(x) = x^2$ by drawing the graphs of both functions in the following viewing rectangles.

(a) $[0, 3]$ by $[0, 8]$

(b) $[0, 6]$ by $[0, 25]$

(c) $[0, 20]$ by $[0, 1000]$

▼ **SOLUTION**

(a) Figure 4(a) shows that the graph of $g(x) = x^2$ catches up with, and becomes higher than, the graph of $f(x) = 2^x$ at $x = 2$.

(b) The larger viewing rectangle in Figure 4(b) shows that the graph of $f(x) = 2^x$ overtakes that of $g(x) = x^2$ when $x = 4$.

(c) Figure 4(c) gives a more global view and shows that when x is large, $f(x) = 2^x$ is much larger than $g(x) = x^2$.

(a)

(b)

(c)

FIGURE 4

The Natural Exponential Function

Any positive number can be used as the base for an exponential function, but some bases are used more frequently than others. We will see in the remaining sections of this chapter that the bases 2 and 10 are convenient for certain applications, but the most important base is the number denoted by the letter e.

The number e is defined as the value that $(1 + 1/n)^n$ approaches as n becomes large. (In calculus this idea is made more precise through the concept of a limit.) The table in the margin shows the values of the expression $(1 + 1/n)^n$ for increasingly large values of n. It appears that, correct to five decimal places, $e \approx 2.71828$; in fact, the approximate value to 20 decimal places is

$$e \approx 2.71828182845904523536$$

It can be shown that e is an irrational number, so we cannot write its exact value in decimal form.

Why use such a strange base for an exponential function? It might seem at first that a base such as 10 is easier to work with. We will see, however, that in certain applications the number e is the best possible base. In this section we study how e occurs in the description of compound interest.

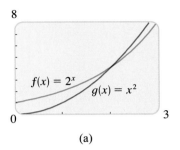

n	$\left(1 + \dfrac{1}{n}\right)^n$
1	2.00000
5	2.48832
10	2.59374
100	2.70481
1000	2.71692
10,000	2.71815
100,000	2.71827
1,000,000	2.71828

The notation e was chosen by Leonhard Euler, probably because it is the first letter of the word *exponential*.

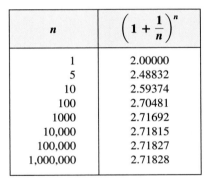

FIGURE 5 Graph of the natural exponential function

> ### THE NATURAL EXPONENTIAL FUNCTION
>
> The **natural exponential function** is the exponential function
> $$f(x) = e^x$$
> with base e. It is often referred to as *the* exponential function.

Since $2 < e < 3$, the graph of the natural exponential function lies between the graphs of $y = 2^x$ and $y = 3^x$, as shown in Figure 5.

Scientific calculators have a special key for the function $f(x) = e^x$. We use this key in the next example.

▶ **EXAMPLE 6** | Evaluating the Exponential Function

Evaluate each expression correct to five decimal places.

(a) e^3 **(b)** $2e^{-0.53}$ **(c)** $e^{4.8}$

▼ **SOLUTION**   We use the $\boxed{e^x}$ key on a calculator to evaluate the exponential function.

(a) $e^3 \approx 20.08554$ **(b)** $2e^{-0.53} \approx 1.17721$ **(c)** $e^{4.8} \approx 121.51042$ ▲

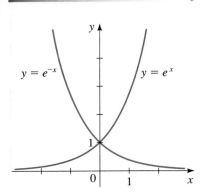

FIGURE 6

▶ **EXAMPLE 7** | Transformations of the Exponential Function

Sketch the graph of each function.

(a) $f(x) = e^{-x}$

(b) $g(x) = 3e^{0.5x}$

▼ **SOLUTION**

(a) We start with the graph of $y = e^x$ and reflect in the y-axis to obtain the graph of $y = e^{-x}$ as in Figure 6.

(b) We calculate several values, plot the resulting points, then connect the points with a smooth curve. The graph is shown in Figure 7.

x	$f(x) = 3e^{0.5x}$
-3	0.67
-2	1.10
-1	1.82
0	3.00
1	4.95
2	8.15
3	13.45

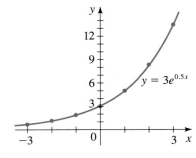

FIGURE 7

▶ **EXAMPLE 8** | An Exponential Model for the Spread of a Virus

An infectious disease begins to spread in a small city of population 10,000. After t days, the number of people who have succumbed to the virus is modeled by the function

$$v(t) = \frac{10{,}000}{5 + 1245e^{-0.97t}}$$

(a) How many infected people are there initially (at time $t = 0$)?

(b) Find the number of infected people after one day, two days, and five days.

(c) Graph the function v, and describe its behavior.

▼ **SOLUTION**

(a) Since $v(0) = 10{,}000/(5 + 1245e^0) = 10{,}000/1250 = 8$, we conclude that 8 people initially have the disease.

(b) Using a calculator, we evaluate $v(1)$, $v(2)$, and $v(5)$ and then round off to obtain the following values.

Days	Infected people
1	21
2	54
5	678

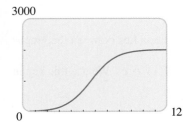

FIGURE 8

$$v(t) = \frac{10{,}000}{5 + 1245e^{-0.97t}}$$

(c) From the graph in Figure 8 we see that the number of infected people first rises slowly, then rises quickly between day 3 and day 8, and then levels off when about 2000 people are infected.

The graph in Figure 8 is called a *logistic curve* or a *logistic growth model*. Curves like it occur frequently in the study of population growth.

■ Compound Interest

Exponential functions occur in calculating compound interest. If an amount of money P, called the **principal**, is invested at an interest rate i per time period, then after one time period the interest is Pi, and the amount A of money is

$$A = P + Pi = P(1 + i)$$

If the interest is reinvested, then the new principal is $P(1 + i)$, and the amount after another time period is $A = P(1 + i)(1 + i) = P(1 + i)^2$. Similarly, after a third time period the amount is $A = P(1 + i)^3$. In general, after k periods the amount is

$$A = P(1 + i)^k$$

Notice that this is an exponential function with base $1 + i$.

If the annual interest rate is r and if interest is compounded n times per year, then in each time period the interest rate is $i = r/n$, and there are nt time periods in t years. This leads to the following formula for the amount after t years.

COMPOUND INTEREST

Compound interest is calculated by the formula

$$A(t) = P\left(1 + \frac{r}{n}\right)^{nt}$$

where $A(t)$ = amount after t years

P = principal

r = interest rate per year

n = number of times interest is compounded per year

t = number of years

---

r is often referred to as the nominal annual interest rate.

▶ **EXAMPLE 9** | Calculating Compound Interest

A sum of $1000 is invested at an interest rate of 12% per year. Find the amounts in the account after 3 years if interest is compounded annually, semiannually, quarterly, monthly, and daily.

▼ **SOLUTION**   We use the compound interest formula with $P = \$1000$, $r = 0.12$, and $t = 3$.

Compounding	n	Amount after 3 years	
Annual	1	$1000\left(1 + \dfrac{0.12}{1}\right)^{1(3)}$	$= \$1404.93$
Semiannual	2	$1000\left(1 + \dfrac{0.12}{2}\right)^{2(3)}$	$= \$1418.52$
Quarterly	4	$1000\left(1 + \dfrac{0.12}{4}\right)^{4(3)}$	$= \$1425.76$
Monthly	12	$1000\left(1 + \dfrac{0.12}{12}\right)^{12(3)}$	$= \$1430.77$
Daily	365	$1000\left(1 + \dfrac{0.12}{365}\right)^{365(3)}$	$= \$1433.24$

We see from Example 9 that the interest paid increases as the number of compounding periods n increases. Let's see what happens as n increases indefinitely. If we let $m = n/r$, then

$$A(t) = P\left(1 + \frac{r}{n}\right)^{nt} = P\left[\left(1 + \frac{r}{n}\right)^{n/r}\right]^{rt} = P\left[\left(1 + \frac{1}{m}\right)^{m}\right]^{rt}$$

Recall that as m becomes large, the quantity $(1 + 1/m)^m$ approaches the number e. Thus, the amount approaches $A = Pe^{rt}$. This expression gives the amount when the interest is compounded at "every instant."

CONTINUOUSLY COMPOUNDED INTEREST

Continuously compounded interest is calculated by the formula

$$A(t) = Pe^{rt}$$

where $A(t)$ = amount after t years

P = principal

r = interest rate per year

t = number of years

▶ **EXAMPLE 10** | Calculating Continuously Compounded Interest

Find the amount after 3 years if $1000 is invested at an interest rate of 12% per year, compounded continuously.

▼ **SOLUTION**   We use the formula for continuously compounded interest with $P = 1000, $r = 0.12$, and $t = 3$ to get

$$A(3) = 1000e^{(0.12)3} = 1000e^{0.36} = \$1433.33$$

Compare this amount with the amounts in Example 9.   ▲

| 5.2 | Logarithmic Functions |

LEARNING OBJECTIVES

After completing this section, you will be able to:

- Evaluate logarithmic functions
- Change between logarithmic and exponential forms of an expression
- Use basic properties of logarithms
- Graph logarithmic functions
- Use common and natural logarithms

In this section we study the inverses of exponential functions.

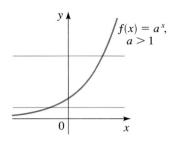

FIGURE 1 $f(x) = a^x$ is one-to-one

Logarithmic Functions

Every exponential function $f(x) = a^x$, with $a > 0$ and $a \neq 1$, is a one-to-one function by the Horizontal Line Test (see Figure 1 for the case $a > 1$) and therefore has an inverse function. The inverse function f^{-1} is called the *logarithmic function with base a* and is denoted by $\log_a$. Recall from Section 3.7 that f^{-1} is defined by

$$f^{-1}(x) = y \quad \Leftrightarrow \quad f(y) = x$$

This leads to the following definition of the logarithmic function.

> **DEFINITION OF THE LOGARITHMIC FUNCTION**
>
> Let a be a positive number with $a \neq 1$. The **logarithmic function with base a**, denoted by **$\log_a$**, is defined by
>
> $$\log_a x = y \quad \Leftrightarrow \quad a^y = x$$
>
> So $\log_a x$ is the *exponent* to which the base a must be raised to give x.

We read $\log_a x = y$ as "log base a of x is y."

By tradition the name of the logarithmic function is $\log_a$, not just a single letter. Also, we usually omit the parentheses in the function notation and write

$$\log_a(x) = \log_a x$$

When we use the definition of logarithms to switch back and forth between the **logarithmic form** $\log_a x = y$ and the **exponential form** $a^y = x$, it is helpful to notice that, in both forms, the base is the same:

Logarithmic form	**Exponential form**
Exponent	Exponent
$\log_a x = y$	$a^y = x$
Base	Base

▷ **EXAMPLE 1** | Logarithmic and Exponential Forms

The logarithmic and exponential forms are equivalent equations: If one is true, then so is the other. So we can switch from one form to the other as in the following illustrations.

x	$\log_{10} x$
10^4	4
10^3	3
10^2	2
10	1
1	0
10^{-1}	-1
10^{-2}	-2
10^{-3}	-3
10^{-4}	-4

Logarithmic form	**Exponential form**
$\log_{10} 100{,}000 = 5$	$10^5 = 100{,}000$
$\log_2 8 = 3$	$2^3 = 8$
$\log_2\left(\frac{1}{8}\right) = -3$	$2^{-3} = \frac{1}{8}$
$\log_5 s = r$	$5^r = s$

▲

It is important to understand that $\log_a x$ is an *exponent*. For example, the numbers in the right column of the table in the margin are the logarithms (base 10) of the numbers in the left column. This is the case for all bases, as the following example illustrates.

▶ **EXAMPLE 2 |** Evaluating Logarithms

(a) $\log_{10} 1000 = 3$ because $10^3 = 1000$

(b) $\log_2 32 = 5$ because $2^5 = 32$

(c) $\log_{10} 0.1 = -1$ because $10^{-1} = 0.1$

(d) $\log_{16} 4 = \frac{1}{2}$ because $16^{1/2} = 4$ ▲

Inverse Function Property:

$$f^{-1}(f(x)) = x$$

$$f(f^{-1}(x)) = x$$

When we apply the Inverse Function Property described on page 176 to $f(x) = a^x$ and $f^{-1}(x) = \log_a x$, we get

$$\log_a(a^x) = x \qquad x \in \mathbb{R}$$

$$a^{\log_a x} = x \qquad x > 0$$

We list these and other properties of logarithms discussed in this section.

PROPERTIES OF LOGARITHMS

Property	Reason
1. $\log_a 1 = 0$	We must raise a to the power 0 to get 1.
2. $\log_a a = 1$	We must raise a to the power 1 to get a.
3. $\log_a a^x = x$	We must raise a to the power x to get a^x.
4. $a^{\log_a x} = x$	$\log_a x$ is the power to which a must be raised to get x.

▶ **EXAMPLE 3 |** Applying Properties of Logarithms

We illustrate the properties of logarithms when the base is 5.

$\log_5 1 = 0$ *Property 1* $\log_5 5 = 1$ *Property 2*

$\log_5 5^8 = 8$ *Property 3* $5^{\log_5 12} = 12$ *Property 4*   ▲

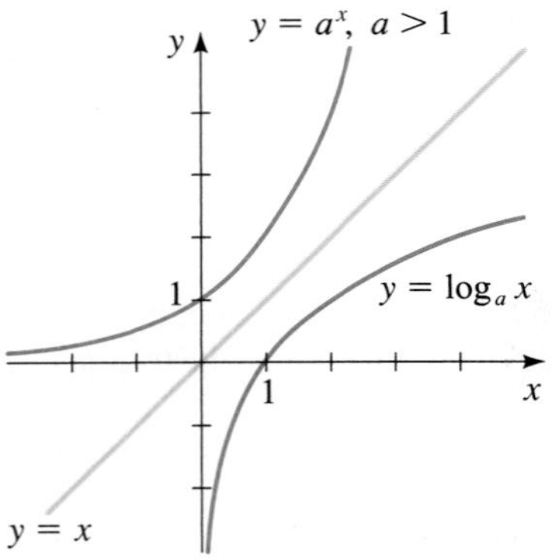

■ Graphs of Logarithmic Functions

$y = a^x, \ a > 1$

$y = \log_a x$

$y = x$

FIGURE 2 Graph of the logarithmic function $f(x) = \log_a x$

Arrow notation is explained on page 225.

Recall that if a one-to-one function f has domain A and range B, then its inverse function f^{-1} has domain B and range A. Since the exponential function $f(x) = a^x$ with $a \neq 1$ has domain $\mathbb{R}$ and range $(0, \infty)$, we conclude that its inverse function, $f^{-1}(x) = \log_a x$, has domain $(0, \infty)$ and range $\mathbb{R}$.

The graph of $f^{-1}(x) = \log_a x$ is obtained by reflecting the graph of $f(x) = a^x$ in the line $y = x$. Figure 2 shows the case $a > 1$. The fact that $y = a^x$ (for $a > 1$) is a very rapidly increasing function for $x > 0$ implies that $y = \log_a x$ is a very slowly increasing function for $x > 1$.

Since $\log_a 1 = 0$, the x-intercept of the function $y = \log_a x$ is 1. The y-axis is a vertical asymptote of $y = \log_a x$ because $\log_a x \to -\infty$ as $x \to 0^+$.

▶ **EXAMPLE 4 |** Graphing a Logarithmic Function by Plotting Points

Sketch the graph of $f(x) = \log_2 x$.

▼ **SOLUTION**   To make a table of values, we choose the x-values to be powers of 2 so that we can easily find their logarithms. We plot these points and connect them with a smooth curve as in Figure 3.

x	$\log_2 x$
2^3	3
2^2	2
2	1
1	0
2^{-1}	-1
2^{-2}	-2
2^{-3}	-3
2^{-4}	-4

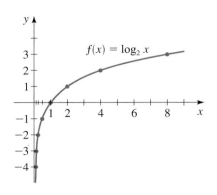

FIGURE 3

Figure 4 shows the graphs of the family of logarithmic functions with bases 2, 3, 5, and 10. These graphs are drawn by reflecting the graphs of $y = 2^x$, $y = 3^x$, $y = 5^x$, and $y = 10^x$ (see Figure 2 in Section 5.1) in the line $y = x$. We can also plot points as an aid to sketching these graphs, as illustrated in Example 4.

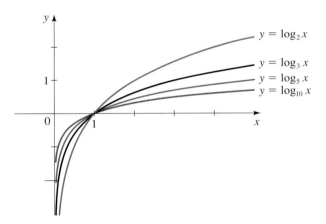

FIGURE 4   A family of logarithmic functions

In the next two examples we graph logarithmic functions by starting with the basic graphs in Figure 4 and using the transformations of Section 3.5.

▶ **EXAMPLE 5** | Reflecting Graphs of Logarithmic Functions

Sketch the graph of each function.

(a) $g(x) = -\log_2 x$

(b) $h(x) = \log_2(-x)$

▼ **SOLUTION**

(a) We start with the graph of $f(x) = \log_2 x$ and reflect in the x-axis to get the graph of $g(x) = -\log_2 x$ in Figure 5(a) on the next page.

(b) We start with the graph of $f(x) = \log_2 x$ and reflect in the y-axis to get the graph of $h(x) = \log_2(-x)$ in Figure 5(b) on the next page.

FIGURE 5   (a)   (b)

▶ **EXAMPLE 6** | Shifting Graphs of Logarithmic Functions

Find the domain of each function, and sketch the graph.

(a) $g(x) = 2 + \log_5 x$

(b) $h(x) = \log_{10}(x - 3)$

▼ **SOLUTION**

(a) The graph of g is obtained from the graph of $f(x) = \log_5 x$ (Figure 4) by shifting upward 2 units (see Figure 6). The domain of f is $(0, \infty)$.

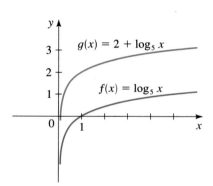

FIGURE 6

(b) The graph of h is obtained from the graph of $f(x) = \log_{10} x$ (Figure 4) by shifting to the right 3 units (see Figure 7 below). The line $x = 3$ is a vertical asymptote. Since $\log_{10} x$ is defined only when $x > 0$, the domain of $h(x) = \log_{10}(x - 3)$ is

$$\{x \mid x - 3 > 0\} = \{x \mid x > 3\} = (3, \infty)$$

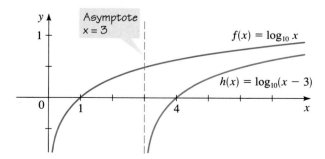

FIGURE 7

■ Common Logarithms

We now study logarithms with base 10.

> **COMMON LOGARITHM**
>
> The logarithm with base 10 is called the **common logarithm** and is denoted by omitting the base:
>
> $$\log x = \log_{10} x$$

From the definition of logarithms we can easily find that

$$\log 10 = 1 \quad \text{and} \quad \log 100 = 2$$

But how do we find log 50? We need to find the exponent y such that $10^y = 50$. Clearly, 1 is too small and 2 is too large. So

$$1 < \log 50 < 2$$

To get a better approximation, we can experiment to find a power of 10 closer to 50. Fortunately, scientific calculators are equipped with a $\boxed{\text{LOG}}$ key that directly gives values of common logarithms.

▶ **EXAMPLE 7** | Evaluating Common Logarithms

Use a calculator to find appropriate values of $f(x) = \log x$ and use the values to sketch the graph.

▼ **SOLUTION**   We make a table of values, using a calculator to evaluate the function at those values of x that are not powers of 10. We plot those points and connect them by a smooth curve as in Figure 8.

x	$\log x$
0.01	−2
0.1	−1
0.5	−0.301
1	0
4	0.602
5	0.699
10	1

FIGURE 8

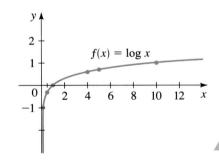

Human response to sound and light intensity is logarithmic.

Scientists model human response to stimuli (such as sound, light, or pressure) using logarithmic functions. For example, the intensity of a sound must be increased manyfold before we "feel" that the loudness has simply doubled. The psychologist Gustav Fechner formulated the law as

$$S = k \log\left(\frac{I}{I_0}\right)$$

where S is the subjective intensity of the stimulus, I is the physical intensity of the stimulus, I_0 stands for the threshold physical intensity, and k is a constant that is different for each sensory stimulus.

We study the decibel scale in more detail in Section 5.5.

► **EXAMPLE 8** | Common Logarithms and Sound

The perception of the loudness B (in decibels, dB) of a sound with physical intensity I (in W/m^2) is given by

$$B = 10 \log\left(\frac{I}{I_0}\right)$$

where I_0 is the physical intensity of a barely audible sound. Find the decibel level (loudness) of a sound whose physical intensity I is 100 times that of I_0.

▼ **SOLUTION**   We find the decibel level B by using the fact that $I = 100I_0$.

$$B = 10 \log\left(\frac{I}{I_0}\right) \qquad \text{Definition of } B$$

$$= 10 \log\left(\frac{100I_0}{I_0}\right) \qquad I = 100I_0$$

$$= 10 \log 100 \qquad \text{Cancel } I_0$$

$$= 10 \cdot 2 = 20 \qquad \text{Definition of log}$$

The loudness of the sound is 20 dB.   ▲

Natural Logarithms

Of all possible bases a for logarithms, it turns out that the most convenient choice for the purposes of calculus is the number e, which we defined in Section 5.1.

The notation ln is an abbreviation for the Latin name *logarithmus naturalis*.

NATURAL LOGARITHM

The logarithm with base e is called the **natural logarithm** and is denoted by **ln**:

$$\ln x = \log_e x$$

The natural logarithmic function $y = \ln x$ is the inverse function of the exponential function $y = e^x$. Both functions are graphed in Figure 9. By the definition of inverse functions we have

$$\ln x = y \quad \Leftrightarrow \quad e^y = x$$

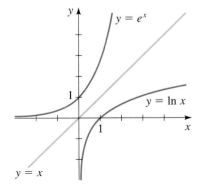

FIGURE 9 Graph of the natural logarithmic function

If we substitute $a = e$ and write "ln" for "$\log_e$" in the properties of logarithms mentioned earlier, we obtain the following properties of natural logarithms.

PROPERTIES OF NATURAL LOGARITHMS

Property	Reason
1. $\ln 1 = 0$	We must raise e to the power 0 to get 1.
2. $\ln e = 1$	We must raise e to the power 1 to get e.
3. $\ln e^x = x$	We must raise e to the power x to get e^x.
4. $e^{\ln x} = x$	$\ln x$ is the power to which e must be raised to get x.

Calculators are equipped with an $\boxed{\text{LN}}$ key that directly gives the values of natural logarithms.

▶ **EXAMPLE 9** | Evaluating the Natural Logarithm Function

(a) $\ln e^8 = 8$ *Definition of natural logarithm*

(b) $\ln\left(\dfrac{1}{e^2}\right) = \ln e^{-2} = -2$ *Definition of natural logarithm*

(c) $\ln 5 \approx 1.609$ *Use* $\boxed{\text{LN}}$ *key on calculator*   ▲

▶ **EXAMPLE 10** | Finding the Domain of a Logarithmic Function

Find the domain of the function $f(x) = \ln(4 - x^2)$.

▼ **SOLUTION**   As with any logarithmic function, $\ln x$ is defined when $x > 0$. Thus, the domain of f is

$$\{x \mid 4 - x^2 > 0\} = \{x \mid x^2 < 4\} = \{x \mid |x| < 2\}$$

$$= \{x \mid -2 < x < 2\} = (-2, 2)$$ ▲

▶ **EXAMPLE 11** | Drawing the Graph of a Logarithmic Function

Draw the graph of the function $y = x \ln(4 - x^2)$, and use it to find the asymptotes and local maximum and minimum values.

▼ **SOLUTION**   As in Example 10 the domain of this function is the interval $(-2, 2)$, so we choose the viewing rectangle $[-3, 3]$ by $[-3, 3]$. The graph is shown in Figure 10, and from it we see that the lines $x = -2$ and $x = 2$ are vertical asymptotes.

The function has a local maximum point to the right of $x = 1$ and a local minimum point to the left of $x = -1$. By zooming in and tracing along the graph with the cursor, we find that the local maximum value is approximately 1.13 and this occurs when $x \approx 1.15$. Similarly (or by noticing that the function is odd), we find that the local minimum value is about -1.13, and it occurs when $x \approx -1.15$.   ▲

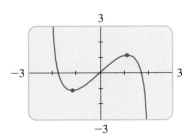

FIGURE 10
$y = x \ln(4 - x^2)$

| 5.3 | Laws of Logarithms |

LEARNING OBJECTIVES

After completing this section, you will be able to:

■ Use the Laws of Logarithms to evaluate logarithmic expressions
■ Use the Laws of Logarithms to expand logarithmic expressions
■ Use the Laws of Logarithms to combine logarithmic expressions
■ Use the Change of Base Formula

In this section we study properties of logarithms. These properties give logarithmic functions a wide range of applications, as we will see in Section 5.5.

■ Laws of Logarithms

Since logarithms are exponents, the Laws of Exponents give rise to the Laws of Logarithms.

LAWS OF LOGARITHMS

Let a be a positive number, with $a \neq 1$. Let A, B, and C be any real numbers with $A > 0$ and $B > 0$.

Law	Description
1. $\log_a(AB) = \log_a A + \log_a B$	The logarithm of a product of numbers is the sum of the logarithms of the numbers.
2. $\log_a\left(\dfrac{A}{B}\right) = \log_a A - \log_a B$	The logarithm of a quotient of numbers is the difference of the logarithms of the numbers.
3. $\log_a(A^C) = C \log_a A$	The logarithm of a power of a number is the exponent times the logarithm of the number.

▼ **PROOF** We make use of the property $\log_a a^x = x$ from Section 5.2.

Law 1   Let $\log_a A = u$ and $\log_a B = v$. When written in exponential form, these equations become

$$a^u = A \qquad \text{and} \qquad a^v = B$$

Thus

$$\log_a(AB) = \log_a(a^u a^v) = \log_a(a^{u+v})$$
$$= u + v = \log_a A + \log_a B$$

Law 2   Using Law 1, we have

$$\log_a A = \log_a\left[\left(\frac{A}{B}\right)B\right] = \log_a\left(\frac{A}{B}\right) + \log_a B$$

so

$$\log_a\left(\frac{A}{B}\right) = \log_a A - \log_a B$$

Law 3   Let $\log_a A = u$. Then $a^u = A$, so

$$\log_a(A^C) = \log_a(a^u)^C = \log_a(a^{uC}) = uC = C \log_a A \qquad \blacktriangle$$

▶ **EXAMPLE 1** | Using the Laws of Logarithms to Evaluate Expressions

Evaluate each expression.

(a) $\log_4 2 + \log_4 32$

(b) $\log_2 80 - \log_2 5$

(c) $-\frac{1}{3}\log 8$

▼ **SOLUTION**

(a) $\log_4 2 + \log_4 32 = \log_4(2 \cdot 32)$ Law 1

$\qquad\qquad\qquad\quad = \log_4 64 = 3$ Because $64 = 4^3$

(b) $\log_2 80 - \log_2 5 = \log_2\left(\frac{80}{5}\right)$ Law 2

$\qquad\qquad\qquad\quad = \log_2 16 = 4$ Because $16 = 2^4$

(c) $-\frac{1}{3}\log 8 = \log 8^{-1/3}$ Law 3

$\qquad\qquad\quad = \log\left(\frac{1}{2}\right)$ Property of negative exponents

$\qquad\qquad\quad \approx -0.301$ Calculator   ▲

■ Expanding and Combining Logarithmic Expressions

The Laws of Logarithms allow us to write the logarithm of a product or a quotient as the sum or difference of logarithms. This process, called *expanding* a logarithmic expression, is illustrated in the next example.

▶ **EXAMPLE 2** | Expanding Logarithmic Expressions

Use the Laws of Logarithms to expand each expression.

(a) $\log_2(6x)$ **(b)** $\log_5(x^3 y^6)$ **(c)** $\ln\left(\dfrac{ab}{\sqrt[3]{c}}\right)$

▼ **SOLUTION**

(a) $\log_2(6x) = \log_2 6 + \log_2 x$ Law 1

(b) $\log_5(x^3 y^6) = \log_5 x^3 + \log_5 y^6$ Law 1

$\qquad\qquad\quad = 3\log_5 x + 6\log_5 y$ Law 3

(c) $\ln\left(\dfrac{ab}{\sqrt[3]{c}}\right) = \ln(ab) - \ln\sqrt[3]{c}$ Law 2

$\qquad\qquad\quad = \ln a + \ln b - \ln c^{1/3}$ Law 1

$\qquad\qquad\quad = \ln a + \ln b - \frac{1}{3}\ln c$ Law 3   ▲

The Laws of Logarithms also allow us to reverse the process of expanding that was done in Example 2. That is, we can write sums and differences of logarithms as a single logarithm. This process, called *combining* logarithmic expressions, is illustrated in the next example.

▶ **EXAMPLE 3** | Combining Logarithmic Expressions

Combine $3\log x + \frac{1}{2}\log(x + 1)$ into a single logarithm.

▼ **SOLUTION**

$$3 \log x + \tfrac{1}{2}\log(x + 1) = \log x^3 + \log(x + 1)^{1/2} \qquad \text{Law 3}$$

$$= \log(x^3(x + 1)^{1/2}) \qquad \text{Law 1} \qquad \blacktriangle$$

▶ **EXAMPLE 4** | Combining Logarithmic Expressions

Combine $3 \ln s + \tfrac{1}{2}\ln t - 4 \ln(t^2 + 1)$ into a single logarithm.

▼ **SOLUTION**

$$3 \ln s + \tfrac{1}{2}\ln t - 4 \ln(t^2 + 1) = \ln s^3 + \ln t^{1/2} - \ln(t^2 + 1)^4 \qquad \text{Law 3}$$

$$= \ln(s^3 t^{1/2}) - \ln(t^2 + 1)^4 \qquad \text{Law 1}$$

$$= \ln\!\left(\frac{s^3 \sqrt{t}}{(t^2 + 1)^4}\right) \qquad \text{Law 2} \qquad \blacktriangle$$

WARNING   Although the Laws of Logarithms tell us how to compute the logarithm of a product or a quotient, *there is no corresponding rule for the logarithm of a sum or a difference*. For instance,

$$\log_a(x + y) \neq \log_a x + \log_a y$$

In fact, we know that the right side is equal to $\log_a(xy)$. Also, don't improperly simplify quotients or powers of logarithms. For instance,

$$\frac{\log 6}{\log 2} \neq \log\!\left(\frac{6}{2}\right) \qquad \text{and} \qquad (\log_2 x)^3 \neq 3 \log_2 x$$

Logarithmic functions are used to model a variety of situations involving human behavior. One such behavior is how quickly we forget things we have learned. For example, if you learn algebra at a certain performance level (say, 90% on a test) and then don't use algebra for a while, how much will you retain after a week, a month, or a year? Hermann Ebbinghaus (1850–1909) studied this phenomenon and formulated the law described in the next example.

Forgetting what we've learned depends logarithmically on how long ago we learned it.

▶ **EXAMPLE 5** | The Law of Forgetting

Ebbinghaus' Law of Forgetting states that if a task is learned at a performance level P_0, then after a time interval t the performance level P satisfies

$$\log P = \log P_0 - c \log(t + 1)$$

where c is a constant that depends on the type of task and t is measured in months.

(a) Solve for P.

(b) If your score on a history test is 90, what score would you expect to get on a similar test after two months? After a year? (Assume that $c = 0.2$.)

▼ **SOLUTION**

(a) We first combine the right-hand side.

$$\log P = \log P_0 - c \log(t + 1) \qquad \textit{Given equation}$$

$$\log P = \log P_0 - \log(t + 1)^c \qquad \textit{Law 3}$$

$$\log P = \log \frac{P_0}{(t + 1)^c} \qquad \textit{Law 2}$$

$$P = \frac{P_0}{(t + 1)^c} \qquad \textit{Because log is one-to-one}$$

(b) Here $P_0 = 90$, $c = 0.2$, and t is measured in months.

$$\text{In two months:} \qquad t = 2 \qquad \text{and} \qquad P = \frac{90}{(2 + 1)^{0.2}} \approx 72$$

$$\text{In one year:} \qquad t = 12 \qquad \text{and} \qquad P = \frac{90}{(12 + 1)^{0.2}} \approx 54$$

Your expected scores after two months and one year are 72 and 54, respectively. ▲

■ Change of Base

For some purposes, we find it useful to change from logarithms in one base to logarithms in another base. Suppose we are given $\log_a x$ and want to find $\log_b x$. Let

$$y = \log_b x$$

We write this in exponential form and take the logarithm, with base a, of each side.

$$b^y = x \qquad\qquad \text{Exponential form}$$

$$\log_a(b^y) = \log_a x \qquad \text{Take } \log_a \text{ of each side}$$

$$y \log_a b = \log_a x \qquad \text{Law 3}$$

$$y = \frac{\log_a x}{\log_a b} \qquad \text{Divide by } \log_a b$$

We may write the Change of Base Formula as

$$\log_b x = \left(\frac{1}{\log_a b}\right) \log_a x$$

So $\log_b x$ is just a constant multiple of $\log_a x$; the constant is $\frac{1}{\log_a b}$.

This proves the following formula.

CHANGE OF BASE FORMULA

$$\log_b x = \frac{\log_a x}{\log_a b}$$

In particular, if we put $x = a$, then $\log_a a = 1$, and this formula becomes

$$\log_b a = \frac{1}{\log_a b}$$

We can now evaluate a logarithm to *any* base by using the Change of Base Formula to express the logarithm in terms of common logarithms or natural logarithms and then using a calculator.

▶ **EXAMPLE 6** | Evaluating Logarithms with the Change of Base Formula

Use the Change of Base Formula and common or natural logarithms to evaluate each logarithm, correct to five decimal places.

(a) $\log_8 5$ **(b)** $\log_9 20$

▼ **SOLUTION**

We get the same answer whether we use $\log_{10}$ or ln:

$$\log_8 5 = \frac{\ln 5}{\ln 8} \approx 0.77398$$

(a) We use the Change of Base Formula with $b = 8$ and $a = 10$:

$$\log_8 5 = \frac{\log_{10} 5}{\log_{10} 8} \approx 0.77398$$

(b) We use the Change of Base Formula with $b = 9$ and $a = e$:

$$\log_9 20 = \frac{\ln 20}{\ln 9} \approx 1.36342$$ ▲

▶ **EXAMPLE 7** | Using the Change of Base Formula to Graph a Logarithmic Function

Use a graphing calculator to graph $f(x) = \log_6 x$.

▼ **SOLUTION** Calculators don't have a key for $\log_6$, so we use the Change of Base Formula to write

$$f(x) = \log_6 x = \frac{\ln x}{\ln 6}$$

Since calculators do have an $\boxed{\text{LN}}$ key, we can enter this new form of the function and graph it. The graph is shown in Figure 1. ▲

FIGURE 1 $f(x) = \log_6 x = \dfrac{\ln x}{\ln 6}$

5.4 | Exponential and Logarithmic Equations

LEARNING OBJECTIVES

After completing this section, you will be able to:

- ▨ Solve exponential equations
- ▨ Solve logarithmic equations
- ▨ Solve problems involving compound interest
- ▨ Calculate annual percentage yield

In this section we solve equations that involve exponential or logarithmic functions. The techniques that we develop here will be used in the next section for solving applied problems.

▨ Exponential Equations

An *exponential equation* is one in which the variable occurs in the exponent. For example,

$$2^x = 7$$

The variable x presents a difficulty because it is in the exponent. To deal with this difficulty, we take the logarithm of each side and then use the Laws of Logarithms to "bring down x" from the exponent.

$2^x = 7$	*Given equation*
$\ln 2^x = \ln 7$	*Take ln of each side*
$x \ln 2 = \ln 7$	*Law 3 (bring down exponent)*
$x = \dfrac{\ln 7}{\ln 2}$	*Solve for x*
≈ 2.807	*Calculator*

Recall that Law 3 of the Laws of Logarithms says that $\log_a A^C = C \log_a A$.

The method that we used to solve $2^x = 7$ is typical of how we solve exponential equations in general.

> **GUIDELINES FOR SOLVING EXPONENTIAL EQUATIONS**
>
> **1.** Isolate the exponential expression on one side of the equation.
> **2.** Take the logarithm of each side, then use the Laws of Logarithms to "bring down the exponent."
> **3.** Solve for the variable.

▶ **EXAMPLE 1** | Solving an Exponential Equation

Find the solution of the equation $3^{x+2} = 7$, correct to six decimal places.

▼ **SOLUTION**   We take the common logarithm of each side and use Law 3.

We could have used natural logarithms instead of common logarithms. In fact, using the same steps, we get

$$x = \frac{\ln 7}{\ln 3} - 2 \approx -0.228756$$

$3^{x+2} = 7$	*Given equation*
$\log(3^{x+2}) = \log 7$	*Take log of each side*
$(x + 2)\log 3 = \log 7$	*Law 3 (bring down exponent)*
$x + 2 = \dfrac{\log 7}{\log 3}$	*Divide by log 3*
$x = \dfrac{\log 7}{\log 3} - 2$	*Subtract 2*
≈ -0.228756	*Calculator*

▲

Check Your Answer

Substituting $x = -0.228756$ into the original equation and using a calculator, we get

$$3^{(-0.228756)+2} \approx 7 \quad ✔$$

▶ **EXAMPLE 2** | Solving an Exponential Equation

Solve the equation $8e^{2x} = 20$.

▼ **SOLUTION**   We first divide by 8 to isolate the exponential term on one side of the equation.

$8e^{2x} = 20$	*Given equation*
$e^{2x} = \frac{20}{8}$	*Divide by 8*
$\ln e^{2x} = \ln 2.5$	*Take ln of each side*
$2x = \ln 2.5$	*Property of ln*
$x = \dfrac{\ln 2.5}{2}$	*Divide by 2*
≈ 0.458	*Calculator*

▲

Check Your Answer

Substituting $x = 0.458$ into the original equation and using a calculator, we get

$$8e^{2(0.458)} \approx 20 \quad ✔$$

▶ **EXAMPLE 3** │ Solving an Exponential Equation Algebraically and Graphically

Solve the equation $e^{3-2x} = 4$ algebraically and graphically.

▼ **SOLUTION 1:** Algebraic

Since the base of the exponential term is e, we use natural logarithms to solve this equation.

$e^{3-2x} = 4$	Given equation
$\ln(e^{3-2x}) = \ln 4$	Take ln of each side
$3 - 2x = \ln 4$	Property of ln
$-2x = -3 + \ln 4$	Subtract 3
$x = \frac{1}{2}(3 - \ln 4) \approx 0.807$	Multiply by $-\frac{1}{2}$

You should check that this answer satisfies the original equation.

▼ **SOLUTION 2:** Graphical

We graph the equations $y = e^{3-2x}$ and $y = 4$ in the same viewing rectangle as in Figure 1. The solutions occur where the graphs intersect. Zooming in on the point of intersection of the two graphs, we see that $x \approx 0.81$.   ▲

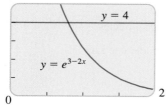

FIGURE 1

▶ **EXAMPLE 4** │ An Exponential Equation of Quadratic Type

Solve the equation $e^{2x} - e^x - 6 = 0$.

▼ **SOLUTION**   To isolate the exponential term, we factor.

If we let $w = e^x$, we get the quadratic equation

$$w^2 - w - 6 = 0$$

which factors as

$$(w - 3)(w + 2) = 0$$

$e^{2x} - e^x - 6 = 0$	Given equation
$(e^x)^2 - e^x - 6 = 0$	Law of Exponents
$(e^x - 3)(e^x + 2) = 0$	Factor (a quadratic in e^x)
$e^x - 3 = 0 \quad$ or $\quad e^x + 2 = 0$	Zero-Product Property
$e^x = 3 \qquad\qquad e^x = -2$	

The equation $e^x = 3$ leads to $x = \ln 3$. But the equation $e^x = -2$ has no solution because $e^x > 0$ for all x. Thus, $x = \ln 3 \approx 1.0986$ is the only solution. You should check that this answer satisfies the original equation.   ▲

▶ **EXAMPLE 5** │ Solving an Exponential Equation

Solve the equation $3xe^x + x^2e^x = 0$.

▼ **SOLUTION**   First we factor the left side of the equation.

Check Your Answers

$x = 0$:

$3(0)e^0 + 0^2e^0 = 0$ ✔

$x = -3$:

$3(-3)e^{-3} + (-3)^2e^{-3}$

$= -9e^{-3} + 9e^{-3} = 0$ ✔

$3xe^x + x^2e^x = 0$	Given equation
$x(3 + x)e^x = 0$	Factor out common factors
$x(3 + x) = 0$	Divide by e^x (because $e^x \neq 0$)
$x = 0 \quad$ or $\quad 3 + x = 0$	Zero-Product Property

Thus, the solutions are $x = 0$ and $x = -3$.   ▲

■ Logarithmic Equations

A *logarithmic equation* is one in which a logarithm of the variable occurs. For example,

$$\log_2(x + 2) = 5$$

To solve for x, we write the equation in exponential form.

$$x + 2 = 2^5 \qquad \text{Exponential form}$$

$$x = 32 - 2 = 30 \qquad \text{Solve for x}$$

Another way of looking at the first step is to raise the base, 2, to each side of the equation.

$$2^{\log_2(x+2)} = 2^5 \qquad \text{Raise 2 to each side}$$

$$x + 2 = 2^5 \qquad \text{Property of logarithms}$$

$$x = 32 - 2 = 30 \qquad \text{Solve for x}$$

The method used to solve this simple problem is typical. We summarize the steps as follows.

GUIDELINES FOR SOLVING LOGARITHMIC EQUATIONS

1. Isolate the logarithmic term on one side of the equation; you might first need to combine the logarithmic terms.

2. Write the equation in exponential form (or raise the base to each side of the equation).

3. Solve for the variable.

▶ **EXAMPLE 6** | Solving Logarithmic Equations

Solve each equation for x.

(a) $\ln x = 8$ (b) $\log_2(25 - x) = 3$

▼ **SOLUTION**

(a)
$$\ln x = 8 \qquad \text{Given equation}$$
$$x = e^8 \qquad \text{Exponential form}$$

Therefore, $x = e^8 \approx 2981$.
    We can also solve this problem another way:

$$\ln x = 8 \qquad \text{Given equation}$$
$$e^{\ln x} = e^8 \qquad \text{Raise e to each side}$$
$$x = e^8 \qquad \text{Property of ln}$$

(b) The first step is to rewrite the equation in exponential form.

$$\log_2(25 - x) = 3 \qquad \text{Given equation}$$
$$25 - x = 2^3 \qquad \text{Exponential form (or raise 2 to each side)}$$
$$25 - x = 8$$
$$x = 25 - 8 = 17$$

Check Your Answer

If $x = 17$, we get

$$\log_2(25 - 17) = \log_2 8 = 3 \qquad ✔$$

EXAMPLE 7 | Solving a Logarithmic Equation

Solve the equation $4 + 3\log(2x) = 16$.

SOLUTION We first isolate the logarithmic term. This allows us to write the equation in exponential form.

$4 + 3\log(2x) = 16$	Given equation
$3\log(2x) = 12$	Subtract 4
$\log(2x) = 4$	Divide by 3
$2x = 10^4$	Exponential form (or raise 10 to each side)
$x = 5000$	Divide by 2

Check Your Answer

If $x = 5000$, we get

$$\begin{aligned} 4 + 3\log 2(5000) &= 4 + 3\log 10{,}000 \\ &= 4 + 3(4) \\ &= 16 \quad ✓ \end{aligned}$$

EXAMPLE 8 | Solving a Logarithmic Equation Algebraically and Graphically

Solve the equation $\log(x + 2) + \log(x - 1) = 1$ algebraically and graphically.

SOLUTION 1: Algebraic

We first combine the logarithmic terms, using the Laws of Logarithms.

$\log[(x + 2)(x - 1)] = 1$	Law 1
$(x + 2)(x - 1) = 10$	Exponential form (or raise 10 to each side)
$x^2 + x - 2 = 10$	Expand left side
$x^2 + x - 12 = 0$	Subtract 10
$(x + 4)(x - 3) = 0$	Factor
$x = -4$ or $x = 3$	

We check these potential solutions in the original equation and find that $x = -4$ is not a solution (because logarithms of negative numbers are undefined), but $x = 3$ is a solution. (See *Check Your Answers*.)

Check Your Answers

$x = -4$:

$$\begin{aligned} &\log(-4 + 2) + \log(-4 - 1) \\ &\quad = \log(-2) + \log(-5) \\ &\quad\quad \text{undefined} \quad ✗ \end{aligned}$$

$x = 3$:

$$\begin{aligned} &\log(3 + 2) + \log(3 - 1) \\ &\quad = \log 5 + \log 2 = \log(5 \cdot 2) \\ &\quad = \log 10 = 1 \quad ✓ \end{aligned}$$

SOLUTION 2: Graphical

We first move all terms to one side of the equation:

$$\log(x + 2) + \log(x - 1) - 1 = 0$$

Then we graph

$$y = \log(x + 2) + \log(x - 1) - 1$$

as in Figure 2. The solutions are the x-intercepts of the graph. Thus, the only solution is $x \approx 3$.

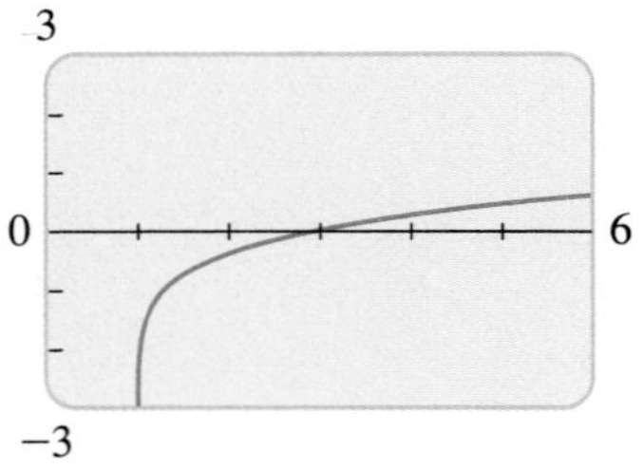

FIGURE 2

EXAMPLE 9 | Solving a Logarithmic Equation Graphically

Solve the equation $x^2 = 2\ln(x + 2)$.

In Example 9 it's not possible to isolate x algebraically, so we must solve the equation graphically.

SOLUTION We first move all terms to one side of the equation

$$x^2 - 2\ln(x + 2) = 0$$

Then we graph

$$y = x^2 - 2\ln(x + 2)$$

as in Figure 3. The solutions are the *x*-intercepts of the graph. Zooming in on the *x*-intercepts, we see that there are two solutions:

$$x \approx -0.71 \qquad \text{and} \qquad x \approx 1.60$$

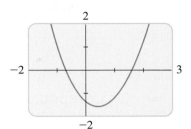

FIGURE 3

Logarithmic equations are used in determining the amount of light that reaches various depths in a lake. (This information helps biologists to determine the types of life a lake can support.) As light passes through water (or other transparent materials such as glass or plastic), some of the light is absorbed. It's easy to see that the murkier the water, the more light is absorbed. The exact relationship between light absorption and the distance light travels in a material is described in the next example.

The intensity of light in a lake diminishes with depth.

▶ **EXAMPLE 10** | Transparency of a Lake

If I_0 and I denote the intensity of light before and after going through a material and x is the distance (in feet) the light travels in the material, then according to the **Beer-Lambert Law,**

$$-\frac{1}{k}\ln\left(\frac{I}{I_0}\right) = x$$

where k is a constant depending on the type of material.

(a) Solve the equation for I.

(b) For a certain lake $k = 0.025$, and the light intensity is $I_0 = 14$ lumens (lm). Find the light intensity at a depth of 20 ft.

▼ **SOLUTION**

(a) We first isolate the logarithmic term.

$$-\frac{1}{k}\ln\left(\frac{I}{I_0}\right) = x \qquad \text{\textit{Given equation}}$$

$$\ln\left(\frac{I}{I_0}\right) = -kx \qquad \text{\textit{Multiply by }} -k$$

$$\frac{I}{I_0} = e^{-kx} \qquad \text{\textit{Exponential form}}$$

$$I = I_0 e^{-kx} \qquad \text{\textit{Multiply by }} I_0$$

(b) We find I using the formula from part (a).

$$I = I_0e^{-kx} \qquad \text{From part (a)}$$

$$= 14e^{(-0.025)(20)} \qquad I_0 = 14, k = 0.025, x = 20$$

$$\approx 8.49 \qquad \text{Calculator}$$

The light intensity at a depth of 20 ft is about 8.5 lm.

■ Compound Interest

Recall the formulas for interest that we found in Section 5.1. If a principal P is invested at an interest rate r for a period of t years, then the amount A of the investment is given by

$$A = P(1 + r) \qquad \text{Simple interest (for one year)}$$

$$A(t) = P\left(1 + \frac{r}{n}\right)^{nt} \qquad \text{Interest compounded } n \text{ times per year}$$

$$A(t) = Pe^{rt} \qquad \text{Interest compounded continuously}$$

We can use logarithms to determine the time it takes for the principal to increase to a given amount.

▶ **EXAMPLE 11** | Finding the Term for an Investment to Double

A sum of $5000 is invested at an interest rate of 5% per year. Find the time required for the money to double if the interest is compounded according to the following method.

(a) Semiannual

(b) Continuous

▼ **SOLUTION**

(a) We use the formula for compound interest with $P = \$5000$, $A(t) = \$10,000$, $r = 0.05$, and $n = 2$ and solve the resulting exponential equation for t.

$$5000\left(1 + \frac{0.05}{2}\right)^{2t} = 10,000 \qquad P\left(1 + \frac{r}{n}\right)^{nt} = A$$

$$(1.025)^{2t} = 2 \qquad \text{Divide by 5000}$$

$$\log 1.025^{2t} = \log 2 \qquad \text{Take log of each side}$$

$$2t \log 1.025 = \log 2 \qquad \text{Law 3 (bring down the exponent)}$$

$$t = \frac{\log 2}{2 \log 1.025} \qquad \text{Divide by 2 log 1.025}$$

$$t \approx 14.04 \qquad \text{Calculator}$$

The money will double in 14.04 years.

(b) We use the formula for continuously compounded interest with $P = \$5000$, $A(t) = \$10,000$, and $r = 0.05$ and solve the resulting exponential equation for t.

$$5000e^{0.05t} = 10,000 \qquad Pe^{rt} = A$$

$$e^{0.05t} = 2 \qquad \text{Divide by 5000}$$

$$\ln e^{0.05t} = \ln 2 \qquad \text{Take ln of each side}$$

$$0.05t = \ln 2 \qquad \text{Property of ln}$$

$$t = \frac{\ln 2}{0.05} \qquad \text{Divide by 0.05}$$

$$t \approx 13.86 \qquad \text{Calculator}$$

The money will double in 13.86 years. ▲

▶ **EXAMPLE 12** | Time Required to Grow an Investment

A sum of $1000 is invested at an interest rate of 4% per year. Find the time required for the amount to grow to $4000 if interest is compounded continuously.

▼ **SOLUTION**   We use the formula for continuously compounded interest with $P = \$1000$, $A(t) = \$4000$, and $r = 0.04$ and solve the resulting exponential equation for t.

$$1000e^{0.04t} = 4000 \qquad Pe^{rt} = A$$

$$e^{0.04t} = 4 \qquad \text{Divide by 1000}$$

$$0.04t = \ln 4 \qquad \text{Take ln of each side}$$

$$t = \frac{\ln 4}{0.04} \qquad \text{Divide by 0.04}$$

$$t \approx 34.66 \qquad \text{Calculator}$$

The amount will be $4000 in about 34 years and 8 months. ▲

If an investment earns compound interest, then the **annual percentage yield** (APY) is the *simple* interest rate that yields the same amount at the end of one year.

▶ **EXAMPLE 13** | Calculating the Annual Percentage Yield

Find the annual percentage yield for an investment that earns interest at a rate of 6% per year, compounded daily.

▼ **SOLUTION**   After one year, a principal P will grow to the amount

$$A = P\left(1 + \frac{0.06}{365}\right)^{365} = P(1.06183)$$

The formula for simple interest is

$$A = P(1 + r)$$

Comparing, we see that $1 + r = 1.06183$, so $r = 0.06183$. Thus, the annual percentage yield is 6.183%. ▲

5.5 | Modeling with Exponential and Logarithmic Functions

LEARNING OBJECTIVES

After completing this section, you will be able to:

▪ Find exponential models of population growth
▪ Find exponential models of radioactive decay
▪ Find models using Newton's Law of Cooling
▪ Use logarithmic scales (pH, Richter, and decibel scales)

Many processes that occur in nature, such as population growth, radioactive decay, heat diffusion, and numerous others, can be modeled by using exponential functions. Logarithmic functions are used in models for the loudness of sounds, the intensity of earthquakes, and many other phenomena. In this section we study exponential and logarithmic models.

▪ Exponential Models of Population Growth

Biologists have observed that the population of a species doubles its size in a fixed period of time. For example, under ideal conditions a certain population of bacteria doubles in size every 3 hours. If the culture is started with 1000 bacteria, then after 3 hours there will be 2000 bacteria, after another 3 hours there will be 4000, and so on. If we let $n = n(t)$ be the number of bacteria after t hours, then

$$n(0) = 1000$$

$$n(3) = 1000 \cdot 2$$

$$n(6) = (1000 \cdot 2) \cdot 2 = 1000 \cdot 2^2$$

$$n(9) = (1000 \cdot 2^2) \cdot 2 = 1000 \cdot 2^3$$

$$n(12) = (1000 \cdot 2^3) \cdot 2 = 1000 \cdot 2^4$$

From this pattern it appears that the number of bacteria after t hours is modeled by the function

$$n(t) = 1000 \cdot 2^{t/3}$$

In general, suppose that the initial size of a population is n_0 and the doubling period is a. Then the size of the population at time t is modeled by

$$n(t) = n_0 2^{ct}$$

where $c = 1/a$. If we knew the tripling time b, then the formula would be $n(t) = n_0 3^{ct}$, where $c = 1/b$. These formulas indicate that the growth of the bacteria is modeled by an exponential function. But what base should we use? The answer is e, because then it can be shown (using calculus) that the population is modeled by

$$n(t) = n_0 e^{rt}$$

where r is the *relative rate of growth of population, expressed as a proportion of the population at any time.* For instance, if $r = 0.02$, then at any time t the growth rate is 2% of the population at time t.

Notice that the formula for population growth is the same as that for continuously compounded interest. In fact, the same principle is at work in both cases: The growth of a population (or an investment) per time period is proportional to the size of the population (or the amount of the investment). A population of 1,000,000 will increase more in one year than a population of 1000; in exactly the same way, an investment of $1,000,000 will increase more in one year than an investment of $1000.

EXPONENTIAL GROWTH MODEL

A population that experiences **exponential growth** increases according to the model

$$n(t) = n_0 e^{rt}$$

where $n(t)$ = population at time t
         n_0 = initial size of the population
         r = relative rate of growth (expressed as a proportion of the population)
         t = time

In the following examples we assume that the populations grow exponentially.

▶ **EXAMPLE 1** | Predicting the Size of a Population

The initial bacterium count in a culture is 500. A biologist later makes a sample count of bacteria in the culture and finds that the relative rate of growth is 40% per hour.

(a) Find a function that models the number of bacteria after t hours.

(b) What is the estimated count after 10 hours?

(c) Sketch the graph of the function $n(t)$.

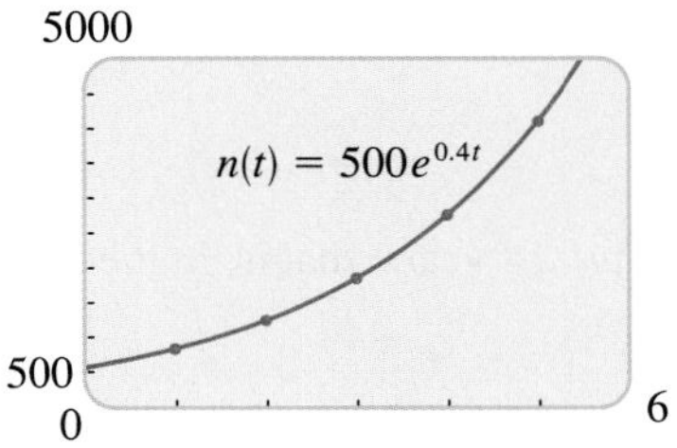

▼ **SOLUTION**

(a) We use the exponential growth model with $n_0 = 500$ and $r = 0.4$ to get

$$n(t) = 500 e^{0.4t}$$

where t is measured in hours.

(b) Using the function in part (a), we find that the bacterium count after 10 hours is

$$n(10) = 500 e^{0.4(10)} = 500 e^4 \approx 27{,}300$$

(c) The graph is shown in Figure 1.   ▲

FIGURE 1

$n(t) = 500 e^{0.4t}$

▶ **EXAMPLE 2** | Comparing Different Rates of Population Growth

In 2000 the population of the world was 6.1 billion, and the relative rate of growth was 1.4% per year. It is claimed that a rate of 1.0% per year would make a significant difference in the total population in just a few decades. Test this claim by estimating the population of the world in the year 2050 using a relative rate of growth of (a) 1.4% per year and (b) 1.0% per year.

Graph the population functions for the next 100 years for the two relative growth rates in the same viewing rectangle.

The relative growth of world population has been declining over the past few decades—from 2% in 1995 to 1.3% in 2006.

▼ **SOLUTION**

(a) By the exponential growth model we have

$$n(t) = 6.1 e^{0.014t}$$

where $n(t)$ is measured in billions and t is measured in years since 2000. Because the year 2050 is 50 years after 2000, we find

$$n(50) = 6.1 e^{0.014(50)} = 6.1 e^{0.7} \approx 12.3$$

The estimated population in the year 2050 is about 12.3 billion.

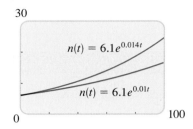

FIGURE 2

(b) We use the function

$$n(t) = 6.1e^{0.010t}$$

and find

$$n(50) = 6.1e^{0.010(50)} = 6.1e^{0.50} \approx 10.1$$

The estimated population in the year 2050 is about 10.1 billion.

The graphs in Figure 2 show that a small change in the relative rate of growth will, over time, make a large difference in population size.   ▲

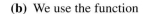

▶ **EXAMPLE 3** | World Population Projections

The population of the world in 2000 was 6.1 billion, and the estimated relative growth rate was 1.4% per year. If the population continues to grow at this rate, when will it reach 122 billion?

▼ **SOLUTION**   We use the population growth function with $n_0 = 6.1$ billion, $r = 0.014$, and $n(t) = 122$ billion. This leads to an exponential equation, which we solve for t.

$$6.1e^{0.014t} = 122 \qquad n_0 e^{rt} = n(t)$$

$$e^{0.014t} = 20 \qquad \text{Divide by 6.1}$$

$$\ln e^{0.014t} = \ln 20 \qquad \text{Take ln of each side}$$

$$0.014t = \ln 20 \qquad \text{Property of ln}$$

$$t = \frac{\ln 20}{0.014} \qquad \text{Divide by 0.014}$$

$$t \approx 213.98 \qquad \text{Calculator}$$

Thus, the population will reach 122 billion in approximately 214 years, that is, in the year $2000 + 214 = 2214$.   ▲

> **Standing Room Only**
>
> The population of the world was about 6.1 billion in 2000 and was increasing at 1.4% per year. Assuming that each person occupies an average of 4 ft² of the surface of the earth, the exponential model for population growth projects that by the year 2801 there will be standing room only! (The total land surface area of the world is about 1.8×10^{15} ft².)

▶ **EXAMPLE 4** | Finding the Initial Population

A certain breed of rabbit was introduced onto a small island about 8 years ago. The current rabbit population on the island is estimated to be 4100, with a relative growth rate of 55% per year.

(a) What was the initial size of the rabbit population?

(b) Estimate the population 12 years from now.

▼ **SOLUTION**

(a) From the exponential growth model we have

$$n(t) = n_0 e^{0.55t}$$

and we know that the population at time $t = 8$ is $n(8) = 4100$. We substitute what we know into the equation and solve for n_0:

$$4100 = n_0 e^{0.55(8)}$$

$$n_0 = \frac{4100}{e^{0.55(8)}} \approx \frac{4100}{81.45} \approx 50$$

Thus, we estimate that 50 rabbits were introduced onto the island.

(b) Now that we know n_0, we can write a formula for population growth:

$$n(t) = 50e^{0.55t}$$

Twelve years from now, $t = 8 + 12 = 20$ and

$$n(20) = 50e^{0.55(20)} \approx 2{,}993{,}707$$

We estimate that the rabbit population on the island 12 years from now will be about 3 million. ▲

Another way to solve part (b) is to let t be the number of years from now. In this case $n_0 = 4100$ (the current population), and the population 12 years from now will be

$$n(12) = 4100e^{0.55(12)} \approx 3 \text{ million}$$

Can the rabbit population in Example 4(b) actually reach such a high number? In reality, as the island becomes overpopulated with rabbits, the rabbit population growth will be slowed because of food shortage and other factors.

▶ **EXAMPLE 5** | The Number of Bacteria in a Culture

A culture starts with 10,000 bacteria, and the number doubles every 40 minutes.

(a) Find a function that models the number of bacteria at time t.

(b) Find the number of bacteria after one hour.

(c) After how many minutes will there be 50,000 bacteria?

(d) Sketch a graph of the number of bacteria at time t.

▼ **SOLUTION**

(a) To find the function that models this population growth, we need to find the rate r. To do this, we use the formula for population growth with $n_0 = 10{,}000$, $t = 40$, and $n(t) = 20{,}000$ and then solve for r.

$$10{,}000e^{r(40)} = 20{,}000 \qquad n_0 e^{rt} = n(t)$$

$$e^{40r} = 2 \qquad \text{Divide by 10,000}$$

$$\ln e^{40r} = \ln 2 \qquad \text{Take ln of each side}$$

$$40r = \ln 2 \qquad \text{Property of ln}$$

$$r = \frac{\ln 2}{40} \qquad \text{Divide by 40}$$

$$r \approx 0.01733 \qquad \text{Calculator}$$

Now that we know that $r \approx 0.01733$, we can write the function for the population growth:

$$n(t) = 10{,}000e^{0.01733t}$$

(b) Using the function that we found in part (a) with $t = 60$ min (one hour), we get

$$n(60) = 10{,}000e^{0.01733(60)} \approx 28{,}287$$

Thus, the number of bacteria after one hour is approximately 28,000.

(c) We use the function we found in part (a) with $n(t) = 50{,}000$ and solve the resulting exponential equation for t.

$$10{,}000e^{0.01733t} = 50{,}000 \qquad n_0 e^{rt} = n(t)$$

$$e^{0.01733t} = 5 \qquad \text{Divide by 10,000}$$

$$\ln e^{0.01733t} = \ln 5 \qquad \text{Take ln of each side}$$

$$0.01733t = \ln 5 \qquad \text{Property of ln}$$

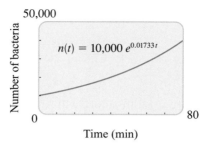

FIGURE 3

The half-lives of **radioactive elements** vary from very long to very short. Here are some examples.

Element	Half-life
Thorium-232	14.5 billion years
Uranium-235	4.5 billion years
Thorium-230	80,000 years
Plutonium-239	24,360 years
Carbon-14	5,730 years
Radium-226	1,600 years
Cesium-137	30 years
Strontium-90	28 years
Polonium-210	140 days
Thorium-234	25 days
Iodine-135	8 days
Radon-222	3.8 days
Lead-211	3.6 minutes
Krypton-91	10 seconds

$$t = \frac{\ln 5}{0.01733} \qquad \text{Divide by 0.01733}$$

$$t \approx 92.9 \qquad \text{Calculator}$$

The bacterium count will reach 50,000 in approximately 93 min.

(d) The graph of the function $n(t) = 10,000e^{0.01733t}$ is shown in Figure 3. ▲

■ Radioactive Decay

Radioactive substances decay by spontaneously emitting radiation. The rate of decay is directly proportional to the mass of the substance. This is analogous to population growth, except that the mass of radioactive material *decreases*. It can be shown that the mass $m(t)$ remaining at time t is modeled by the function

$$m(t) = m_0 e^{-rt}$$

where r is the rate of decay expressed as a proportion of the mass and m_0 is the initial mass. Physicists express the rate of decay in terms of **half-life**, the time required for half the mass to decay. We can obtain the rate r from this as follows. If h is the half-life, then a mass of 1 unit becomes $\frac{1}{2}$ unit when $t = h$. Substituting this into the model, we get

$$\tfrac{1}{2} = 1 \cdot e^{-rh} \qquad m(t) = m_0 e^{-rt}$$

$$\ln\left(\tfrac{1}{2}\right) = -rh \qquad \text{Take ln of each side}$$

$$r = -\frac{1}{h}\ln(2^{-1}) \qquad \text{Solve for } r$$

$$r = \frac{\ln 2}{h} \qquad \ln 2^{-1} = -\ln 2 \text{ by Law 3}$$

This last equation allows us to find the rate r from the half-life h.

RADIOACTIVE DECAY MODEL

If m_0 is the initial mass of a radioactive substance with half-life h, then the mass remaining at time t is modeled by the function

$$m(t) = m_0 e^{-rt}$$

where $r = \dfrac{\ln 2}{h}$.

▶ **EXAMPLE 6** | Radioactive Decay

Polonium-210 (^{210}Po) has a half-life of 140 days. Suppose a sample of this substance has a mass of 300 mg.

(a) Find a function that models the amount of the sample remaining at time t.

(b) Find the mass remaining after one year.

(c) How long will it take for the sample to decay to a mass of 200 mg?

(d) Draw a graph of the sample mass as a function of time.

▼ **SOLUTION**

(a) Using the model for radioactive decay with $m_0 = 300$ and $r = (\ln 2/140) \approx 0.00495$, we have

$$m(t) = 300e^{-0.00495t}$$

(b) We use the function we found in part (a) with $t = 365$ (one year).

$$m(365) = 300e^{-0.00495(365)} \approx 49.256$$

Thus, approximately 49 mg of ^{210}Po remains after one year.

(c) We use the function that we found in part (a) with $m(t) = 200$ and solve the resulting exponential equation for t.

$$300e^{-0.00495t} = 200 \qquad m(t) = m_o e^{-rt}$$
$$e^{-0.00495t} = \tfrac{2}{3} \qquad \text{Divided by 300}$$
$$\ln e^{-0.00495t} = \ln \tfrac{2}{3} \qquad \text{Take ln of each side}$$
$$-0.00495t = \ln \tfrac{2}{3} \qquad \text{Property of ln}$$
$$t = -\frac{\ln \tfrac{2}{3}}{0.00495} \qquad \text{Divide by } -0.00495$$
$$t \approx 81.9 \qquad \text{Calculator}$$

The time required for the sample to decay to 200 mg is about 82 days.

(d) A graph of the function $m(t) = 300e^{-0.00495t}$ is shown in Figure 4. ▲

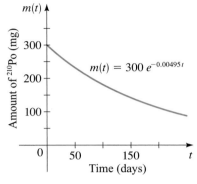

$m(t) = 300\,e^{-0.00495t}$

FIGURE 4

■ Newton's Law of Cooling

Newton's Law of Cooling states that the rate of cooling of an object is proportional to the temperature difference between the object and its surroundings, provided that the temperature difference is not too large. Using calculus, the following model can be deduced from this law.

NEWTON'S LAW OF COOLING

If D_0 is the initial temperature difference between an object and its surroundings, and if its surroundings have temperature T_s, then the temperature of the object at time t is modeled by the function

$$T(t) = T_s + D_0 e^{-kt}$$

where k is a positive constant that depends on the type of object.

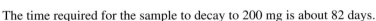

▶ **EXAMPLE 7** | Newton's Law of Cooling

A cup of coffee has a temperature of 200°F and is placed in a room that has a temperature of 70°F. After 10 min the temperature of the coffee is 150°F.

(a) Find a function that models the temperature of the coffee at time t.

(b) Find the temperature of the coffee after 15 min.

(c) When will the coffee have cooled to 100°F?

(d) Illustrate by drawing a graph of the temperature function.

▼ **SOLUTION**

(a) The temperature of the room is $T_s = 70°F$, and the initial temperature difference is

$$D_0 = 200 - 70 = 130°F$$

So by Newton's Law of Cooling, the temperature after t minutes is modeled by the function

$$T(t) = 70 + 130e^{-kt}$$

We need to find the constant k associated with this cup of coffee. To do this, we use the fact that when $t = 10$, the temperature is $T(10) = 150$. So we have

$$70 + 130e^{-10k} = 150 \qquad \text{\small $T_s + D_o e^{-kt} = T(t)$}$$

$$130e^{-10k} = 80 \qquad \text{\small Subtract 70}$$

$$e^{-10k} = \tfrac{8}{13} \qquad \text{\small Divide by 130}$$

$$-10k = \ln \tfrac{8}{13} \qquad \text{\small Take ln of each side}$$

$$k = -\tfrac{1}{10} \ln \tfrac{8}{13} \qquad \text{\small Divide by −10}$$

$$k \approx 0.04855 \qquad \text{\small Calculator}$$

Substituting this value of k into the expression for $T(t)$, we get

$$T(t) = 70 + 130e^{-0.04855t}$$

(b) We use the function that we found in part (a) with $t = 15$.

$$T(15) = 70 + 130e^{-0.04855(15)} \approx 133°F$$

(c) We use the function that we found in part (a) with $T(t) = 100$ and solve the resulting exponential equation for t.

$$70 + 130e^{-0.04855t} = 100 \qquad \text{\small $T_s + D_o e^{-kt} = T(t)$}$$

$$130e^{-0.04855t} = 30 \qquad \text{\small Subtract 70}$$

$$e^{-0.04855t} = \tfrac{3}{13} \qquad \text{\small Divide by 130}$$

$$-0.04855t = \ln \tfrac{3}{13} \qquad \text{\small Take ln of each side}$$

$$t = \frac{\ln \tfrac{3}{13}}{-0.04855} \qquad \text{\small Divide by −0.04855}$$

$$t \approx 30.2 \qquad \text{\small Calculator}$$

The coffee will have cooled to 100°F after about half an hour.

(d) The graph of the temperature function is sketched in Figure 5. Notice that the line $t = 70$ is a horizontal asymptote. (Why?) ▲

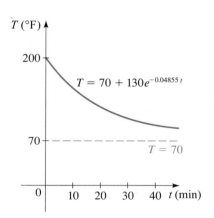

FIGURE 5 Temperature of coffee after t minutes

■ Logarithmic Scales

When a physical quantity varies over a very large range, it is often convenient to take its logarithm in order to have a more manageable set of numbers. We discuss three such situations: the pH scale, which measures acidity; the Richter scale, which measures the intensity of earthquakes; and the decibel scale, which measures the loudness of sounds. Other quantities that are measured on logarithmic scales are light intensity, information capacity, and radiation.

The pH Scale Chemists measured the acidity of a solution by giving its hydrogen ion concentration until Søren Peter Lauritz Sørensen, in 1909, proposed a more convenient measure. He defined

$$\text{pH} = -\log[\text{H}^+]$$

pH for Some Common Substances

Substance	pH
Milk of magnesia	10.5
Seawater	8.0–8.4
Human blood	7.3–7.5
Crackers	7.0–8.5
Hominy	6.9–7.9
Cow's milk	6.4–6.8
Spinach	5.1–5.7
Tomatoes	4.1–4.4
Oranges	3.0–4.0
Apples	2.9–3.3
Limes	1.3–2.0
Battery acid	1.0

where $[H^+]$ is the concentration of hydrogen ions measured in moles per liter (M). He did this to avoid very small numbers and negative exponents. For instance,

$$\text{if} \quad [H^+] = 10^{-4}\,M, \quad \text{then} \quad pH = -\log_{10}(10^{-4}) = -(-4) = 4$$

Solutions with a pH of 7 are defined as *neutral*, those with pH < 7 are *acidic*, and those with pH > 7 are *basic*. Notice that when the pH increases by one unit, $[H^+]$ decreases by a factor of 10.

▶ **EXAMPLE 8** | pH Scale and Hydrogen Ion Concentration

(a) The hydrogen ion concentration of a sample of human blood was measured to be $[H^+] = 3.16 \times 10^{-8}$ M. Find the pH and classify the blood as acidic or basic.

(b) The most acidic rainfall ever measured occurred in Scotland in 1974; its pH was 2.4. Find the hydrogen ion concentration.

▼ **SOLUTION**

(a) A calculator gives

$$pH = -\log[H^+] = -\log(3.16 \times 10^{-8}) \approx 7.5$$

Since this is greater than 7, the blood is basic.

(b) To find the hydrogen ion concentration, we need to solve for $[H^+]$ in the logarithmic equation

$$\log[H^+] = -pH$$

So we write it in exponential form.

$$[H^+] = 10^{-pH}$$

In this case pH = 2.4, so

$$[H^+] = 10^{-2.4} \approx 4.0 \times 10^{-3}\,M$$ ▲

The Richter Scale   In 1935 the American geologist Charles Richter (1900–1984) defined the magnitude M of an earthquake to be

$$M = \log \frac{I}{S}$$

where I is the intensity of the earthquake (measured by the amplitude of a seismograph reading taken 100 km from the epicenter of the earthquake) and S is the intensity of a "standard" earthquake (whose amplitude is 1 micron $= 10^{-4}$ cm). The magnitude of a standard earthquake is

$$M = \log \frac{S}{S} = \log 1 = 0$$

Richter studied many earthquakes that occurred between 1900 and 1950. The largest had magnitude 8.9 on the Richter scale, and the smallest had magnitude 0. This corresponds to a ratio of intensities of 800,000,000, so the Richter scale provides more manageable numbers to work with. For instance, an earthquake of magnitude 6 is ten times stronger than an earthquake of magnitude 5.

Largest Earthquakes

Location	Date	Magnitude
Chile	1960	9.5
Alaska	1964	9.2
Sumatra	2004	9.1
Alaska	1957	9.1
Kamchatka	1952	9.0
Ecuador	1906	8.8
Alaska	1965	8.7
Sumatra	2005	8.7
Tibet	1950	8.6
Kamchatka	1923	8.5
Indonesia	1938	8.5
Kuril Islands	1963	8.5

▶ **EXAMPLE 9** | Magnitude of Earthquakes

The 1906 earthquake in San Francisco had an estimated magnitude of 8.3 on the Richter scale. In the same year a powerful earthquake occurred on the Colombia-Ecuador border that was four times as intense. What was the magnitude of the Colombia-Ecuador earthquake on the Richter scale?

▼ **SOLUTION** If I is the intensity of the San Francisco earthquake, then from the definition of magnitude we have

$$M = \log \frac{I}{S} = 8.3$$

The intensity of the Colombia-Ecuador earthquake was $4I$, so its magnitude was

$$M = \log \frac{4I}{S} = \log 4 + \log \frac{I}{S} = \log 4 + 8.3 \approx 8.9$$ ▲

▶ **EXAMPLE 10** | Intensity of Earthquakes

The 1989 Loma Prieta earthquake that shook San Francisco had a magnitude of 7.1 on the Richter scale. How many times more intense was the 1906 earthquake (see Example 9) than the 1989 event?

▼ **SOLUTION** If I_1 and I_2 are the intensities of the 1906 and 1989 earthquakes, then we are required to find I_1/I_2. To relate this to the definition of magnitude, we divide the numerator and denominator by S.

$$\log \frac{I_1}{I_2} = \log \frac{I_1/S}{I_2/S} \qquad \text{Divide numerator and denominator by } S$$

$$= \log \frac{I_1}{S} - \log \frac{I_2}{S} \qquad \text{Law 2 of logarithms}$$

$$= 8.3 - 7.1 = 1.2 \qquad \text{Definition of earthquake magnitude}$$

Therefore,

$$\frac{I_1}{I_2} = 10^{\log(I_1/I_2)} = 10^{1.2} \approx 16$$

The 1906 earthquake was about 16 times as intense as the 1989 earthquake. ▲

The Decibel Scale The ear is sensitive to an extremely wide range of sound intensities. We take as a reference intensity $I_0 = 10^{-12}$ W/m² (watts per square meter) at a frequency of 1000 hertz, which measures a sound that is just barely audible (the threshold of hearing). The psychological sensation of loudness varies with the logarithm of the intensity (the Weber-Fechner Law), so the **intensity level** B, measured in decibels (dB), is defined as

$$B = 10 \log \frac{I}{I_0}$$

The intensity level of the barely audible reference sound is

$$B = 10 \log \frac{I_0}{I_0} = 10 \log 1 = 0 \text{ dB}$$

© Roger Ressmeyer/CORBIS

The **intensity levels of sounds** that we can hear vary from very loud to very soft. Here are some examples of the decibel levels of commonly heard sounds.

Source of sound	B (dB)
Jet takeoff	140
Jackhammer	130
Rock concert	120
Subway	100
Heavy traffic	80
Ordinary traffic	70
Normal conversation	50
Whisper	30
Rustling leaves	10–20
Threshold of hearing	0

▶ **EXAMPLE 11** | Sound Intensity of a Jet Takeoff

Find the decibel intensity level of a jet engine during takeoff if the intensity was measured at 100 W/m^2.

▼ **SOLUTION**   From the definition of intensity level we see that

$$B = 10 \log \frac{I}{I_0} = 10 \log \frac{10^2}{10^{-12}} = 10 \log 10^{14} = 140 \text{ dB}$$

Thus, the intensity level is 140 dB.   ▲

The table in the margin on page 278 lists decibel intensity levels for some common sounds ranging from the threshold of human hearing to the jet takeoff of Example 11. The threshold of pain is about 120 dB.

▶ CHAPTER 5 | REVIEW

▼ PROPERTIES AND FORMULAS

Exponential Functions (pp. 242–245)

The **exponential function** f with base a (where $a > 0$, $a \neq 1$) is defined for all real numbers x by

$$f(x) = a^x$$

The domain of f is $\mathbb{R}$, and the range of f is $(0, \infty)$. The graph of f has one of the following shapes, depending on the value of a:

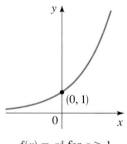

$f(x) = a^x$ for $a > 1$

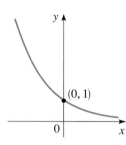

$f(x) = a^x$ for $0 < a < 1$

The Natural Exponential Function (p. 247)

The **natural exponential function** is the exponential function with base e:

$$f(x) = e^x$$

The number e is defined to be the number that the expression $(1 + 1/n)^n$ approaches as $n \to \infty$. An approximate value for the irrational number e is

$$e \approx 2.7182818284590\ldots$$

Compound Interest (pp. 249–250)

If a principal P is invested in an account paying an annual interest rate r, compounded n times a year, then after t years the **amount** $A(t)$ in the account is

$$A(t) = P\left(1 + \frac{r}{n}\right)^{nt}$$

If the interest is compounded **continuously**, then the amount is

$$A(t) = Pe^{rt}$$

Logarithmic Functions (p. 251)

The **logarithmic function** $\log_a$ with base a (where $a > 0$, $a \neq 1$) is defined for $x > 0$ by

$$\log_a x = y \iff a^y = x$$

So $\log_a x$ is the exponent to which the base a must be raised to give y.

The domain of $\log_a$ is $(0, \infty)$, and the range is $\mathbb{R}$. For $a > 1$, the graph of the function $\log_a$ has the following shape:

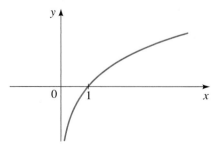

$y = \log_a x$, $a > 1$

Properties of Logarithms (p. 252)

1. $\log_a 1 = 0$ 2. $\log_a a = 1$

3. $\log_a a^x = x$ 4. $a^{\log_a x} = x$

Common and Natural Logarithms (pp. 255–257)

The logarithm function with base 10 is called the **common logarithm** and is denoted **log**. So

$$\log x = \log_{10} x$$

The logarithm function with base e is called the **natural logarithm** and is denoted **ln**. So

$$\ln x = \log_e x$$

Laws of Logarithms (p. 258)

Let a be a logarithm base $(a > 0, a \neq 1)$, and let A, B, and C be any real numbers or algebraic expressions that represent real numbers, with $A > 0$ and $B > 0$. Then:

1. $\log_a(AB) = \log_a A + \log_a B$

2. $\log_a(A/B) = \log_a A - \log_a B$

3. $\log_a(A^C) = C \log_a A$

Change of Base Formula (p. 261)

$$\log_b x = \frac{\log_a x}{\log_a b}$$

Guidelines for Solving Exponential Equations (p. 263)

1. Isolate the exponential term on one side of the equation.

2. Take the logarithm of each side, and use the Laws of Logarithms to "bring down the exponent."

3. Solve for the variable.

Guidelines for Solving Logarithmic Equations (p. 265)

1. Isolate the logarithmic term(s) on one side of the equation, and use the Laws of Logarithms to combine logarithmic terms if necessary.

2. Rewrite the equation in exponential form.

3. Solve for the variable.

Exponential Growth Model (p. 271)

A population experiences **exponential growth** if it can be modeled by the exponential function

$$n(t) = n_0 e^{rt}$$

where $n(t)$ is the population at time t, n_0 is the initial population (at time $t = 0$), and r is the relative growth rate (expressed as a proportion of the population).

Radioactive Decay Model (p. 274)

If a **radioactive substance** with half-life h has initial mass m_0, then at time t the mass $m(t)$ of the substance that remains is modeled by the exponential function

$$m(t) = m_0 e^{-rt}$$

where $r = \dfrac{\ln 2}{h}$.

Newton's Law of Cooling (p. 275)

If an object has an initial temperature that is D_0 degrees warmer than the surrounding temperature T_s, then at time t the temperature $T(t)$ of the object is modeled by the function

$$T(t) = T_s + D_0 e^{-kt}$$

where the constant $k > 0$ depends on the size and type of the object.

Logarithmic Scales (pp. 276–279)

The **pH scale** measures the acidity of a solution:

$$pH = -\log[H^+]$$

The **Richter scale** measures the intensity of earthquakes:

$$M = \log \frac{I}{S}$$

The **decibel scale** measures the intensity of sound:

$$B = 10 \log \frac{I}{I_0}$$

▼ CONCEPT SUMMARY

Section 5.1

	Review Exercises
▪ Evaluate exponential functions	1–4
▪ Graph exponential functions	5–8, 73–74, 80
▪ Evaluate and graph the natural exponential function	4, 13–14
▪ Find compound interest	89–90
▪ Find continuously compounded interest	89(d), 90(c), 92

Section 5.2

	Review Exercises
▪ Evaluate logarithmic functions	29–44
▪ Change between logarithmic and exponential forms of an expression	21–28
▪ Use basic properties of logarithms	30, 31, 33, 36
▪ Graph logarithmic functions	9–12, 15–16, 75–76
▪ Use common and natural logarithms	17–20, 23, 24, 27, 28

Section 5.3

Review Exercises

- Use the Laws of Logarithms to evaluate logarithmic expressions — 29, 32, 34, 35, 37–44
- Use the Laws of Logarithms to expand logarithmic expressions — 45–50
- Use the Laws of Logarithms to combine logarithmic expressions — 51–56
- Use the Change of Base Formula — 83–86

Section 5.4

Review Exercises

- Solve exponential equations — 57–64, 69–72, 78
- Solve logarithmic equations — 65–68, 77
- Solve problems involving compound interest — 91–92
- Calculate annual percentage yield — 93–94

Section 5.5

Review Exercises

- Find exponential models of population growth — 95–96, 101
- Find exponential models of radioactive decay — 97–100
- Find models using Newton's Law of Cooling — 102
- Use logarithmic scales (pH, Richter, and decibel scales) — 103–106

▼ EXERCISES

1–4 ▪ Use a calculator to find the indicated values of the exponential function, correct to three decimal places.

1. $f(x) = 5^x$; $f(-1.5), f(\sqrt{2}), f(2.5)$

2. $f(x) = 3 \cdot 2^x$; $f(-2.2), f(\sqrt{7}), f(5.5)$

3. $g(x) = 4 \cdot \left(\frac{2}{3}\right)^{x-2}$; $g(-0.7), g(e), g(\pi)$

4. $g(x) = \frac{7}{4}e^{x+1}$; $g(-2), g(\sqrt{3}), g(3.6)$

5–16 ▪ Sketch the graph of the function. State the domain, range, and asymptote.

5. $f(x) = 2^{-x+1}$

6. $f(x) = 3^{x-2}$

7. $g(x) = 3 + 2^x$

8. $g(x) = 5^{-x} - 5$

9. $f(x) = \log_3(x - 1)$

10. $g(x) = \log(-x)$

11. $f(x) = 2 - \log_2 x$

12. $f(x) = 3 + \log_5(x + 4)$

13. $F(x) = e^x - 1$

14. $G(x) = \frac{1}{2}e^{x-1}$

15. $g(x) = 2 \ln x$

16. $g(x) = \ln(x^2)$

17–20 ▪ Find the domain of the function.

17. $f(x) = 10^{x^2} + \log(1 - 2x)$

18. $g(x) = \log(2 + x - x^2)$

19. $h(x) = \ln(x^2 - 4)$

20. $k(x) = \ln|x|$

21–24 ▪ Write the equation in exponential form.

21. $\log_2 1024 = 10$

22. $\log_6 37 = x$

23. $\log x = y$

24. $\ln c = 17$

25–28 ▪ Write the equation in logarithmic form.

25. $2^6 = 64$

26. $49^{-1/2} = \frac{1}{7}$

27. $10^x = 74$

28. $e^k = m$

29–44 ▪ Evaluate the expression without using a calculator.

29. $\log_2 128$

30. $\log_8 1$

31. $10^{\log 45}$

32. $\log 0.000001$

33. $\ln(e^6)$

34. $\log_4 8$

35. $\log_3\left(\frac{1}{27}\right)$

36. $2^{\log_2 13}$

37. $\log_5 \sqrt{5}$

38. $e^{2\ln 7}$

39. $\log 25 + \log 4$

40. $\log_3 \sqrt{243}$

41. $\log_2 16^{23}$

42. $\log_5 250 - \log_5 2$

43. $\log_8 6 - \log_8 3 + \log_8 2$

44. $\log \log 10^{100}$

45–50 ▪ Expand the logarithmic expression.

45. $\log(AB^2C^3)$

46. $\log_2(x \sqrt{x^2 + 1})$

47. $\ln\sqrt{\dfrac{x^2 - 1}{x^2 + 1}}$

48. $\log\left(\dfrac{4x^3}{y^2(x - 1)^5}\right)$

49. $\log_5\left(\dfrac{x^2(1 - 5x)^{3/2}}{\sqrt{x^3 - x}}\right)$

50. $\ln\left(\dfrac{\sqrt[3]{x^4 + 12}}{(x + 16)\sqrt{x - 3}}\right)$

51–56 ▪ Combine into a single logarithm.

51. $\log 6 + 4 \log 2$

52. $\log x + \log(x^2 y) + 3 \log y$

53. $\frac{3}{2}\log_2(x - y) - 2\log_2(x^2 + y^2)$

54. $\log_5 2 + \log_5(x + 1) - \frac{1}{3}\log_5(3x + 7)$

55. $\log(x - 2) + \log(x + 2) - \frac{1}{2}\log(x^2 + 4)$

56. $\frac{1}{2}[\ln(x - 4) + 5 \ln(x^2 + 4x)]$

57–68 ▪ Solve the equation. Find the exact solution if possible; otherwise, use a calculator to approximate to two decimals.

57. $3^{2x-7} = 27$

58. $5^{4-x} = \frac{1}{125}$

59. $2^{3x-5} = 7$

60. $10^{6-3x} = 18$

61. $4^{1-x} = 3^{2x+5}$

62. $e^{3x/4} = 10$

63. $x^2 e^{2x} + 2x e^{2x} = 8 e^{2x}$ **64.** $3^{2x} - 3^x - 6 = 0$

65. $\log_2(1 - x) = 4$

66. $\log x + \log(x + 1) = \log 12$

67. $\log_8(x + 5) - \log_8(x - 2) = 1$

68. $\ln(2x - 3) + 1 = 0$

69–72 ▪ Use a calculator to find the solution of the equation, correct to six decimal places.

69. $5^{-2x/3} = 0.63$ **70.** $2^{3x-5} = 7$

71. $5^{2x+1} = 3^{4x-1}$ **72.** $e^{-15k} = 10,000$

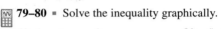 **73–76** ▪ Draw a graph of the function and use it to determine the asymptotes and the local maximum and minimum values.

73. $y = e^{x/(x+2)}$ **74.** $y = 10^x - 5^x$

75. $y = \log(x^3 - x)$ **76.** $y = 2x^2 - \ln x$

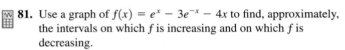 **77–78** ▪ Find the solutions of the equation, correct to two decimal places.

77. $3 \log x = 6 - 2x$ **78.** $4 - x^2 = e^{-2x}$

79–80 ▪ Solve the inequality graphically.

79. $\ln x > x - 2$ **80.** $e^x < 4x^2$

81. Use a graph of $f(x) = e^x - 3e^{-x} - 4x$ to find, approximately, the intervals on which f is increasing and on which f is decreasing.

82. Find an equation of the line shown in the figure.

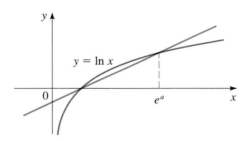

83–86 ▪ Use the Change of Base Formula to evaluate the logarithm, correct to six decimal places.

83. $\log_4 15$ **84.** $\log_7(\frac{3}{4})$

85. $\log_9 0.28$ **86.** $\log_{100} 250$

87. Which is larger, $\log_4 258$ or $\log_5 620$?

88. Find the inverse of the function $f(x) = 2^{3^x}$, and state its domain and range.

89. If $12,000 is invested at an interest rate of 10% per year, find the amount of the investment at the end of 3 years for each compounding method.
(a) Semiannual   (b) Monthly
(c) Daily   (d) Continuous

90. A sum of $5000 is invested at an interest rate of $8\frac{1}{2}\%$ per year, compounded semiannually.
(a) Find the amount of the investment after $1\frac{1}{2}$ years.

(b) After what period of time will the investment amount to $7000?
(c) If interest were compounded continously instead of semi-annually, how long would it take for the amount to grow to $7000?

91. A money market account pays 5.2% annual interest, compounded daily. If $100,000 is invested in this account, how long will it take for the account to accumulate $10,000 in interest?

92. A retirement savings plan pays 4.5% interest, compounded continuously. How long will it take for an investment in this plan to double?

93–94 ▪ Determine the annual percentage yield (APY) for the given nominal annual interest rate and compounding frequency.

93. 4.25%;   daily   **94.** 3.2%;   monthly

95. The stray-cat population in a small town grows exponentially. In 1999 the town had 30 stray cats, and the relative growth rate was 15% per year.
(a) Find a function that models the stray-cat population $n(t)$ after t years.
(b) Find the projected population after 4 years.
(c) Find the number of years required for the stray-cat population to reach 500.

96. A culture contains 10,000 bacteria initially. After an hour the bacteria count is 25,000.
(a) Find the doubling period.
(b) Find the number of bacteria after 3 hours.

97. Uranium-234 has a half-life of 2.7×10^5 years.
(a) Find the amount remaining from a 10-mg sample after a thousand years.
(b) How long will it take this sample to decompose until its mass is 7 mg?

98. A sample of bismuth-210 decayed to 33% of its original mass after 8 days.
(a) Find the half-life of this element.
(b) Find the mass remaining after 12 days.

99. The half-life of radium-226 is 1590 years.
(a) If a sample has a mass of 150 mg, find a function that models the mass that remains after t years.
(b) Find the mass that will remain after 1000 years.
(c) After how many years will only 50 mg remain?

100. The half-life of palladium-100 is 4 days. After 20 days a sample has been reduced to a mass of 0.375 g.
(a) What was the initial mass of the sample?
(b) Find a function that models the mass remaining after t days.
(c) What is the mass after 3 days?
(d) After how many days will only 0.15 g remain?

101. The graph shows the population of a rare species of bird, where t represents years since 1999 and $n(t)$ is measured in thousands.
(a) Find a function that models the bird population at time t in the form $n(t) = n_0 e^{rt}$.

(b) What is the bird population expected to be in the year 2010?

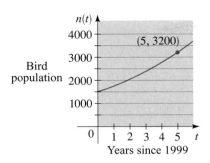

102. A car engine runs at a temperature of 190°F. When the engine is turned off, it cools according to Newton's Law of Cooling with constant $k = 0.0341$, where the time is measured in minutes. Find the time needed for the engine to cool to 90°F if the surrounding temperature is 60°F.

103. The hydrogen ion concentration of fresh egg whites was measured as

$$[H^+] = 1.3 \times 10^{-8} \text{ M}$$

Find the pH, and classify the substance as acidic or basic.

104. The pH of lime juice is 1.9. Find the hydrogen ion concentration.

105. If one earthquake has magnitude 6.5 on the Richter scale, what is the magnitude of another quake that is 35 times as intense?

106. The drilling of a jackhammer was measured at 132 dB. The sound of whispering was measured at 28 dB. Find the ratio of the intensity of the drilling to that of the whispering.

1. Sketch the graph of each function, and state its domain, range, and asymptote. Show the x- and y-intercepts on the graph.

   (a) $f(x) = 2^{-x} + 4$ (b) $g(x) = \log_3(x + 3)$

2. (a) Write the equation $6^{2x} = 25$ in logarithmic form.

   (b) Write the equation $\ln A = 3$ in exponential form.

3. Find the exact value of each expression.

   (a) $10^{\log 36}$ (b) $\ln e^3$

   (c) $\log_3 \sqrt{27}$ (d) $\log_2 80 - \log_2 10$

   (e) $\log_8 4$ (f) $\log_6 4 + \log_6 9$

4. Use the Laws of Logarithms to expand the expression.

$$\log \sqrt[3]{\frac{x + 2}{x^4(x^2 + 4)}}$$

5. Combine into a single logarithm: $\ln x - 2\ln(x^2 + 1) + \frac{1}{2}\ln(3 - x^4)$

6. Find the solution of the equation, correct to two decimal places.

   (a) $2^{x-1} = 10$ (b) $5\ln(3 - x) = 4$

   (c) $10^{x+3} = 6^{2x}$ (d) $\log_2(x + 2) + \log_2(x - 1) = 2$

7. The initial size of a culture of bacteria is 1000. After one hour the bacteria count is 8000.

   (a) Find a function that models the population after t hours.

   (b) Find the population after 1.5 hours.

   (c) When will the population reach 15,000?

   (d) Sketch the graph of the population function.

8. Suppose that $12,000 is invested in a savings account paying 5.6% interest per year.

   (a) Write the formula for the amount in the account after t years if interest is compounded monthly.

   (b) Find the amount in the account after 3 years if interest is compounded daily.

   (c) How long will it take for the amount in the account to grow to $20,000 if interest is compounded semiannually?

9. The half-life of krypton-91 (^{91}Kr) is 10 seconds. At time $t = 0$ a heavy canister contains 3 g of this radioactive gas.

   (a) Find a function that models the amount $A(t)$ of ^{91}Kr remaining in the canister after t seconds.

   (b) How much ^{91}Kr remains after one minute?

   (c) When will the amount of ^{91}Kr remaining be reduced to 1 μg (1 microgram, or 10^{-6} g)?

10. An earthquake measuring 6.4 on the Richter scale struck Japan in July 2007, causing extensive damage. Earlier that year, a minor earthquake measuring 3.1 on the Richter scale was felt in parts of Pennsylvania. How many times more intense was the Japanese earthquake than the Pennsylvania earthquake?

1. Let $f(x) = x^2 - 4x$ and $g(x) = \sqrt{x + 4}$. Find each of the following:

   (a) The domain of f

   (b) The domain of g

   (c) $f(-2), f(0), f(4), g(0), g(8), g(-6)$

   (d) $f(x + 2), g(x + 2), f(2 + h)$

   (e) The average rate of change of g between $x = 5$ and $x = 21$

   (f) $f \circ g, g \circ f, f(g(12)), g(f(12))$

   (g) The inverse of g

2. Let $f(x) = \begin{cases} 4 & \text{if } x \leq 2 \\ x - 3 & \text{if } x > 2 \end{cases}$

   (a) Evaluate $f(0), f(1), f(2), f(3),$ and $f(4)$.

   (b) Sketch the graph of f.

3. Let f be the quadratic function $f(x) = -2x^2 + 8x + 5$.

   (a) Express f in standard form.

   (b) Find the maximum or minimum value of f.

   (c) Sketch the graph of f.

   (d) Find the interval on which f is increasing and the interval on which f is decreasing.

   (e) How is the graph of $g(x) = -2x^2 + 8x + 10$ obtained from the graph of f?

   (f) How is the graph of $h(x) = -2(x + 3)^2 + 8(x + 3) + 5$ obtained from the graph of f?

4. Without using a graphing calculator, match each of the following functions to the graphs below. Give reasons for your choices.

$$f(x) = x^3 - 8x \qquad g(x) = -x^4 + 8x^2 \qquad r(x) = \frac{2x + 3}{x^2 - 9}$$

$$s(x) = \frac{2x - 3}{x^2 + 9} \qquad h(x) = 2^x - 5 \qquad k(x) = 2^{-x} + 3$$

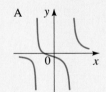
A

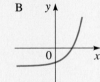
B

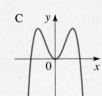
C

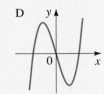
D

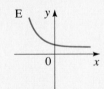
E

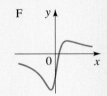
F

5. Let $P(x) = 2x^3 - 11x^2 + 10x + 8$.

   (a) List all possible rational zeros of P.

   (b) Determine which of the numbers you listed in part (a) actually are zeros of P.

   (c) Factor P completely.

   (d) Sketch a graph of P.

6. Let $Q(x) = x^5 - 3x^4 + 3x^3 + x^2 - 4x + 2$.

   (a) Find all zeros of Q, real and complex, and state their multiplicities.

   (b) Factor Q completely.

   (c) Factor Q into linear and irreducible quadratic factors.

7. Let $r(x) = \dfrac{3x^2 + 6x}{x^2 - x - 2}$. Find the x- and y-intercepts and the horizontal and vertical asymptotes. Then sketch the graph of r.

8. Sketch graphs of the following functions on the same coordinate plane.

    (a) $f(x) = 2 - e^x$ **(b)** $g(x) = \ln(x + 1)$

9. (a) Find the exact value of $\log_3 16 - 2 \log_3 36$.

    (b) Use the Laws of Logarithms to expand the expression

$$\log\left(\frac{x^5 \sqrt{x - 1}}{2x - 3} \right)$$

10. Solve the equations.

    (a) $\log_2 x + \log_2(x - 2) = 3$

    (b) $2e^{3x} - 11e^{2x} + 10e^x + 8 = 0$ [*Hint:* Compare to the polynomial in problem 5.]

11. A sum of $25,000 is deposited into an account paying 5.4% interest per year, compounded daily.

    (a) What will the amount in the account be after 3 years?

    (b) When will the account have grown to $35,000?

    (c) How long will it take for the initial deposit to double?

12. After a shipwreck, 120 rats manage to swim from the wreckage to a deserted island. The rat population on the island grows exponentially, and after 15 months there are 280 rats on the island.

    (a) Find a function that models the population t months after the arrival of the rats.

    (b) What will the population be 3 years after the shipwreck?

    (c) When will the population reach 2000?

Systems of Equations and Inequalities

Many real-world problems involve several changing quantities. Such problems are modeled using several equations that contain several variables, rather than by just one equation in one variable. Such a set of equations is called a *system of equations*. In this chapter we learn how to find solutions for such systems. We also study systems of inequalities.

The following Warm-Up Exercises will help you review the basic algebra skills that you will need for this chapter.

WARM-UP EXERCISES

1. Consider the following equation: $x^2e^x - xe^x - 2e^x - x^2 + x = -2$. Which of the following are solutions to this equation?

   $-2 \quad -1 \quad 0 \quad 1 \quad 2$?

2. Solve the following equation for x: $3x + 2y = 4$.

3. Which two of the following lines are parallel?

$$y = \frac{3}{2}x + 2$$
$$y - 4 = 4(x - 2)$$
$$y = -\frac{2}{3}x - 4$$
$$2y - 3x = 6$$
$$y = 2x + 3$$

4. A pitcher contains 3 quarts of lemonade that is 15% lemon juice. How many quarts of lemon juice are in the pitcher? What would be the percentage of lemon juice if 1 quart of water were added to the pitcher?

5. Add the following rational expressions:

$$\frac{12n}{3n - 2} + \frac{8}{2 - 3n}$$

6. Add the following rational expressions:

   (a) $\dfrac{3}{x + 2} + \dfrac{4}{x - 2}$ **(b)** $\dfrac{6}{(x - 1)} + \dfrac{7}{(x - 1)^2}$

7. Solve the following inequalities:

   (a) $3x - 6 \leq -9$ **(b)** $-2x + 4 > 10$

8. Graph the circle whose equation is $x^2 - 4x + y^2 - 2y = 4$.

6.1 | Systems of Equations

LEARNING OBJECTIVES

After completing this section, you will be able to:

- Solve a system of equations using the substitution method
- Solve a system of equations using the elimination method
- Solve a system of equations using the graphical method

We have already seen how a real-world situation can be modeled by an equation (Section 1.2). But many such situations involve too many variables to be modeled by a *single* equation. For example, weather depends on the relationship among many variables, including temperature, wind speed, air pressure, and humidity. So to model (and forecast) the weather, scientists use many equations, each having many variables. Such collections of equations, called systems of equations, *work together* to describe the weather. Systems of equations with hundreds or even thousands of variables are used extensively by airlines to establish consistent flight schedules and by telecommunications companies to find efficient routings for telephone calls. In this chapter we learn how to solve systems of equations that consist of several equations in several variables.

In this section we learn how to solve systems of two equations in two variables. We learn three different methods of solving such systems: by substitution, by elimination, and graphically.

■ Systems of Equations and Their Solutions

A **system of equations** is a set of equations that involve the same variables. A **solution** of a system is an assignment of values for the variables that makes *each* equation in the system true. To **solve** a system means to find all solutions of the system.

Here is an example of a system of two equations in two variables:

$$\begin{cases} 2x - y = 5 & \text{Equation 1} \\ x + 4y = 7 & \text{Equation 2} \end{cases}$$

We can check that $x = 3$ and $y = 1$ is a solution of this system.

Equation 1	**Equation 2**
$2x - y = 5$	$x + 4y = 7$
$2(3) - 1 = 5$ ✔	$3 + 4(1) = 7$ ✔

The solution can also be written as the ordered pair $(3, 1)$.

Note that the graphs of Equations 1 and 2 are lines (see Figure 1). Since the solution $(3, 1)$ satisfies each equation, the point $(3, 1)$ lies on each line. So it is the point of intersection of the two lines.

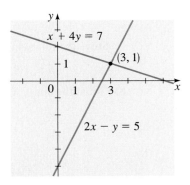

FIGURE 1

■ Substitution Method

In the **substitution method** we start with one equation in the system and solve for one variable in terms of the other variable. The following box describes the procedure.

SUBSTITUTION METHOD

1. **Solve for One Variable.** Choose one equation, and solve for one variable in terms of the other variable.

2. **Substitute.** Substitute the expression you found in Step 1 into the other equation to get an equation in one variable, then solve for that variable.

3. **Back-Substitute.** Substitute the value you found in Step 2 back into the expression found in Step 1 to solve for the remaining variable.

▶ **EXAMPLE 1** | Substitution Method

Find all solutions of the system.

$$\begin{cases} 2x + y = 1 & \text{Equation 1} \\ 3x + 4y = 14 & \text{Equation 2} \end{cases}$$

▼ **SOLUTION** We solve for y in the first equation.

Solve for one variable ▶
$$y = 1 - 2x \qquad \textit{Solve for y in Equation 1}$$

Now we substitute for y in the second equation and solve for x:

Substitute ▶
$$3x + 4(1 - 2x) = 14 \qquad \textit{Substitute y = 1 − 2x into Equation 2}$$
$$3x + 4 - 8x = 14 \qquad \textit{Expand}$$
$$-5x + 4 = 14 \qquad \textit{Simplify}$$
$$-5x = 10 \qquad \textit{Subtract 4}$$
$$x = -2 \qquad \textit{Solve for x}$$

Next we back-substitute $x = -2$ into the equation $y = 1 - 2x$:

Back-substitute ▶
$$y = 1 - 2(-2) = 5 \qquad \textit{Back-substitute}$$

Thus, $x = -2$ and $y = 5$, so the solution is the ordered pair $(-2, 5)$. Figure 2 shows that the graphs of the two equations intersect at the point $(-2, 5)$.

Check Your Answer

$x = -2, y = 5$:

$$\begin{cases} 2(-2) + 5 = 1 \\ 3(-2) + 4(5) = 14 \end{cases} ✔$$

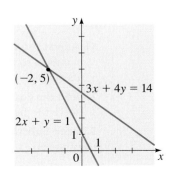

FIGURE 2

▶ **EXAMPLE 2** | Substitution Method

Find all solutions of the system.

$$\begin{cases} x^2 + y^2 = 100 & \text{Equation 1} \\ 3x - y = 10 & \text{Equation 2} \end{cases}$$

▼ **SOLUTION**   We start by solving for y in the second equation.

Solve for one variable ▶
$$y = 3x - 10 \qquad \text{Solve for } y \text{ in Equation 2}$$

Next we substitute for y in the first equation and solve for x:

Substitute ▶
$$x^2 + (3x - 10)^2 = 100 \qquad \text{Substitute } y = 3x - 10 \text{ into Equation 1}$$

$$x^2 + (9x^2 - 60x + 100) = 100 \qquad \text{Expand}$$

$$10x^2 - 60x = 0 \qquad \text{Simplify}$$

$$10x(x - 6) = 0 \qquad \text{Factor}$$

$$x = 0 \quad \text{or} \quad x = 6 \qquad \text{Solve for } x$$

Now we back-substitute these values of x into the equation $y = 3x - 10$.

Back-substitute ▶
$$\text{For} \quad x = 0: \quad y = 3(0) - 10 = -10 \qquad \text{Back-substitute}$$

$$\text{For} \quad x = 6: \quad y = 3(6) - 10 = 8 \qquad \text{Back-substitute}$$

So we have two solutions: $(0, -10)$ and $(6, 8)$.

The graph of the first equation is a circle, and the graph of the second equation is a line; Figure 3 shows that the graphs intersect at the two points $(0, -10)$ and $(6, 8)$.

Check Your Answers

$x = 0, y = -10$:

$$\begin{cases} (0)^2 + (-10)^2 = 100 \\ 3(0) - (-10) = 10 \end{cases} \quad ✔$$

$x = 6, y = 8$:

$$\begin{cases} (6)^2 + (8)^2 = 36 + 64 = 100 \\ 3(6) - (8) = 18 - 8 = 10 \end{cases} \quad ✔$$

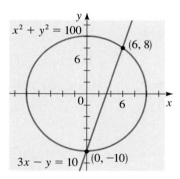

FIGURE 3

Elimination Method

To solve a system using the **elimination method**, we try to combine the equations using sums or differences so as to eliminate one of the variables.

ELIMINATION METHOD

1. **Adjust the Coefficients.**   Multiply one or more of the equations by appropriate numbers so that the coefficient of one variable in one equation is the negative of its coefficient in the other equation.

2. **Add the Equations.**   Add the two equations to eliminate one variable, then solve for the remaining variable.

3. **Back-Substitute.**   Substitute the value that you found in Step 2 back into one of the original equations, and solve for the remaining variable.

▶ **EXAMPLE 3** | Elimination Method

Find all solutions of the system.

$$\begin{cases} 3x + 2y = 14 & \text{Equation 1} \\ x - 2y = 2 & \text{Equation 2} \end{cases}$$

▼ **SOLUTION**   Since the coefficients of the y-terms are negatives of each other, we can add the equations to eliminate y.

$$\begin{cases} 3x + 2y = 14 \\ x - 2y = 2 \end{cases} \quad \text{System}$$

$$\overline{\quad 4x \qquad = 16} \quad \text{Add}$$

$$x = 4 \quad \text{Solve for } x$$

Now we back-substitute $x = 4$ into one of the original equations and solve for y. Let's choose the second equation because it looks simpler.

$$x - 2y = 2 \qquad \text{Equation 2}$$

$$4 - 2y = 2 \qquad \text{Back-substitute } x = 4 \text{ into Equation 2}$$

$$-2y = -2 \qquad \text{Subtract 4}$$

$$y = 1 \qquad \text{Solve for } y$$

The solution is $(4, 1)$. Figure 4 shows that the graphs of the equations in the system intersect at the point $(4, 1)$. ▲

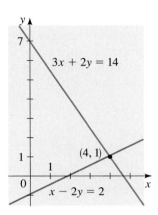

FIGURE 4

▶ **EXAMPLE 4** | Elimination Method

Find all solutions of the system.

$$\begin{cases} 3x^2 + 2y = 26 & \text{Equation 1} \\ 5x^2 + 7y = 3 & \text{Equation 2} \end{cases}$$

▼ **SOLUTION**   We choose to eliminate the x-term, so we multiply the first equation by 5 and the second equation by -3. Then we add the two equations and solve for y.

$$\begin{cases} 15x^2 + 10y = 130 & 5 \times \text{Equation 1} \\ -15x^2 - 21y = -9 & (-3) \times \text{Equation 2} \end{cases}$$

$$\overline{\qquad\quad -11y = 121} \quad \text{Add}$$

$$y = -11 \quad \text{Solve for } y$$

Now we back-substitute $y = -11$ into one of the original equations, say $3x^2 + 2y = 26$, and solve for x:

$$3x^2 + 2(-11) = 26 \qquad \text{Back-substitute } y = -11 \text{ into Equation 1}$$

$$3x^2 = 48 \qquad \text{Add 22}$$

$$x^2 = 16 \qquad \text{Divide by 3}$$

$$x = -4 \quad \text{or} \quad x = 4 \qquad \text{Solve for } x$$

So we have two solutions: $(-4, -11)$ and $(4, -11)$.

The graphs of both equations are parabolas (see Section 4.1). Figure 5 shows that the graphs intersect at the two points $(-4, -11)$ and $(4, -11)$.

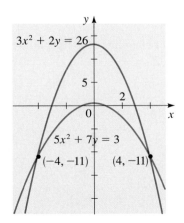

FIGURE 5

Check Your Answers

$x = -4, y = -11:$

$$\begin{cases} 3(-4)^2 + 2(-11) = 26 \\ 5(-4)^2 + 7(-11) = 3 \end{cases} \quad ✔$$

$x = 4, y = -11:$

$$\begin{cases} 3(4)^2 + 2(-11) = 26 \\ 5(4)^2 + 7(-11) = 3 \end{cases} \quad ✔$$

▲

■ Graphical Method

In the **graphical method** we use a graphing device to solve the system of equations. Note that with many graphing devices, any equation must first be expressed in terms of one or more functions of the form $y = f(x)$ before we can use the calculator to graph it. Not all equations can be readily expressed in this way, so not all systems can be solved by this method.

GRAPHICAL METHOD

1. **Graph Each Equation.**   Express each equation in a form suitable for the graphing calculator by solving for y as a function of x. Graph the equations on the same screen.

2. **Find the Intersection Points.**   The solutions are the x- and y-coordinates of the points of intersection.

It may be more convenient to solve for x in terms of y in the equations. In that case, in Step 1 graph x as a function of y instead.

▶ **EXAMPLE 5** | Graphical Method

Find all solutions of the system.

$$\begin{cases} x^2 - y = 2 \\ 2x - y = -1 \end{cases}$$

Graph each equation ▶   ▼ **SOLUTION**   Solving for y in terms of x, we get the equivalent system

$$\begin{cases} y = x^2 - 2 \\ y = 2x + 1 \end{cases}$$

Find intersection points ▶   Figure 6 shows that the graphs of these equations intersect at two points. Zooming in, we see that the solutions are

$$(-1, -1) \quad \text{and} \quad (3, 7)$$

Check Your Answers

$x = -1, y = -1:$

$$\begin{cases} (-1)^2 - (-1) = 2 \\ 2(-1) - (-1) = -1 \end{cases} \quad ✔$$

$x = 3, y = 7:$

$$\begin{cases} 3^2 - 7 = 2 \\ 2(3) - 7 = -1 \end{cases} \quad ✔$$

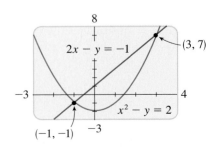

FIGURE 6

▶ **EXAMPLE 6** | Solving a System of Equations Graphically

Find all solutions of the system, correct to one decimal place.

$$\begin{cases} x^2 + y^2 = 12 & \text{Equation 1} \\ y = 2x^2 - 5x & \text{Equation 2} \end{cases}$$

Graph each equation ▶

▼ **SOLUTION** The graph of the first equation is a circle, and the graph of the second is a parabola. To graph the circle on a graphing calculator, we must first solve for y in terms of x (see Section 2.3).

$$x^2 + y^2 = 12$$
$$y^2 = 12 - x^2 \qquad \text{Isolate } y^2 \text{ on LHS}$$
$$y = \pm\sqrt{12 - x^2} \qquad \text{Take square roots}$$

To graph the circle, we must graph both functions:

$$y = \sqrt{12 - x^2} \quad \text{and} \quad y = -\sqrt{12 - x^2}$$

Find intersection points ▶

In Figure 7 the graph of the circle is shown in red and the parabola in blue. The graphs intersect in Quadrants I and II. Zooming in, or using the `Intersect` command, we see that the intersection points are $(-0.559, 3.419)$ and $(2.847, 1.974)$. There also appears to be an intersection point in quadrant IV. However, when we zoom in, we see that the curves come close to each other but don't intersect (see Figure 8). Thus, the system has two solutions; correct to the nearest tenth, they are

$$(-0.6, 3.4) \quad \text{and} \quad (2.8, 2.0)$$

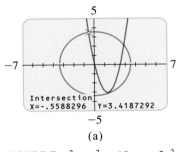

(a)

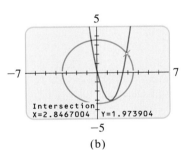

(b)

FIGURE 7 $x^2 + y^2 = 12$, $y = 2x^2 - 5x$

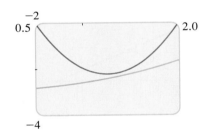

FIGURE 8 Zooming in

6.2 | Systems of Linear Equations in Two Variables

LEARNING OBJECTIVES

After completing this section, you will be able to:

▪ Solve a system of two linear equations in two variables
▪ Determine whether a system of two linear equations in two variables has one solution, infinitely many solutions, or no solution
▪ Model with linear systems

Recall that an equation of the form $Ax + By = C$ is called linear because its graph is a line (see Section 2.4). In this section we study systems of two linear equations in two variables.

■ Systems of Linear Equations in Two Variables

A **system of two linear equations in two variables** has the form

$$\begin{cases} a_1x + b_1y = c_1 \\ a_2x + b_2y = c_2 \end{cases}$$

We can use either the substitution method or the elimination method to solve such systems algebraically. But since the elimination method is usually easier for linear systems, we use elimination rather than substitution in our examples.

The graph of a linear system in two variables is a pair of lines, so to solve the system graphically, we must find the intersection point(s) of the lines. Two lines may intersect in a single point, they may be parallel, or they may coincide, as shown in Figure 1. So there are three possible outcomes in solving such a system.

NUMBER OF SOLUTIONS OF A LINEAR SYSTEM IN TWO VARIABLES

For a system of linear equations in two variables, exactly one of the following is true. (See Figure 1.)

1. The system has exactly one solution.

2. The system has no solution.

3. The system has infinitely many solutions.

A system that has no solution is said to be **inconsistent**. A system with infinitely many solutions is called **dependent**.

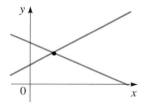

(a) Linear system with one solution. Lines intersect at a single point.

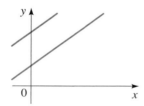

(b) Linear system with no solution. Lines are parallel—they do not intersect.

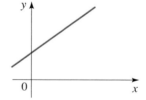

(c) Linear system with infinitely many solutions. Lines coincide—equations are for the same line.

FIGURE 1

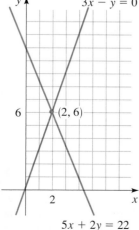

FIGURE 2

▶ **EXAMPLE 1** | A Linear System with One Solution

Solve the system and graph the lines.

$$\begin{cases} 3x - y = 0 & \text{Equation 1} \\ 5x + 2y = 22 & \text{Equation 2} \end{cases}$$

▼ **SOLUTION**   We eliminate y from the equations and solve for x.

$$\begin{cases} 6x - 2y = 0 & \text{2 × Equation 1} \\ 5x + 2y = 22 \end{cases}$$

$$\overline{11x = 22} \quad \text{Add}$$

$$x = 2 \quad \text{Solve for } x$$

Now we back-substitute into the first equation and solve for y:

$$6(2) - 2y = 0 \quad \text{Back-substitute } x = 2$$

$$-2y = -12 \quad \text{Subtract } 6 \times 2 = 12$$

$$y = 6 \quad \text{Solve for } y$$

Check Your Answer

$x = 2, \quad y = 6$:

$$\begin{cases} 3(2) - (6) = 0 \\ 5(2) + 2(6) = 22 \end{cases} \checkmark$$

The solution of the system is the ordered pair $(2, 6)$, that is,

$$x = 2, \qquad y = 6$$

The graph in Figure 2 shows that the lines in the system intersect at the point $(2, 6)$. ▲

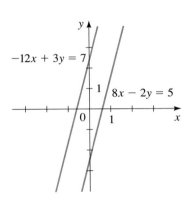

FIGURE 3

▶ **EXAMPLE 2** | A Linear System with No Solution

Solve the system.

$$\begin{cases} 8x - 2y = 5 & \text{Equation 1} \\ -12x + 3y = 7 & \text{Equation 2} \end{cases}$$

▼ **SOLUTION**   This time we try to find a suitable combination of the two equations to eliminate the variable y. Multiplying the first equation by 3 and the second equation by 2 gives

$$\begin{cases} 24x - 6y = 15 & \text{3 × Equation 1} \\ -24x + 6y = 14 & \text{2 × Equation 2} \end{cases}$$
$$\overline{\qquad 0 = 29 \qquad \text{Add}}$$

Adding the two equations eliminates *both x and y* in this case, and we end up with $0 = 29$, which is obviously false. No matter what values we assign to x and y, we cannot make this statement true, so the system has *no solution*. Figure 3 shows that the lines in the system are parallel and do not intersect. The system is inconsistent. ▲

▶ **EXAMPLE 3** | A Linear System with Infinitely Many Solutions

Solve the system.

$$\begin{cases} 3x - 6y = 12 & \text{Equation 1} \\ 4x - 8y = 16 & \text{Equation 2} \end{cases}$$

▼ **SOLUTION**   We multiply the first equation by 4 and the second by 3 to prepare for subtracting the equations to eliminate x. The new equations are

$$\begin{cases} 12x - 24y = 48 & \text{4 × Equation 1} \\ 12x - 24y = 48 & \text{3 × Equation 2} \end{cases}$$

We see that the two equations in the original system are simply different ways of expressing the equation of one single line. The coordinates of any point on this line give a solution of the system. Writing the equation in slope-intercept form, we have $y = \frac{1}{2}x - 2$. So if we let t represent any real number, we can write the solution as

$$x = t$$
$$y = \tfrac{1}{2}t - 2$$

We can also write the solution in ordered-pair form as

$$\left(t, \tfrac{1}{2}t - 2\right)$$

where t is any real number. The system has infinitely many solutions (see Figure 4). ▲

In Example 3, to get specific solutions, we have to assign values to t. For instance, if $t = 1$, we get the solution $\left(1, -\frac{3}{2}\right)$. If $t = 4$, we get the solution $(4, 0)$. For every value of t we get a different solution. (See Figure 4.)

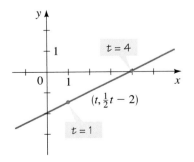

FIGURE 4

■ Modeling with Linear Systems

Frequently, when we use equations to solve problems in the sciences or in other areas, we obtain systems like the ones we've been considering. When modeling with systems of equations, we use the following guidelines, which are similar to those in Section 1.2.

GUIDELINES FOR MODELING WITH SYSTEMS OF EQUATIONS

1. **Identify the Variables.**   Identify the quantities that the problem asks you to find. These are usually determined by a careful reading of the question posed at the end of the problem. Introduce notation for the variables (call them x and y or some other letters).

2. **Express All Unknown Quantities in Terms of the Variables.**   Read the problem again, and express all the quantities mentioned in the problem in terms of the variables you defined in Step 1.

3. **Set Up a System of Equations.**   Find the crucial facts in the problem that give the relationships between the expressions you found in Step 2. Set up a system of equations (or a model) that expresses these relationships.

4. **Solve the System and Interpret the Results.**   Solve the system you found in Step 3, check your solutions, and state your final answer as a sentence that answers the question posed in the problem.

The next two examples illustrate how to model with systems of equations.

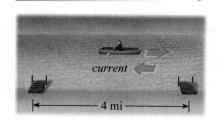

EXAMPLE 4 | A Distance-Speed-Time Problem

A woman rows a boat upstream from one point on a river to another point 4 mi away in $1\frac{1}{2}$ hours. The return trip, traveling with the current, takes only 45 min. How fast does she row relative to the water, and at what speed is the current flowing?

▼ **SOLUTION**   We are asked to find the rowing speed and the speed of the current, so we let

$$x = \text{rowing speed (mi/h)}$$

Identify the variables ▶

$$y = \text{current speed (mi/h)}$$

The woman's speed when she rows upstream is her rowing speed minus the speed of the current; her speed downstream is her rowing speed plus the speed of the current. Now we translate this information into the language of algebra.

Express unknown quantities in terms of the variable ▶

In Words	In Algebra
Rowing speed	x
Current speed	y
Speed upstream	$x - y$
Speed downstream	$x + y$

The distance upstream and downstream is 4 mi, so using the fact that speed × time = distance for both legs of the trip, we get

$$\text{speed upstream} \times \text{time upstream} = \text{distance traveled}$$

$$\text{speed downstream} \times \text{time downstream} = \text{distance traveled}$$

In algebraic notation this translates into the following equations.

Set up a system of equations ▶

$$(x - y)\tfrac{3}{2} = 4 \qquad \text{Equation 1}$$

$$(x + y)\tfrac{3}{4} = 4 \qquad \text{Equation 2}$$

(The times have been converted to hours, since we are expressing the speeds in miles per *hour*.) We multiply the equations by 2 and 4, respectively, to clear the denominators.

Solve the system ▶

$$\begin{cases} 3x - 3y = 8 & \text{2 × Equation 1} \\ 3x + 3y = 16 & \text{4 × Equation 2} \end{cases}$$

$$\begin{array}{ll} 6x \phantom{{}- 3y} = 24 & \text{Add} \\ x \phantom{{}- 3y} = 4 & \text{Solve for } x \end{array}$$

Back-substituting this value of x into the first equation (the second works just as well) and solving for y gives

$$\begin{array}{ll} 3(4) - 3y = 8 & \text{Back-substitute } x = 4 \\ -3y = 8 - 12 & \text{Subtract 12} \\ y = \tfrac{4}{3} & \text{Solve for } y \end{array}$$

The woman rows at 4 mi/h, and the current flows at $1\tfrac{1}{3}$ mi/h.

Check Your Answer

Speed upstream is

$$\frac{\text{distance}}{\text{time}} = \frac{4 \text{ mi}}{1\tfrac{1}{2}\text{ h}} = 2\tfrac{2}{3}\text{ mi/h}$$

and this should equal

rowing speed − current flow

$$= 4 \text{ mi/h} - \tfrac{4}{3}\text{ mi/h} = 2\tfrac{2}{3}\text{ mi/h}$$

Speed downstream is

$$\frac{\text{distance}}{\text{time}} = \frac{4 \text{ mi}}{\tfrac{3}{4}\text{ h}} = 5\tfrac{1}{3}\text{ mi/h}$$

and this should equal

rowing speed + current flow

$$= 4 \text{ mi/h} + \tfrac{4}{3}\text{ mi/h} = 5\tfrac{1}{3}\text{ mi/h} \qquad ✔$$

▲

▶ **EXAMPLE 5** | A Mixture Problem

A vintner fortifies wine that contains 10% alcohol by adding a 70% alcohol solution to it. The resulting mixture has an alcoholic strength of 16% and fills 1000 one-liter bottles. How many liters (L) of the wine and of the alcohol solution does the vintner use?

▼ **SOLUTION**   Since we are asked for the amounts of wine and alcohol, we let

Identify the variables ▶

$$x = \text{amount of wine used (L)}$$

$$y = \text{amount of alcohol solution used (L)}$$

From the fact that the wine contains 10% alcohol and the solution contains 70% alcohol, we get the following.

Express all unknown quantities in terms of the variable ▶

In Words	In Algebra
Amount of wine used (L)	x
Amount of alcohol solution used (L)	y
Amount of alcohol in wine (L)	$0.10x$
Amount of alcohol in solution (L)	$0.70y$

The volume of the mixture must be the total of the two volumes the vintner is adding together, so

$$x + y = 1000$$

Also, the amount of alcohol in the mixture must be the total of the alcohol contributed by the wine and by the alcohol solution, that is,

$$0.10x + 0.70y = (0.16)1000$$

$$0.10x + 0.70y = 160 \qquad \text{Simplify}$$

$$x + 7y = 1600 \qquad \text{Multiply by 10 to clear decimals}$$

Thus, we get the system

Set up a system of equations ▶

$$\begin{cases} x + y = 1000 & \text{Equation 1} \\ x + 7y = 1600 & \text{Equation 2} \end{cases}$$

Subtracting the first equation from the second eliminates the variable x, and we get

Solve the system ▶

$$6y = 600 \qquad \text{Subtract Equation 1 from Equation 2}$$

$$y = 100 \qquad \text{Solve for } y$$

We now back-substitute $y = 100$ into the first equation and solve for x.

$$x + 100 = 1000 \qquad \text{Back-substitute } y = 100$$

$$x = 900 \qquad \text{Solve for } x$$

The vintner uses 900 L of wine and 100 L of the alcohol solution. ▲

6.3 | Systems of Linear Equations in Several Variables

LEARNING OBJECTIVES

After completing this section, you will be able to:

■ Use Gaussian elimination to solve a system of three (or more) linear equations
■ Determine whether a system of three (or more) linear equations has one solution, infinitely many solutions, or no solution
■ Model with linear systems in three (or more) variables

A **linear equation in n variables** is an equation that can be put in the form

$$a_1x_1 + a_2x_2 + \cdots + a_nx_n = c$$

where $a_1, a_2, \ldots, a_n$ and c are real numbers, and $x_1, x_2, \ldots, x_n$ are the variables. If we have only three or four variables, we generally use x, y, z, and w instead of x_1, x_2, x_3, and x_4. Such equations are called *linear* because if we have just two variables, the equation is $a_1x + a_2y = c$, which is the equation of a line. Here are some examples of equations in three variables that illustrate the difference between linear and nonlinear equations.

Linear equations	**Nonlinear equations**	
$6x_1 - 3x_2 + \sqrt{5}x_3 = 10$	$x^2 + 3y - \sqrt{z} = 5$	Not linear because it contains the square and the square root of a variable
$x + y + z = 2w - \frac{1}{2}$	$x_1x_2 + 6x_3 = -6$	Not linear because it contains a product of variables

In this section we study systems of linear equations in three or more variables.

▨ Solving a Linear System

The following are two examples of systems of linear equations in three variables. The second system is in **triangular form**; that is, the variable x doesn't appear in the second equation, and the variables x and y do not appear in the third equation.

<table>
<tr><td align="center">A system of linear equations</td><td align="center">A system in triangular form</td></tr>
<tr><td>

$$\begin{cases} x - 2y - z = 1 \\ -x + 3y + 3z = 4 \\ 2x - 3y + z = 10 \end{cases}$$

</td><td>

$$\begin{cases} x - 2y - z = 1 \\ y + 2z = 5 \\ z = 3 \end{cases}$$

</td></tr>
</table>

It's easy to solve a system that is in triangular form by using back-substitution. So our goal in this section is to start with a system of linear equations and change it to a system in triangular form that has the same solutions as the original system. We begin by showing how to use back-substitution to solve a system that is already in triangular form.

▶ **EXAMPLE 1** | Solving a Triangular System Using Back-Substitution

Solve the system using back-substitution:

$$\begin{cases} x - 2y - z = 1 & \text{Equation 1} \\ y + 2z = 5 & \text{Equation 2} \\ z = 3 & \text{Equation 3} \end{cases}$$

▼ **SOLUTION**   From the last equation we know that $z = 3$. We back-substitute this into the second equation and solve for y.

$$y + 2(3) = 5 \qquad \text{Back-substitute } z = 3 \text{ into Equation 2}$$

$$y = -1 \qquad \text{Solve for } y$$

Then we back-substitute $y = -1$ and $z = 3$ into the first equation and solve for x.

$$x - 2(-1) - (3) = 1 \qquad \text{Back-substitute } y = -1 \text{ and } z = 3 \text{ into Equation 1}$$

$$x = 2 \qquad \text{Solve for } x$$

The solution of the system is $x = 2$, $y = -1$, $z = 3$. We can also write the solution as the ordered triple $(2, -1, 3)$. ▲

To change a system of linear equations to an **equivalent system** (that is, a system with the same solutions as the original system), we use the elimination method. This means that we can use the following operations.

OPERATIONS THAT YIELD AN EQUIVALENT SYSTEM

1. Add a nonzero multiple of one equation to another.

2. Multiply an equation by a nonzero constant.

3. Interchange the positions of two equations.

To solve a linear system, we use these operations to change the system to an equivalent triangular system. Then we use back-substitution as in Example 1. This process is called **Gaussian elimination**.

▶ **EXAMPLE 2** | Solving a System of Three Equations in Three Variables

Solve the system using Gaussian elimination.

$$\begin{cases} x - 2y + 3z = 1 & \text{Equation 1} \\ x + 2y - z = 13 & \text{Equation 2} \\ 3x + 2y - 5z = 3 & \text{Equation 3} \end{cases}$$

▼ **SOLUTION** We need to change this to a triangular system, so we begin by eliminating the x-term from the second equation.

$$\begin{array}{ll} x + 2y - z = 13 & \text{Equation 2} \\ \underline{x - 2y + 3z = 1} & \text{Equation 1} \\ 4y - 4z = 12 & \text{Equation 2} + (-1) \times \text{Equation 1} = \text{new Equation 2} \end{array}$$

This gives us a new, equivalent system that is one step closer to triangular form:

$$\begin{cases} x - 2y + 3z = 1 & \text{Equation 1} \\ 4y - 4z = 12 & \text{Equation 2} \\ 3x + 2y - 5z = 3 & \text{Equation 3} \end{cases}$$

Now we eliminate the x-term from the third equation.

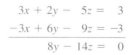

$$\begin{cases} x - 2y + 3z = 1 \\ 4y - 4z = 12 \\ 8y - 14z = 0 \end{cases} \quad \text{Equation 3} + (-3) \times \text{Equation 1} = \text{new Equation 3}$$

Then we eliminate the y-term from the third equation.

$$\begin{array}{l} 8y - 14z = 0 \\ \underline{-8y + 8z = -24} \\ -6z = -24 \end{array}$$

$$\begin{cases} x - 2y + 3z = 1 \\ 4y - 4z = 12 \\ -6z = -24 \end{cases} \quad \text{Equation 3} + (-2) \times \text{Equation 2} = \text{new Equation 3}$$

The system is now in triangular form, but it will be easier to work with if we divide the second and third equations by the common factors of each term.

$$\begin{cases} x - 2y + 3z = 1 \\ y - z = 3 & \tfrac{1}{4} \times \text{Equation 2} = \text{new Equation 2} \\ z = 4 & -\tfrac{1}{6} \times \text{Equation 3} = \text{new Equation 3} \end{cases}$$

Intersection of Three Planes

When you study calculus or linear algebra, you will learn that the graph of a linear equation in three variables is a plane in a three-dimensional coordinate system. For a system of three equations in three variables, the following situations arise:

1. The three planes intersect in a single point.

The system has a unique solution.

2. The three planes intersect in more than one point.

The system has infinitely many solutions.

3. The three planes have no point in common.

The system has no solution.

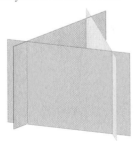

Now we use back-substitution to solve the system. From the third equation we get $z = 4$. We back-substitute this into the second equation and solve for y.

$$y - (4) = 3 \qquad \text{Back-substitute } z = 4 \text{ into Equation 2}$$
$$y = 7 \qquad \text{Solve for } y$$

Now we back-substitute $y = 7$ and $z = 4$ into the first equation and solve for x.

$$x - 2(7) + 3(4) = 1 \qquad \text{Back-substitute } y = 7 \text{ and } z = 4 \text{ into Equation 1}$$
$$x = 3 \qquad \text{Solve for } x$$

The solution of the system is $x = 3$, $y = 7$, $z = 4$, which we can write as the ordered triple $(3, 7, 4)$.

Check Your Answer

$x = 3, y = 7, z = 4$:

$$
\begin{aligned}
(3) - 2(7) + 3(4) &= 1 \\
(3) + 2(7) - (4) &= 13 \\
3(3) + 2(7) - 5(4) &= 3 \qquad ✔
\end{aligned}
$$

■ The Number of Solutions of a Linear System

Just as in the case of two variables, a system of equations in several variables may have one solution, no solution, or infinitely many solutions. The graphical interpretation of the solutions of a linear system is analogous to that for systems of equations in two variables (see the margin note).

NUMBER OF SOLUTIONS OF A LINEAR SYSTEM
For a system of linear equations, exactly one of the following is true.   **1.** The system has exactly one solution.   **2.** The system has no solution.   **3.** The system has infinitely many solutions.

A system with no solutions is said to be **inconsistent**, and a system with infinitely many solutions is said to be **dependent**. As we see in the next example, a linear system has no solution if we end up with a *false equation* after applying Gaussian elimination to the system.

▶ **EXAMPLE 3** | A System with No Solution

Solve the following system.

$$
\begin{cases}
x + 2y - 2z = 1 & \text{Equation 1} \\
2x + 2y - z = 6 & \text{Equation 2} \\
3x + 4y - 3z = 5 & \text{Equation 3}
\end{cases}
$$

▼ **SOLUTION**   To put this in triangular form, we begin by eliminating the x-terms from the second equation and the third equation.

$$\begin{cases} x + 2y - 2z = 1 \\ \quad\;\; -2y + 3z = 4 \\ 3x + 4y + 3z = 5 \end{cases}$$

Equation 2 + (−2) × Equation 1 = new Equation 2

$$\begin{cases} x + 2y - 2z = 1 \\ \quad\;\; -2y + 3z = 4 \\ \quad\;\; -2y + 3z = 2 \end{cases}$$

Equation 3 + (−3) × Equation 1 = new Equation 3

Now we eliminate the y-term from the third equation.

$$\begin{cases} x + 2y - 2z = 1 \\ \quad\;\; -2y + 3z = 4 \\ \qquad\qquad 0 = 2 \end{cases}$$

Equation 3 + (−1) × Equation 2 = new Equation 3

The system is now in triangular form, but the third equation says $0 = 2$, which is false. No matter what values we assign to x, y, and z, the third equation will never be true. This means that the system has *no solution*. ▲

▶ **EXAMPLE 4** | A System with Infinitely Many Solutions

Solve the following system.

$$\begin{cases} x - \;\, y + 5z = -2 \quad & \text{Equation 1} \\ 2x + \;\, y + 4z = \quad 2 & \text{Equation 2} \\ 2x + 4y - 2z = \quad 8 & \text{Equation 3} \end{cases}$$

▼ **SOLUTION** To put this in triangular form, we begin by eliminating the x-terms from the second equation and the third equation.

$$\begin{cases} x - \;\, y + 5z = -2 \\ \qquad 3y - 6z = \quad 6 \\ 2x + 4y - 2z = \quad 8 \end{cases}$$

Equation 2 + (−2) × Equation 1 = new Equation 2

$$\begin{cases} x - y + \;\, 5z = -2 \\ \quad\;\; 3y - \;\, 6z = \quad 6 \\ \quad\;\; 6y - 12z = \;\; 12 \end{cases}$$

Equation 3 + (−2) × Equation 1 = new Equation 3

Now we eliminate the y-term from the third equation.

$$\begin{cases} x - y + 5z = -2 \\ \quad\;\; 3y - 6z = \quad 6 \\ \qquad\qquad 0 = \quad 0 \end{cases}$$

Equation 3 + (−2) × Equation 2 = new Equation 3

The new third equation is true, but it gives us no new information, so we can drop it from the system. Only two equations are left. We can use them to solve for x and y in terms of z, but z can take on any value, so there are infinitely many solutions.

To find the complete solution of the system, we begin by solving for y in terms of z, using the new second equation.

$$3y - 6z = 6 \qquad \text{Equation 2}$$

$$y - 2z = 2 \qquad \text{Multiply by } \tfrac{1}{3}$$

$$y = 2z + 2 \qquad \text{Solve for } y$$

Then we solve for x in terms of z, using the first equation.

$$x - (2z + 2) + 5z = -2 \qquad \text{Substitute } y = 2z + 2 \text{ into Equation 1}$$

$$x + 3z - 2 = -2 \qquad \text{Simplify}$$

$$x = -3z \qquad \text{Solve for } x$$

To describe the complete solution, we let t represent any real number. The solution is

$$x = -3t$$

$$y = 2t + 2$$

$$z = t$$

We can also write this as the ordered triple $(-3t, 2t + 2, t)$.   ▲

In the solution of Example 4 the variable t is called a **parameter**. To get a specific solution, we give a specific value to the parameter t. For instance, if we set $t = 2$, we get

$$x = -3(2) = -6$$

$$y = 2(2) + 2 = 6$$

$$z = 2$$

Thus, $(-6, 6, 2)$ is a solution of the system. Here are some other solutions of the system obtained by substituting other values for the parameter t.

Parameter t	Solution $(-3t, 2t + 2, t)$
-1	$(3, 0, -1)$
0	$(0, 2, 0)$
3	$(-9, 8, 3)$
10	$(-30, 22, 10)$

You should check that these points satisfy the original equations. There are infinitely many choices for the parameter t, so the system has infinitely many solutions.

■ Modeling Using Linear Systems

Linear systems are used to model situations that involve several varying quantities. In the next example we consider an application of linear systems to finance.

▶ **EXAMPLE 5** | Modeling a Financial Problem Using a Linear System

Jason receives an inheritance of $50,000. His financial advisor suggests that he invest this in three mutual funds: a money-market fund, a blue-chip stock fund, and a high-tech stock fund. The advisor estimates that the money-market fund will return 5% over the next year, the blue-chip fund 9%, and the high-tech fund 16%. Jason wants a total first-year return of $4000. To avoid excessive risk, he decides to invest three times as much in the money-market fund as in the high-tech stock fund. How much should he invest in each fund?

▼ **SOLUTION**

Let $x = $ amount invested in the money-market fund

$y = $ amount invested in the blue-chip stock fund

$z = $ amount invested in the high-tech stock fund

We convert each fact given in the problem into an equation.

$$x + y + z = 50{,}000 \qquad \text{Total amount invested is \$50,000}$$
$$0.05x + 0.09y + 0.16z = 4000 \qquad \text{Total investment return is \$4000}$$
$$x = 3z \qquad \text{Money-market amount is 3 × high-tech amount}$$

Multiplying the second equation by 100 and rewriting the third gives the following system, which we solve using Gaussian elimination.

$$\begin{cases} x + y + z = 50{,}000 \\ 5x + 9y + 16z = 400{,}000 \qquad \text{100 × Equation 2} \\ x \qquad - 3z = 0 \qquad \text{Subtract 3z} \end{cases}$$

$$\begin{cases} x + y + z = 50{,}000 \\ 4y + 11z = 150{,}000 \qquad \text{Equation 2 + (−5) × Equation 1 = new Equation 2} \\ -y - 4z = -50{,}000 \qquad \text{Equation 3 + (−1) × Equation 1 = new Equation 3} \end{cases}$$

$$\begin{cases} x + y + z = 50{,}000 \\ -5z = -50{,}000 \qquad \text{Equation 2 + 4 × Equation 3 = new Equation 3} \\ -y - 4z = -50{,}000 \end{cases}$$

$$\begin{cases} x + y + z = 50{,}000 \\ z = 10{,}000 \qquad (-\tfrac{1}{5}) \times \text{Equation 2} \\ y + 4z = 50{,}000 \qquad (-1) \times \text{Equation 3} \end{cases}$$

$$\begin{cases} x + y + z = 50{,}000 \\ y + 4z = 50{,}000 \qquad \text{Interchange Equations 2 and 3} \\ z = 10{,}000 \end{cases}$$

Now that the system is in triangular form, we use back-substitution to find that $x = 30{,}000$, $y = 10{,}000$, and $z = 10{,}000$. This means that Jason should invest

$30,000 in the money-market fund

$10,000 in the blue-chip stock fund

$10,000 in the high-tech stock fund

6.4 | Partial Fractions

LEARNING OBJECTIVES

After completing this section, you will be able to:

- Find the form of the partial fraction decomposition of a rational expression in the following cases.
  - —Denominator contains distinct linear factors
  - —Denominator contains repeated linear factors
  - —Denominator contains distinct quadratic factors
  - —Denominator contains repeated quadratic factors
- Find the partial fraction decomposition of a rational expression in the above cases

To write a sum or difference of fractional expressions as a single fraction, we bring them to a common denominator. For example,

◼ Common denominator ➡

$$\frac{1}{x - 1} + \frac{1}{2x + 1} = \frac{3x}{2x^2 - x - 1}$$

⬅ ◼ Partial fractions ◼

$$\frac{1}{x - 1} + \frac{1}{2x + 1} = \frac{(2x + 1) + (x - 1)}{(x - 1)(2x + 1)} = \frac{3x}{2x^2 - x - 1}$$

But for some applications of algebra to calculus we must reverse this process—that is, we must express a fraction such as $3x/(2x^2 - x - 1)$ as the sum of the simpler fractions $1/(x - 1)$ and $1/(2x + 1)$. These simpler fractions are called *partial fractions*; we learn how to find them in this section.

Let r be the rational function

$$r(x) = \frac{P(x)}{Q(x)}$$

where the degree of P is less than the degree of Q. By the Linear and Quadratic Factors Theorem in Section 4.5, every polynomial with real coefficients can be factored completely into linear and irreducible quadratic factors, that is, factors of the form $ax + b$ and $ax^2 + bx + c$, where a, b, and c are real numbers. For instance,

$$x^4 - 1 = (x^2 - 1)(x^2 + 1) = (x - 1)(x + 1)(x^2 + 1)$$

After we have completely factored the denominator Q of r, we can express $r(x)$ as a sum of **partial fractions** of the form

$$\frac{A}{(ax + b)^i} \quad \text{and} \quad \frac{Ax + B}{(ax^2 + bx + c)^j}$$

This sum is called the **partial fraction decomposition** of r. Let's examine the details of the four possible cases.

CASE 1: THE DENOMINATOR IS A PRODUCT OF DISTINCT LINEAR FACTORS

Suppose that we can factor $Q(x)$ as

$$Q(x) = (a_1x + b_1)(a_2x + b_2) \cdots (a_nx + b_n)$$

with no factor repeated. In this case the partial fraction decomposition of $P(x)/Q(x)$ takes the form

$$\frac{P(x)}{Q(x)} = \frac{A}{a_1x + b_1} + \frac{A_2}{a_2x + b_2} + \cdots + \frac{A_n}{a_nx + b_n}$$

The constants $A_1, A_2, \ldots, A_n$ are determined as in the following example.

▶ **EXAMPLE 1** | Distinct Linear Factors

Find the partial fraction decomposition of $\dfrac{5x + 7}{x^3 + 2x^2 - x - 2}$.

▼ **SOLUTION** The denominator factors as follows:

$$x^3 + 2x^2 - x - 2 = x^2(x + 2) - (x + 2) = (x^2 - 1)(x + 2)$$
$$= (x - 1)(x + 1)(x + 2)$$

This gives us the partial fraction decomposition

$$\frac{5x + 7}{x^3 + 2x^2 - x - 2} = \frac{A}{x - 1} + \frac{B}{x + 1} + \frac{C}{x + 2}$$

Multiplying each side by the common denominator, $(x - 1)(x + 1)(x + 2)$, we get

$$5x + 7 = A(x + 1)(x + 2) + B(x - 1)(x + 2) + C(x - 1)(x + 1)$$

$$= A(x^2 + 3x + 2) + B(x^2 + x - 2) + C(x^2 - 1) \qquad \text{Expand}$$

$$= (A + B + C)x^2 + (3A + B)x + (2A - 2B - C) \qquad \text{Combine like terms}$$

If two polynomials are equal, then their coefficients are equal. Thus, since $5x + 7$ has no x^2-term, we have $A + B + C = 0$. Similarly, by comparing the coefficients of x, we see that $3A + B = 5$, and by comparing constant terms, we get $2A - 2B - C = 7$. This leads to the following system of linear equations for A, B, and C.

$$\begin{cases} A + B + C = 0 & \text{Equation 1: Coefficients of } x^2 \\ 3A + B = 5 & \text{Equation 2: Coefficients of } x \\ 2A - 2B - C = 7 & \text{Equation 3: Constant coefficients} \end{cases}$$

We use Gaussian elimination to solve this system.

$$\begin{cases} A + B + C = 0 & \\ -2B - 3C = 5 & \text{Equation 2 + (−3) × Equation 1} \\ -4B - 3C = 7 & \text{Equation 3 + (−2) × Equation 1} \end{cases}$$

$$\begin{cases} A + B + C = 0 & \\ -2B - 3C = 5 & \\ 3C = -3 & \text{Equation 3 + (−2) × Equation 2} \end{cases}$$

From the third equation we get $C = -1$. Back-substituting, we find that $B = -1$ and $A = 2$. So the partial fraction decomposition is

$$\frac{5x + 7}{x^3 + 2x^2 - x - 2} = \frac{2}{x - 1} + \frac{-1}{x + 1} + \frac{-1}{x + 2} \qquad \blacktriangle$$

The same approach works in the remaining cases. We set up the partial fraction decomposition with the unknown constants A, B, C, ... Then we multiply each side of the resulting equation by the common denominator, simplify the right-hand side of the equation, and equate coefficients. This gives a set of linear equations that will always have a unique solution (provided that the partial fraction decomposition has been set up correctly).

CASE 2: THE DENOMINATOR IS A PRODUCT OF LINEAR FACTORS, SOME OF WHICH ARE REPEATED

Suppose the complete factorization of $Q(x)$ contains the linear factor $ax + b$ repeated k times; that is, $(ax + b)^k$ is a factor of $Q(x)$. Then, corresponding to each such factor, the partial fraction decomposition for $P(x)/Q(x)$ contains

$$\frac{A_1}{ax + b} + \frac{A_2}{(ax + b)^2} + \cdots + \frac{A_k}{(ax + b)^k}$$

▶ **EXAMPLE 2** | Repeated Linear Factors

Find the partial fraction decomposition of $\dfrac{x^2 + 1}{x(x - 1)^3}$.

▼ **SOLUTION** Because the factor $x - 1$ is repeated three times in the denominator, the partial fraction decomposition has the form

$$\frac{x^2 + 1}{x(x - 1)^3} = \frac{A}{x} + \frac{B}{x - 1} + \frac{C}{(x - 1)^2} + \frac{D}{(x - 1)^3}$$

Multiplying each side by the common denominator, $x(x - 1)^3$, gives

$$\begin{aligned} x^2 + 1 &= A(x - 1)^3 + Bx(x - 1)^2 + Cx(x - 1) + Dx \\ &= A(x^3 - 3x^2 + 3x - 1) + B(x^3 - 2x^2 + x) + C(x^2 - x) + Dx \qquad \text{Expand} \\ &= (A + B)x^3 + (-3A - 2B + C)x^2 + (3A + B - C + D)x - A \qquad \text{Combine like terms} \end{aligned}$$

Equating coefficients, we get the following equations.

$$\begin{cases} A + B & = 0 & \text{Coefficients of } x^3 \\ -3A - 2B + C & = 1 & \text{Coefficients of } x^2 \\ 3A + B - C + D = 0 & & \text{Coefficients of } x \\ -A & = 1 & \text{Constant coefficients} \end{cases}$$

If we rearrange these equations by putting the last one in the first position, we can easily see (using substitution) that the solution to the system is $A = -1, B = 1, C = 0, D = 2$, so the partial fraction decomposition is

$$\frac{x^2 + 1}{x(x - 1)^3} = \frac{-1}{x} + \frac{1}{x - 1} + \frac{2}{(x - 1)^3}$$

▲

CASE 3: THE DENOMINATOR HAS IRREDUCIBLE QUADRATIC FACTORS, NONE OF WHICH IS REPEATED

Suppose the complete factorization of $Q(x)$ contains the quadratic factor $ax^2 + bx + c$ (which can't be factored further). Then, corresponding to this, the partial fraction decomposition of $P(x)/Q(x)$ will have a term of the form

$$\frac{Ax + B}{ax^2 + bx + c}$$

▶ **EXAMPLE 3** | Distinct Quadratic Factors

Find the partial fraction decomposition of $\dfrac{2x^2 - x + 4}{x^3 + 4x}$.

▼ **SOLUTION** Since $x^3 + 4x = x(x^2 + 4)$, which can't be factored further, we write

$$\frac{2x^2 - x + 4}{x^3 + 4x} = \frac{A}{x} + \frac{Bx + C}{x^2 + 4}$$

Multiplying by $x(x^2 + 4)$, we get

$$\begin{aligned} 2x^2 - x + 4 &= A(x^2 + 4) + (Bx + C)x \\ &= (A + B)x^2 + Cx + 4A \end{aligned}$$

Equating coefficients gives us the equations

$$\begin{cases} A + B = 2 & \text{Coefficients of } x^2 \\ C = -1 & \text{Coefficients of } x \\ 4A = 4 & \text{Constant coefficient} \end{cases}$$

so $A = 1$, $B = 1$, and $C = -1$. The required partial fraction decomposition is

$$\frac{2x^2 - x + 4}{x^3 + 4x} = \frac{1}{x} + \frac{x - 1}{x^2 + 4}$$

▲

> **CASE 4: THE DENOMINATOR HAS A REPEATED IRREDUCIBLE QUADRATIC FACTOR**
>
> Suppose the complete factorization of $Q(x)$ contains the factor $(ax^2 + bx + c)^k$, where $ax^2 + bx + c$ can't be factored further. Then the partial fraction decomposition of $P(x)/Q(x)$ will have the terms
>
> $$\frac{A_1x + B_1}{ax^2 + bx + c} + \frac{A_2x + B_2}{(ax^2 + bx + c)^2} + \cdots + \frac{A_kx + B_k}{(ax^2 + bx + c)^k}$$

▶ **EXAMPLE 4** | Repeated Quadratic Factors

Write the form of the partial fraction decomposition of

$$\frac{x^5 - 3x^2 + 12x - 1}{x^3(x^2 + x + 1)(x^2 + 2)^3}$$

▼ **SOLUTION**

$$\frac{x^5 - 3x^2 + 12x - 1}{x^3(x^2 + x + 1)(x^2 + 2)^3}$$

$$= \frac{A}{x} + \frac{B}{x^2} + \frac{C}{x^3} + \frac{Dx + E}{x^2 + x + 1} + \frac{Fx + G}{x^2 + 2} + \frac{Hx + I}{(x^2 + 2)^2} + \frac{Jx + K}{(x^2 + 2)^3}$$

▲

To find the values of A, B, C, D, E, F, G, H, I, J, and K in Example 4, we would have to solve a system of 11 linear equations. Although possible, this would certainly involve a great deal of work!

The techniques that we have described in this section apply only to rational functions $P(x)/Q(x)$ in which the degree of P is less than the degree of Q. If this isn't the case, we must first use long division to divide Q into P.

▶ **EXAMPLE 5** | Using Long Division to Prepare for Partial Fractions

Find the partial fraction decomposition of

$$\frac{2x^4 + 4x^3 - 2x^2 + x + 7}{x^3 + 2x^2 - x - 2}$$

▼ **SOLUTION** Since the degree of the numerator is larger than the degree of the denominator, we use long division to obtain

$$\frac{2x^4 + 4x^3 - 2x^2 + x + 7}{x^3 + 2x^2 - x - 2} = 2x + \frac{5x + 7}{x^3 + 2x^2 - x - 2}$$

$$
\begin{array}{r}
2x \\
x^3 + 2x^2 - x - 2 \overline{)2x^4 + 4x^3 - 2x^2 + x + 7} \\
\underline{2x^4 + 4x^3 - 2x^2 - 4x} \\
5x + 7
\end{array}
$$

The remainder term now satisfies the requirement that the degree of the numerator is less than the degree of the denominator. At this point we proceed as in Example 1 to obtain the decomposition

$$\frac{2x^4 + 4x^3 - 2x^2 + x + 7}{x^3 + 2x^2 - x - 2} = 2x + \frac{2}{x - 1} + \frac{-1}{x + 1} + \frac{-1}{x + 2}$$

▲

6.5 | Systems of Inequalities

LEARNING OBJECTIVES

After completing this section, you will be able to:

- Graph the solution of an inequality
- Graph the solution of a system of inequalities
- Graph the solution of a system of linear inequalities

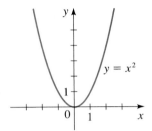

FIGURE 1

In this section we study systems of inequalities in two variables from a graphical point of view.

■ Graphing an Inequality

We begin by considering the graph of a single inequality. We already know that the graph of $y = x^2$, for example, is the *parabola* in Figure 1. If we replace the equal sign by the symbol $\geq$, we obtain the *inequality*

$$y \geq x^2$$

Its graph consists of not just the parabola in Figure 1, but also every point whose y-coordinate is *larger* than x^2. We indicate the solution in Figure 2(a) by shading the points *above* the parabola.

Similarly, the graph of $y \leq x^2$ in Figure 2(b) consists of all points on and *below* the parabola. However, the graphs of $y > x^2$ and $y < x^2$ do not include the points on the parabola itself, as indicated by the dashed curves in Figures 2(c) and 2(d).

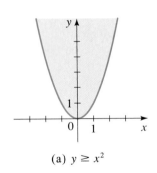

(a) $y \geq x^2$

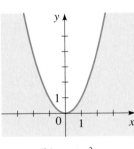

(b) $y \leq x^2$

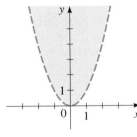

(c) $y > x^2$

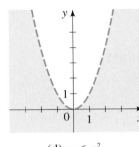

(d) $y < x^2$

FIGURE 2

The graph of an inequality, in general, consists of a region in the plane whose boundary is the graph of the equation obtained by replacing the inequality sign ($\geq$, $\leq$, $>$, or $<$) with an equal sign. To determine which side of the graph gives the solution set of the inequality, we need only check **test points**.

GRAPHING INEQUALITIES

To graph an inequality, we carry out the following steps.

1. **Graph Equation.** Graph the equation corresponding to the inequality. Use a dashed curve for $>$ or $<$ and a solid curve for $\leq$ or $\geq$.

2. **Test Points.** Test one point in each region formed by the graph in Step 1. If the point satisfies the inequality, then all the points in that region satisfy the inequality. (In that case, shade the region to indicate that it is part of the graph.) If the test point does not satisfy the inequality, then the region isn't part of the graph.

▶ **EXAMPLE 1** | Graphs of Inequalities

Graph each inequality.

(a) $x^2 + y^2 < 25$ **(b)** $x + 2y \geq 5$

▼ **SOLUTION**

(a) The graph of $x^2 + y^2 = 25$ is a circle of radius 5 centered at the origin. The points on the circle itself do not satisfy the inequality because it is of the form $<$, so we graph the circle with a dashed curve, as shown in Figure 3.

To determine whether the inside or the outside of the circle satisfies the inequality, we use the test points $(0, 0)$ on the inside and $(6, 0)$ on the outside. To do this, we substitute the coordinates of each point into the inequality and check whether the result satisfies the inequality. (Note that *any* point inside or outside the circle can serve as a test point. We have chosen these points for simplicity.)

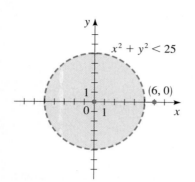

FIGURE 3

Test point	$x^2 + y^2 < 25$	Conclusion
$(0, 0)$	$0^2 + 0^2 = 0 < 25$	Part of graph
$(6, 0)$	$6^2 + 0^2 = 36 \not< 25$	Not part of graph

Thus, the graph of $x^2 + y^2 < 25$ is the set of all points *inside* the circle (see Figure 3).

(b) The graph of $x + 2y = 5$ is the line shown in Figure 4. We use the test points $(0, 0)$ and $(5, 5)$ on opposite sides of the line.

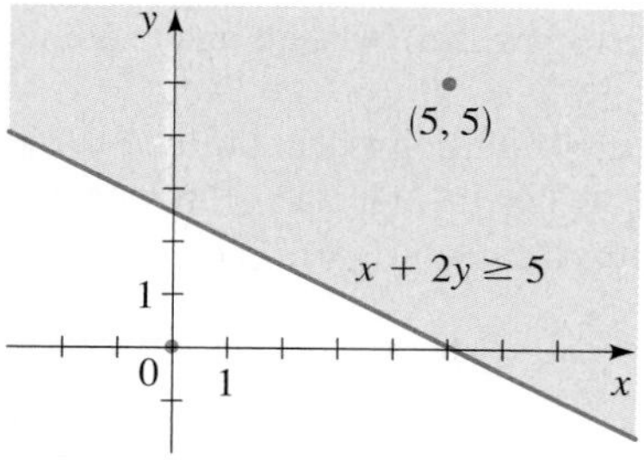

Test point	$x + 2y \geq 5$	Conclusion
$(0, 0)$	$0 + 2(0) = 0 \not\geq 5$	Not part of graph
$(5, 5)$	$5 + 2(5) = 15 \geq 5$	Part of graph

FIGURE 4

Our check shows that the points *above* the line satisfy the inequality.

Alternatively, we could put the inequality into slope-intercept form and graph it directly:

$$x + 2y \geq 5$$
$$2y \geq -x + 5$$
$$y \geq -\tfrac{1}{2}x + \tfrac{5}{2}$$

From this form we see that the graph includes all points whose y-coordinates are *greater* than those on the line $y = -\tfrac{1}{2}x + \tfrac{5}{2}$; that is, the graph consists of the points *on or above* this line, as shown in Figure 4. ▲

■ Systems of Inequalities

We now consider *systems* of inequalities. The solution of such a system is the set of all points in the coordinate plane that satisfy every inequality in the system.

▶ **EXAMPLE 2** | A System of Two Inequalities

Graph the solution of the system of inequalities, and label its vertices.

$$\begin{cases} x^2 + y^2 < 25 \\ x + 2y \geq 5 \end{cases}$$

▼ **SOLUTION** These are the two inequalities of Example 1. In this example we wish to graph only those points that simultaneously satisfy both inequalities. The solution consists of the intersection of the graphs in Example 1. On the next page in Figure 5(a) we show the two regions on the same coordinate plane (in different colors), and in Figure 5(b) we show their intersection.

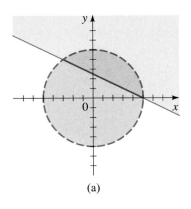

(a)

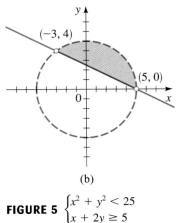

(b)

FIGURE 5 $\begin{cases} x^2 + y^2 < 25 \\ x + 2y \geq 5 \end{cases}$

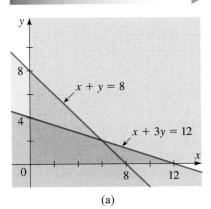

(a)

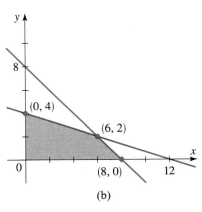

(b)

FIGURE 6

Vertices   The points $(-3, 4)$ and $(5, 0)$ in Figure 5(b) are the **vertices** of the solution set. They are obtained by solving the system of *equations*

$$\begin{cases} x^2 + y^2 = 25 \\ x + 2y = 5 \end{cases}$$

We solve this system of equations by substitution. Solving for x in the second equation gives $x = 5 - 2y$, and substituting this into the first equation gives

$$(5 - 2y)^2 + y^2 = 25 \qquad \text{Substitute } x = 5 - 2y$$
$$(25 - 20y + 4y^2) + y^2 = 25 \qquad \text{Expand}$$
$$-20y + 5y^2 = 0 \qquad \text{Simplify}$$
$$-5y(4 - y) = 0 \qquad \text{Factor}$$

Thus, $y = 0$ or $y = 4$. When $y = 0$, we have $x = 5 - 2(0) = 5$, and when $y = 4$, we have $x = 5 - 2(4) = -3$. So the points of intersection of these curves are $(5, 0)$ and $(-3, 4)$.

Note that in this case the vertices are not part of the solution set, since they don't satisfy the inequality $x^2 + y^2 < 25$ (so they are graphed as open circles in the figure). They simply show where the "corners" of the solution set lie.   ▲

■ Systems of Linear Inequalities

An inequality is **linear** if it can be put into one of the following forms:

$$ax + by \geq c \qquad ax + by \leq c \qquad ax + by > c \qquad ax + by < c$$

In the next example we graph the solution set of a system of linear inequalities.

▶ **EXAMPLE 3** | A System of Four Linear Inequalities

Graph the solution set of the system, and label its vertices.

$$\begin{cases} x + 3y \leq 12 \\ x + y \leq 8 \\ x \geq 0 \\ y \geq 0 \end{cases}$$

▼ **SOLUTION**   In Figure 6 we first graph the lines given by the equations that correspond to each inequality. To determine the graphs of the linear inequalities, we need to check only one test point. For simplicity let's use the point $(0, 0)$.

Inequality	Test point $(0, 0)$	Conclusion
$x + 3y \leq 12$	$0 + 3(0) = 0 \leq 12$	Satisfies inequality
$x + y \leq 8$	$0 + 0 = 0 \leq 8$	Satisfies inequality

Since $(0, 0)$ is below the line $x + 3y = 12$, our check shows that the region on or below the line must satisfy the inequality. Likewise, since $(0, 0)$ is below the line $x + y = 8$, our check shows that the region on or below this line must satisfy the inequality. The inequalities $x \geq 0$ and $y \geq 0$ say that x and y are nonnegative. These regions are sketched in Figure 6(a), and the intersection—the solution set—is sketched in Figure 6(b).

Vertices   The coordinates of each vertex are obtained by simultaneously solving the equations of the lines that intersect at that vertex. From the system

$$\begin{cases} x + 3y = 12 \\ x + y = 8 \end{cases}$$

we get the vertex $(6, 2)$. The origin $(0, 0)$ is also clearly a vertex. The other two vertices are at the x- and y-intercepts of the corresponding lines: $(8, 0)$ and $(0, 4)$. In this case all the vertices *are* part of the solution set. ▲

▶ **EXAMPLE 4** | A System of Linear Inequalities

Graph the solution set of the system.

$$\begin{cases} x + 2y \geq 8 \\ -x + 2y \leq 4 \\ 3x - 2y \leq 8 \end{cases}$$

▼ **SOLUTION** We must graph the lines that correspond to these inequalities and then shade the appropriate regions, as in Example 3. We will use a graphing calculator, so we must first isolate y on the left-hand side of each inequality.

$$\begin{cases} y \geq -\frac{1}{2}x + 4 \\ y \leq \frac{1}{2}x + 2 \\ y \geq \frac{3}{2}x - 4 \end{cases}$$

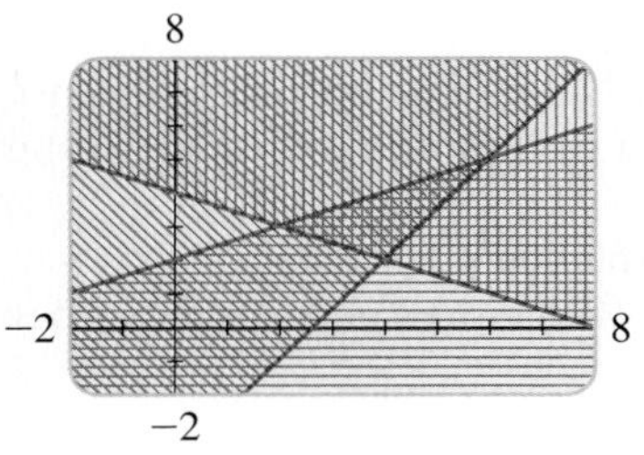

Using the shading feature of the calculator, we obtain the graph in Figure 7. The solution set is the triangular region that is shaded in all three patterns. We then use TRACE or the Intersect command to find the vertices of the region. The solution set is graphed in Figure 8. ▲

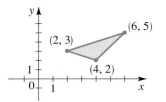

FIGURE 7

FIGURE 8

When a region in the plane can be covered by a (sufficiently large) circle, it is said to be **bounded**. A region that is not bounded is called **unbounded**. For example, the regions graphed in Figures 3, 5(b), 6(b), and 8 are bounded, whereas those in Figures 2 and 4 are unbounded. An unbounded region cannot be "fenced in"—it extends infinitely far in at least one direction.

■ **Application: Feasible Regions**

Many applied problems involve *constraints* on the variables. For instance, a factory manager has only a certain number of workers that can be assigned to perform jobs on the factory floor. A farmer deciding what crops to cultivate has only a certain amount of land that can be seeded. Such constraints or limitations can usually be expressed as systems of inequalities. When dealing with applied inequalities, we usually refer to the solution set of a system as a *feasible region*, because the points in the solution set represent feasible (or possible) values for the quantities being studied.

▶ **EXAMPLE 5** | Restricting Pollutant Outputs

A factory produces two agricultural pesticides, A and B. For every barrel of A, the factory emits 0.25 kg of carbon monoxide (CO) and 0.60 kg of sulfur dioxide (SO_2); and for every barrel of B, it emits 0.50 kg of CO and 0.20 kg of SO_2. Pollution laws restrict the factory's output of CO to a maximum of 75 kg and SO_2 to a maximum of 90 kg per day.

(a) Find a system of inequalities that describes the number of barrels of each pesticide the factory can produce and still satisfy the pollution laws. Graph the feasible region.

(b) Would it be legal for the factory to produce 100 barrels of A and 80 barrels of B per day?

(c) Would it be legal for the factory to produce 60 barrels of A and 160 barrels of B per day?

▼ **SOLUTION**

(a) To set up the required inequalities, it is helpful to organize the given information into a table.

	A	B	Maximum
CO (kg)	0.25	0.50	75
SO_2 (kg)	0.60	0.20	90

We let

$$x = \text{number of barrels of A produced per day}$$

$$y = \text{number of barrels of B produced per day}$$

From the data in the table and the fact that x and y can't be negative, we obtain the following inequalities.

$$\begin{cases} 0.25x + 0.50y \le 75 & \text{CO inequality} \\ 0.60x + 0.20y \le 90 & \text{SO}_2 \text{ inequality} \\ x \ge 0, \quad y \ge 0 \end{cases}$$

Multiplying the first inequality by 4 and the second by 5 simplifies this to

$$\begin{cases} x + 2y \le 300 \\ 3x + y \le 450 \\ x \ge 0, \quad y \ge 0 \end{cases}$$

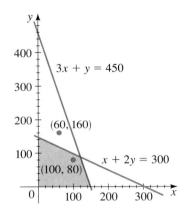

The feasible region is the solution of this system of inequalities, shown in Figure 9.

(b) Since the point $(100, 80)$ lies inside the feasible region, this production plan is legal (see Figure 9).

(c) Since the point $(60, 160)$ lies outside the feasible region, this production plan is not legal. It violates the CO restriction, although it does not violate the SO_2 restriction (see Figure 9).

FIGURE 9

▶ CHAPTER 6 | REVIEW

▼ PROPERTIES AND FORMULAS

Substitution Method (p. 289)

To solve a pair of equations in two variables by substitution:

1. Solve for one variable in terms of the other variable in one equation.

2. **Substitute** into the other equation to get an equation in one variable, and solve for this variable.

3. Substitute the value(s) of the variable you have found into either original equation, and solve for the remaining variable.

Elimination Method (p. 290)

To solve a pair of equations in two variables by elimination:

1. Multiply the equations by appropriate constants so that the term(s) involving one of the variables are of opposite sign in the equations.

2. Add the equations to **eliminate** that one variable; this gives an equation in the other variable. Solve for this variable.

3. Substitute the value(s) of the variable that you have found into either original equation, and solve for the remaining variable.

Graphical Method (p. 292)

To solve a pair of equations in two variables graphically:

1. Put each equation in function form, $y = f(x)$

2. Use a graphing calculator to **graph** the equations on a common screen.

3. Find the points of intersection of the graphs. The solutions are the x- and y-coordinates of the points of intersection.

Gaussian Elimination (p. 300)

When we use **Gaussian elimination** to solve a system of linear equations, we use the following operations to change the system to an **equivalent** simpler system:

1. Add a nonzero multiple of one equation to another.

2. Multiply an equation by a nonzero constant.

3. Interchange the position of two equations in the system.

Number of Solutions of a System of Linear Equations (pp. 294, 301)

A system of linear equations can have:

1. A unique solution for each variable.

2. No solution, in which case the system is **inconsistent**.

3. Infinitely many solutions, in which case the system is **dependent**.

How to Determine the Number of Solutions of a Linear System (p. 301)

When we use **Gaussian elimination** to solve a system of linear equations, then we can tell that the system has:

1. **No solution** (is *inconsistent*) if we arrive at a false equation of the form $0 = c$, where c is nonzero.

2. **Infinitely many solutions** (is *dependent*) if the system is consistent but we end up with fewer equations than variables (after discarding redundant equations of the form $0 = 0$).

Partial Fractions (pp. 304–308)

The *partial fraction decomposition* of a rational function

$$r(x) = \frac{P(x)}{Q(x)}$$

(where the degree of P is less than the degree of Q) is a sum of simpler fractional expressions that equal $r(x)$ when brought to a common denominator. The denominator of each simpler fraction is either a linear or quadratic factor of $Q(x)$ or a power of such a linear or quadratic factor. So to find the terms of the partial fraction decomposition, we first factor $Q(x)$ into linear and irreducible quadratic factors. The terms then have the following forms, depending on the factors of $Q(x)$.

1. For every **distinct linear factor** $ax + b$, there is a term of the form

$$\frac{A}{ax + b}$$

2. For every **repeated linear factor** $(ax + b)^m$, there are terms of the form

$$\frac{A_1}{ax + b} + \frac{A_2}{(ax + b)^2} + \cdots + \frac{A_m}{(ax + b)^m}$$

3. For every **distinct quadratic factor** $ax^2 + bx + c$, there is a term of the form

$$\frac{Ax + B}{ax^2 + bx + c}$$

4. For every **repeated quadratic factor** $(ax^2 + bx + c)^m$, there are terms of the form

$$\frac{A_1x + B_1}{ax^2 + bx + c} + \frac{A_2x + B_2}{(ax^2 + bx + c)^2} + \cdots + \frac{A_mx + B_m}{(ax^2 + bx + c)^m}$$

Graphing Inequalities (p. 309)

To graph an inequality:

1. Graph the equation that corresponds to the inequality. This "boundary curve" divides the coordinate plane into separate regions.

2. Use **test points** to determine which region(s) satisfy the inequality.

3. Shade the region(s) that satisfy the inequality, and use a solid line for the boundary curve if it satisfies the inequality ($\leq$ or $\geq$) and a dashed line if it does not ($<$ or $>$).

Graphing Systems of Inequalities (pp. 310–311)

To graph the solution of a system of inequalities (or **feasible region** determined by the inequalities):

1. Graph all the inequalities on the same coordinate plane.

2. The solution is the intersection of the solutions of all the inequalities, so shade the region that satisfies all the inequalities.

3. Determine the coordinates of the intersection points of all the boundary curves that touch the solution set of the system. These points are the **vertices** of the solution.

▼ CONCEPT SUMMARY

Section 6.1   **Review Exercises**

- Solve a system of equations using the substitution method   1–6, 11, 12
- Solve a system of equations using the elimination method   7–10, 13, 14
- Solve a system of equations using the graphical method   15–18

Section 6.2   **Review Exercises**

- Solve a system of two linear equations in two variables   5–8
- Determine whether a system of two linear equations in two variables has one solution, infinitely many solutions, or no solution   5–8
- Model with linear systems in two variables   27–28

Section 6.3

Section 6.4

Section 6.5

▼ EXERCISES

1–4 ■ Two equations and their graphs are given. Find the intersection point(s) of the graphs by solving the system.

1. $\begin{cases} 2x + 3y = 7 \\ x - 2y = 0 \end{cases}$

2. $\begin{cases} 3x + y = 8 \\ y = x^2 - 5x \end{cases}$

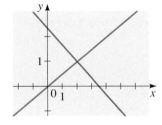

3. $\begin{cases} x^2 + y = 2 \\ x^2 - 3x - y = 0 \end{cases}$

4. $\begin{cases} x - y = -2 \\ x^2 + y^2 - 4y = 4 \end{cases}$

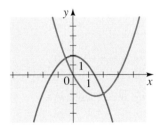

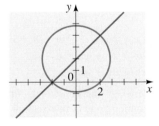

5–10 ■ Solve the system of equations and graph the lines.

5. $\begin{cases} 3x - y = 5 \\ 2x + y = 5 \end{cases}$

6. $\begin{cases} y = 2x + 6 \\ y = -x + 3 \end{cases}$

7. $\begin{cases} 2x - 7y = 28 \\ y = \frac{2}{7}x - 4 \end{cases}$

8. $\begin{cases} 6x - 8y = 15 \\ -\frac{3}{2}x + 2y = -4 \end{cases}$

9. $\begin{cases} 2x - y = 1 \\ x + 3y = 10 \\ 3x + 4y = 15 \end{cases}$

10. $\begin{cases} 2x + 5y = 9 \\ -x + 3y = 1 \\ 7x - 2y = 14 \end{cases}$

11–14 ■ Solve the system of equations.

11. $\begin{cases} y = x^2 + 2x \\ y = 6 + x \end{cases}$

12. $\begin{cases} x^2 + y^2 = 8 \\ y = x + 2 \end{cases}$

13. $\begin{cases} 3x + \dfrac{4}{y} = 6 \\ x - \dfrac{8}{y} = 4 \end{cases}$

14. $\begin{cases} x^2 + y^2 = 10 \\ x^2 + 2y^2 - 7y = 0 \end{cases}$

15–18 ■ Use a graphing device to solve the system, correct to the nearest hundredth.

15. $\begin{cases} 0.32x + 0.43y = 0 \\ 7x - 12y = 341 \end{cases}$

16. $\begin{cases} \sqrt{12}\,x - 3\sqrt{2}\,y = 660 \\ 7137x + 3931y = 20{,}000 \end{cases}$

17. $\begin{cases} x - y^2 = 10 \\ x = \frac{1}{22}y + 12 \end{cases}$

18. $\begin{cases} y = 5^x + x \\ y = x^5 + 5 \end{cases}$

19–26 ■ Find the complete solution of the system, or show that the system has no solution.

19. $\begin{cases} x + y + 2z = 6 \\ 2x + 5z = 12 \\ x + 2y + 3z = 9 \end{cases}$

20. $\begin{cases} x - 2y + 3z = 1 \\ x - 3y - z = 0 \\ 2x - 6z = 6 \end{cases}$

21. $\begin{cases} x - 2y + 3z = 1 \\ 2x - y + z = 3 \\ 2x - 7y + 11z = 2 \end{cases}$

22. $\begin{cases} x + y + z + w = 2 \\ 2x - 3z = 5 \\ x - 2y + 4w = 9 \\ x + y + 2z + 3w = 5 \end{cases}$

23. $\begin{cases} x - 3y + z = 4 \\ 4x - y + 15z = 5 \end{cases}$

24. $\begin{cases} 2x - 3y + 4z = 3 \\ 4x - 5y + 9z = 13 \\ 2x + 7z = 0 \end{cases}$

25. $\begin{cases} -x + 4y + z = 8 \\ 2x - 6y + z = -9 \\ x - 6y - 4z = -15 \end{cases}$

26. $\begin{cases} x \quad\quad - z + w = 2 \\ 2x + y \quad\quad - 2w = 12 \\ 3y + z + w = 4 \\ x + y - z \quad\quad = 10 \end{cases}$

27. Eleanor has two children, Kieran and Siobhan. Kieran is 4 years older than Siobhan, and the sum of their ages is 22. How old are the children?

28. A man invests his savings in two accounts, one paying 6% interest per year and the other paying 7%. He has twice as much invested in the 7% account as in the 6% account, and his annual interest income is $600. How much is invested in each account?

29. A piggy bank contains 50 coins, all of them nickels, dimes, or quarters. The total value of the coins is $5.60, and the value of the dimes is five times the value of the nickels. How many coins of each type are there?

30. Tornie is a commercial fisherman who trolls for salmon on the British Columbia coast. One day he catches a total of 25 fish of three salmon species: coho, sockeye, and pink. He catches three more coho than the other two species combined; moreover, he catches twice as many coho as sockeye. How many fish of each species has he caught?

31–38 ▪ Find the partial fraction decomposition of the rational function.

31. $\dfrac{3x + 1}{x^2 - 2x - 15}$

32. $\dfrac{8}{x^3 - 4x}$

33. $\dfrac{2x - 4}{x(x - 1)^2}$

34. $\dfrac{6x - 4}{x^3 - 2x^2 - 4x + 8}$

35. $\dfrac{2x - 1}{x^3 + x}$

36. $\dfrac{5x^2 - 3x + 10}{x^4 + x^2 - 2}$

37. $\dfrac{3x^2 - x + 6}{(x^2 + 2)^2}$

38. $\dfrac{x^2 + x + 1}{x(x^2 + 1)^2}$

39–40 ▪ An equation and its graph are given. Find an inequality whose solution is the shaded region.

39. $x + y^2 = 4$

40. $x^2 + y^2 = 8$

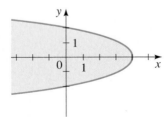

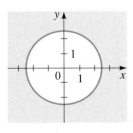

41–44 ▪ Graph the inequality.

41. $3x + y \le 6$

42. $y \ge x^2 - 3$

43. $x^2 + y^2 > 9$

44. $x - y^2 < 4$

45–48 ▪ The figure shows the graphs of the equations corresponding to the given inequalities. Shade the solution set of the system of inequalities.

45. $\begin{cases} y \ge x^2 - 3x \\ y \le \frac{1}{3}x - 1 \end{cases}$

46. $\begin{cases} y \ge x - 1 \\ x^2 + y^2 \le 1 \end{cases}$

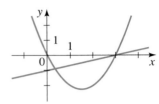

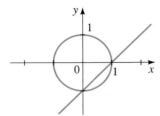

47. $\begin{cases} x + y \ge 2 \\ y - x \le 2 \\ x \le 3 \end{cases}$

48. $\begin{cases} y \ge -2x \\ y \le 2x \\ y \le -\frac{1}{2}x + 2 \end{cases}$

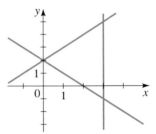

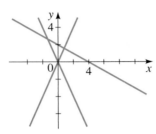

49–52 ▪ Graph the solution set of the system of inequalities. Find the coordinates of all vertices, and determine whether the solution set is bounded or unbounded.

49. $\begin{cases} x^2 + y^2 < 9 \\ x + y < 0 \end{cases}$

50. $\begin{cases} y - x^2 \ge 4 \\ y < 20 \end{cases}$

51. $\begin{cases} x \ge 0, y \ge 0 \\ x + 2y \le 12 \\ y \le x + 4 \end{cases}$

52. $\begin{cases} x \ge 4 \\ x + y \ge 24 \\ x \le 2y + 12 \end{cases}$

53–54 ▪ Solve for x, y, and z in terms of a, b, and c.

53. $\begin{cases} -x + y + z = a \\ x - y + z = b \\ x + y - z = c \end{cases}$

54. $\begin{cases} ax + by + cz = a - b + c \\ bx + by + cz = c \\ cx + cy + cz = c \end{cases}$ $(a \ne b, b \ne c, c \ne 0)$

55. For what values of k do the following three lines have a common point of intersection?

$$x + y = 12$$
$$kx - y = 0$$
$$y - x = 2k$$

56. For what value of k does the following system have infinitely many solutions?

$$\begin{cases} kx + y + z = 0 \\ x + 2y + kz = 0 \\ -x + \quad 3z = 0 \end{cases}$$

1–2 ▪ A system of equations is given.

(a) Determine whether the system is linear or nonlinear.

(b) Find all solutions of the system.

1. $\begin{cases} 3x + 5y = 4 \\ x - 4y = 7 \end{cases}$

2. $\begin{cases} 10x - y^2 = 4 \\ 2x + y = 2 \end{cases}$

3. Use a graphing device to find all solutions of the system correct to two decimal places.

$$\begin{cases} x - 2y = 1 \\ y = x^3 - 2x^2 \end{cases}$$

4. In $2\frac{1}{2}$ hours an airplane travels 600 km against the wind. It takes 50 min to travel 300 km with the wind. Find the speed of the wind and the speed of the airplane in still air.

5–8 ▪ A system of linear equations is given.

(a) Find the complete solution of the system, or show that there is no solution.

(b) State whether the system is inconsistent, dependent, or neither.

5. $\begin{cases} x + 2y + z = 3 \\ x + 3y + 2z = 3 \\ 2x + 3y - z = 8 \end{cases}$

6. $\begin{cases} x - y + 9z = -8 \\ \qquad\; -4z = 7 \\ 3x - y + z = 5 \end{cases}$

7. $\begin{cases} 2x - y + z = 0 \\ 3x + 2y - 3z = 1 \\ x - 4y + 5z = -1 \end{cases}$

8. $\begin{cases} x + y - 2z = 8 \\ 2x - y = 20 \\ 2x + 2y - 5z = 15 \end{cases}$

9. Anne, Barry, and Cathy enter a coffee shop. Anne orders two coffees, one juice, and two doughnuts and pays $6.25. Barry orders one coffee and three doughnuts and pays $3.75. Cathy orders three coffees, one juice, and four doughnuts and pays $9.25. Find the price of coffee, juice, and doughnuts at this coffee shop.

10–11 ▪ Graph the solution set of the system of inequalities. Label the vertices with their coordinates.

10. $\begin{cases} 2x + y \le 8 \\ x - y \ge -2 \\ x + 2y \ge 4 \end{cases}$

11. $\begin{cases} x^2 + y \le 5 \\ y \ge 2x + 5 \end{cases}$

12–13 ▪ Find the partial fraction decomposition of the rational function.

12. $\dfrac{4x - 1}{(x - 1)^2(x + 2)}$

13. $\dfrac{2x - 3}{x^3 + 3x}$

Matrices and Determinants

In this chapter we learn new methods of solving systems of linear equations. Such a system can be represented by a rectangular array of numbers called a matrix. Matrices can be thought of as single objects that can be added, subtracted, or multiplied together, just like real numbers. These operations on matrices allow us to write a system of equations as a single *matrix equation*, thus providing us with a powerful method of solving such systems. We also study other methods of solving linear systems (including a method called Cramer's Rule).

The following Warm-Up Exercises will help you review the basic algebra skills that you will need for this chapter.

WARM-UP EXERCISES

1. Solve the following system using Gaussian elimination:

$$x + y + z = 0$$
$$2x - 2y + 3z = -1$$
$$x + 2y + 2z = 4$$

2. Solve the following system:

$$x - 3y = 2$$
$$3x - 9y = 4$$

3. State the property of real numbers being used:

(a) $3x - y = -y + 3x$ (b) $a(b^2 x) = (ab^2)x$ (c) $9(6 + e) = 54 + 9e$

4. Factor: $abx^2 - 3bx^2$.

5. Find the multiplicative inverse, if it exists, of the following quantities:

(a) 3 (b) $\dfrac{x + 3}{x - 2}$ (c) $\dfrac{a^{1/3} - \sqrt[3]{a}}{a + 5}$

7.1 | Matrices and Systems of Linear Equations

LEARNING OBJECTIVES
After completing this section, you will be able to:

▪ Find the augmented matrix of a linear system
▪ Solve a linear system using elementary row operations
▪ Solve a linear system using the row-echelon form of its matrix
▪ Solve a system using the reduced row-echelon form of its matrix
▪ Determine the number of solutions of a linear system from the row-echelon form of its matrix
▪ Model using linear systems

A *matrix* is simply a rectangular array of numbers. Matrices* are used to organize information into categories that correspond to the rows and columns of the matrix. For example, a scientist might organize information on a population of endangered whales as follows:

$$
\begin{array}{c c c c}
 & \text{Immature} & \text{Juvenile} & \text{Adult} \\
\text{Male} & \begin{bmatrix} 12 & 52 & 18 \\ \text{Female} & 15 & 42 & 11 \end{bmatrix}
\end{array}
$$

This is a compact way of saying there are 12 immature males, 15 immature females, 18 adult males, and so on.

In this section we represent a linear system by a matrix, called the *augmented matrix* of the system:

Linear system   **Augmented matrix**

$$
\begin{cases} 2x - y = 5 & \text{Equation 1} \\ x + 4y = 7 & \text{Equation 2} \end{cases}
\qquad
\begin{bmatrix} 2 & -1 & 5 \\ 1 & 4 & 7 \end{bmatrix}
$$
$$
 x \quad y
$$

The augmented matrix contains the same information as the system, but in a simpler form. The operations we learned for solving systems of equations can now be performed on the augmented matrix.

▪ Matrices

We begin by defining the various elements that make up a matrix.

---

*The plural of *matrix* is *matrices*.

> **DEFINITION OF MATRIX**
>
> An $m \times n$ **matrix** is a rectangular array of numbers with m **rows** and n **columns**.
>
> $$\begin{bmatrix} a_{11} & a_{12} & a_{13} & \cdots & a_{1n} \\ a_{21} & a_{22} & a_{23} & \cdots & a_{2n} \\ a_{31} & a_{32} & a_{33} & \cdots & a_{3n} \\ \vdots & \vdots & \vdots & \ddots & \vdots \\ a_{m1} & a_{m2} & a_{m3} & \cdots & a_{mn} \end{bmatrix} \left. \begin{matrix} \leftarrow \\ \leftarrow \\ \leftarrow \\ \\ \leftarrow \end{matrix} \right\} m \text{ rows}$$
>
> $$\underbrace{\uparrow \qquad \uparrow \qquad \uparrow \qquad\qquad \uparrow}_{n \text{ columns}}$$
>
> We say that the matrix has **dimension** $m \times n$. The numbers a_{ij} are the **entries** of the matrix. The subscript on the entry a_{ij} indicates that it is in the ith row and the jth column.

Here are some examples of matrices.

Matrix	**Dimension**	
$\begin{bmatrix} 1 & 3 & 0 \\ 2 & 4 & -1 \end{bmatrix}$	2×3	2 rows by 3 columns
$\begin{bmatrix} 6 & -5 & 0 & 1 \end{bmatrix}$	1×4	1 row by 4 columns

■ The Augmented Matrix of a Linear System

We can write a system of linear equations as a matrix, called the **augmented matrix** of the system, by writing only the coefficients and constants that appear in the equations. Here is an example.

Linear system   **Augmented matrix**

$$\begin{cases} 3x - 2y + z = 5 \\ x + 3y - z = 0 \\ -x + 4z = 11 \end{cases} \qquad \begin{bmatrix} 3 & -2 & 1 & 5 \\ 1 & 3 & -1 & 0 \\ -1 & 0 & 4 & 11 \end{bmatrix}$$

Notice that a missing variable in an equation corresponds to a 0 entry in the augmented matrix.

▶ **EXAMPLE 1** | Finding the Augmented Matrix of a Linear System

Write the augmented matrix of the system of equations.

$$\begin{cases} 6x - 2y - z = 4 \\ x + 3z = 1 \\ 7y + z = 5 \end{cases}$$

▼ **SOLUTION** First we write the linear system with the variables lined up in columns.

$$\begin{cases} 6x - 2y - z = 4 \\ x + 3z = 1 \\ 7y + z = 5 \end{cases}$$

The augmented matrix is the matrix whose entries are the coefficients and the constants in this system.

$$\begin{bmatrix} 6 & -2 & -1 & 4 \\ 1 & 0 & 3 & 1 \\ 0 & 7 & 1 & 5 \end{bmatrix}$$

▲

Elementary Row Operations

The operations that we used in Section 6.3 to solve linear systems correspond to operations on the rows of the augmented matrix of the system. For example, adding a multiple of one equation to another corresponds to adding a multiple of one row to another.

ELEMENTARY ROW OPERATIONS

1. Add a multiple of one row to another.

2. Multiply a row by a nonzero constant.

3. Interchange two rows.

Note that performing any of these operations on the augmented matrix of a system does not change its solution. We use the following notation to describe the elementary row operations:

Symbol	Description
$R_i + kR_j \rightarrow R_i$	Change the ith row by adding k times row j to it, and then put the result back in row i.
kR_i	Multiply the ith row by k.
$R_i \leftrightarrow R_j$	Interchange the ith and jth rows.

In the next example we compare the two ways of writing systems of linear equations.

▶ **EXAMPLE 2** | Using Elementary Row Operations to Solve a Linear System

Solve the system of linear equations.

$$\begin{cases} x - y + 3z = 4 \\ x + 2y - 2z = 10 \\ 3x - y + 5z = 14 \end{cases}$$

▼ **SOLUTION**   Our goal is to eliminate the x-term from the second equation and the x- and y-terms from the third equation. For comparison, we write both the system of equations and its augmented matrix.

System		**Augmented matrix**

$$\begin{cases} x - y + 3z = 4 \\ x + 2y - 2z = 10 \\ 3x - y + 5z = 14 \end{cases} \qquad \begin{bmatrix} 1 & -1 & 3 & 4 \\ 1 & 2 & -2 & 10 \\ 3 & -1 & 5 & 14 \end{bmatrix}$$

Add $(-1) \times$ Equation 1 to Equation 2.
Add $(-3) \times$ Equation 1 to Equation 3.

$$\begin{cases} x - y + 3z = 4 \\ 3y - 5z = 6 \\ 2y - 4z = 2 \end{cases} \qquad \xrightarrow[R_3 - 3R_1 \rightarrow R_3]{R_2 - R_1 \rightarrow R_2} \qquad \begin{bmatrix} 1 & -1 & 3 & 4 \\ 0 & 3 & -5 & 6 \\ 0 & 2 & -4 & 2 \end{bmatrix}$$

Multiply Equation 3 by $\frac{1}{2}$.

$$\begin{cases} x - y + 3z = 4 \\ 3y - 5z = 6 \\ y - 2z = 1 \end{cases}$$

$\xrightarrow{\frac{1}{2}R_3}$

$$\begin{bmatrix} 1 & -1 & 3 & 4 \\ 0 & 3 & -5 & 6 \\ 0 & 1 & -2 & 1 \end{bmatrix}$$

Add $(-3) \times$ Equation 3 to Equation 2 (to eliminate y from Equation 2).

$$\begin{cases} x - y + 3z = 4 \\ z = 3 \\ y - 2z = 1 \end{cases}$$

$\xrightarrow{R_2 - 3R_3 \to R_2}$

$$\begin{bmatrix} 1 & -1 & 3 & 4 \\ 0 & 0 & 1 & 3 \\ 0 & 1 & -2 & 1 \end{bmatrix}$$

Interchange Equations 2 and 3.

$$\begin{cases} x - y + 3z = 4 \\ y - 2z = 1 \\ z = 3 \end{cases}$$

$\xrightarrow{R_2 \leftrightarrow R_3}$

$$\begin{bmatrix} 1 & -1 & 3 & 4 \\ 0 & 1 & -2 & 1 \\ 0 & 0 & 1 & 3 \end{bmatrix}$$

Now we use back-substitution to find that $x = 2$, $y = 7$, and $z = 3$. The solution is $(2, 7, 3)$. ▲

Gaussian Elimination

In general, to solve a system of linear equations using its augmented matrix, we use elementary row operations to arrive at a matrix in a certain form. This form is described in the following box.

> **ROW-ECHELON FORM AND REDUCED ROW-ECHELON FORM OF A MATRIX**
>
> A matrix is in **row-echelon form** if it satisfies the following conditions.
>
> **1.** The first nonzero number in each row (reading from left to right) is 1. This is called the **leading entry**.
>
> **2.** The leading entry in each row is to the right of the leading entry in the row immediately above it.
>
> **3.** All rows consisting entirely of zeros are at the bottom of the matrix.
>
> A matrix is in **reduced row-echelon** form if it is in row-echelon form and also satisfies the following condition.
>
> **4.** Every number above and below each leading entry is a 0.

In the following matrices, the first one is not in row-echelon form. The second one *is* in row-echelon form, and the third one is in reduced row-echelon form. The entries in red are the leading entries.

Not in row-echelon form

$$\begin{bmatrix} 0 & 1 & -\frac{1}{2} & 0 & 6 \\ 1 & 0 & 3 & 4 & -5 \\ 0 & 0 & 0 & 1 & 0.4 \\ 0 & 1 & 1 & 0 & 0 \end{bmatrix}$$

Leading 1's do not shift to the right in successive rows.

Row-echelon form

$$\begin{bmatrix} 1 & 3 & -6 & 10 & 0 \\ 0 & 0 & 1 & 4 & -3 \\ 0 & 0 & 0 & 1 & \frac{1}{2} \\ 0 & 0 & 0 & 0 & 0 \end{bmatrix}$$

Leading 1's shift to the right in successive rows.

Reduced row-echelon form

$$\begin{bmatrix} 1 & 3 & 0 & 0 & 0 \\ 0 & 0 & 1 & 0 & -3 \\ 0 & 0 & 0 & 1 & \frac{1}{2} \\ 0 & 0 & 0 & 0 & 0 \end{bmatrix}$$

Leading 1's have 0's above and below them.

Here is a systematic way to put a matrix in row-echelon form using elementary row operations:

- Start by obtaining 1 in the top left corner. Then obtain zeros below that 1 by adding appropriate multiples of the first row to the rows below it.

- Next, obtain a leading 1 in the next row, and then obtain zeros below that 1.
- At each stage make sure that every leading entry is to the right of the leading entry in the row above it—rearrange the rows if necessary.
- Continue this process until you arrive at a matrix in row-echelon form.

This is how the process might work for a 3×4 matrix:

Once an augmented matrix is in row-echelon form, we can solve the corresponding linear system using back-substitution. This technique is called **Gaussian elimination**, in honor of its inventor, the German mathematician C. F. Gauss.

> ### SOLVING A SYSTEM USING GAUSSIAN ELIMINATION
>
> 1. **Augmented Matrix.** Write the augmented matrix of the system.
> 2. **Row-Echelon Form.** Use elementary row operations to change the augmented matrix to row-echelon form.
> 3. **Back-Substitution.** Write the new system of equations that corresponds to the row-echelon form of the augmented matrix and solve by back-substitution.

▶ **EXAMPLE 3** | Solving a System Using Row-Echelon Form

Solve the system of linear equations using Gaussian elimination.

$$\begin{cases} 4x + 8y - 4z = 4 \\ 3x + 8y + 5z = -11 \\ -2x + y + 12z = -17 \end{cases}$$

▼ **SOLUTION** We first write the augmented matrix of the system, and then use elementary row operations to put it in row-echelon form.

Need a 1 here.

Augmented matrix ▶
$$\begin{bmatrix} ④ & 8 & -4 & 4 \\ 3 & 8 & 5 & -11 \\ -2 & 1 & 12 & -17 \end{bmatrix}$$

$\xrightarrow{\frac{1}{4}R_1}$
$$\begin{bmatrix} 1 & 2 & -1 & 1 \\ ③ & 8 & 5 & -11 \\ ⊖② & 1 & 12 & -17 \end{bmatrix}$$
Need 0's here.

$\xrightarrow[R_3 + 2R_1 \to R_3]{R_2 - 3R_1 \to R_2}$
$$\begin{bmatrix} 1 & 2 & -1 & 1 \\ 0 & ② & 8 & -14 \\ 0 & 5 & 10 & -15 \end{bmatrix}$$
Need a 1 here.

$\xrightarrow{\frac{1}{2}R_2}$
$$\begin{bmatrix} 1 & 2 & -1 & 1 \\ 0 & 1 & 4 & -7 \\ 0 & ⑤ & 10 & -15 \end{bmatrix}$$
Need a 0 here.

$\xrightarrow{R_3 - 5R_2 \to R_3}$
$$\begin{bmatrix} 1 & 2 & -1 & 1 \\ 0 & 1 & 4 & -7 \\ 0 & 0 & ⊖10 & 20 \end{bmatrix}$$
Need a 1 here.

Reduced row-echelon form ▶

$$\xrightarrow{-\frac{1}{10}R_3} \begin{bmatrix} 1 & 2 & -1 & 1 \\ 0 & 1 & 4 & -7 \\ 0 & 0 & 1 & -2 \end{bmatrix}$$

We now have an equivalent matrix in row-echelon form, and the corresponding system of equations is

$$\begin{cases} x + 2y - z = 1 \\ \quad\quad y + 4z = -7 \\ \quad\quad\quad\quad z = -2 \end{cases}$$

Back-substitute ▶ We use back-substitution to solve the system.

$$y + 2(-2) = -3 \qquad \text{Back-substitute } z = -2 \text{ into Equation 2}$$
$$y = 1 \qquad \text{Solve for } y$$
$$x + 2(1) - (-2) = 1 \qquad \text{Back-substitute } y = 1 \text{ and } z = -2 \text{ into Equation 1}$$
$$x = -3 \qquad \text{Solve for } x$$

So the solution of the system is $(-3, 1, -2)$.   ▲

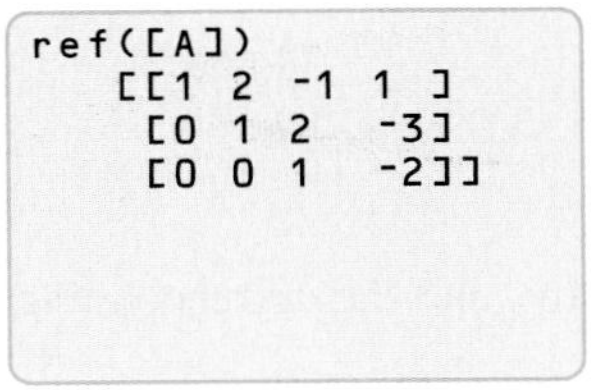

```
ref([A])
   [[1  2  -1  1 ]
    [0  1  2  -3]
    [0  0  1  -2]]
```

FIGURE 1

Graphing calculators have a "row-echelon form" command that puts a matrix in row-echelon form. (On the TI-83 this command is `ref`.) For the augmented matrix in Example 3 the `ref` command gives the output shown in Figure 1. Notice that the row-echelon form that is obtained by the calculator differs from the one we got in Example 3. This is because the calculator used different row operations than we did. You should check that your calculator's row-echelon form leads to the same solution as ours.

▨ Gauss-Jordan Elimination

If we put the augmented matrix of a linear system in *reduced* row-echelon form, then we don't need to back-substitute to solve the system. To put a matrix in reduced row-echelon form, we use the following steps.

- Use the elementary row operations to put the matrix in row-echelon form.
- Obtain zeros above each leading entry by adding multiples of the row containing that entry to the rows above it. Begin with the last leading entry and work up.

Here is how the process works for a 3 × 4 matrix:

$$\begin{bmatrix} 1 & \blacksquare & \blacksquare & \blacksquare \\ 0 & 1 & \blacksquare & \blacksquare \\ 0 & 0 & 1 & \blacksquare \end{bmatrix} \rightarrow \begin{bmatrix} 1 & \blacksquare & 0 & \blacksquare \\ 0 & 1 & 0 & \blacksquare \\ 0 & 0 & 1 & \blacksquare \end{bmatrix} \rightarrow \begin{bmatrix} 1 & 0 & 0 & \blacksquare \\ 0 & 1 & 0 & \blacksquare \\ 0 & 0 & 1 & \blacksquare \end{bmatrix}$$

Using the reduced row-echelon form to solve a system is called **Gauss-Jordan elimination**. The process is illustrated in the next example.

▶ **EXAMPLE 4** | Solving a System Using Reduced Row-Echelon Form

Solve the system of linear equations, using Gauss-Jordan elimination.

$$\begin{cases} 4x + 8y - 4z = 4 \\ 3x + 8y + 5z = -11 \\ -2x + y + 12z = -17 \end{cases}$$

▼ **SOLUTION**   In Example 3 we used Gaussian elimination on the augmented matrix of this system to arrive at an equivalent matrix in row-echelon form. We continue using elementary row operations on the last matrix in Example 3 to arrive at an equivalent matrix in reduced row-echelon form.

$$\begin{bmatrix} 1 & 2 & \boxed{-1} & 1 \\ 0 & 1 & \boxed{4} & -7 \\ 0 & 0 & 1 & -2 \end{bmatrix}$$

Need 0's here.

$$\xrightarrow[R_1 + R_3 \to R_1]{R_2 - 4R_3 \to R_2} \begin{bmatrix} 1 & \boxed{2} & 0 & -1 \\ 0 & 1 & 0 & 1 \\ 0 & 0 & 1 & -2 \end{bmatrix}$$

Need a 0 here.

$$\xrightarrow{R_1 - 2R_2 \to R_1} \begin{bmatrix} 1 & 0 & 0 & -3 \\ 0 & 1 & 0 & 1 \\ 0 & 0 & 1 & -2 \end{bmatrix}$$

We now have an equivalent matrix in reduced row-echelon form, and the corresponding system of equations is

Since the system in reduced row-echelon form, back-subsitution is not required to get the solution

$$\begin{cases} x = -3 \\ y = 1 \\ z = -2 \end{cases}$$

Hence we immediately arrive at the solution $(-3, 1, -2)$.   ▲

Graphing calculators also have a command that puts a matrix in reduced row-echelon form. (On the TI-83 this command is `rref`.) For the augmented matrix in Example 4, the `rref` command gives the output shown in Figure 2. The calculator gives the same reduced row-echelon form as the one we got in Example 4. This is because every matrix has a *unique* reduced row-echelon form.

```
rref([A])
      [[1 0 0 -3]
       [0 1 0 1 ]
       [0 0 1 -2]]
```

FIGURE 2

■ Inconsistent and Dependent Systems

The systems of linear equations that we considered in Examples 1–4 had exactly one solution. But as we know from Section 6.3, a linear system may have one solution, no solution, or infinitely many solutions. Fortunately, the row-echelon form of a system allows us to determine which of these cases applies, as described in the following box.

First we need some terminology. A **leading variable** in a linear system is one that corresponds to a leading entry in the row-echelon form of the augmented matrix of the system.

THE SOLUTIONS OF A LINEAR SYSTEM IN ROW-ECHELON FORM

Suppose the augmented matrix of a system of linear equations has been transformed by Gaussian elimination into row-echelon form. Then exactly one of the following is true.

1. **No solution.** If the row-echelon form contains a row that represents the equation $0 = c$, where c is not zero, then the system has no solution. A system with no solution is called **inconsistent**.

2. **One solution.** If each variable in the row-echelon form is a leading variable, then the system has exactly one solution, which we find using back-substitution or Gauss-Jordan elimination.

3. **Infinitely many solutions.** If the variables in the row-echelon form are not all leading variables and if the system is not inconsistent, then it has infinitely many solutions. In this case the system is called **dependent**. We solve the system by putting the matrix in reduced row-echelon form and then expressing the leading variables in terms of the nonleading variables. The nonleading variables may take on any real numbers as their values.

The matrices below, all in row-echelon form, illustrate the three cases described above.

No solution	**One solution**	**Infinitely many solutions**
$\begin{bmatrix} 1 & 2 & 5 & 7 \\ 0 & 1 & 3 & 4 \\ 0 & 0 & 0 & 1 \end{bmatrix}$	$\begin{bmatrix} 1 & 6 & -1 & 3 \\ 0 & 1 & 2 & -2 \\ 0 & 0 & 1 & 8 \end{bmatrix}$	$\begin{bmatrix} 1 & 2 & -3 & 1 \\ 0 & 1 & 5 & -2 \\ 0 & 0 & 0 & 0 \end{bmatrix}$
Last equation says $0 = 1$.	Each variable is a leading variable.	z is not a leading variable.

▶ **EXAMPLE 5** | A System with No Solution

Solve the system.

$$\begin{cases} x - 3y + 2z = 12 \\ 2x - 5y + 5z = 14 \\ x - 2y + 3z = 20 \end{cases}$$

▼ **SOLUTION** We transform the system into row-echelon form.

$$\begin{bmatrix} 1 & -3 & 2 & 12 \\ 2 & -5 & 5 & 14 \\ 1 & -2 & 3 & 20 \end{bmatrix} \xrightarrow[\substack{R_3 - R_1 \to R_3}]{R_2 - 2R_1 \to R_2} \begin{bmatrix} 1 & -3 & 2 & 12 \\ 0 & 1 & 1 & -10 \\ 0 & 1 & 1 & 8 \end{bmatrix}$$

$$\xrightarrow{R_3 - R_2 \to R_3} \begin{bmatrix} 1 & -3 & 2 & 12 \\ 0 & 1 & 1 & -10 \\ 0 & 0 & 0 & 18 \end{bmatrix} \xrightarrow{\frac{1}{18}R_3} \begin{bmatrix} 1 & -3 & 2 & 12 \\ 0 & 1 & 1 & -10 \\ 0 & 0 & 0 & 1 \end{bmatrix}$$

This last matrix is in row-echelon form, so we can stop the Gaussian elimination process. Now if we translate the last row back into equation form, we get $0x + 0y + 0z = 1$, or $0 = 1$, which is false. No matter what values we pick for x, y, and z, the last equation will never be a true statement. This means that the system *has no solution.* ▲

Figure 3 shows the row-echelon form produced by a TI-83 calculator for the augmented matrix in Example 5. You should check that this gives the same solution.

```
ref([A])
[[1 -2.5 2.5 7  ]
 [0  1    1  -10]
 [0  0    0   1 ]]
```

FIGURE 3

▶ **EXAMPLE 6** | A System with Infinitely Many Solutions

Find the complete solution of the system.

$$\begin{cases} -3x - 5y + 36z = 10 \\ -x \qquad\quad + 7z = 5 \\ x + y - 10z = -4 \end{cases}$$

▼ **SOLUTION**   We transform the system into reduced row-echelon form.

$$\begin{bmatrix} -3 & -5 & 36 & 10 \\ -1 & 0 & 7 & 5 \\ 1 & 1 & -10 & -4 \end{bmatrix} \xrightarrow{R_1 \leftrightarrow R_3} \begin{bmatrix} 1 & 1 & -10 & -4 \\ -1 & 0 & 7 & 5 \\ -3 & -5 & 36 & 10 \end{bmatrix}$$

$$\xrightarrow[R_3 + 3R_1 \to R_3]{R_2 + R_1 \to R_2} \begin{bmatrix} 1 & 1 & -10 & -4 \\ 0 & 1 & -3 & 1 \\ 0 & -2 & 6 & -2 \end{bmatrix} \xrightarrow{R_3 + 2R_2 \to R_3} \begin{bmatrix} 1 & 1 & -10 & -4 \\ 0 & 1 & -3 & 1 \\ 0 & 0 & 0 & 0 \end{bmatrix}$$

$$\xrightarrow{R_1 - R_2 \to R_1} \begin{bmatrix} 1 & 0 & -7 & -5 \\ 0 & 1 & -3 & 1 \\ 0 & 0 & 0 & 0 \end{bmatrix}$$

The third row corresponds to the equation $0 = 0$. This equation is always true, no matter what values are used for x, y, and z. Since the equation adds no new information about the variables, we can drop it from the system. So the last matrix corresponds to the system

$$\begin{cases} x \qquad - 7z = -5 & \text{Equation 1} \\ \quad y - 3z = \ \ 1 & \text{Equation 2} \end{cases}$$

Leading variables

Reduced row-echelon form on the TI-83 calculator:

```
rref([A])
   [[1 0 -7 -5]
    [0 1 -3 1 ]
    [0 0 0  0 ]]
```

Now we solve for the leading variables x and y in terms of the nonleading variable z:

$$x = 7z - 5 \qquad \text{Solve for x in Equation 1}$$

$$y = 3z + 1 \qquad \text{Solve for y in Equation 2}$$

To obtain the complete solution, we let t represent any real number, and we express x, y, and z in terms of t:

$$x = 7t - 5$$
$$y = 3t + 1$$
$$z = t$$

We can also write the solution as the ordered triple $(7t - 5, 3t + 1, t)$, where t is any real number.   ▲

In Example 6, to get specific solutions, we give a specific value to t. For example, if $t = 1$, then

$$x = 7(1) - 5 = 2$$
$$y = 3(1) + 1 = 4$$
$$z = 1$$

Here are some other solutions of the system obtained by substituting other values for the parameter t.

Parameter t	Solution $(7t - 5, 3t + 1, t)$
-1	$(-12, -2, -1)$
0	$(-5, 1, 0)$
2	$(9, 7, 2)$
5	$(30, 16, 5)$

▶ **EXAMPLE 7** | A System with Infinitely Many Solutions

Find the complete solution of the system.

$$\begin{cases} x + 2y - 3z - 4w = 10 \\ x + 3y - 3z - 4w = 15 \\ 2x + 2y - 6z - 8w = 10 \end{cases}$$

▼ **SOLUTION**   We transform the system into reduced row-echelon form.

$$\begin{bmatrix} 1 & 2 & -3 & -4 & 10 \\ 1 & 3 & -3 & -4 & 15 \\ 2 & 2 & -6 & -8 & 10 \end{bmatrix} \xrightarrow[R_3 - 2R_1 \to R_3]{R_2 - R_1 \to R_2} \begin{bmatrix} 1 & 2 & -3 & -4 & 10 \\ 0 & 1 & 0 & 0 & 5 \\ 0 & -2 & 0 & 0 & -10 \end{bmatrix}$$

$$\xrightarrow{R_3 + 2R_2 \to R_3} \begin{bmatrix} 1 & 2 & -3 & -4 & 10 \\ 0 & 1 & 0 & 0 & 5 \\ 0 & 0 & 0 & 0 & 0 \end{bmatrix} \xrightarrow{R_1 - 2R_2 \to R_1} \begin{bmatrix} 1 & 0 & -3 & -4 & 0 \\ 0 & 1 & 0 & 0 & 5 \\ 0 & 0 & 0 & 0 & 0 \end{bmatrix}$$

This is in reduced row-echelon form. Since the last row represents the equation $0 = 0$, we may discard it. So the last matrix corresponds to the system

$$\begin{cases} x \quad\quad -3z - 4w = 0 \\ \quad y \quad\quad\quad\quad = 5 \end{cases}$$

Leading variables

To obtain the complete solution, we solve for the leading variables x and y in terms of the nonleading variables z and w, and we let z and w be any real numbers. Thus, the complete solution is

$$\begin{aligned} x &= 3s + 4t \\ y &= 5 \\ z &= s \\ w &= t \end{aligned}$$

where s and t are any real numbers.   ▲

    Note that s and t do *not* have to be the *same* real number in the solution for Example 7. We can choose arbitrary values for each if we wish to construct a specific solution to the system. For example, if we let $s = 1$ and $t = 2$, then we get the solution $(11, 5, 1, 2)$. You should check that this does indeed satisfy all three of the original equations in Example 7.

Examples 6 and 7 illustrate this general fact: If a system in row-echelon form has n nonzero equations in m variables ($m > n$), then the complete solution will have $m - n$ nonleading variables. For instance, in Example 6 we arrived at *two* nonzero equations in the *three* variables x, y, and z, which gave us $3 - 2 = 1$ nonleading variable.

■ Modeling with Linear Systems

Linear equations, often containing hundreds or even thousands of variables, occur frequently in the applications of algebra to the sciences and to other fields. For now, let's consider an example that involves only three variables.

▶ **EXAMPLE 8** | Nutritional Analysis Using a System of Linear Equations

A nutritionist is performing an experiment on student volunteers. He wishes to feed one of his subjects a daily diet that consists of a combination of three commercial diet foods: MiniCal, LiquiFast, and SlimQuick. For the experiment it is important that the subject consume exactly 500 mg of potassium, 75 g of protein, and 1150 units of vitamin D every day. The amounts of these nutrients in one ounce of each food are given in the table. How many ounces of each food should the subject eat every day to satisfy the nutrient requirements exactly?

	MiniCal	LiquiFast	SlimQuick
Potassium (mg)	50	75	10
Protein (g)	5	10	3
Vitamin D (units)	90	100	50

▼ **SOLUTION** Let x, y, and z represent the number of ounces of MiniCal, LiquiFast, and SlimQuick, respectively, that the subject should eat every day. This means that he will get $50x$ mg of potassium from MiniCal, $75y$ mg from LiquiFast, and $10z$ mg from SlimQuick, for a total of $50x + 75y + 10z$ mg potassium in all. Since the potassium requirement is 500 mg, we get the first equation below. Similar reasoning for the protein and vitamin D requirements leads to the system

$$\begin{cases} 50x + 75y + 10z = 500 & \text{Potassium} \\ 5x + 10y + 3z = 75 & \text{Protein} \\ 90x + 100y + 50z = 1150 & \text{Vitamin D} \end{cases}$$

Dividing the first equation by 5 and the third one by 10 gives the system

$$\begin{cases} 10x + 15y + 2z = 100 \\ 5x + 10y + 3z = 75 \\ 9x + 10y + 5z = 115 \end{cases}$$

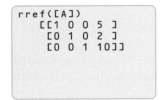

FIGURE 4

We can solve this system using Gaussian elimination, or we can use a graphing calculator to find the reduced row-echelon form of the augmented matrix of the system. Using the `rref` command on the TI-83, we get the output in Figure 4. From the reduced row-echelon form we see that $x = 5$, $y = 2$, $z = 10$. The subject should be fed 5 oz of MiniCal, 2 oz of LiquiFast, and 10 oz of SlimQuick every day. ▲

Check Your Answer

$x = 5$, $y = 2$, $z = 10$:

$$\begin{cases} 10(5) + 15(2) + 2(10) = 100 \\ 5(5) + 10(2) + 3(10) = 75 \\ 9(5) + 10(2) + 5(10) = 115 \end{cases} ✔$$

A more practical application might involve dozens of foods and nutrients rather than just three. Such problems lead to systems with large numbers of variables and equations. Computers or graphing calculators are essential for solving such large systems.

7.2 | The Algebra of Matrices

LEARNING OBJECTIVES

After completing this section, you will be able to:

- Determine whether two matrices are equal
- Use addition, subtraction, and scalar multiplication of matrices
- Multiply matrices
- Write a linear system in matrix form

Thus far, we have used matrices simply for notational convenience when solving linear systems. Matrices have many other uses in mathematics and the sciences, and for most of these applications a knowledge of matrix algebra is essential. Like numbers, matrices can be added, subtracted, multiplied, and divided. In this section we learn how to perform these algebraic operations on matrices.

■ Equality of Matrices

Two matrices are equal if they have the same entries in the same positions.

Equal matrices

$$\begin{bmatrix} \sqrt{4} & 2^2 & e^0 \\ 0.5 & 1 & 1-1 \end{bmatrix} = \begin{bmatrix} 2 & 4 & 1 \\ \frac{1}{2} & \frac{2}{2} & 0 \end{bmatrix}$$

Unequal matrices

$$\begin{bmatrix} 1 & 2 \\ 3 & 4 \\ 5 & 6 \end{bmatrix} \neq \begin{bmatrix} 1 & 3 & 5 \\ 2 & 4 & 6 \end{bmatrix}$$

> **EQUALITY OF MATRICES**
>
> The matrices $A = [a_{ij}]$ and $B = [b_{ij}]$ are **equal** if and only if they have the same dimension $m \times n$, and corresponding entries are equal, that is,
>
> $$a_{ij} = b_{ij}$$
>
> for $i = 1, 2, \ldots, m$ and $j = 1, 2, \ldots, n$.

▶ **EXAMPLE 1** | Equal Matrices

Find a, b, c, and d, if

$$\begin{bmatrix} a & b \\ c & d \end{bmatrix} = \begin{bmatrix} 1 & 3 \\ 5 & 2 \end{bmatrix}$$

▼ **SOLUTION**   Since the two matrices are equal, corresponding entries must be the same. So we must have $a = 1$, $b = 3$, $c = 5$, and $d = 2$.   ▲

■ Addition, Subtraction, and Scalar Multiplication of Matrices

Two matrices can be added or subtracted if they have the same dimension. (Otherwise, their sum or difference is undefined.) We add or subtract the matrices by adding or subtracting corresponding entries. To multiply a matrix by a number, we multiply every element of the matrix by that number. This is called the *scalar product*.

SUM, DIFFERENCE, AND SCALAR PRODUCT OF MATRICES

Let $A = [a_{ij}]$ and $B = [b_{ij}]$ be matrices of the same dimension $m \times n$, and let c be any real number.

1. The **sum** $A + B$ is the $m \times n$ matrix obtained by adding corresponding entries of A and B.

$$A + B = [a_{ij} + b_{ij}]$$

2. The **difference** $A - B$ is the $m \times n$ matrix obtained by subtracting corresponding entries of A and B.

$$A - B = [a_{ij} - b_{ij}]$$

3. The **scalar product** cA is the $m \times n$ matrix obtained by multiplying each entry of A by c.

$$cA = [ca_{ij}]$$

▶ **EXAMPLE 2** | Performing Algebraic Operations on Matrices

Let

$$A = \begin{bmatrix} 2 & -3 \\ 0 & 5 \\ 7 & -\frac{1}{2} \end{bmatrix} \quad B = \begin{bmatrix} 1 & 0 \\ -3 & 1 \\ 2 & 2 \end{bmatrix}$$

$$C = \begin{bmatrix} 7 & -3 & 0 \\ 0 & 1 & 5 \end{bmatrix} \quad D = \begin{bmatrix} 6 & 0 & -6 \\ 8 & 1 & 9 \end{bmatrix}$$

Carry out each indicated operation, or explain why it cannot be performed.

(a) $A + B$ **(b)** $C - D$ **(c)** $C + A$ **(d)** $5A$

▼ **SOLUTION**

(a) $A + B = \begin{bmatrix} 2 & -3 \\ 0 & 5 \\ 7 & -\frac{1}{2} \end{bmatrix} + \begin{bmatrix} 1 & 0 \\ -3 & 1 \\ 2 & 2 \end{bmatrix} = \begin{bmatrix} 3 & -3 \\ -3 & 6 \\ 9 & \frac{3}{2} \end{bmatrix}$

(b) $C - D = \begin{bmatrix} 7 & -3 & 0 \\ 0 & 1 & 5 \end{bmatrix} - \begin{bmatrix} 6 & 0 & -6 \\ 8 & 1 & 9 \end{bmatrix} = \begin{bmatrix} 1 & -3 & 6 \\ -8 & 0 & -4 \end{bmatrix}$

(c) $C + A$ is undefined because we can't add matrices of different dimensions.

(d) $5A = 5\begin{bmatrix} 2 & -3 \\ 0 & 5 \\ 7 & -\frac{1}{2} \end{bmatrix} = \begin{bmatrix} 10 & -15 \\ 0 & 25 \\ 35 & -\frac{5}{2} \end{bmatrix}$

▲

The properties in the box on the next page follow from the definitions of matrix addition and scalar multiplication and the corresponding properties of real numbers.

> **PROPERTIES OF ADDITION AND SCALAR MULTIPLICATION OF MATRICES**
>
> Let A, B, and C be $m \times n$ matrices and let c and d be scalars.
>
> | | |
> |---|---|
> | $A + B = B + A$ | Commutative Property of Matrix Addition |
> | $(A + B) + C = A + (B + C)$ | Associative Property of Matrix Addition |
> | $c(dA) = cdA$ | Associative Property of Scalar Multiplication |
> | $(c + d)A = cA + dA$ $c(A + B) = cA + cB$ | Distributive Properties of Scalar Multiplication |

▶ **EXAMPLE 3** | Solving a Matrix Equation

Solve the matrix equation

$$2X - A = B$$

for the unknown matrix X, where

$$A = \begin{bmatrix} 2 & 3 \\ -5 & 1 \end{bmatrix} \qquad B = \begin{bmatrix} 4 & -1 \\ 1 & 3 \end{bmatrix}$$

▼ **SOLUTION** We use the properties of matrices to solve for X.

$$2X - A = B \qquad \textit{Given equation}$$

$$2X = B + A \qquad \textit{Add the matrix A to each side}$$

$$X = \tfrac{1}{2}(B + A) \qquad \textit{Multiply each side by the scalar } \tfrac{1}{2}$$

So

$$X = \frac{1}{2}\left(\begin{bmatrix} 4 & -1 \\ 1 & 3 \end{bmatrix} + \begin{bmatrix} 2 & 3 \\ -5 & 1 \end{bmatrix} \right) \qquad \textit{Substitute the matrices A and B}$$

$$= \frac{1}{2}\begin{bmatrix} 6 & 2 \\ -4 & 4 \end{bmatrix} \qquad \textit{Add matrices}$$

$$= \begin{bmatrix} 3 & 1 \\ -2 & 2 \end{bmatrix} \qquad \textit{Multiply by the scalar } \tfrac{1}{2}$$

▲

■ Multiplication of Matrices

Multiplying two matrices is more difficult to describe than other matrix operations. In later examples we will see why taking the matrix product involves a rather complex procedure, which we now describe.

First, the product AB (or $A \cdot B$) of two matrices A and B is defined only when the number of columns in A is equal to the number of rows in B. This means that if we write their dimensions side by side, the two inner numbers must match:

Matrices	A	B
Dimensions	$m \times n$	$n \times k$

Columns in A   Rows in B

If the dimensions of A and B match in this fashion, then the product AB is a matrix of dimension $m \times k$. Before describing the procedure for obtaining the elements of AB, we define the *inner product* of a row of A and a column of B.

If $\begin{bmatrix} a_1 & a_2 & \cdots & a_n \end{bmatrix}$ is a row of A, and if $\begin{bmatrix} b_1 \\ b_2 \\ \vdots \\ b_n \end{bmatrix}$ is a column of B, then their **inner product**

is the number $a_1 b_1 + a_2 b_2 + \cdots + a_n b_n$. For example, taking the inner product of

$$\begin{bmatrix} 2 & -1 & 0 & 4 \end{bmatrix} \text{ and } \begin{bmatrix} 5 \\ 4 \\ -3 \\ \frac{1}{2} \end{bmatrix} \text{ gives}$$

$$2 \cdot 5 + (-1) \cdot 4 + 0 \cdot (-3) + 4 \cdot \tfrac{1}{2} = 8$$

We now define the **product** AB of two matrices.

MATRIX MULTIPLICATION

If $A = \begin{bmatrix} a_{ij} \end{bmatrix}$ is an $m \times n$ matrix and $B = \begin{bmatrix} b_{ij} \end{bmatrix}$ an $n \times k$ matrix, then their product is the $m \times k$ matrix

$$C = \begin{bmatrix} c_{ij} \end{bmatrix}$$

where c_{ij} is the inner product of the ith row of A and the jth column of B. We write the product as

$$C = AB$$

This definition of matrix product says that each entry in the matrix AB is obtained from a *row* of A and a *column* of B as follows: The entry c_{ij} in the ith row and jth column of the matrix AB is obtained by multiplying the entries in the ith row of A with the corresponding entries in the jth column of B and adding the results.

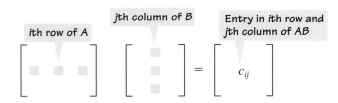

▶ **EXAMPLE 4** | Multiplying Matrices

Let

$$A = \begin{bmatrix} 1 & 3 \\ -1 & 0 \end{bmatrix} \quad \text{and} \quad B = \begin{bmatrix} -1 & 5 & 2 \\ 0 & 4 & 7 \end{bmatrix}$$

Calculate, if possible, the products AB and BA.

Inner numbers match, so product is defined.

$$2 \times 2 \quad 2 \times 3$$

Outer numbers give dimension of product: 2 × 3.

▼ **SOLUTION**   Since A has dimension 2×2 and B has dimension 2×3, the product AB is defined and has dimension 2×3. We can therefore write

$$AB = \begin{bmatrix} 1 & 3 \\ -1 & 0 \end{bmatrix} \begin{bmatrix} -1 & 5 & 2 \\ 0 & 4 & 7 \end{bmatrix} = \begin{bmatrix} ? & ? & ? \\ ? & ? & ? \end{bmatrix}$$

where the question marks must be filled in using the rule defining the product of two matrices. If we define $C = AB = [c_{ij}]$, then the entry c_{11} is the inner product of the first row of A and the first column of B:

$$\begin{bmatrix} 1 & 3 \\ -1 & 0 \end{bmatrix} \begin{bmatrix} -1 & 5 & 2 \\ 0 & 4 & 7 \end{bmatrix} \qquad 1 \cdot (-1) + 3 \cdot 0 = -1$$

Similarly, we calculate the remaining entries of the product as follows.

Entry	Inner product of:	Value	Product matrix
c_{12}	$\begin{bmatrix} 1 & 3 \\ -1 & 0 \end{bmatrix} \begin{bmatrix} -1 & 5 & 2 \\ 0 & 4 & 7 \end{bmatrix}$	$1 \cdot 5 + 3 \cdot 4 = 17$	$\begin{bmatrix} -1 & 17 & \\ & & \end{bmatrix}$
c_{13}	$\begin{bmatrix} 1 & 3 \\ -1 & 0 \end{bmatrix} \begin{bmatrix} -1 & 5 & 2 \\ 0 & 4 & 7 \end{bmatrix}$	$1 \cdot 2 + 3 \cdot 7 = 23$	$\begin{bmatrix} -1 & 17 & 23 \\ & & \end{bmatrix}$
c_{21}	$\begin{bmatrix} 1 & 3 \\ -1 & 0 \end{bmatrix} \begin{bmatrix} -1 & 5 & 2 \\ 0 & 4 & 7 \end{bmatrix}$	$(-1) \cdot (-1) + 0 \cdot 0 = 1$	$\begin{bmatrix} -1 & 17 & 23 \\ 1 & & \end{bmatrix}$
c_{22}	$\begin{bmatrix} 1 & 3 \\ -1 & 0 \end{bmatrix} \begin{bmatrix} -1 & 5 & 2 \\ 0 & 4 & 7 \end{bmatrix}$	$(-1) \cdot 5 + 0 \cdot 4 = -5$	$\begin{bmatrix} -1 & 17 & 23 \\ 1 & -5 & \end{bmatrix}$
c_{23}	$\begin{bmatrix} 1 & 3 \\ -1 & 0 \end{bmatrix} \begin{bmatrix} -1 & 5 & 2 \\ 0 & 4 & 7 \end{bmatrix}$	$(-1) \cdot 2 + 0 \cdot 7 = -2$	$\begin{bmatrix} -1 & 17 & 23 \\ 1 & -5 & -2 \end{bmatrix}$

Thus, we have

$$AB = \begin{bmatrix} -1 & 17 & 23 \\ 1 & -5 & -2 \end{bmatrix}$$

The product BA is not defined, however, because the dimensions of B and A are

$$2 \times 3 \qquad \text{and} \qquad 2 \times 2$$

The inner two numbers are not the same, so the rows and columns won't match up when we try to calculate the product. ▲

Not equal, so product not defined.

$2 \times 3 \quad 2 \times 2$

Graphing calculators and computers are capable of performing matrix algebra. For instance, if we enter the matrices in Example 4 into the matrix variables [A] and [B] on a TI-83 calculator, then the calculator finds their product as shown in Figure 1.

```
[A]*[B]
    [[-1  17  23]
     [1   -5  -2]]
```

FIGURE 1

■ Properties of Matrix Multiplication

Although matrix multiplication is not commutative, it does obey the Associative and Distributive Properties.

PROPERTIES OF MATRIX MULTIPLICATION

Let A, B, and C be matrices for which the following products are defined. Then

$$A(BC) = (AB)C \qquad \text{Associative Property}$$

$$A(B + C) = AB + AC$$
$$(B + C)A = BA + CA \qquad \text{Distributive Property}$$

The next example shows that even when both AB and BA are defined, they aren't necessarily equal. This result proves that matrix multiplication is *not* commutative.

▶ **EXAMPLE 5** | Matrix Multiplication Is Not Commutative

Let
$$A = \begin{bmatrix} 5 & 7 \\ -3 & 0 \end{bmatrix} \quad \text{and} \quad B = \begin{bmatrix} 1 & 2 \\ 9 & -1 \end{bmatrix}$$

Calculate the products AB and BA.

▼ **SOLUTION**   Since both matrices A and B have dimension 2×2, both products AB and BA are defined, and each product is also a 2×2 matrix.

$$AB = \begin{bmatrix} 5 & 7 \\ -3 & 0 \end{bmatrix}\begin{bmatrix} 1 & 2 \\ 9 & -1 \end{bmatrix} = \begin{bmatrix} 5\cdot1 + 7\cdot9 & 5\cdot2 + 7\cdot(-1) \\ (-3)\cdot1 + 0\cdot9 & (-3)\cdot2 + 0\cdot(-1) \end{bmatrix} = \begin{bmatrix} 68 & 3 \\ -3 & -6 \end{bmatrix}$$

$$BA = \begin{bmatrix} 1 & 2 \\ 9 & -1 \end{bmatrix}\begin{bmatrix} 5 & 7 \\ -3 & 0 \end{bmatrix} = \begin{bmatrix} 1\cdot5 + 2\cdot(-3) & 1\cdot7 + 2\cdot0 \\ 9\cdot5 + (-1)\cdot(-3) & 9\cdot7 + (-1)\cdot0 \end{bmatrix} = \begin{bmatrix} -1 & 7 \\ 48 & 63 \end{bmatrix}$$

This shows that, in general, $AB \neq BA$. In fact, in this example AB and BA don't even have an entry in common.   ▲

■ Applications of Matrix Multiplication

We now consider some applied examples that give some indication of why mathematicians chose to define the matrix product in such an apparently bizarre fashion. Example 6 shows how our definition of matrix product allows us to express a system of linear equations as a single matrix equation.

▶ **EXAMPLE 6** | Writing a Linear System as a Matrix Equation

Show that the following matrix equation is equivalent to the system of equations in Example 2 of Section 7.1.

$$\begin{bmatrix} 1 & -1 & 3 \\ 1 & 2 & -2 \\ 3 & -1 & 5 \end{bmatrix}\begin{bmatrix} x \\ y \\ z \end{bmatrix} = \begin{bmatrix} 4 \\ 10 \\ 14 \end{bmatrix}$$

Matrix equations like this one are described in more detail on page 343.

▼ **SOLUTION**   If we perform matrix multiplication on the left side of the equation, we get

$$\begin{bmatrix} x - y + 3z \\ x + 2y - 2z \\ 3x - y + 5z \end{bmatrix} = \begin{bmatrix} 4 \\ 10 \\ 14 \end{bmatrix}$$

Because two matrices are equal only if their corresponding entries are equal, we equate entries to get

$$\begin{cases} x - y + 3z = 4 \\ x + 2y - 2z = 10 \\ 3x - y + 5z = 14 \end{cases}$$

This is exactly the system of equations in Example 2 of Section 7.1.   ▲

▶ **EXAMPLE 7** | Representing Demographic Data by Matrices

In a certain city the proportions of voters in each age group who are registered as Democrats, Republicans, or Independents are given by the following matrix.

$$
\begin{array}{c}
\qquad\qquad\quad \textbf{Age} \\
\qquad \textbf{18--30}\ \ \textbf{31--50}\ \ \textbf{Over 50}
\end{array}
$$

$$
\begin{array}{r}
\textbf{Democrat} \\
\textbf{Republican} \\
\textbf{Independent}
\end{array}
\begin{bmatrix}
0.30 & 0.60 & 0.50 \\
0.50 & 0.35 & 0.25 \\
0.20 & 0.05 & 0.25
\end{bmatrix} = A
$$

The next matrix gives the distribution, by age and sex, of the voting population of this city.

$$
\begin{array}{ccc}
& & \textbf{Male} \qquad \textbf{Female}
\end{array}
$$

$$
\textbf{Age}\
\begin{array}{r}
\textbf{18--30} \\
\textbf{31--50} \\
\textbf{Over 50}
\end{array}
\begin{bmatrix}
5{,}000 & 6{,}000 \\
10{,}000 & 12{,}000 \\
12{,}000 & 15{,}000
\end{bmatrix} = B
$$

For this problem, let's make the (highly unrealistic) assumption that within each age group, political preference is not related to gender. That is, the percentage of Democrat males in the 18–30 group, for example, is the same as the percentage of Democrat females in this group.

(a) Calculate the product AB.

(b) How many males are registered as Democrats in this city?

(c) How many females are registered as Republicans?

▼ **SOLUTION**

(a) $AB = \begin{bmatrix} 0.30 & 0.60 & 0.50 \\ 0.50 & 0.35 & 0.25 \\ 0.20 & 0.05 & 0.25 \end{bmatrix} \begin{bmatrix} 5{,}000 & 6{,}000 \\ 10{,}000 & 12{,}000 \\ 12{,}000 & 15{,}000 \end{bmatrix} = \begin{bmatrix} 13{,}500 & 16{,}500 \\ 9{,}000 & 10{,}950 \\ 4{,}500 & 5{,}550 \end{bmatrix}$

(b) When we take the inner product of a row in A with a column in B, we are adding the number of people in each age group who belong to the category in question. For example, the entry c_{21} of AB (the 9000) is obtained by taking the inner product of the Republican row in A with the Male column in B. This number is therefore the total number of male Republicans in this city. We can label the rows and columns of AB as follows.

$$
\begin{array}{c}
\qquad\qquad \textbf{Male} \quad \textbf{Female}
\end{array}
$$

$$
\begin{array}{r}
\textbf{Democrat} \\
\textbf{Republican} \\
\textbf{Independent}
\end{array}
\begin{bmatrix}
13{,}500 & 16{,}500 \\
9{,}000 & 10{,}950 \\
4{,}500 & 5{,}550
\end{bmatrix} = AB
$$

Thus, 13,500 males are registered as Democrats in this city.

(c) There are 10,950 females registered as Republicans.   ▲

In Example 7 the entries in each column of A add up to 1. (Can you see why this has to be true, given what the matrix describes?) A matrix with this property is called **stochastic**. Stochastic matrices are used extensively in statistics, where they arise frequently in situations like the one described here.

■ Computer Graphics

One important use of matrices is in the digital representation of images. A digital camera or a scanner converts an image into a matrix by dividing the image into a rectangular array of elements called pixels. Each pixel is assigned a value that represents the color, brightness, or some other feature of that location. For example, in a 256-level gray-scale image each pixel is assigned a value between 0 and 255, where 0 represents white, 255 represents black, and the numbers in between represent increasing gradations of gray. The gradations of a much simpler 8-level gray scale are shown in Figure 2. We use this 8-level gray scale to illustrate the process.

To digitize the black and white image in Figure 3(a), we place a grid over the picture as shown in Figure 3(b). Each cell in the grid is compared to the gray scale, and then assigned a value between 0 and 7 depending on which gray square in the scale most closely matches the "darkness" of the cell. (If the cell is not uniformly gray, an average value is assigned.) The values are stored in the matrix shown in Figure 3(c). The digital image corresponding to this matrix is shown in Figure 3(d). Obviously the grid that we have used is far too coarse to provide good image resolution. In practice, currently available high-resolution digital cameras use matrices with dimension as large as 2048×2048.

FIGURE 2

(a) Original image

(b) 10×10 grid

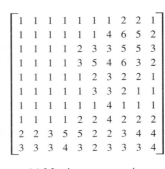

(c) Matrix representation

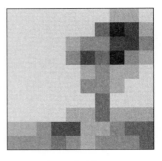

(d) Digital image

FIGURE 3

$$\begin{bmatrix} 1 & 1 & 1 & 1 & 1 & 1 & 1 & 2 & 2 & 1 \\ 1 & 1 & 1 & 1 & 1 & 1 & 4 & 6 & 5 & 2 \\ 1 & 1 & 1 & 1 & 2 & 3 & 3 & 5 & 5 & 3 \\ 1 & 1 & 1 & 1 & 3 & 5 & 4 & 6 & 3 & 2 \\ 1 & 1 & 1 & 1 & 1 & 2 & 3 & 2 & 2 & 1 \\ 1 & 1 & 1 & 1 & 1 & 3 & 3 & 2 & 1 & 1 \\ 1 & 1 & 1 & 1 & 1 & 1 & 4 & 1 & 1 & 1 \\ 1 & 1 & 1 & 1 & 2 & 2 & 4 & 2 & 2 & 2 \\ 2 & 2 & 3 & 5 & 5 & 2 & 2 & 3 & 4 & 4 \\ 3 & 3 & 3 & 4 & 3 & 2 & 3 & 3 & 3 & 4 \end{bmatrix}$$

Once the image is stored as a matrix, it can be manipulated by using matrix operations. For example, to darken the image, we add a constant to each entry in the matrix; to lighten the image, we subtract a constant. To increase the contrast, we darken the darker areas and lighten the lighter areas, so we could add 1 to each entry that is 4, 5, or 6 and subtract 1 from each entry that is 1, 2, or 3. (Note that we cannot darken an entry of 7 or lighten a 0.) Applying this process to the matrix in Figure 3(c) produces the new matrix in Figure 4(a). This generates the high-contrast image shown in Figure 4(b).

$$\begin{bmatrix} 0 & 0 & 0 & 0 & 0 & 0 & 0 & 1 & 1 & 0 \\ 0 & 0 & 0 & 0 & 0 & 0 & 5 & 7 & 6 & 1 \\ 0 & 0 & 0 & 0 & 1 & 2 & 2 & 6 & 6 & 2 \\ 0 & 0 & 0 & 0 & 2 & 6 & 5 & 7 & 2 & 1 \\ 0 & 0 & 0 & 0 & 0 & 1 & 2 & 1 & 1 & 0 \\ 0 & 0 & 0 & 0 & 0 & 2 & 2 & 1 & 0 & 0 \\ 0 & 0 & 0 & 0 & 0 & 0 & 5 & 0 & 0 & 0 \\ 0 & 0 & 0 & 0 & 1 & 1 & 5 & 1 & 1 & 1 \\ 1 & 1 & 2 & 6 & 6 & 1 & 1 & 2 & 5 & 5 \\ 2 & 2 & 2 & 5 & 2 & 1 & 2 & 2 & 2 & 5 \end{bmatrix}$$

(a) Matrix modified to
    increase contrast

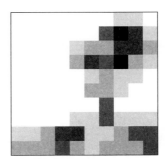

(b) High contrast image

FIGURE 4

7.3 | Inverses of Matrices and Matrix Equations

LEARNING OBJECTIVES
After completing this section, you will be able to:

- Determine when two matrices are inverses of each other
- Find the inverse of a 2 × 2 matrix
- Find the inverse of an $n \times n$ matrix
- Solve a linear system by writing it as a matrix equation
- Model using matrix equations

In the preceding section we saw that when the dimensions are appropriate, matrices can be added, subtracted, and multiplied. In this section we investigate division of matrices. With this operation we can solve equations that involve matrices.

■ The Inverse of a Matrix

First, we define *identity matrices*, which play the same role for matrix multiplication as the number 1 does for ordinary multiplication of numbers; that is, $1 \cdot a = a \cdot 1 = a$ for all numbers a. In the following definition the term **main diagonal** refers to the entries of a square matrix whose row and column numbers are the same. These entries stretch diagonally down the matrix, from top left to bottom right.

> The **identity matrix** I_n is the $n \times n$ matrix for which each main diagonal entry is a 1 and for which all other entries are 0.

Thus, the 2 × 2, 3 × 3, and 4 × 4 identity matrices are

$$I_2 = \begin{bmatrix} 1 & 0 \\ 0 & 1 \end{bmatrix} \qquad I_3 = \begin{bmatrix} 1 & 0 & 0 \\ 0 & 1 & 0 \\ 0 & 0 & 1 \end{bmatrix} \qquad I_4 = \begin{bmatrix} 1 & 0 & 0 & 0 \\ 0 & 1 & 0 & 0 \\ 0 & 0 & 1 & 0 \\ 0 & 0 & 0 & 1 \end{bmatrix}$$

Identity matrices behave like the number 1 in the sense that

$$A \cdot I_n = A \qquad \text{and} \qquad I_n \cdot B = B$$

whenever these products are defined.

▶ **EXAMPLE 1** | Identity Matrices

The following matrix products show how multiplying a matrix by an identity matrix of the appropriate dimension leaves the matrix unchanged.

$$\begin{bmatrix} 1 & 0 \\ 0 & 1 \end{bmatrix} \begin{bmatrix} 3 & 5 & 6 \\ -1 & 2 & 7 \end{bmatrix} = \begin{bmatrix} 3 & 5 & 6 \\ -1 & 2 & 7 \end{bmatrix}$$

$$\begin{bmatrix} -1 & 7 & \frac{1}{2} \\ 12 & 1 & 3 \\ -2 & 0 & 7 \end{bmatrix} \begin{bmatrix} 1 & 0 & 0 \\ 0 & 1 & 0 \\ 0 & 0 & 1 \end{bmatrix} = \begin{bmatrix} -1 & 7 & \frac{1}{2} \\ 12 & 1 & 3 \\ -2 & 0 & 7 \end{bmatrix}$$

If A and B are $n \times n$ matrices, and if $AB = BA = I_n$, then we say that B is the *inverse* of A, and we write $B = A^{-1}$. The concept of the inverse of a matrix is analogous to that of the reciprocal of a real number.

INVERSE OF A MATRIX

Let A be a square $n \times n$ matrix. If there exists an $n \times n$ matrix A^{-1} with the property that

$$AA^{-1} = A^{-1}A = I_n$$

then we say that A^{-1} is the **inverse** of A.

▶ **EXAMPLE 2** | Verifying That a Matrix Is an Inverse

Verify that B is the inverse of A, where

$$A = \begin{bmatrix} 2 & 1 \\ 5 & 3 \end{bmatrix} \quad \text{and} \quad B = \begin{bmatrix} 3 & -1 \\ -5 & 2 \end{bmatrix}$$

▼ **SOLUTION**   We perform the matrix multiplications to show that $AB = I$ and $BA = I$:

$$\begin{bmatrix} 2 & 1 \\ 5 & 3 \end{bmatrix}\begin{bmatrix} 3 & -1 \\ -5 & 2 \end{bmatrix} = \begin{bmatrix} 2 \cdot 3 + 1(-5) & 2(-1) + 1 \cdot 2 \\ 5 \cdot 3 + 3(-5) & 5(-1) + 3 \cdot 2 \end{bmatrix} = \begin{bmatrix} 1 & 0 \\ 0 & 1 \end{bmatrix}$$

$$\begin{bmatrix} 3 & -1 \\ -5 & 2 \end{bmatrix}\begin{bmatrix} 2 & 1 \\ 5 & 3 \end{bmatrix} = \begin{bmatrix} 3 \cdot 2 + (-1)5 & 3 \cdot 1 + (-1)3 \\ (-5)2 + 2 \cdot 5 & (-5)1 + 2 \cdot 3 \end{bmatrix} = \begin{bmatrix} 1 & 0 \\ 0 & 1 \end{bmatrix} \quad ▲$$

■ Finding the Inverse of a 2 × 2 Matrix

The following rule provides a simple way for finding the inverse of a 2×2 matrix, when it exists. For larger matrices there is a more general procedure for finding inverses, which we consider later in this section.

INVERSE OF A 2 × 2 MATRIX

If $A = \begin{bmatrix} a & b \\ c & d \end{bmatrix}$ then $A^{-1} = \dfrac{1}{ad - bc}\begin{bmatrix} d & -b \\ -c & a \end{bmatrix}$.

If $ad - bc = 0$, then A has no inverse.

▶ **EXAMPLE 3** | Finding the Inverse of a 2 × 2 Matrix

Let

$$A = \begin{bmatrix} 4 & 5 \\ 2 & 3 \end{bmatrix}$$

Find A^{-1} and verify that $AA^{-1} = A^{-1}A = I_2$.

▼ **SOLUTION**   Using the rule for the inverse of a 2×2 matrix, we get

$$A^{-1} = \frac{1}{4 \cdot 3 - 5 \cdot 2} \begin{bmatrix} 3 & -5 \\ -2 & 4 \end{bmatrix} = \frac{1}{2} \begin{bmatrix} 3 & -5 \\ -2 & 4 \end{bmatrix} = \begin{bmatrix} \frac{3}{2} & -\frac{5}{2} \\ -1 & 2 \end{bmatrix}$$

To verify that this is indeed the inverse of A, we calculate AA^{-1} and $A^{-1}A$:

$$AA^{-1} = \begin{bmatrix} 4 & 5 \\ 2 & 3 \end{bmatrix} \begin{bmatrix} \frac{3}{2} & -\frac{5}{2} \\ -1 & 2 \end{bmatrix} = \begin{bmatrix} 4 \cdot \frac{3}{2} + 5(-1) & 4(-\frac{5}{2}) + 5 \cdot 2 \\ 2 \cdot \frac{3}{2} + 3(-1) & 2(-\frac{5}{2}) + 3 \cdot 2 \end{bmatrix} = \begin{bmatrix} 1 & 0 \\ 0 & 1 \end{bmatrix}$$

$$A^{-1}A = \begin{bmatrix} \frac{3}{2} & -\frac{5}{2} \\ -1 & 2 \end{bmatrix} \begin{bmatrix} 4 & 5 \\ 2 & 3 \end{bmatrix} = \begin{bmatrix} \frac{3}{2} \cdot 4 + (-\frac{5}{2})2 & \frac{3}{2} \cdot 5 + (-\frac{5}{2})3 \\ (-1)4 + 2 \cdot 2 & (-1)5 + 2 \cdot 3 \end{bmatrix} = \begin{bmatrix} 1 & 0 \\ 0 & 1 \end{bmatrix} \quad \blacktriangle$$

The quantity $ad - bc$ that appears in the rule for calculating the inverse of a 2×2 matrix is called the **determinant** of the matrix. If the determinant is 0, then the matrix does not have an inverse (since we cannot divide by 0).

■ Finding the Inverse of an $n \times n$ Matrix

For 3×3 and larger square matrices the following technique provides the most efficient way to calculate their inverses. If A is an $n \times n$ matrix, we first construct the $n \times 2n$ matrix that has the entries of A on the left and of the identity matrix I_n on the right:

$$\left[\begin{array}{cccc|cccc} a_{11} & a_{12} & \cdots & a_{1n} & 1 & 0 & \cdots & 0 \\ a_{21} & a_{22} & \cdots & a_{2n} & 0 & 1 & \cdots & 0 \\ \vdots & \vdots & \ddots & \vdots & \vdots & \vdots & \ddots & \vdots \\ a_{n1} & a_{n2} & \cdots & a_{nn} & 0 & 0 & \cdots & 1 \end{array} \right]$$

We then use the elementary row operations on this new large matrix to change the left side into the identity matrix. (This means that we are changing the large matrix to reduced row-echelon form.) The right side is transformed automatically into A^{-1}. (We omit the proof of this fact.)

▶ **EXAMPLE 4** | Finding the Inverse of a 3×3 Matrix

Let A be the matrix

$$A = \begin{bmatrix} 1 & -2 & -4 \\ 2 & -3 & -6 \\ -3 & 6 & 15 \end{bmatrix}$$

(a) Find A^{-1}.

(b) Verify that $AA^{-1} = A^{-1}A = I_3$.

▼ **SOLUTION**

(a) We begin with the 3×6 matrix whose left half is A and whose right half is the identity matrix.

$$\left[\begin{array}{ccc|ccc} 1 & -2 & -4 & 1 & 0 & 0 \\ 2 & -3 & -6 & 0 & 1 & 0 \\ -3 & 6 & 15 & 0 & 0 & 1 \end{array} \right]$$

We then transform the left half of this new matrix into the identity matrix by performing the following sequence of elementary row operations on the *entire* new matrix:

$$\xrightarrow[\begin{array}{c}R_2 - 2R_1 \to R_2\\ R_3 + 3R_1 \to R_3\end{array}]{} \left[\begin{array}{ccc|ccc} 1 & -2 & -4 & 1 & 0 & 0 \\ 0 & 1 & 2 & -2 & 1 & 0 \\ 0 & 0 & 3 & 3 & 0 & 1 \end{array}\right]$$

$$\xrightarrow{\frac{1}{3}R_3} \left[\begin{array}{ccc|ccc} 1 & -2 & -4 & 1 & 0 & 0 \\ 0 & 1 & 2 & -2 & 1 & 0 \\ 0 & 0 & 1 & 1 & 0 & \frac{1}{3} \end{array}\right]$$

$$\xrightarrow{R_1 + 2R_2 \to R_1} \left[\begin{array}{ccc|ccc} 1 & 0 & 0 & -3 & 2 & 0 \\ 0 & 1 & 2 & -2 & 1 & 0 \\ 0 & 0 & 1 & 1 & 0 & \frac{1}{3} \end{array}\right]$$

$$\xrightarrow{R_2 - 2R_3 \to R_2} \left[\begin{array}{ccc|ccc} 1 & 0 & 0 & -3 & 2 & 0 \\ 0 & 1 & 0 & -4 & 1 & -\frac{2}{3} \\ 0 & 0 & 1 & 1 & 0 & \frac{1}{3} \end{array}\right]$$

We have now transformed the left half of this matrix into an identity matrix. (This means that we have put the entire matrix in reduced row-echelon form.) Note that to do this in as systematic a fashion as possible, we first changed the elements below the main diagonal to zeros, just as we would if we were using Gaussian elimination. We then changed each main diagonal element to a 1 by multiplying by the appropriate constant(s). Finally, we completed the process by changing the remaining entries on the left side to zeros.

The right half is now A^{-1}.

$$A^{-1} = \left[\begin{array}{ccc} -3 & 2 & 0 \\ -4 & 1 & -\frac{2}{3} \\ 1 & 0 & \frac{1}{3} \end{array}\right]$$

(b) We calculate AA^{-1} and $A^{-1}A$ and verify that both products give the identity matrix I_3.

$$AA^{-1} = \left[\begin{array}{ccc} 1 & -2 & -4 \\ 2 & -3 & -6 \\ -3 & 6 & 15 \end{array}\right]\left[\begin{array}{ccc} -3 & 2 & 0 \\ -4 & 1 & -\frac{2}{3} \\ 1 & 0 & \frac{1}{3} \end{array}\right] = \left[\begin{array}{ccc} 1 & 0 & 0 \\ 0 & 1 & 0 \\ 0 & 0 & 1 \end{array}\right]$$

$$A^{-1}A = \left[\begin{array}{ccc} -3 & 2 & 0 \\ -4 & 1 & -\frac{2}{3} \\ 1 & 0 & \frac{1}{3} \end{array}\right]\left[\begin{array}{ccc} 1 & -2 & -4 \\ 2 & -3 & -6 \\ -3 & 6 & 15 \end{array}\right] = \left[\begin{array}{ccc} 1 & 0 & 0 \\ 0 & 1 & 0 \\ 0 & 0 & 1 \end{array}\right]$$ ▲

Graphing calculators are also able to calculate matrix inverses. On the TI-82 and TI-83 calculators, matrices are stored in memory using names such as `[A]`, `[B]`, `[C]`, ... To find the inverse of `[A]`, we key in

$$\boxed{[A]} \quad \boxed{x^{-1}} \quad \boxed{\text{ENTER}}$$

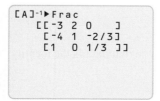

FIGURE 1

For the matrix of Example 4 this results in the output shown in Figure 1 (where we have also used the ▶ F r a c command to display the output in fraction form rather than in decimal form).

The next example shows that not every square matrix has an inverse.

▶ **EXAMPLE 5** | A Matrix That Does Not Have an Inverse

Find the inverse of the matrix.

$$\begin{bmatrix} 2 & -3 & -7 \\ 1 & 2 & 7 \\ 1 & 1 & 4 \end{bmatrix}$$

▼ **SOLUTION** We proceed as follows.

$$\begin{bmatrix} 2 & -3 & -7 & | & 1 & 0 & 0 \\ 1 & 2 & 7 & | & 0 & 1 & 0 \\ 1 & 1 & 4 & | & 0 & 0 & 1 \end{bmatrix} \xrightarrow{R_1 \longleftrightarrow R_2} \begin{bmatrix} 1 & 2 & 7 & | & 0 & 1 & 0 \\ 2 & -3 & -7 & | & 1 & 0 & 0 \\ 1 & 1 & 4 & | & 0 & 0 & 1 \end{bmatrix}$$

$$\xrightarrow[R_3 - R_1 \to R_3]{R_2 - 2R_1 \to R_2} \begin{bmatrix} 1 & 2 & 7 & | & 0 & 1 & 0 \\ 0 & -7 & -21 & | & 1 & -2 & 0 \\ 0 & -1 & -3 & | & 0 & -1 & 1 \end{bmatrix}$$

$$\xrightarrow{-\frac{1}{7}R_2} \begin{bmatrix} 1 & 2 & 7 & | & 0 & 1 & 0 \\ 0 & 1 & 3 & | & -\frac{1}{7} & \frac{2}{7} & 0 \\ 0 & -1 & -3 & | & 0 & -1 & 1 \end{bmatrix}$$

$$\xrightarrow[R_1 - 2R_2 \to R_1]{R_3 + R_2 \to R_3} \begin{bmatrix} 1 & 0 & 1 & | & \frac{2}{7} & \frac{3}{7} & 0 \\ 0 & 1 & 3 & | & -\frac{1}{7} & \frac{2}{7} & 0 \\ 0 & 0 & 0 & | & -\frac{1}{7} & -\frac{5}{7} & 1 \end{bmatrix}$$

At this point we would like to change the 0 in the $(3, 3)$ position of this matrix to a 1 without changing the zeros in the $(3, 1)$ and $(3, 2)$ positions. But there is no way to accomplish this, because no matter what multiple of rows 1 and/or 2 we add to row 3, we can't change the third zero in row 3 without changing the first or second zero as well. Thus, we cannot change the left half to the identity matrix, so the original matrix doesn't have an inverse. ▲

ERR:SINGULAR MAT
1:Quit
2:Goto

FIGURE 2

If we encounter a row of zeros on the left when trying to find an inverse, as in Example 5, then the original matrix does not have an inverse. If we try to calculate the inverse of the matrix from Example 5 on a TI-83 calculator, we get the error message shown in Figure 2. (A matrix that has no inverse is called *singular*.)

■ **Matrix Equations**

We saw in Example 6 in Section 7.2 that a system of linear equations can be written as a single matrix equation. For example, the system

$$\begin{cases} x - 2y - 4z = 7 \\ 2x - 3y - 6z = 5 \\ -3x + 6y + 15z = 0 \end{cases}$$

is equivalent to the matrix equation

$$\underbrace{\begin{bmatrix} 1 & -2 & -4 \\ 2 & -3 & -6 \\ -3 & 6 & 15 \end{bmatrix}}_{A} \underbrace{\begin{bmatrix} x \\ y \\ z \end{bmatrix}}_{X} = \underbrace{\begin{bmatrix} 7 \\ 5 \\ 0 \end{bmatrix}}_{B}$$

If we let

$$A = \begin{bmatrix} 1 & -2 & -4 \\ 2 & -3 & -6 \\ -3 & 6 & 15 \end{bmatrix} \qquad X = \begin{bmatrix} x \\ y \\ z \end{bmatrix} \qquad B = \begin{bmatrix} 7 \\ 5 \\ 0 \end{bmatrix}$$

then this matrix equation can be written as

$$AX = B$$

The matrix A is called the **coefficient matrix**.

We solve this matrix equation by multiplying each side by the inverse of A (provided that this inverse exists):

Solving the matrix equation $AX = B$ is very similar to solving the simple real-number equation

$$3x = 12$$

which we do by multiplying each side by the reciprocal (or inverse) of 3:

$$\tfrac{1}{3}(3x) = \tfrac{1}{3}(12)$$

$$x = 4$$

$$AX = B$$

$$A^{-1}(AX) = A^{-1}B \qquad \text{Multiply on left by } A^{-1}$$

$$(A^{-1}A)X = A^{-1}B \qquad \text{Associative Property}$$

$$I_3X = A^{-1}B \qquad \text{Property of inverses}$$

$$X = A^{-1}B \qquad \text{Property of identity matrix}$$

In Example 4 we showed that

$$A^{-1} = \begin{bmatrix} -3 & 2 & 0 \\ -4 & 1 & -\tfrac{2}{3} \\ 1 & 0 & \tfrac{1}{3} \end{bmatrix}$$

So from $X = A^{-1}B$ we have

$$\underbrace{\begin{bmatrix} x \\ y \\ z \end{bmatrix}}_{X} = \underbrace{\begin{bmatrix} -3 & 2 & 0 \\ -4 & 1 & -\tfrac{2}{3} \\ 1 & 0 & \tfrac{1}{3} \end{bmatrix}}_{A^{-1}} \underbrace{\begin{bmatrix} 7 \\ 5 \\ 0 \end{bmatrix}}_{B} = \begin{bmatrix} -11 \\ -23 \\ 7 \end{bmatrix}$$

Thus, $x = -11$, $y = -23$, $z = 7$ is the solution of the original system.

We have proved that the matrix equation $AX = B$ can be solved by the following method.

SOLVING A MATRIX EQUATION

If A is a square $n \times n$ matrix that has an inverse A^{-1} and if X is a variable matrix and B a known matrix, both with n rows, then the solution of the matrix equation

$$AX = B$$

is given by

$$X = A^{-1}B$$

▶ **EXAMPLE 6** | Solving a System Using a Matrix Inverse

(a) Write the system of equations as a matrix equation.

(b) Solve the system by solving the matrix equation.

$$\begin{cases} 2x - 5y = 15 \\ 3x - 6y = 36 \end{cases}$$

▼ **SOLUTION**

(a) We write the system as a matrix equation of the form $AX = B$:

$$\begin{bmatrix} 2 & -5 \\ 3 & -6 \end{bmatrix} \begin{bmatrix} x \\ y \end{bmatrix} = \begin{bmatrix} 15 \\ 36 \end{bmatrix}$$

$$\underset{A}{} \quad \underset{X}{} = \underset{B}{}$$

(b) Using the rule for finding the inverse of a 2×2 matrix, we get

$$A^{-1} = \begin{bmatrix} 2 & -5 \\ 3 & -6 \end{bmatrix}^{-1} = \frac{1}{2(-6) - (-5)3} \begin{bmatrix} -6 & -(-5) \\ -3 & 2 \end{bmatrix} = \frac{1}{3}\begin{bmatrix} -6 & 5 \\ -3 & 2 \end{bmatrix}$$

Multiplying each side of the matrix equation by this inverse matrix, we get

$$\begin{bmatrix} x \\ y \end{bmatrix} = \frac{1}{3}\begin{bmatrix} -6 & 5 \\ -3 & 2 \end{bmatrix}\begin{bmatrix} 15 \\ 36 \end{bmatrix} = \begin{bmatrix} 30 \\ 9 \end{bmatrix}$$

$$\underset{X}{} = \underset{A^{-1}}{} \quad \underset{B}{}$$

So $x = 30$ and $y = 9$.   ▲

■ Modeling with Matrix Equations

Suppose we need to solve several systems of equations with the same coefficient matrix. Then converting the systems to matrix equations provides an efficient way to obtain the solutions, because we need to find the inverse of the coefficient matrix only once. This procedure is particularly convenient if we use a graphing calculator to perform the matrix operations, as in the next example.

▶ **EXAMPLE 7** | Modeling Nutritional Requirements Using Matrix Equations

A pet-store owner feeds his hamsters and gerbils different mixtures of three types of rodent food: KayDee Food, Pet Pellets, and Rodent Chow. He wishes to feed his animals the correct amount of each brand to satisfy their daily requirements for protein, fat, and carbohydrates exactly. Suppose that hamsters require 340 mg of protein, 280 mg of fat, and 440 mg of carbohydrates, and gerbils need 480 mg of protein, 360 mg of fat, and 680 mg of carbohydrates each day. The amount of each nutrient (in mg) in one gram of each brand is given in the following table. How many grams of each food should the storekeeper feed his hamsters and gerbils daily to satisfy their nutrient requirements?

	KayDee Food	Pet Pellets	Rodent Chow
Protein (mg)	10	0	20
Fat (mg)	10	20	10
Carbohydrates (mg)	5	10	30

▼ SOLUTION   We let x_1, x_2, and x_3 be the respective amounts (in grams) of KayDee Food, Pet Pellets, and Rodent Chow that the hamsters should eat and y_1, y_2, and y_3 be the corresponding amounts for the gerbils. Then we want to solve the matrix equations

$$\begin{bmatrix} 10 & 0 & 20 \\ 10 & 20 & 10 \\ 5 & 10 & 30 \end{bmatrix} \begin{bmatrix} x_1 \\ x_2 \\ x_3 \end{bmatrix} = \begin{bmatrix} 340 \\ 280 \\ 440 \end{bmatrix} \qquad \text{Hamster equation}$$

$$\begin{bmatrix} 10 & 0 & 20 \\ 10 & 20 & 10 \\ 5 & 10 & 30 \end{bmatrix} \begin{bmatrix} y_1 \\ y_2 \\ y_3 \end{bmatrix} = \begin{bmatrix} 480 \\ 360 \\ 680 \end{bmatrix} \qquad \text{Gerbil equation}$$

Let

$$A = \begin{bmatrix} 10 & 0 & 20 \\ 10 & 20 & 10 \\ 5 & 10 & 30 \end{bmatrix} \quad B = \begin{bmatrix} 340 \\ 280 \\ 440 \end{bmatrix} \quad C = \begin{bmatrix} 480 \\ 360 \\ 680 \end{bmatrix} \quad X = \begin{bmatrix} x_1 \\ x_2 \\ x_3 \end{bmatrix} \quad Y = \begin{bmatrix} y_1 \\ y_2 \\ y_3 \end{bmatrix}$$

Then we can write these matrix equations as

$$AX = B \qquad \text{Hamster equation}$$

$$AY = C \qquad \text{Gerbil equation}$$

We want to solve for X and Y, so we multiply both sides of each equation by A^{-1}, the inverse of the coefficient matrix. We could find A^{-1} by hand, but it is more convenient to use a graphing calculator as shown in Figure 3.

(a)   (b)

FIGURE 3

So

$$X = A^{-1}B = \begin{bmatrix} 10 \\ 3 \\ 12 \end{bmatrix} \qquad Y = A^{-1}C = \begin{bmatrix} 8 \\ 4 \\ 20 \end{bmatrix}$$

Thus, each hamster should be fed 10 g of KayDee Food, 3 g of Pet Pellets, and 12 g of Rodent Chow, and each gerbil should be fed 8 g of KayDee Food, 4 g of Pet Pellets, and 20 g of Rodent Chow daily. ▲

| **7.4** | # Determinants and Cramer's Rule |

LEARNING OBJECTIVES

After completing this section, you will be able to:

- Find the determinant of a 2 × 2 matrix
- Find the determinant of an $n \times n$ matrix
- Use the Invertibility Criterion
- Use row and column transformations in finding the determinant of a matrix
- Use Cramer's Rule to solve a linear system
- Use determinants to find the area of a triangle in the coordinate plane

If a matrix is **square** (that is, if it has the same number of rows as columns), then we can assign to it a number called its *determinant*. Determinants can be used to solve systems of linear equations, as we will see later in this section. They are also useful in determining whether a matrix has an inverse.

■ Determinant of a 2 × 2 Matrix

We denote the determinant of a square matrix A by the symbol $\det(A)$ or $|A|$. We first define $\det(A)$ for the simplest cases. If $A = [a]$ is a 1×1 matrix, then $\det(A) = a$. The following box gives the definition of a 2×2 determinant.

> We will use both notations, $\det(A)$ and $|A|$, for the determinant of A. Although the symbol $|A|$ looks like the absolute value symbol, it will be clear from the context which meaning is intended.

DETERMINANT OF A 2 × 2 MATRIX

The **determinant** of the 2 × 2 matrix $A = \begin{bmatrix} a & b \\ c & d \end{bmatrix}$ is

$$\det(A) = |A| = \begin{vmatrix} a & b \\ c & d \end{vmatrix} = ad - bc$$

▶ **EXAMPLE 1** | Determinant of a 2 × 2 Matrix

Evaluate $|A|$ for $A = \begin{bmatrix} 6 & -3 \\ 2 & 3 \end{bmatrix}$.

▼ **SOLUTION**

> To evaluate a 2 × 2 determinant, we take the product of the diagonal from top left to bottom right and subtract the product from top right to bottom left, as indicated by the arrows.

$$\begin{vmatrix} 6 & -3 \\ 2 & 3 \end{vmatrix} = 6 \cdot 3 - (-3)2 = 18 - (-6) = 24 \quad ▲$$

■ Determinant of an $n \times n$ Matrix

To define the concept of determinant for an arbitrary $n \times n$ matrix, we need the following terminology.

Let A be an $n \times n$ matrix.

1. The **minor** M_{ij} of the element a_{ij} is the determinant of the matrix obtained by deleting the ith row and jth column of A.

2. The **cofactor** A_{ij} of the element a_{ij} is

$$A_{ij} = (-1)^{i+j} M_{ij}$$

For example, if A is the matrix

$$\begin{bmatrix} 2 & 3 & -1 \\ 0 & 2 & 4 \\ -2 & 5 & 6 \end{bmatrix}$$

then the minor M_{12} is the determinant of the matrix obtained by deleting the first row and second column from A. Thus,

$$M_{12} = \begin{vmatrix} 2 & 3 & -1 \\ 0 & 2 & 4 \\ -2 & 5 & 6 \end{vmatrix} = \begin{vmatrix} 0 & 4 \\ -2 & 6 \end{vmatrix} = 0(6) - 4(-2) = 8$$

So the cofactor $A_{12} = (-1)^{1+2} M_{12} = -8$. Similarly,

$$M_{33} = \begin{vmatrix} 2 & 3 & -1 \\ 0 & 2 & 4 \\ -2 & 5 & 6 \end{vmatrix} = \begin{vmatrix} 2 & 3 \\ 0 & 2 \end{vmatrix} = 2 \cdot 2 - 3 \cdot 0 = 4$$

So $A_{33} = (-1)^{3+3} M_{33} = 4$.

Note that the cofactor of a_{ij} is simply the minor of a_{ij} multiplied by either 1 or -1, depending on whether $i + j$ is even or odd. Thus, in a 3×3 matrix we obtain the cofactor of any element by prefixing its minor with the sign obtained from the following checkerboard pattern.

$$\begin{bmatrix} + & - & + \\ - & + & - \\ + & - & + \end{bmatrix}$$

We are now ready to define the determinant of any square matrix.

THE DETERMINANT OF A SQUARE MATRIX

If A is an $n \times n$ matrix, then the **determinant** of A is obtained by multiplying each element of the first row by its cofactor, and then adding the results. In symbols.

$$\det(A) = |A| = \begin{vmatrix} a_{11} & a_{12} & \cdots & a_{1n} \\ a_{21} & a_{22} & \cdots & a_{2n} \\ \vdots & \vdots & \ddots & \vdots \\ a_{n1} & a_{n2} & \cdots & a_{nn} \end{vmatrix} = a_{11}A_{11} + a_{12}A_{12} + \cdots + a_{1n}A_{1n}$$

▶ **EXAMPLE 2** | Determinant of a 3 × 3 Matrix

Evaluate the determinant of the matrix.

$$A = \begin{bmatrix} 2 & 3 & -1 \\ 0 & 2 & 4 \\ -2 & 5 & 6 \end{bmatrix}$$

▼ **SOLUTION**

$$\det(A) = \begin{vmatrix} 2 & 3 & -1 \\ 0 & 2 & 4 \\ -2 & 5 & 6 \end{vmatrix} = 2\begin{vmatrix} 2 & 4 \\ 5 & 6 \end{vmatrix} - 3\begin{vmatrix} 0 & 4 \\ -2 & 6 \end{vmatrix} + (-1)\begin{vmatrix} 0 & 2 \\ -2 & 5 \end{vmatrix}$$

$$= 2(2\cdot 6 - 4\cdot 5) - 3[0\cdot 6 - 4(-2)] - [0\cdot 5 - 2(-2)]$$

$$= -16 - 24 - 4$$

$$= -44 \qquad\qquad ▲$$

In our definition of the determinant we used the cofactors of elements in the first row only. This is called **expanding the determinant by the first row**. In fact, *we can expand the determinant by any row or column in the same way, and obtain the same result in each case* (although we won't prove this). The next example illustrates this principle.

▶ **EXAMPLE 3** | Expanding a Determinant About a Row and a Column

Let A be the matrix of Example 2. Evaluate the determinant of A by expanding

(a) by the second row

(b) by the third column

Verify that each expansion gives the same value.

▼ **SOLUTION**

(a) Expanding by the second row, we get

$$\det(A) = \begin{vmatrix} 2 & 3 & -1 \\ 0 & 2 & 4 \\ -2 & 5 & 6 \end{vmatrix} = -0\begin{vmatrix} 3 & -1 \\ 5 & 6 \end{vmatrix} + 2\begin{vmatrix} 2 & -1 \\ -2 & 6 \end{vmatrix} - 4\begin{vmatrix} 2 & 3 \\ -2 & 5 \end{vmatrix}$$

$$= 0 + 2[2\cdot 6 - (-1)(-2)] - 4[2\cdot 5 - 3(-2)]$$

$$= 0 + 20 - 64 = -44$$

(b) Expanding by the third column gives

Graphing calculators are capable of computing determinants. Here is the output when the TI-83 is used to calculate the determinant in Example 3:

```
[A]
        [[2   3  -1]
         [0   2   4 ]
         [-2  5   6 ]]
det([A])
                   -44
```

$$\det(A) = \begin{vmatrix} 2 & 3 & -1 \\ 0 & 2 & 4 \\ -2 & 5 & 6 \end{vmatrix}$$

$$= -1\begin{vmatrix} 0 & 2 \\ -2 & 5 \end{vmatrix} - 4\begin{vmatrix} 2 & 3 \\ -2 & 5 \end{vmatrix} + 6\begin{vmatrix} 2 & 3 \\ 0 & 2 \end{vmatrix}$$

$$= -[0\cdot 5 - 2(-2)] - 4[2\cdot 5 - 3(-2)] + 6(2\cdot 2 - 3\cdot 0)$$

$$= -4 - 64 + 24 = -44$$

In both cases we obtain the same value for the determinant as when we expanded by the first row in Example 2. ▲

The following criterion allows us to determine whether a square matrix has an inverse without actually calculating the inverse. This is one of the most important uses of the determinant in matrix algebra, and it is the reason for the name *determinant*.

INVERTIBILITY CRITERION

If A is a square matrix, then A has an inverse if and only if $\det(A) \neq 0$.

We will not prove this fact, but from the formula for the inverse of a 2×2 matrix (page 340) you can see why it is true in the 2×2 case.

▶ **EXAMPLE 4** | Using the Determinant to Show That a Matrix Is Not Invertible

Show that the matrix A has no inverse.

$$A = \begin{bmatrix} 1 & 2 & 0 & 4 \\ 0 & 0 & 0 & 3 \\ 5 & 6 & 2 & 6 \\ 2 & 4 & 0 & 9 \end{bmatrix}$$

▼ **SOLUTION** We begin by calculating the determinant of A. Since all but one of the elements of the second row is zero, we expand the determinant by the second row. If we do this, we see from the following equation that only the cofactor A_{24} will have to be calculated.

$$\det(A) = \begin{vmatrix} 1 & 2 & 0 & 4 \\ 0 & 0 & 0 & 3 \\ 5 & 6 & 2 & 6 \\ 2 & 4 & 0 & 9 \end{vmatrix}$$

$$= -0 \cdot A_{21} + 0 \cdot A_{22} - 0 \cdot A_{23} + 3 \cdot A_{24} = 3A_{24}$$

$$= 3\begin{vmatrix} 1 & 2 & 0 \\ 5 & 6 & 2 \\ 2 & 4 & 0 \end{vmatrix} \qquad \text{Expand this by column 3}$$

$$= 3(-2)\begin{vmatrix} 1 & 2 \\ 2 & 4 \end{vmatrix}$$

$$= 3(-2)(1 \cdot 4 - 2 \cdot 2) = 0$$

Since the determinant of A is zero, A cannot have an inverse, by the Invertibility Criterion. ▲

■ Row and Column Transformations

The preceding example shows that if we expand a determinant about a row or column that contains many zeros, our work is reduced considerably because we don't have to evaluate the cofactors of the elements that are zero. The following principle often simplifies the process of finding a determinant by introducing zeros into it without changing its value.

ROW AND COLUMN TRANSFORMATIONS OF A DETERMINANT

If A is a square matrix and if the matrix B is obtained from A by adding a multiple of one row to another or a multiple of one column to another, then $\det(A) = \det(B)$.

▶ **EXAMPLE 5** | Using Row and Column Transformations to Calculate a Determinant

Find the determinant of the matrix A. Does it have an inverse?

$$A = \begin{bmatrix} 8 & 2 & -1 & -4 \\ 3 & 5 & -3 & 11 \\ 24 & 6 & 1 & -12 \\ 2 & 2 & 7 & -1 \end{bmatrix}$$

▼ **SOLUTION** If we add -3 times row 1 to row 3, we change all but one element of row 3 to zeros:

$$\begin{bmatrix} 8 & 2 & -1 & -4 \\ 3 & 5 & -3 & 11 \\ 0 & 0 & 4 & 0 \\ 2 & 2 & 7 & -1 \end{bmatrix}$$

This new matrix has the same determinant as A, and if we expand its determinant by the third row, we get

$$\det(A) = 4 \begin{vmatrix} 8 & 2 & -4 \\ 3 & 5 & 11 \\ 2 & 2 & -1 \end{vmatrix}$$

Now, adding 2 times column 3 to column 1 in this determinant gives us

$$\det(A) = 4 \begin{vmatrix} 0 & 2 & -4 \\ 25 & 5 & 11 \\ 0 & 2 & -1 \end{vmatrix} \qquad \text{Expand this by column 1}$$

$$= 4(-25) \begin{vmatrix} 2 & -4 \\ 2 & -1 \end{vmatrix}$$

$$= 4(-25)[2(-1) - (-4)2] = -600$$

Since the determinant of A is not zero, A does have an inverse. ▲

■ Cramer's Rule

The solutions of linear equations can sometimes be expressed by using determinants. To illustrate, let's solve the following pair of linear equations for the variable x.

$$\begin{cases} ax + by = r \\ cx + dy = s \end{cases}$$

To eliminate the variable y, we multiply the first equation by d and the second by b, and subtract.

$$\begin{aligned} adx + bdy &= rd \\ bcx + bdy &= bs \\ \hline adx - bcx &= rd - bs \end{aligned}$$

Factoring the left-hand side, we get $(ad - bc)x = rd - bs$. Assuming that $ad - bc \neq 0$, we can now solve this equation for x:

$$x = \frac{rd - bs}{ad - bc}$$

Similarly, we find

$$y = \frac{as - cr}{ad - bc}$$

The numerator and denominator of the fractions for x and y are determinants of 2×2 matrices. So we can express the solution of the system using determinants as follows.

CRAMER'S RULE FOR SYSTEMS IN TWO VARIABLES

The linear system

$$\begin{cases} ax + by = r \\ cx + dy = s \end{cases}$$

has the solution

$$x = \frac{\begin{vmatrix} r & b \\ s & d \end{vmatrix}}{\begin{vmatrix} a & b \\ c & d \end{vmatrix}} \qquad y = \frac{\begin{vmatrix} a & r \\ c & s \end{vmatrix}}{\begin{vmatrix} a & b \\ c & d \end{vmatrix}}$$

provided that $\begin{vmatrix} a & b \\ c & d \end{vmatrix} \neq 0$.

Using the notation

$$D = \begin{bmatrix} a & b \\ c & d \end{bmatrix} \qquad D_x = \begin{bmatrix} r & b \\ s & d \end{bmatrix} \qquad D_y = \begin{bmatrix} a & r \\ c & s \end{bmatrix}$$

| Coefficient matrix | Replace first column of D by r and s. | Replace second column of D by r and s. |

we can write the solution of the system as

$$x = \frac{|D_x|}{|D|} \quad \text{and} \quad y = \frac{|D_y|}{|D|}$$

▶ **EXAMPLE 6** | Using Cramer's Rule to Solve a System with Two Variables

Use Cramer's Rule to solve the system.

$$\begin{cases} 2x + 6y = -1 \\ x + 8y = 2 \end{cases}$$

▼ **SOLUTION**   For this system we have

$$|D| = \begin{vmatrix} 2 & 6 \\ 1 & 8 \end{vmatrix} = 2 \cdot 8 - 6 \cdot 1 = 10$$

$$|D_x| = \begin{vmatrix} -1 & 6 \\ 2 & 8 \end{vmatrix} = (-1)8 - 6 \cdot 2 = -20$$

$$|D_y| = \begin{vmatrix} 2 & -1 \\ 1 & 2 \end{vmatrix} = 2 \cdot 2 - (-1)1 = 5$$

The solution is

$$x = \frac{|D_x|}{|D|} = \frac{-20}{10} = -2$$

$$y = \frac{|D_y|}{|D|} = \frac{5}{10} = \frac{1}{2}$$

Cramer's Rule can be extended to apply to any system of n linear equations in n variables in which the determinant of the coefficient matrix is not zero. As we saw in the preceding section, any such system can be written in matrix form as

$$\begin{bmatrix} a_{11} & a_{12} & \cdots & a_{1n} \\ a_{21} & a_{22} & \cdots & a_{2n} \\ \vdots & \vdots & \ddots & \vdots \\ a_{n1} & a_{n2} & \cdots & a_{nn} \end{bmatrix} \begin{bmatrix} x_1 \\ x_2 \\ \vdots \\ x_n \end{bmatrix} = \begin{bmatrix} b_1 \\ b_2 \\ \vdots \\ b_n \end{bmatrix}$$

By analogy with our derivation of Cramer's Rule in the case of two equations in two unknowns, we let D be the coefficient matrix in this system, and D_{x_i} be the matrix obtained by replacing the ith column of D by the numbers $b_1, b_2, \ldots, b_n$ that appear to the right of the equal sign. The solution of the system is then given by the following rule.

CRAMER'S RULE

If a system of n linear equations in the n variables $x_1, x_2, \ldots, x_n$ is equivalent to the matrix equation $DX = B$, and if $|D| \neq 0$, then its solutions are

$$x_1 = \frac{|D_{x_1}|}{|D|} \qquad x_2 = \frac{|D_{x_2}|}{|D|} \qquad \cdots \qquad x_n = \frac{|D_{x_n}|}{|D|}$$

where D_{x_i} is the matrix obtained by replacing the ith column of D by the $n \times 1$ matrix B.

▶ **EXAMPLE 7** | Using Cramer's Rule to Solve a System with Three Variables

Use Cramer's Rule to solve the system.

$$\begin{cases} 2x - 3y + 4z = 1 \\ x \qquad\quad + 6z = 0 \\ 3x - 2y \qquad\quad = 5 \end{cases}$$

▼ **SOLUTION** First, we evaluate the determinants that appear in Cramer's Rule. Note that D is the coefficient matrix and that D_x, D_y, and D_z are obtained by replacing the first, second, and third columns of D by the constant terms.

$$|D| = \begin{vmatrix} 2 & -3 & 4 \\ 1 & 0 & 6 \\ 3 & -2 & 0 \end{vmatrix} = -38 \qquad |D_x| = \begin{vmatrix} 1 & -3 & 4 \\ 0 & 0 & 6 \\ 5 & -2 & 0 \end{vmatrix} = -78$$

$$|D_y| = \begin{vmatrix} 2 & 1 & 4 \\ 1 & 0 & 6 \\ 3 & 5 & 0 \end{vmatrix} = -22 \qquad |D_z| = \begin{vmatrix} 2 & -3 & 1 \\ 1 & 0 & 0 \\ 3 & -2 & 5 \end{vmatrix} = 13$$

Now we use Cramer's Rule to get the solution:

$$x = \frac{|D_x|}{|D|} = \frac{-78}{-38} = \frac{39}{19}$$

$$y = \frac{|D_y|}{|D|} = \frac{-22}{-38} = \frac{11}{19}$$

$$z = \frac{|D_z|}{|D|} = \frac{13}{-38} = -\frac{13}{38}$$

▲

Solving the system in Example 7 using Gaussian elimination would involve matrices whose elements are fractions with fairly large denominators. Thus, in cases like Examples 6 and 7, Cramer's Rule gives us an efficient way to solve systems of linear equations. But in systems with more than three equations, evaluating the various determinants that are involved is usually a long and tedious task (unless you are using a graphing calculator). Moreover, the rule doesn't apply if $|D| = 0$ or if D is not a square matrix. So Cramer's Rule is a useful alternative to Gaussian elimination, but only in some situations.

■ Areas of Triangles Using Determinants

Determinants provide a simple way to calculate the area of a triangle in the coordinate plane.

AREA OF A TRIANGLE

If a triangle in the coordinate plane has vertices (a_1, b_1), (a_2, b_2), and (a_3, b_3), then its area is

$$\text{area} = \pm\frac{1}{2}\begin{vmatrix} a_1 & b_1 & 1 \\ a_2 & b_2 & 1 \\ a_3 & b_3 & 1 \end{vmatrix}$$

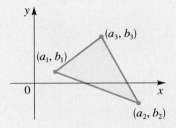

where the sign is chosen to make the area positive.

You are asked to prove this formula in the exercise set.

▶ **EXAMPLE 8** | Area of a Triangle

Find the area of the triangle shown in Figure 1.

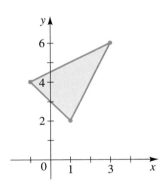

FIGURE 1

We can calculate the determinant by hand or by using a graphing calculator.

```
[A]
        [[-1  4  1]
         [3   6  1]
         [1   2  1]]
det([A])
            -12
```

▼ **SOLUTION** The vertices are $(1, 2)$, $(3, 6)$, and $(-1, 4)$. Using the formula in the preceding box, we get

$$\text{area} = \pm\frac{1}{2}\begin{vmatrix} -1 & 4 & 1 \\ 3 & 6 & 1 \\ 1 & 2 & 1 \end{vmatrix} = \pm\frac{1}{2}(-12)$$

To make the area positive, we choose the negative sign in the formula. Thus, the area of the triangle is

$$\text{area} = -\frac{1}{2}(-12) = 6$$

▲

▶ **CHAPTER 7** | REVIEW

▼ PROPERTIES AND FORMULAS

Matrices (p. 321)

A **matrix** A of **dimension** $m \times n$ is a rectangular array of numbers with m **rows** and n **columns**:

$$A = \begin{bmatrix} a_{11} & a_{12} & \cdots & a_{1n} \\ a_{21} & a_{22} & \cdots & a_{2n} \\ \vdots & \vdots & \ddots & \vdots \\ a_{m1} & a_{m2} & \cdots & a_{mn} \end{bmatrix}$$

Augmented Matrix of a System (p. 321)

The **augmented matrix** of a system of linear equations is the matrix consisting of the coefficients and the constant terms. For example, for the two-variable system

$$a_{11}x + a_{12}x = b_1$$
$$a_{21}x + a_{22}x = b_2$$

the augmented matrix is

$$\begin{bmatrix} a_{11} & a_{12} & b_1 \\ a_{21} & a_{22} & b_2 \end{bmatrix}$$

Elementary Row Operations (p. 322)

To solve a system of linear equations using the augmented matrix of the system, the following operations can be used to transform the rows of the matrix:

1. Add a nonzero multiple of one row to another.
2. Multiply a row by a nonzero constant.
3. Interchange two rows.

Row-Echelon Form of a Matrix (p. 323)

A matrix is in **row-echelon form** if its entries satisfy the following conditions:

1. The first nonzero entry in each row (the **leading entry**) is the number 1.
2. The leading entry of each row is to the right of the leading entry in the row above it.
3. All rows consisting entirely of zeros are at the bottom of the matrix.

If the matrix also satisfies the following condition, it is in **reduced row-echelon** form:

4. If a column contains a leading entry, then every other entry in that column is a 0.

Number of Solutions of a Linear System (p. 327)

If the augmented matrix of a system of linear equations has been reduced to row-echelon form using elementary row operations, then the system has:

1. **No solution** if the row-echelon form contains a row that represents the equation $0 = 1$. In this case the system is **inconsistent**.
2. **One solution** if each variable in the row-echelon form is a **leading variable**.

3. **Infinitely many solutions** if the system is not inconsistent but not every variable is a leading variable. In this case the system is **dependent**.

Operations on Matrices (p. 332)

If A and B are $m \times n$ matrices and c is a scalar (real number), then:

1. The **sum** $A + B$ is the $m \times n$ matrix that is obtained by adding corresponding entries of A and B.
2. The **difference** $A - B$ is the $m \times n$ matrix that is obtained by subtracting corresponding entries of A and B.
3. The **scalar product** cA is the $m \times n$ matrix that is obtained by multiplying each entry of A by c.

Multiplication of Matrices (p. 334)

If A is an $m \times n$ matrix and B is an $n \times k$ matrix (so the number of columns of A is the same as the number of rows of B), then the **matrix product** AB is the $m \times k$ matrix whose ij-entry is the inner product of the ith row of A and the jth column of B.

Properties of Matrix Operations (pp. 333, 335)

If A, B, and C are matrices of compatible dimensions then the following properties hold:

1. **Commutativity of addition:**

$$A + B = B + A$$

2. **Associativity:**

$$(A + B) + C = A + (B + C)$$
$$(AB)C = A(BC)$$

3. **Distributivity:**

$$A(B + C) = AB + AC$$
$$(B + C)A = BA + CA$$

(Note that matrix *multiplication* is *not* commutative.)

Identity Matrix (p. 339)

The **identity matrix** I_n is the $n \times n$ matrix whose main diagonal entries are all 1 and whose other entries are all 0:

$$I_n = \begin{bmatrix} 1 & 0 & \cdots & 0 \\ 0 & 1 & \cdots & 0 \\ \vdots & \vdots & \ddots & \vdots \\ 0 & 0 & \cdots & 1 \end{bmatrix}$$

If A is an $m \times n$ matrix, then

$$AI_n = A \qquad \text{and} \qquad I_m A = A$$

Inverse of a Matrix (p. 340)

If A is an $n \times n$ matrix, then the inverse of A is the $n \times n$ matrix A^{-1} with the following properties:

$$A^{-1}A = I_n \qquad \text{and} \qquad AA^{-1} = I_n$$

To find the inverse of a matrix, we use a procedure involving elementary row operations (explained on page 322). (Note that *some* square matrices do not have an inverse.)

Inverse of a 2 × 2 Matrix (p. 340)

For 2 × 2 matrices the following special rule provides a shortcut for finding the inverse:

$$A = \begin{bmatrix} a & b \\ c & d \end{bmatrix} \quad \Rightarrow \quad A^{-1} = \frac{1}{ad - bc} \begin{bmatrix} d & -b \\ -c & a \end{bmatrix}$$

Writing a Linear System as a Matrix Equation (pp. 343–344)

A system of n linear equations in n variables can be written as a single matrix equation

$$AX = B$$

where A is the $n \times n$ matrix of coefficients, X is the $n \times 1$ matrix of the variables, and B is the $n \times 1$ matrix of the constants. For example, the linear system of two equations in two variables

$$a_{11}x + a_{12}x = b_1$$
$$a_{21}x + a_{22}x = b_2$$

can be expressed as

$$\begin{bmatrix} a_{11} & a_{12} \\ a_{21} & a_{22} \end{bmatrix} \begin{bmatrix} x \\ y \end{bmatrix} = \begin{bmatrix} b_1 \\ b_2 \end{bmatrix}$$

Solving Matrix Equations (p. 344)

If A is an invertible $n \times n$ matrix, X is an $n \times 1$ variable matrix, and B is an $n \times 1$ constant matrix, then the matrix equation

$$AX = B$$

has the unique solution

$$X = A^{-1}B$$

Determinant of a 2 × 2 Matrix (p. 347)

The **determinant** of the matrix

$$A = \begin{bmatrix} a & b \\ c & d \end{bmatrix}$$

is the *number*

$$\det(A) = |A| = ad - bc$$

Minors and Cofactors (p. 348)

If $A = |a_{ij}|$ is an $n \times n$ matrix, then the **minor** M_{ij} of the entry a_{ij} is the determinant of the matrix obtained by deleting the ith row and the jth column of A.

The **cofactor** A_{ij} of the entry a_{ij} is

$$A_{ij} = (-1)^{i+j}M_{ij}$$

(Thus, the minor and the cofactor of each entry either are the same or are negatives of each other.)

Determinant of an $n \times n$ Matrix (p. 348)

To find the **determinant** of the $n \times n$ matrix

$$A = \begin{bmatrix} a_{11} & a_{12} & \cdots & a_{1n} \\ a_{21} & a_{22} & \cdots & a_{2n} \\ \vdots & \vdots & \ddots & \vdots \\ a_{n1} & a_{n2} & \cdots & a_{nn} \end{bmatrix}$$

we choose a row or column to **expand**, and then we calculate the number that is obtained by multiplying each element of that row or column by its cofactor and then adding the resulting products. For example, if we choose to expand about the first row, we get

$$\det(A) = |A| = a_{11}A_{11} + a_{12}A_{12} + \cdots + a_{1n}A_{1n}$$

Invertibility Criterion (p. 350)

A square matrix has an inverse if and only if its determinant is not 0.

Row and Column Transformations (p. 351)

If we add a nonzero multiple of one row to another row in a square matrix or a nonzero multiple of one column to another column, then the determinant of the matrix is unchanged.

Cramer's Rule (pp. 352–354)

If a system of n linear equations in the n variables $x_1, x_2, \ldots, x_n$ is equivalent to the matrix equation $DX = B$ and if $|D| \neq 0$, then the solutions of the system are

$$x_1 = \frac{|D_{x_1}|}{|D|} \qquad x_2 = \frac{|D_{x_2}|}{|D|} \qquad \cdots \qquad x_n = \frac{|D_{x_n}|}{|D|}$$

where D_{x_i} is the matrix that is obtained from D by replacing its ith column by the constant matrix B.

Area of a Triangle Using Determinants (p. 355)

If a triangle in the coordinate plane has vertices (a_1, b_1), (a_2, b_2), and (a_3, b_3), then the area of the triangle is given by

$$\text{area} = \pm\frac{1}{2}\begin{vmatrix} a_1 & b_1 & 1 \\ a_2 & b_2 & 1 \\ a_3 & b_3 & 1 \end{vmatrix}$$

where the sign is chosen to make the area positive.

▼ CONCEPT SUMMARY

Section 7.1	Review Exercises
▪ Find the augmented matrix of a linear system	1–20
▪ Solve a linear system using elementary row operations	7–20
▪ Solve a linear system using the row-echelon form of its matrix	7–12
▪ Solve a system using the reduced row-echelon form of its matrix	13–20
▪ Determine the number of solutions of a linear system from the row-echelon form of its matrix	7–20
▪ Model using linear systems	63–64

Section 7.2

- Determine whether two matrices are equal
- Use addition, subtraction, and scalar multiplication of matrices
- Multiply matrices
- Write a linear system in matrix form

Section 7.3

- Determine when two matrices are inverses of each other
- Find the inverse of a 2×2 matrix
- Find the inverse of an $n \times n$ matrix
- Solve a linear system by writing it as a matrix equation
- Model using matrix equations

Section 7.4

- Find the determinant of a 2×2 matrix
- Find the determinant of an $n \times n$ matrix
- Use the Invertibility Criterion
- Use row and column transformations in computing the determinant of a matrix
- Use Cramer's Rule to solve a linear system
- Use determinants to find the area of a triangle in the coordinate plane

▼ EXERCISES

1–6 ■ A matrix is given. **(a)** State the dimension of the matrix. **(b)** Is the matrix in row-echelon form? **(c)** Is the matrix in reduced row-echelon form? **(d)** Write the system of equations for which the given matrix is the augmented matrix.

1. $\begin{bmatrix} 1 & 2 & -5 \\ 0 & 1 & 3 \end{bmatrix}$ **2.** $\begin{bmatrix} 1 & 0 & 6 \\ 0 & 1 & 0 \end{bmatrix}$

3. $\begin{bmatrix} 1 & 0 & 8 & 0 \\ 0 & 1 & 5 & -1 \\ 0 & 0 & 0 & 0 \end{bmatrix}$ **4.** $\begin{bmatrix} 1 & 3 & 6 & 2 \\ 2 & 1 & 0 & 5 \\ 0 & 0 & 1 & 0 \end{bmatrix}$

5. $\begin{bmatrix} 0 & 1 & -3 & 4 \\ 1 & 1 & 0 & 7 \\ 1 & 2 & 1 & 2 \end{bmatrix}$ **6.** $\begin{bmatrix} 1 & 8 & 6 & -4 \\ 0 & 1 & -3 & 5 \\ 0 & 0 & 2 & -7 \\ 1 & 1 & 1 & 0 \end{bmatrix}$

7–12 ■ Use Gaussian elimination to find the complete solution of the system, or show that no solution exists.

7. $\begin{cases} x + 2y + 2z = 6 \\ x - y = -1 \\ 2x + y + 3z = 7 \end{cases}$ **8.** $\begin{cases} x - y + z = 2 \\ x + y + 3z = 6 \\ 2y + 3z = 5 \end{cases}$

9. $\begin{cases} x - 2y + 3z = -2 \\ 2x - y + z = 2 \\ 2x - 7y + 11z = -9 \end{cases}$ **10.** $\begin{cases} x - y + z = 2 \\ x + y + 3z = 6 \\ 3x - y + 5z = 10 \end{cases}$

11. $\begin{cases} x + y + z + w = 0 \\ x - y - 4z - w = -1 \\ x - 2y + 4w = -7 \\ 2x + 2y + 3z + 4w = -3 \end{cases}$ **12.** $\begin{cases} x + 3z = -1 \\ y - 4w = 5 \\ 2y + z + w = 0 \\ 2x + y + 5z - 4w = 4 \end{cases}$

13–20 ■ Use Gauss-Jordan elimination to find the complete solution of the system, or show that no solution exists.

13. $\begin{cases} x - y + 3z = 2 \\ 2x + y + z = 2 \\ 3x + 4z = 4 \end{cases}$ **14.** $\begin{cases} x - y = 1 \\ x + y + 2z = 3 \\ x - 3y - 2z = -1 \end{cases}$

15. $\begin{cases} x - y + z - w = 0 \\ 3x - y - z - w = 2 \end{cases}$ **16.** $\begin{cases} x - y = 3 \\ 2x + y = 6 \\ x - 2y = 9 \end{cases}$

17. $\begin{cases} x - y + z = 0 \\ 3x + 2y - z = 6 \\ x + 4y - 3z = 3 \end{cases}$ **18.** $\begin{cases} x + 2y + 3z = 2 \\ 2x - y - 5z = 1 \\ 4x + 3y + z = 6 \end{cases}$

19. $\begin{cases} x + y - z - w = 2 \\ x - y + z - w = 0 \\ 2x + 2w = 2 \\ 2x + 4y - 4z - 2w = 6 \end{cases}$ **20.** $\begin{cases} x - y - 2z + 3w = 0 \\ y - z + w = 1 \\ 3x - 2y - 7z + 10w = 2 \end{cases}$

21–22 ■ Determine whether the matrices A and B are equal.

21. $A = \begin{bmatrix} 1 & 2 & 3 \\ 0 & 4 & 6 \\ 0 & 0 & 0 \end{bmatrix}$ $B = \begin{bmatrix} 1 & 2 & 3 \\ 0 & 4 & 6 \end{bmatrix}$

22. $A = \begin{bmatrix} \sqrt{25} & 1 \\ 0 & 2^{-1} \end{bmatrix}$ $B = \begin{bmatrix} 5 & e^0 \\ \log 1 & \frac{1}{2} \end{bmatrix}$

23–34 ▪ Let

$$A = [2 \quad 0 \quad -1] \qquad B = \begin{bmatrix} 1 & 2 & 4 \\ -2 & 1 & 0 \end{bmatrix}$$

$$C = \begin{bmatrix} \frac{1}{2} & 3 \\ 2 & \frac{3}{2} \\ -2 & 1 \end{bmatrix} \qquad D = \begin{bmatrix} 1 & 4 \\ 0 & -1 \\ 2 & 0 \end{bmatrix}$$

$$E = \begin{bmatrix} 2 & -1 \\ -\frac{1}{2} & 1 \end{bmatrix} \qquad F = \begin{bmatrix} 4 & 0 & 2 \\ -1 & 1 & 0 \\ 7 & 5 & 0 \end{bmatrix}$$

$$G = [5]$$

Carry out the indicated operation, or explain why it cannot be performed.

23. $A + B$

24. $C - D$

25. $2C + 3D$

26. $5B - 2C$

27. GA

28. AG

29. BC

30. CB

31. BF

32. FC

33. $(C + D)E$

34. $F(2C - D)$

35–36 ▪ Verify that the matrices A and B are inverses of each other by calculating the products AB and BA.

35. $A = \begin{bmatrix} 2 & -5 \\ -2 & 6 \end{bmatrix} \qquad B = \begin{bmatrix} 3 & \frac{5}{2} \\ 1 & 1 \end{bmatrix}$

36. $A = \begin{bmatrix} 2 & -1 & 3 \\ 2 & -2 & 1 \\ 0 & 1 & 1 \end{bmatrix} \qquad B = \begin{bmatrix} -\frac{3}{2} & 2 & \frac{5}{2} \\ -1 & 1 & 2 \\ 1 & -1 & -1 \end{bmatrix}$

37–42 ▪ Solve the matrix equation for the unknown matrix, X, or show that no solution exists, where

$$A = \begin{bmatrix} 2 & 1 \\ 3 & 2 \end{bmatrix} \quad B = \begin{bmatrix} 1 & -2 \\ -2 & 4 \end{bmatrix} \quad C = \begin{bmatrix} 0 & 1 & 3 \\ -2 & 4 & 0 \end{bmatrix}$$

37. $A + 3X = B$

38. $\frac{1}{2}(X - 2B) = A$

39. $2(X - A) = 3B$

40. $2X + C = 5A$

41. $AX = C$

42. $AX = B$

43–50 ▪ Find the determinant and, if possible, the inverse of the matrix.

43. $\begin{bmatrix} 1 & 4 \\ 2 & 9 \end{bmatrix}$

44. $\begin{bmatrix} 2 & 2 \\ 1 & -3 \end{bmatrix}$

45. $\begin{bmatrix} 4 & -12 \\ -2 & 6 \end{bmatrix}$

46. $\begin{bmatrix} 2 & 4 & 0 \\ -1 & 1 & 2 \\ 0 & 3 & 2 \end{bmatrix}$

47. $\begin{bmatrix} 3 & 0 & 1 \\ 2 & -3 & 0 \\ 4 & -2 & 1 \end{bmatrix}$

48. $\begin{bmatrix} 1 & 2 & 3 \\ 2 & 4 & 5 \\ 2 & 5 & 6 \end{bmatrix}$

49. $\begin{bmatrix} 1 & 0 & 0 & 1 \\ 0 & 2 & 0 & 2 \\ 0 & 0 & 3 & 3 \\ 0 & 0 & 0 & 4 \end{bmatrix}$

50. $\begin{bmatrix} 1 & 0 & 1 & 0 \\ 0 & 1 & 0 & 1 \\ 1 & 1 & 1 & 2 \\ 1 & 2 & 1 & 2 \end{bmatrix}$

51–54 ▪ Express the system of linear equations as a matrix equation. Then solve the matrix equation by multiplying each side by the inverse of the coefficient matrix.

51. $\begin{cases} 12x - 5y = 10 \\ 5x - 2y = 17 \end{cases}$

52. $\begin{cases} 6x - 5y = 1 \\ 8x - 7y = -1 \end{cases}$

53. $\begin{cases} 2x + y + 5z = \frac{1}{3} \\ x + 2y + 2z = \frac{1}{4} \\ x + 3z = \frac{1}{6} \end{cases}$

54. $\begin{cases} 2x + 3z = 5 \\ x + y + 6z = 0 \\ 3x - y + z = 5 \end{cases}$

55. Magda and Ivan grow tomatoes, onions, and zucchini in their backyard and sell them at a roadside stand on Saturdays and Sundays. They price tomatoes at $1.50 per pound, onions at $1.00 per pound, and zucchini at 50 cents per pound. The following table shows the number of pounds of each type of produce that they sold during the last weekend in July.

	Tomatoes	Onions	Zucchini
Saturday	25	16	30
Sunday	14	12	16

(a) Let

$$A = \begin{bmatrix} 25 & 16 & 30 \\ 14 & 12 & 16 \end{bmatrix} \quad \text{and} \quad B = \begin{bmatrix} 1.50 \\ 1.00 \\ 0.50 \end{bmatrix}$$

Compare these matrices to the data given in the problem, and describe what their entries represent.

(b) Only one of the products AB or BA is defined. Calculate the product that *is* defined, and describe what its entries represent.

56. An ATM at a bank in Qualicum Beach, British Columbia, dispenses $20 and $50 bills. Brodie withdraws $600 from this machine and receives a total of 18 bills. Let x be the number of $20 bills and y the number of $50 bills that he receives.

(a) Find a system of two linear equations in x and y that express the information given in the problem.

(b) Write your linear system as a matrix equation of the form $AX = B$.

(c) Find A^{-1}, and use it to solve your matrix equation in part (b). How many bills of each type did Brodie receive?

57–60 ▪ Solve the system using Cramer's Rule.

57. $\begin{cases} 2x + 7y = 13 \\ 6x + 16y = 30 \end{cases}$

58. $\begin{cases} 12x - 11y = 140 \\ 7x + 9y = 20 \end{cases}$

59. $\begin{cases} 2x - y + 5z = 0 \\ -x + 7y = 9 \\ 5x + 4y + 3z = -9 \end{cases}$

60. $\begin{cases} 3x + 4y - z = 10 \\ x - 4z = 20 \\ 2x + y + 5z = 30 \end{cases}$

61–62 ▪ Use the determinant formula for the area of a triangle to find the area of the triangle in the figure.

61.

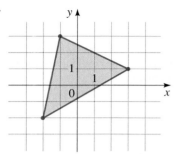

62.

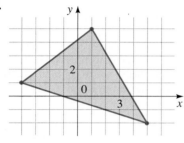

63–64 ▪ Use any of the methods you have learned in this chapter to solve the problem.

63. Clarisse invests $60,000 in money-market accounts at three different banks. Bank A pays 2% interest per year, bank B pays 2.5%, and bank C pays 3%. She decides to invest twice as much in bank B as in the other two banks. After one year, Clarisse has earned $1575 in interest. How much did she invest in each bank?

64. A commercial fisherman fishes for haddock, sea bass, and red snapper. He is paid $1.25 a pound for haddock, $0.75 a pound for sea bass, and $2.00 a pound for red snapper. Yesterday he caught 560 lb of fish worth $575. The haddock and red snapper together are worth $320. How many pounds of each fish did he catch?

1–4 ▪ Determine whether the matrix is in reduced row-echelon form, row-echelon form, or neither.

1. $\begin{bmatrix} 1 & 8 & 0 & 0 \\ 0 & 1 & 7 & 10 \\ 0 & 0 & 0 & 0 \end{bmatrix}$

2. $\begin{bmatrix} 0 & 0 & 0 & 4 \\ 0 & 0 & 2 & 5 \\ 0 & 1 & -2 & 7 \\ 1 & 0 & -3 & 0 \end{bmatrix}$

3. $\begin{bmatrix} 1 & 0 & 0 \\ 0 & 0 & 1 \end{bmatrix}$

4. $\begin{bmatrix} 1 & 0 & 0 & 3 \\ 0 & 1 & 0 & -2 \\ 0 & 0 & 1 & \frac{3}{2} \end{bmatrix}$

5–6 ▪ Use Gaussian elimination to find the complete solution of the system, or show that no solution exists.

5. $\begin{cases} x - y + 2z = 0 \\ 2x - 4y + 5z = -5 \\ 2y - 3z = 5 \end{cases}$

6. $\begin{cases} 2x - 3y + z = 3 \\ x + 2y + 2z = -1 \\ 4x + y + 5z = 4 \end{cases}$

7. Use Gauss-Jordan elimination to find the complete solution of the system.

$$\begin{cases} x + 3y - z = 0 \\ 3x + 4y - 2z = -1 \\ -x + 2y = 1 \end{cases}$$

8–15 ▪ Let

$$A = \begin{bmatrix} 2 & 3 \\ 2 & 4 \end{bmatrix} \quad B = \begin{bmatrix} 2 & 4 \\ -1 & 1 \\ 3 & 0 \end{bmatrix} \quad C = \begin{bmatrix} 1 & 0 & 4 \\ -1 & 1 & 2 \\ 0 & 1 & 3 \end{bmatrix}$$

Carry out the indicated operation, or explain why it cannot be performed.

8. $A + B$ **9.** AB **10.** $BA - 3B$ **11.** CBA

12. A^{-1} **13.** B^{-1} **14.** $\det(B)$ **15.** $\det(C)$

16. (a) Write a matrix equation equivalent to the following system.

$$\begin{cases} 4x - 3y = 10 \\ 3x - 2y = 30 \end{cases}$$

(b) Find the inverse of the coefficient matrix, and use it to solve the system.

17. Only one of the following matrices has an inverse. Find the determinant of each matrix, and use the determinants to identify the one that has an inverse. Then find the inverse.

$$A = \begin{bmatrix} 1 & 4 & 1 \\ 0 & 2 & 0 \\ 1 & 0 & 1 \end{bmatrix} \quad B = \begin{bmatrix} 1 & 4 & 0 \\ 0 & 2 & 0 \\ -3 & 0 & 1 \end{bmatrix}$$

18. Solve using Cramer's Rule:

$$\begin{cases} 2x - z = 14 \\ 3x - y + 5z = 0 \\ 4x + 2y + 3z = -2 \end{cases}$$

19. A shopper buys a mixture of nuts; the almonds cost $4.75 a pound, and the walnuts cost $3.45 a pound. Her total purchase weighs 3 lb and costs $11.91. Use Cramer's Rule to determine how much of each nut she bought.

Conic Sections

In this chapter we study equations for important curves called the conic sections—ellipses, parabolas, and hyperbolas. We encounter conic sections frequently in our everyday lives. For example, a ball thrown upward through the air travels in a path whose shape is a parabola. We'll learn how conic sections can be constructed geometrically, and more importantly, how we can find equations for such curves.

The following Warm-Up Exercises will help you review the basic algebra skills that you will need for this chapter.

WARM-UP EXERCISES

1. Graph the following parabola by putting it in standard form: $y = -2x^2 + 12x - 16$.

2. Find the distance between the points $(2, -4)$ and $(3, 1)$.

3. Graph the circle described by the equation $\dfrac{x^2}{9} + \dfrac{y^2}{9} = 1$.

4. Solve for y, and simplify your answer: $\dfrac{x^2}{a^2} - \dfrac{y^2}{b^2} = 1$.

5. The following is a graph of $y = f(x)$. Sketch a graph of $(y - 1) = f(x - 2)$.

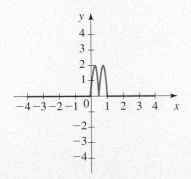

6. Write the equation of a circle with endpoints of a diameter at $(1, 3)$ and $(9, 9)$.

8.1 | Parabolas

LEARNING OBJECTIVES

After completing this section, you will be able to:

- Find geometric properties of a parabola from its equation
- Find the equation of a parabola from some of its geometric properties

---

Conic sections are the curves we get when we make a straight cut in a cone, as shown in the figure. For example, if a cone is cut horizontally, the cross section is a circle. So a circle is a conic section. Other ways of cutting a cone produce parabolas, ellipses, and hyperbolas.

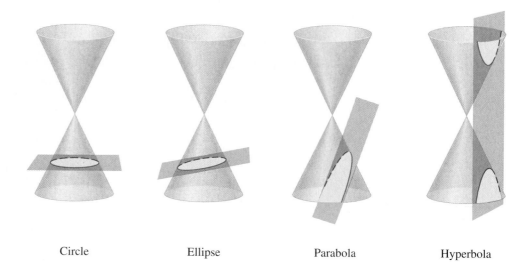

| Circle | Ellipse | Parabola | Hyperbola |

Our goal in this chapter is to find equations whose graphs are the conic sections. We already know from Section 2.2 that the graph of the equation $x^2 + y^2 = r^2$ is a circle. We will find equations for each of the other conic sections by analyzing their *geometric* properties.

■ Geometric Definition of a Parabola

We saw in Section 4.1 that the graph of the equation

$$y = ax^2 + bx + c$$

is a U-shaped curve called a *parabola* that opens either upward or downward, depending on whether the sign of a is positive or negative.

In this section we study parabolas from a geometric rather than an algebraic point of view. We begin with the geometric definition of a parabola and show how this leads to the algebraic formula that we are already familiar with.

GEOMETRIC DEFINITION OF A PARABOLA

A **parabola** is the set of points in the plane that are equidistant from a fixed point F (called the **focus**) and a fixed line l (called the **directrix**).

This definition is illustrated in Figure 1. The **vertex** V of the parabola lies halfway between the focus and the directrix, and the **axis of symmetry** is the line that runs through the focus perpendicular to the directrix.

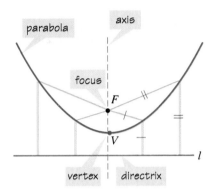

FIGURE 1

In this section we restrict our attention to parabolas that are situated with the vertex at the origin and that have a vertical or horizontal axis of symmetry. (Parabolas in more general positions will be considered in Section 8.4.) If the focus of such a parabola is the point $F(0, p)$, then the axis of symmetry must be vertical, and the directrix has the equation $y = -p$. Figure 2 illustrates the case $p > 0$.

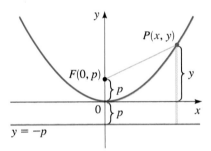

FIGURE 2

If $P(x, y)$ is any point on the parabola, then the distance from P to the focus F (using the Distance Formula) is

$$\sqrt{x^2 + (y - p)^2}$$

The distance from P to the directrix is

$$|y - (-p)| = |y + p|$$

By the definition of a parabola these two distances must be equal:

$$\sqrt{x^2 + (y - p)^2} = |y + p|$$
$$x^2 + (y - p)^2 = |y + p|^2 = (y + p)^2 \qquad \text{Square both sides}$$
$$x^2 + y^2 - 2py + p^2 = y^2 + 2py + p^2 \qquad \text{Expand}$$
$$x^2 - 2py = 2py \qquad \text{Simplify}$$
$$x^2 = 4py$$

If $p > 0$, then the parabola opens upward, but if $p < 0$, it opens downward. When x is replaced by $-x$, the equation remains unchanged, so the graph is symmetric about the y-axis.

■ Equations and Graphs of Parabolas

The following box summarizes what we have just proved about the equation and features of a parabola with a vertical axis.

PARABOLA WITH VERTICAL AXIS

The graph of the equation

$$x^2 = 4py$$

is a parabola with the following properties.

VERTEX	$V(0, 0)$
FOCUS	$F(0, p)$
DIRECTRIX	$y = -p$

The parabola opens upward if $p > 0$ or downward if $p < 0$.

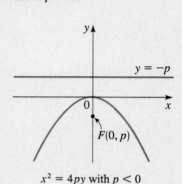

$x^2 = 4py$ with $p > 0$ $x^2 = 4py$ with $p < 0$

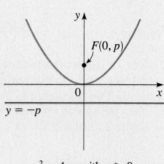

FIGURE 3

▶ **EXAMPLE 1** | Finding the Equation of a Parabola

Find the equation of the parabola with vertex $V(0, 0)$ and focus $F(0, 2)$, and sketch its graph.

▼ **SOLUTION**   Since the focus is $F(0, 2)$, we conclude that $p = 2$ (so the directrix is $y = -2$). Thus, the equation of the parabola is

$$x^2 = 4(2)y \qquad x^2 = 4py \text{ with } p = 2$$
$$x^2 = 8y$$

Since $p = 2 > 0$, the parabola opens upward. See Figure 3.

▲

▶ **EXAMPLE 2** | Finding the Focus and Directrix of a Parabola from Its Equation

Find the focus and directrix of the parabola $y = -x^2$, and sketch the graph.

▼ **SOLUTION**   To find the focus and directrix, we put the given equation in the standard form $x^2 = -y$. Comparing this to the general equation $x^2 = 4py$, we see that $4p = -1$, so $p = -\frac{1}{4}$. Thus, the focus is $F\left(0, -\frac{1}{4}\right)$, and the directrix is $y = \frac{1}{4}$. The graph of the parabola, together with the focus and the directrix, is shown in Figure 4(a). We can also draw the graph using a graphing calculator as shown in Figure 4(b).

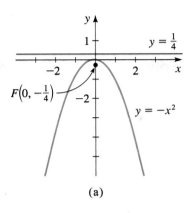

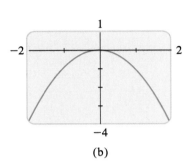

FIGURE 4   (a)   (b)   ▲

Reflecting the graph in Figure 2 about the diagonal line $y = x$ has the effect of interchanging the roles of x and y. This results in a parabola with horizontal axis. By the same method as before, we can prove the following properties.

PARABOLA WITH HORIZONTAL AXIS

The graph of the equation

$$y^2 = 4px$$

is a parabola with the following properties.

VERTEX	$V(0, 0)$
FOCUS	$F(p, 0)$
DIRECTRIX	$x = -p$

The parabola opens to the right if $p > 0$ or to the left if $p < 0$.

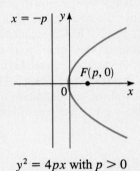

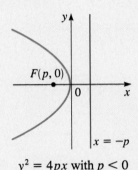

$y^2 = 4px$ with $p > 0$ $y^2 = 4px$ with $p < 0$

▶ **EXAMPLE 3** | A Parabola with Horizontal Axis

A parabola has the equation $6x + y^2 = 0$.

(a) Find the focus and directrix of the parabola, and sketch the graph.

(b) Use a graphing calculator to draw the graph.

▼ **SOLUTION**

(a) To find the focus and directrix, we put the given equation in the standard form $y^2 = -6x$. Comparing this to the general equation $y^2 = 4px$, we see that $4p = -6$, so $p = -\frac{3}{2}$. Thus, the focus is $F\left(-\frac{3}{2}, 0\right)$, and the directrix is $x = \frac{3}{2}$. Since $p < 0$, the parabola opens to the left. The graph of the parabola, together with the focus and the directrix, is shown in Figure 5(a) on the next page.

(b) To draw the graph using a graphing calculator, we need to solve for y.

$$6x + y^2 = 0$$

$$y^2 = -6x \qquad \text{Subtract } 6x$$

$$y = \pm\sqrt{-6x} \qquad \text{Take square roots}$$

To obtain the graph of the parabola, we graph both functions

$$y = \sqrt{-6x} \qquad \text{and} \qquad y = -\sqrt{-6x}$$

as shown in Figure 5(b).

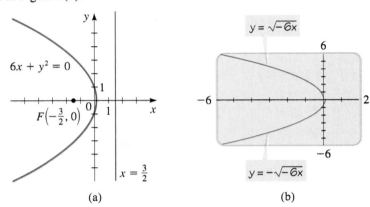

FIGURE 5

(a)   (b)

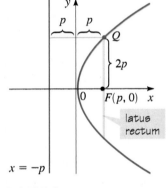

FIGURE 6

The equation $y^2 = 4px$, does not define y as a function of x (see page 148). So to use a graphing calculator to graph a parabola with horizontal axis, we must first solve for y. This leads to two functions, $y = \sqrt{4px}$ and $y = -\sqrt{4px}$. We need to graph both functions to get the complete graph of the parabola. For example, in Figure 5(b) we had to graph both $y = \sqrt{-6x}$ and $y = -\sqrt{-6x}$ to graph the parabola $y^2 = -6x$.

We can use the coordinates of the focus to estimate the "width" of a parabola when sketching its graph. The line segment that runs through the focus perpendicular to the axis, with endpoints on the parabola, is called the **latus rectum**, and its length is the **focal diameter** of the parabola. From Figure 6 we can see that the distance from an endpoint Q of the latus rectum to the directrix is $|2p|$. Thus, the distance from Q to the focus must be $|2p|$ as well (by the definition of a parabola), so the focal diameter is $|4p|$. In the next example we use the focal diameter to determine the "width" of a parabola when graphing it.

▶ **EXAMPLE 4** | The Focal Diameter of a Parabola

Find the focus, directrix, and focal diameter of the parabola $y = \frac{1}{2}x^2$, and sketch its graph.

▼ **SOLUTION** We first put the equation in the form $x^2 = 4py$.

$$y = \tfrac{1}{2}x^2$$

$$x^2 = 2y \qquad \text{Multiply each side by 2}$$

From this equation we see that $4p = 2$, so the focal diameter is 2. Solving for p gives $p = \frac{1}{2}$, so the focus is $\left(0, \frac{1}{2}\right)$ and the directrix is $y = -\frac{1}{2}$. Since the focal diameter is 2, the latus rectum extends 1 unit to the left and 1 unit to the right of the focus. The graph is sketched in Figure 7. ▲

In the next example we graph a family of parabolas, to show how changing the distance between the focus and the vertex affects the "width" of a parabola.

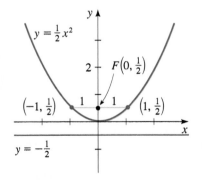

FIGURE 7

▶ **EXAMPLE 5** | A Family of Parabolas

(a) Find equations for the parabolas with vertex at the origin and foci
$F_1(0, \frac{1}{8})$, $F_2(0, \frac{1}{2})$, $F_3(0, 1)$, and $F_4(0, 4)$.

(b) Draw the graphs of the parabolas in part (a). What do you conclude?

▼ **SOLUTION**

(a) Since the foci are on the positive y-axis, the parabolas open upward and have equations of the form $x^2 = 4py$. This leads to the following equations.

Focus	p	Equation $x^2 = 4py$	Form of the equation for graphing calculator
$F_1(0, \frac{1}{8})$	$p = \frac{1}{8}$	$x^2 = \frac{1}{2}y$	$y = 2x^2$
$F_2(0, \frac{1}{2})$	$p = \frac{1}{2}$	$x^2 = 2y$	$y = 0.5x^2$
$F_3(0, 1)$	$p = 1$	$x^2 = 4y$	$y = 0.25x^2$
$F_4(0, 4)$	$p = 4$	$x^2 = 16y$	$y = 0.0625x^2$

(b) The graphs are drawn in Figure 8. We see that the closer the focus to the vertex, the narrower the parabola.

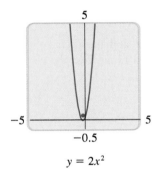

$y = 2x^2$

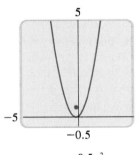

$y = 0.5x^2$

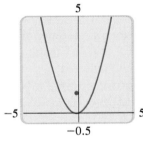

$y = 0.25x^2$

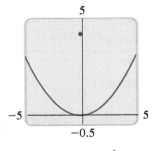

$y = 0.0625x^2$

FIGURE 8 A family of parabolas

■ Applications

Parabolas have an important property that makes them useful as reflectors for lamps and telescopes. Light from a source placed at the focus of a surface with parabolic cross section will be reflected in such a way that it travels parallel to the axis of the parabola (see Figure 9). Thus, a parabolic mirror reflects the light into a beam of parallel rays. Conversely, light approaching the reflector in rays parallel to its axis of symmetry is concentrated to the focus. This *reflection property*, which can be proved by using calculus, is used in the construction of reflecting telescopes.

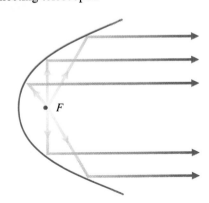

FIGURE 9 Parabolic reflector

▶ **EXAMPLE 6** | Finding the Focal Point of a Searchlight Reflector

A searchlight has a parabolic reflector that forms a "bowl," which is 12 in. wide from rim to rim and 8 in. deep, as shown in Figure 10. If the filament of the light bulb is located at the focus, how far from the vertex of the reflector is it?

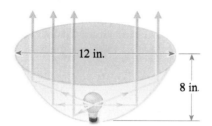

FIGURE 10 A parabolic reflector

▼ **SOLUTION**   We introduce a coordinate system and place a parabolic cross section of the reflector so that its vertex is at the origin and its axis is vertical (see Figure 11). Then the equation of this parabola has the form $x^2 = 4py$. From Figure 11 we see that the point $(6, 8)$ lies on the parabola. We use this to find p.

$$6^2 = 4p(8)   \qquad \text{The point } (6, 8) \text{ satisfies the equation } x^2 = 4py$$

$$36 = 32p$$

$$p = \tfrac{9}{8}$$

The focus is $F\left(0, \tfrac{9}{8}\right)$, so the distance between the vertex and the focus is $\tfrac{9}{8} = 1\tfrac{1}{8}$ in. Because the filament is positioned at the focus, it is located $1\tfrac{1}{8}$ in. from the vertex of the reflector.   ▲

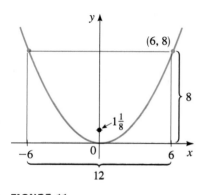

FIGURE 11

8.2 | Ellipses

LEARNING OBJECTIVES
After completing this section, you will be able to:

▪ Find geometric properties of an ellipse from its equation
▪ Find the equation of an ellipse from some of its geometric properties

▪ Geometric Definition of an Ellipse

An ellipse is an oval curve that looks like an elongated circle. More precisely, we have the following definition.

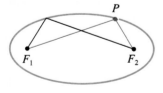

FIGURE 1

> **GEOMETRIC DEFINITION OF AN ELLIPSE**
>
> An **ellipse** is the set of all points in the plane the sum of whose distances from two fixed points F_1 and F_2 is a constant. (See Figure 1.) These two fixed points are the **foci** (plural of **focus**) of the ellipse.

The geometric definition suggests a simple method for drawing an ellipse. Place a sheet of paper on a drawing board, and insert thumbtacks at the two points that are to be the foci of the ellipse. Attach the ends of a string to the tacks, as shown in Figure 2(a). With the point of a pencil, hold the string taut. Then carefully move the pencil around the foci, keeping the string taut at all times. The pencil will trace out an ellipse, because the sum of the distances from the point of the pencil to the foci will always equal the length of the string, which is constant.

If the string is only slightly longer than the distance between the foci, then the ellipse that is traced out will be elongated in shape, as in Figure 2(a), but if the foci are close together relative to the length of the string, the ellipse will be almost circular, as shown in Figure 2(b).

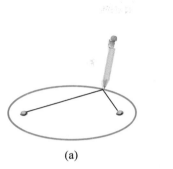
(a)

(b)

FIGURE 2

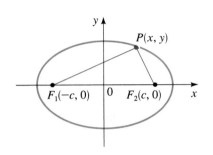

FIGURE 3

To obtain the simplest equation for an ellipse, we place the foci on the x-axis at $F_1(-c, 0)$ and $F_2(c, 0)$ so that the origin is halfway between them (see Figure 3).

For later convenience we let the sum of the distances from a point on the ellipse to the foci be $2a$. Then if $P(x, y)$ is any point on the ellipse, we have

$$d(P, F_1) + d(P, F_2) = 2a$$

So from the Distance Formula

$$\sqrt{(x + c)^2 + y^2} + \sqrt{(x - c)^2 + y^2} = 2a$$

or

$$\sqrt{(x - c)^2 + y^2} = 2a - \sqrt{(x + c)^2 + y^2}$$

Squaring each side and expanding, we get

$$x^2 - 2cx + c^2 + y^2 = 4a^2 - 4a\sqrt{(x + c)^2 + y^2} + (x^2 + 2cx + c^2 + y^2)$$

which simplifies to

$$4a\sqrt{(x + c)^2 + y^2} = 4a^2 + 4cx$$

Dividing each side by 4 and squaring again, we get

$$a^2[(x + c)^2 + y^2] = (a^2 + cx)^2$$

$$a^2x^2 + 2a^2cx + a^2c^2 + a^2y^2 = a^4 + 2a^2cx + c^2x^2$$

$$(a^2 - c^2)x^2 + a^2y^2 = a^2(a^2 - c^2)$$

Since the sum of the distances from P to the foci must be larger than the distance between the foci, we have that $2a > 2c$, or $a > c$. Thus, $a^2 - c^2 > 0$, and we can divide each side of the preceding equation by $a^2(a^2 - c^2)$ to get

$$\frac{x^2}{a^2} + \frac{y^2}{a^2 - c^2} = 1$$

For convenience let $b^2 = a^2 - c^2$ (with $b > 0$). Since $b^2 < a^2$, it follows that $b < a$. The preceding equation then becomes

$$\frac{x^2}{a^2} + \frac{y^2}{b^2} = 1 \quad \text{with } a > b$$

This is the equation of the ellipse. To graph it, we need to know the x- and y-intercepts. Setting $y = 0$, we get

$$\frac{x^2}{a^2} = 1$$

The orbits of the planets are ellipses, with the sun at one focus.

so $x^2 = a^2$, or $x = \pm a$. Thus, the ellipse crosses the x-axis at $(a, 0)$ and $(-a, 0)$, as in Figure 4. These points are called the **vertices** of the ellipse, and the segment that joins them is called the **major axis**. Its length is $2a$.

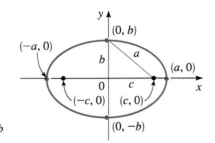

FIGURE 4
$\dfrac{x^2}{a^2} + \dfrac{y^2}{b^2} = 1$ with $a > b$

Similarly, if we set $x = 0$, we get $y = \pm b$, so the ellipse crosses the y-axis at $(0, b)$ and $(0, -b)$. The segment that joins these points is called the **minor axis**, and it has length $2b$. Note that $2a > 2b$, so the major axis is longer than the minor axis. The origin is the **center** of the ellipse.

If the foci of the ellipse are placed on the y-axis at $(0, \pm c)$ rather than on the x-axis, then the roles of x and y are reversed in the preceding discussion, and we get a vertical ellipse.

■ Equations and Graphs of Ellipses

The following box summarizes what we have just proved about the equation and features of an ellipse centered at the origin.

In the standard equation for an ellipse, a^2 is the *larger* denominator and b^2 is the *smaller*. To find c^2, we subtract: larger denominator minus smaller denominator.

ELLIPSE WITH CENTER AT THE ORIGIN

The graph of each of the following equations is an ellipse with center at the origin and having the given properties.

EQUATION	$\dfrac{x^2}{a^2} + \dfrac{y^2}{b^2} = 1$	$\dfrac{x^2}{b^2} + \dfrac{y^2}{a^2} = 1$
	$a > b > 0$	$a > b > 0$
VERTICES	$(\pm a, 0)$	$(0, \pm a)$
MAJOR AXIS	Horizontal, length $2a$	Vertical, length $2a$
MINOR AXIS	Vertical, length $2b$	Horizontal, length $2b$
FOCI	$(\pm c, 0)$, $c^2 = a^2 - b^2$	$(0, \pm c)$, $c^2 = a^2 - b^2$
GRAPH		

▶ **EXAMPLE 1** | Sketching an Ellipse

An ellipse has the equation

$$\frac{x^2}{9} + \frac{y^2}{4} = 1$$

(a) Find the foci, vertices, and the lengths of the major and minor axes, and sketch the graph.

(b) Draw the graph using a graphing calculator.

▼ **SOLUTION**

(a) Since the denominator of x^2 is larger, the ellipse has a horizontal major axis. This gives $a^2 = 9$ and $b^2 = 4$, so $c^2 = a^2 - b^2 = 9 - 4 = 5$. Thus, $a = 3$, $b = 2$, and $c = \sqrt{5}$.

FOCI	$(\pm\sqrt{5}, 0)$
VERTICES	$(\pm 3, 0)$
LENGTH OF MAJOR AXIS	6
LENGTH OF MINOR AXIS	4

The graph is shown in Figure 5(a).

(b) To draw the graph using a graphing calculator, we need to solve for y.

$$\frac{x^2}{9} + \frac{y^2}{4} = 1$$

$$\frac{y^2}{4} = 1 - \frac{x^2}{9} \qquad \text{Subtract } \frac{x^2}{9}$$

$$y^2 = 4\left(1 - \frac{x^2}{9}\right) \qquad \text{Multiply by 4}$$

$$y = \pm 2\sqrt{1 - \frac{x^2}{9}} \qquad \text{Take square roots}$$

Note that the equation of an ellipse does not define y as a function of x (see page 148). That's why we need to graph two functions to graph an ellipse.

To obtain the graph of the ellipse, we graph both functions

$$y = 2\sqrt{1 - x^2/9} \qquad \text{and} \qquad y = -2\sqrt{1 - x^2/9}$$

as shown in Figure 5(b).

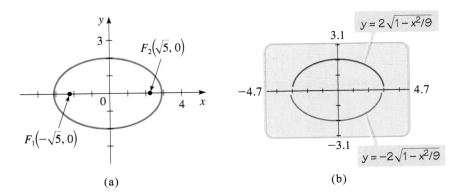

FIGURE 5
$\frac{x^2}{9} + \frac{y^2}{4} = 1$

(a)

(b)

▶ **EXAMPLE 2** | Finding the Foci of an Ellipse

Find the foci of the ellipse $16x^2 + 9y^2 = 144$, and sketch its graph.

▼ **SOLUTION** First we put the equation in standard form. Dividing by 144, we get

$$\frac{x^2}{9} + \frac{y^2}{16} = 1$$

Since $16 > 9$, this is an ellipse with its foci on the y-axis, and with $a = 4$ and $b = 3$. We have

$$c^2 = a^2 - b^2 = 16 - 9 = 7$$

$$c = \sqrt{7}$$

Thus, the foci are $(0, \pm\sqrt{7})$. The graph is shown in Figure 6(a).

We can also draw the graph using a graphing calculator as shown in Figure 6(b).

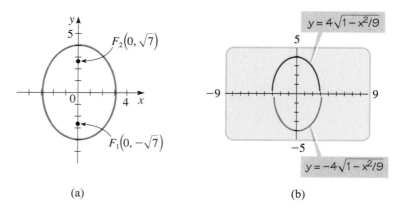

FIGURE 6

$16x^2 + 9y^2 = 144$ (a)   (b)

▶ **EXAMPLE 3** | Finding the Equation of an Ellipse

The vertices of an ellipse are $(\pm4, 0)$, and the foci are $(\pm2, 0)$. Find its equation, and sketch the graph.

▼ **SOLUTION** Since the vertices are $(\pm4, 0)$, we have $a = 4$. The foci are $(\pm2, 0)$, so $c = 2$. To write the equation, we need to find b. Since $c^2 = a^2 - b^2$, we have

$$2^2 = 4^2 - b^2$$

$$b^2 = 16 - 4 = 12$$

Thus, the equation of the ellipse is

$$\frac{x^2}{16} + \frac{y^2}{12} = 1$$

The graph is shown in Figure 7.

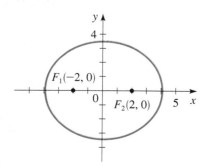

FIGURE 7

$\dfrac{x^2}{16} + \dfrac{y^2}{12} = 1$

■ Eccentricity of an Ellipse

We saw earlier in this section (Figure 2) that if $2a$ is only slightly greater than $2c$, the ellipse is long and thin, whereas if $2a$ is much greater than $2c$, the ellipse is almost circular. We measure the deviation of an ellipse from being circular by the ratio of a and c.

DEFINITION OF ECCENTRICITY

For the ellipse $\dfrac{x^2}{a^2} + \dfrac{y^2}{b^2} = 1$ or $\dfrac{x^2}{b^2} + \dfrac{y^2}{a^2} = 1$ (with $a > b > 0$), the **eccentricity** e is the number

$$e = \frac{c}{a}$$

where $c = \sqrt{a^2 - b^2}$. The eccentricity of every ellipse satisfies $0 < e < 1$.

Thus, if e is close to 1, then c is almost equal to a, and the ellipse is elongated in shape, but if e is close to 0, then the ellipse is close to a circle in shape. The eccentricity is a measure of how "stretched" the ellipse is.

In Figure 8 we show a number of ellipses to demonstrate the effect of varying the eccentricity e.

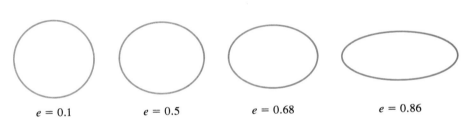

$e = 0.1$ $\qquad$ $e = 0.5$ $\qquad$ $e = 0.68$ $\qquad$ $e = 0.86$

FIGURE 8 Ellipses with various eccentricities

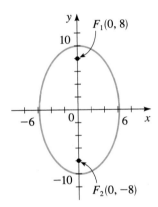

FIGURE 9
$$\frac{x^2}{36} + \frac{y^2}{100} = 1$$

EXAMPLE 4 | Finding the Equation of an Ellipse from Its Eccentricity and Foci

Find the equation of the ellipse with foci $(0, \pm 8)$ and eccentricity $e = \frac{4}{5}$, and sketch its graph.

▼ **SOLUTION** We are given $e = \frac{4}{5}$ and $c = 8$. Thus

$$\frac{4}{5} = \frac{8}{a} \qquad \text{Eccentricity } e = \frac{c}{a}$$

$$4a = 40 \qquad \text{Cross-multiply}$$

$$a = 10$$

To find b, we use the fact that $c^2 = a^2 - b^2$.

$$8^2 = 10^2 - b^2$$

$$b^2 = 10^2 - 8^2 = 36$$

$$b = 6$$

Thus, the equation of the ellipse is

$$\frac{x^2}{36} + \frac{y^2}{100} = 1$$

Because the foci are on the y-axis, the ellipse is oriented vertically. To sketch the ellipse, we find the intercepts: The x-intercepts are ± 6, and the y-intercepts are ± 10. The graph is sketched in Figure 9. ▲

Eccentricities of the Orbits of the Planets

The orbits of the planets are ellipses with the sun at one focus. For most planets these ellipses have very small eccentricity, so they are nearly circular. However, Mercury and Pluto, the innermost and outermost known planets, have visibly elliptical orbits.

Planet	Eccentricity
Mercury	0.206
Venus	0.007
Earth	0.017
Mars	0.093
Jupiter	0.048
Saturn	0.056
Uranus	0.046
Neptune	0.010
Pluto	0.248

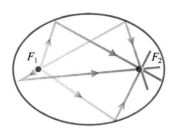

FIGURE 10

Gravitational attraction causes the planets to move in elliptical orbits around the sun with the sun at one focus. This remarkable property was first observed by Johannes Kepler and was later deduced by Isaac Newton from his inverse square law of gravity, using calculus. The orbits of the planets have different eccentricities, but most are nearly circular (see the table in the margin on page 375).

Ellipses, like parabolas, have an interesting *reflection property* that leads to a number of practical applications. If a light source is placed at one focus of a reflecting surface with elliptical cross sections, then all the light will be reflected off the surface to the other focus, as shown in Figure 10. This principle, which works for sound waves as well as for light, is used in *lithotripsy*, a treatment for kidney stones. The patient is placed in a tub of water with elliptical cross sections in such a way that the kidney stone is accurately located at one focus. High-intensity sound waves generated at the other focus are reflected to the stone and destroy it with minimal damage to surrounding tissue. The patient is spared the trauma of surgery and recovers within days instead of weeks.

The reflection property of ellipses is also used in the construction of *whispering galleries*. Sound coming from one focus bounces off the walls and ceiling of an elliptical room and passes through the other focus. In these rooms even quiet whispers spoken at one focus can be heard clearly at the other. Famous whispering galleries include the National Statuary Hall of the U.S. Capitol in Washington, D.C. and the Mormon Tabernacle in Salt Lake City, Utah.

8.3 | Hyperbolas

LEARNING OBJECTIVES

After completing this section, you will be able to:

■ Find geometric properties of a hyperbola from its equation
■ Find the equation of a hyperbola from some of its geometric properties

■ Geometric Definition of a Hyperbola

Although ellipses and hyperbolas have completely different shapes, their definitions and equations are similar. Instead of using the *sum* of distances from two fixed foci, as in the case of an ellipse, we use the *difference* to define a hyperbola.

> **GEOMETRIC DEFINITION OF A HYPERBOLA**
>
> A **hyperbola** is the set of all points in the plane, the difference of whose distances from two fixed points F_1 and F_2 is a constant. (See Figure 1.) These two fixed points are the **foci** of the hyperbola.

As in the case of the ellipse, we get the simplest equation for the hyperbola by placing the foci on the *x*-axis at $(\pm c, 0)$, as shown in Figure 1 on the next page. By definition, if $P(x, y)$ lies on the hyperbola, then either $d(P, F_1) - d(P, F_2)$ or $d(P, F_2) - d(P, F_1)$ must equal some positive constant, which we call $2a$. Thus, we have

$$d(P, F_1) - d(P, F_2) = \pm 2a$$

or
$$\sqrt{(x + c)^2 + y^2} - \sqrt{(x - c)^2 + y^2} = \pm 2a$$

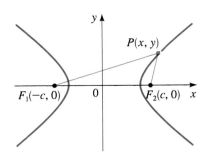

FIGURE 1 P is on the hyperbola if $|d(P, F_1) - d(P, F_2)| = 2a$.

Proceeding as we did in the case of the ellipse (Section 8.2), we simplify this to

$$(c^2 - a^2)x^2 - a^2y^2 = a^2(c^2 - a^2)$$

From triangle PF_1F_2 in Figure 1 we see that $|d(P, F_1) - d(P, F_2)| < 2c$. It follows that $2a < 2c$, or $a < c$. Thus, $c^2 - a^2 > 0$, so we can set $b^2 = c^2 - a^2$. We then simplify the last displayed equation to get

$$\frac{x^2}{a^2} - \frac{y^2}{b^2} = 1$$

This is the *equation of the hyperbola*. If we replace x by $-x$ or y by $-y$ in this equation, it remains unchanged, so the hyperbola is symmetric about both the x- and y-axes and about the origin. The x-intercepts are $\pm a$, and the points $(a, 0)$ and $(-a, 0)$ are the **vertices** of the hyperbola. There is no y-intercept, because setting $x = 0$ in the equation of the hyperbola leads to $-y^2 = b^2$, which has no real solution. Furthermore, the equation of the hyperbola implies that

$$\frac{x^2}{a^2} = \frac{y^2}{b^2} + 1 \geq 1$$

so $x^2/a^2 \geq 1$; thus, $x^2 \geq a^2$, and hence $x \geq a$ or $x \leq -a$. This means that the hyperbola consists of two parts, called its **branches**. The segment joining the two vertices on the separate branches is the **transverse axis** of the hyperbola, and the origin is called its **center**.

If we place the foci of the hyperbola on the y-axis rather than on the x-axis, then this has the effect of reversing the roles of x and y in the derivation of the equation of the hyperbola. This leads to a hyperbola with a vertical transverse axis.

■ Equations and Graphs of Hyperbolas

The main properties of hyperbolas are listed in the following box.

HYPERBOLA WITH CENTER AT THE ORIGIN

The graph of each of the following equations is a hyperbola with center at the origin and having the given properties.

EQUATION	$\dfrac{x^2}{a^2} - \dfrac{y^2}{b^2} = 1 \quad (a > 0, b > 0)$	$\dfrac{y^2}{a^2} - \dfrac{x^2}{b^2} = 1 \quad (a > 0, b > 0)$
VERTICES	$(\pm a, 0)$	$(0, \pm a)$
TRANSVERSE AXIS	Horizontal, length $2a$	Vertical, length $2a$
ASYMPTOTES	$y = \pm \dfrac{b}{a}x$	$y = \pm \dfrac{a}{b}x$
FOCI	$(\pm c, 0), \quad c^2 = a^2 + b^2$	$(0, \pm c), \quad c^2 = a^2 + b^2$
GRAPH	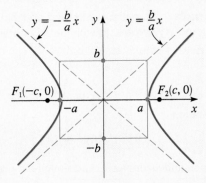	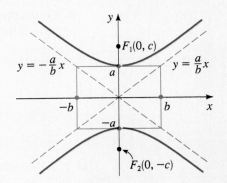

Asymptotes of rational functions are discussed in Section 4.5.

The *asymptotes* mentioned in this box are lines that the hyperbola approaches for large values of x and y. To find the asymptotes in the first case in the box, we solve the equation for y to get

$$y = \pm\frac{b}{a}\sqrt{x^2 - a^2}$$

$$= \pm\frac{b}{a}x\sqrt{1 - \frac{a^2}{x^2}}$$

As x gets large, a^2/x^2 gets closer to zero. In other words, as $x \to \infty$, we have $a^2/x^2 \to 0$. So for large x the value of y can be approximated as $y = \pm(b/a)x$. This shows that these lines are asymptotes of the hyperbola.

Asymptotes are an essential aid for graphing a hyperbola; they help us to determine its shape. A convenient way to find the asymptotes, for a hyperbola with horizontal transverse axis, is to first plot the points $(a, 0)$, $(-a, 0)$, $(0, b)$, and $(0, -b)$. Then sketch horizontal and vertical segments through these points to construct a rectangle, as shown in Figure 2(a) on the next page. We call this rectangle the **central box** of the hyperbola. The slopes of the diagonals of the central box are $\pm b/a$, so by extending them, we obtain the asymptotes $y = \pm(b/a)x$, as sketched in part (b) of the figure. Finally, we plot the vertices and use the asymptotes as a guide in sketching the hyperbola shown in part (c). (A similar procedure applies to graphing a hyperbola that has a vertical transverse axis.)

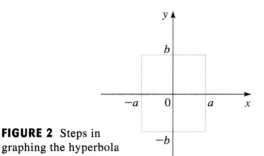

FIGURE 2 Steps in graphing the hyperbola $\dfrac{x^2}{a^2} - \dfrac{y^2}{b^2} = 1$

(a) Central box

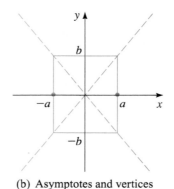

(b) Asymptotes and vertices

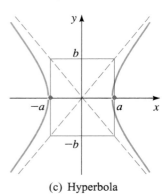

(c) Hyperbola

HOW TO SKETCH A HYPERBOLA

1. **Sketch the Central Box.** This is the rectangle centered at the origin, with sides parallel to the axes, that crosses one axis at $\pm a$, the other at $\pm b$.

2. **Sketch the Asymptotes.** These are the lines obtained by extending the diagonals of the central box.

3. **Plot the Vertices.** These are the two x-intercepts or the two y-intercepts.

4. **Sketch the Hyperbola.** Start at a vertex, and sketch a branch of the hyperbola, approaching the asymptotes. Sketch the other branch in the same way.

▶ **EXAMPLE 1** | A Hyperbola with Horizontal Transverse Axis

A hyperbola has the equation

$$9x^2 - 16y^2 = 144$$

(a) Find the vertices, foci, and asymptotes, and sketch the graph.

(b) Draw the graph using a graphing calculator.

▼ **SOLUTION**

(a) First we divide both sides of the equation by 144 to put it into standard form:

$$\frac{x^2}{16} - \frac{y^2}{9} = 1$$

Because the x^2-term is positive, the hyperbola has a horizontal transverse axis; its vertices and foci are on the x-axis. Since $a^2 = 16$ and $b^2 = 9$, we get $a = 4$, $b = 3$, and $c = \sqrt{16 + 9} = 5$. Thus, we have

VERTICES	$(\pm 4, 0)$
FOCI	$(\pm 5, 0)$
ASYMPTOTES	$y = \pm\frac{3}{4}x$

After sketching the central box and asymptotes, we complete the sketch of the hyperbola as in Figure 3(a).

Note that the equation of a hyperbola does not define y as a function of x (see page 148). That's why we need to graph two functions to graph a hyperbola.

(b) To draw the graph using a graphing calculator, we need to solve for y.

$$9x^2 - 16y^2 = 144$$

$$-16y^2 = -9x^2 + 144 \qquad \text{Subtract } 9x^2$$

$$y^2 = 9\left(\frac{x^2}{16} - 1\right) \qquad \text{Divide by } -16 \text{ and factor } 9$$

$$y = \pm 3\sqrt{\frac{x^2}{16} - 1} \qquad \text{Take square roots}$$

To obtain the graph of the hyperbola, we graph the functions

$$y = 3\sqrt{(x^2/16) - 1} \qquad \text{and} \qquad y = -3\sqrt{(x^2/16) - 1}$$

as shown in Figure 3(b).

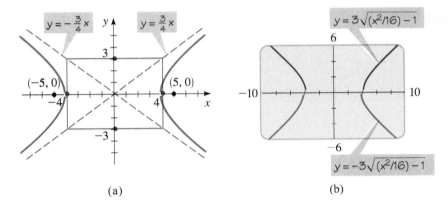

FIGURE 3
$9x^2 - 16y^2 = 144$

(a)

(b)

▶ **EXAMPLE 2** | A Hyperbola with Vertical Transverse Axis

Find the vertices, foci, and asymptotes of the hyperbola, and sketch its graph.

$$x^2 - 9y^2 + 9 = 0$$

▼ **SOLUTION** We begin by writing the equation in the standard form for a hyperbola.

$$x^2 - 9y^2 = -9$$

$$y^2 - \frac{x^2}{9} = 1 \qquad \text{Divide by } -9$$

Because the y^2-term is positive, the hyperbola has a vertical transverse axis; its foci and vertices are on the y-axis. Since $a^2 = 1$ and $b^2 = 9$, we get $a = 1$, $b = 3$, and $c = \sqrt{1 + 9} = \sqrt{10}$. Thus, we have

VERTICES	$(0, \pm 1)$
FOCI	$(0, \pm\sqrt{10})$
ASYMPTOTES	$y = \pm\frac{1}{3}x$

We sketch the central box and asymptotes, then complete the graph, as shown in Figure 4(a).

We can also draw the graph using a graphing calculator, as shown in Figure 4(b).

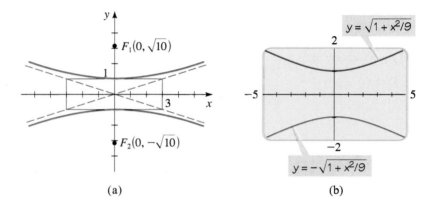

FIGURE 4
$x^2 - 9y^2 + 9 = 0$

(a)　　　　　　　　　　　(b)

▶ **EXAMPLE 3** | Finding the Equation of a Hyperbola from Its Vertices and Foci

Find the equation of the hyperbola with vertices $(\pm 3, 0)$ and foci $(\pm 4, 0)$. Sketch the graph.

▼ **SOLUTION** Since the vertices are on the x-axis, the hyperbola has a horizontal transverse axis. Its equation is of the form

$$\frac{x^2}{3^2} - \frac{y^2}{b^2} = 1$$

We have $a = 3$ and $c = 4$. To find b, we use the relation $a^2 + b^2 = c^2$:

$$3^2 + b^2 = 4^2$$
$$b^2 = 4^2 - 3^2 = 7$$
$$b = \sqrt{7}$$

Thus, the equation of the hyperbola is

$$\frac{x^2}{9} - \frac{y^2}{7} = 1$$

The graph is shown in Figure 5.

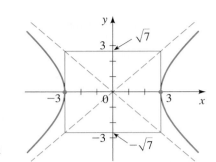

FIGURE 5
$$\frac{x^2}{9} - \frac{y^2}{7} = 1$$

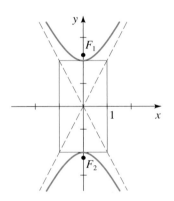

FIGURE 6
$$\frac{y^2}{4} - x^2 = 1$$

EXAMPLE 4 | Finding the Equation of a Hyperbola from Its Vertices and Asymptotes

Find the equation and the foci of the hyperbola with vertices $(0, \pm 2)$ and asymptotes $y = \pm 2x$. Sketch the graph.

▼ **SOLUTION**   Since the vertices are on the y-axis, the hyperbola has a vertical transverse axis with $a = 2$. From the asymptote equation we see that $a/b = 2$. Since $a = 2$, we get $2/b = 2$, and so $b = 1$. Thus, the equation of the hyperbola is

$$\frac{y^2}{4} - x^2 = 1$$

To find the foci, we calculate $c^2 = a^2 + b^2 = 2^2 + 1^2 = 5$, so $c = \sqrt{5}$. Thus, the foci are $(0, \pm\sqrt{5})$. The graph is shown in Figure 6.

Like parabolas and ellipses, hyperbolas have an interesting *reflection property*. Light aimed at one focus of a hyperbolic mirror is reflected toward the other focus, as shown in Figure 7. This property is used in the construction of Cassegrain-type telescopes. A hyperbolic mirror is placed in the telescope tube so that light reflected from the primary parabolic reflector is aimed at one focus of the hyperbolic mirror. The light is then refocused at a more accessible point below the primary reflector (Figure 8).

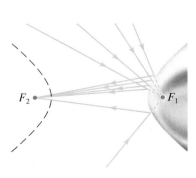

FIGURE 7 Reflection property of hyperbolas

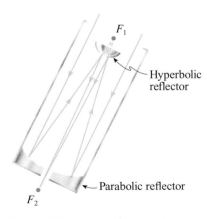

FIGURE 8 Cassegrain-type telescope

The LORAN (LOng RAnge Navigation) system was used until the early 1990s; it has now been superseded by the GPS system. In the LORAN system, hyperbolas are used onboard a ship to determine its location. In Figure 9 radio stations at A and B transmit signals simultaneously for reception by the ship at P. The onboard computer converts the time difference in reception of these signals into a distance difference $d(P, A) - d(P, B)$. From the definition of a hyperbola this locates the ship on one branch of a hyperbola with foci at A and B (sketched in black in the figure). The same procedure is carried out with two other radio stations at C and D, and this locates the ship on a second hyperbola (shown in red in the figure). (In practice, only three stations are needed because one station can be used as a focus for both hyperbolas.) The coordinates of the intersection point of these two hyperbolas, which can be calculated precisely by the computer, give the location of P.

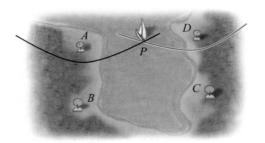

FIGURE 9 LORAN system for finding the location of a ship

8.4 | Shifted Conics

LEARNING OBJECTIVES

After completing this section, you will be able to:

▪ Find geometric properties of a shifted conic from its equation
▪ Find the equation of a shifted conic from some of its geometric properties

In the preceding sections we studied parabolas with vertices at the origin and ellipses and hyperbolas with centers at the origin. We restricted ourselves to these cases because these equations have the simplest form. In this section we consider conics whose vertices and centers are not necessarily at the origin, and we determine how this affects their equations.

▪ Shifting Graphs of Equations

In Section 3.5 we studied transformations of functions that have the effect of shifting their graphs. In general, for any equation in x and y, if we replace x by $x - h$ or by $x + h$, the graph of the new equation is simply the old graph shifted horizontally; if y is replaced by $y - k$ or by $y + k$, the graph is shifted vertically. The following box gives the details.

> ### SHIFTING GRAPHS OF EQUATIONS
>
> If h and k are positive real numbers, then replacing x by $x - h$ or by $x + h$ and replacing y by $y - k$ or by $y + k$ has the following effect(s) on the graph of any equation in x and y.
>
> | Replacement | How the graph is shifted |
> |---|---|
> | **1.** x replaced by $x - h$ | Right h units |
> | **2.** x replaced by $x + h$ | Left h units |
> | **3.** y replaced by $y - k$ | Upward k units |
> | **4.** y replaced by $y + k$ | Downward k units |

■ Shifted Ellipses

Let's apply horizontal and vertical shifting to the ellipse with equation

$$\frac{x^2}{a^2} + \frac{y^2}{b^2} = 1$$

whose graph is shown in Figure 1. If we shift it so that its center is at the point (h, k) instead of at the origin, then its equation becomes

$$\frac{(x - h)^2}{a^2} + \frac{(y - k)^2}{b^2} = 1$$

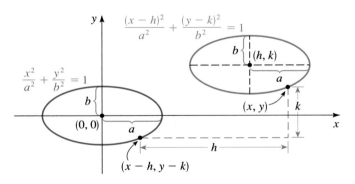

FIGURE 1 Shifted ellipse

▶ **EXAMPLE 1** | Sketching the Graph of a Shifted Ellipse

Sketch the graph of the ellipse

$$\frac{(x + 1)^2}{4} + \frac{(y - 2)^2}{9} = 1$$

and determine the coordinates of the foci.

▼ **SOLUTION**   The ellipse

$$\frac{(x + 1)^2}{4} + \frac{(y - 2)^2}{9} = 1 \qquad \text{Shifted ellipse}$$

is shifted so that its center is at $(-1, 2)$. It is obtained from the ellipse

$$\frac{x^2}{4} + \frac{y^2}{9} = 1 \qquad \text{Ellipse with center at origin}$$

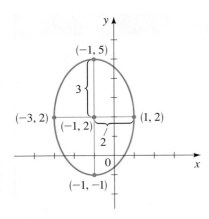

FIGURE 2

$$\frac{(x+1)^2}{4} + \frac{(y-2)^2}{9} = 1$$

by shifting it left 1 unit and upward 2 units. The endpoints of the minor and major axes of the unshifted ellipse are $(2, 0)$, $(-2, 0)$, $(0, 3)$, $(0, -3)$. We apply the required shifts to these points to obtain the corresponding points on the shifted ellipse:

$$(2, 0) \;\rightarrow\; (2 - 1, 0 + 2) = (1, 2)$$

$$(-2, 0) \;\rightarrow\; (-2 - 1, 0 + 2) = (-3, 2)$$

$$(0, 3) \;\rightarrow\; (0 - 1, 3 + 2) = (-1, 5)$$

$$(0, -3) \;\rightarrow\; (0 - 1, -3 + 2) = (-1, -1)$$

This helps us sketch the graph in Figure 2.

To find the foci of the shifted ellipse, we first find the foci of the ellipse with center at the origin. Since $a^2 = 9$ and $b^2 = 4$, we have $c^2 = 9 - 4 = 5$, so $c = \sqrt{5}$. So the foci are $(0, \pm\sqrt{5})$. Shifting left 1 unit and upward 2 units, we get

$$\left(0, \sqrt{5}\right) \;\rightarrow\; \left(0 - 1, \sqrt{5} + 2\right) = \left(-1, 2 + \sqrt{5}\right)$$

$$\left(0, -\sqrt{5}\right) \;\rightarrow\; \left(0 - 1, -\sqrt{5} + 2\right) = \left(-1, 2 - \sqrt{5}\right)$$

Thus, the foci of the shifted ellipse are

$$\left(-1, 2 + \sqrt{5}\right) \quad \text{and} \quad \left(-1, 2 - \sqrt{5}\right) \qquad \blacktriangle$$

■ Shifted Parabolas

Applying shifts to parabolas leads to the equations and graphs shown in Figure 3.

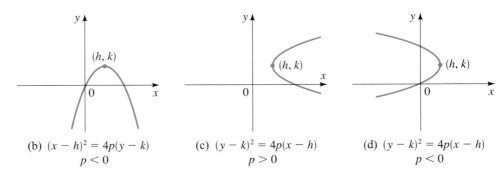

(a) $(x - h)^2 = 4p(y - k)$
$p > 0$

(b) $(x - h)^2 = 4p(y - k)$
$p < 0$

(c) $(y - k)^2 = 4p(x - h)$
$p > 0$

(d) $(y - k)^2 = 4p(x - h)$
$p < 0$

FIGURE 3 Shifted parabolas

▶ **EXAMPLE 2** | Graphing a Shifted Parabola

Determine the vertex, focus, and directrix and sketch the graph of the parabola.

$$x^2 - 4x = 8y - 28$$

▼ **SOLUTION**   We complete the square in x to put this equation into one of the forms in Figure 3.

$$x^2 - 4x + 4 = 8y - 28 + 4 \qquad \textit{Add 4 to complete the square}$$

$$(x - 2)^2 = 8y - 24$$

$$(x - 2)^2 = 8(y - 3) \qquad \textit{Shifted parabola}$$

This parabola opens upward with vertex at $(2, 3)$. It is obtained from the parabola

$$x^2 = 8y \qquad \textit{Parabola with vertex at origin}$$

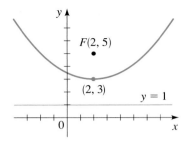

FIGURE 4
$x^2 - 4x = 8y - 28$

by shifting right 2 units and upward 3 units. Since $4p = 8$, we have $p = 2$, so the focus is 2 units above the vertex and the directrix is 2 units below the vertex. Thus, the focus is $(2, 5)$ and the directrix is $y = 1$. The graph is shown in Figure 4.   ▲

Shifted Hyperbolas

Applying shifts to hyperbolas leads to the equations and graphs shown in Figure 5.

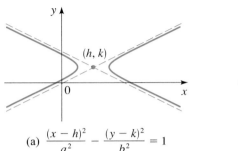

(a) $\dfrac{(x-h)^2}{a^2} - \dfrac{(y-k)^2}{b^2} = 1$

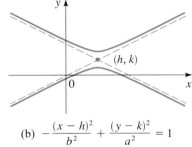

(b) $-\dfrac{(x-h)^2}{b^2} + \dfrac{(y-k)^2}{a^2} = 1$

FIGURE 5 Shifted hyperbolas

▶ **EXAMPLE 3** | Graphing a Shifted Hyperbola

A shifted conic has the equation

$$9x^2 - 72x - 16y^2 - 32y = 16$$

(a) Complete the square in x and y to show that the equation represents a hyperbola.

(b) Find the center, vertices, foci, and asymptotes of the hyperbola, and sketch its graph.

(c) Draw the graph using a graphing calculator.

▼ **SOLUTION**

(a) We complete the squares in both x and y:

$$9(x^2 - 8x \quad\quad) - 16(y^2 + 2y \quad\quad) = 16 \qquad \text{Group terms and factor}$$

$$9(x^2 - 8x + 16) - 16(y^2 + 2y + 1) = 16 + 9 \cdot 16 - 16 \cdot 1 \qquad \text{Complete the squares}$$

$$9(x - 4)^2 - 16(y + 1)^2 = 144 \qquad \text{Divide this by 144}$$

$$\frac{(x-4)^2}{16} - \frac{(y+1)^2}{9} = 1 \qquad \text{Shifted hyperbola}$$

Comparing this to Figure 5(a), we see that this is the equation of a shifted hyperbola.

(b) The shifted hyperbola has center $(4, -1)$ and a horizontal transverse axis.

$$\textbf{CENTER} \quad (4, -1)$$

Its graph will have the same shape as the unshifted hyperbola

$$\frac{x^2}{16} - \frac{y^2}{9} = 1 \qquad \text{Hyperbola with center at origin}$$

Since $a^2 = 16$ and $b^2 = 9$, we have $a = 4$, $b = 3$, and $c = \sqrt{a^2 + b^2} = \sqrt{16 + 9} = 5$. Thus, the foci lie 5 units to the left and to the right of the center, and the vertices lie 4 units to either side of the center.

$$\textbf{FOCI} \qquad (-1, -1) \quad \text{and} \quad (9, -1)$$

$$\textbf{VERTICES} \quad (0, -1) \quad \text{and} \quad (8, -1)$$

The asymptotes of the unshifted hyperbola are $y = \pm\frac{3}{4}x$, so the asymptotes of the shifted hyperbola are found as follows.

$$\textbf{ASYMPTOTES} \quad y + 1 = \pm\tfrac{3}{4}(x - 4)$$
$$y + 1 = \pm\tfrac{3}{4}x \mp 3$$
$$y = \tfrac{3}{4}x - 4 \quad \text{and} \quad y = -\tfrac{3}{4}x + 2$$

To help us sketch the hyperbola, we draw the central box; it extends 4 units left and right from the center and 3 units upward and downward from the center. We then draw the asymptotes and complete the graph of the shifted hyperbola as shown in Figure 6(a).

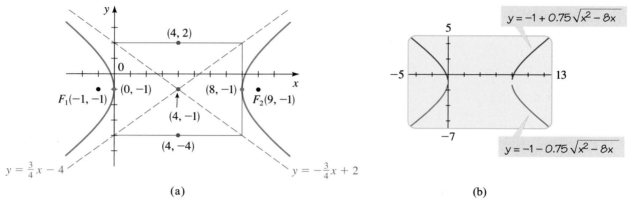

(a)   (b)

FIGURE 6 $9x^2 - 72x - 16y^2 - 32y = 16$

(c) To draw the graph using a graphing calculator, we need to solve for y. The given equation is a quadratic equation in y, so we use the quadratic formula to solve for y. Writing the equation in the form

$$16y^2 + 32y - 9x^2 + 72x + 16 = 0$$

we get

$$y = \frac{-32 \pm \sqrt{32^2 - 4(16)(-9x^2 + 72x + 16)}}{2(16)} \qquad \text{Quadratic Formula}$$

$$= \frac{-32 \pm \sqrt{576x^2 - 4608x}}{32} \qquad \text{Expand}$$

$$= \frac{-32 \pm 24\sqrt{x^2 - 8x}}{32} \qquad \begin{array}{l}\text{Factor 576 from}\\\text{under the radical}\end{array}$$

$$= -1 \pm \tfrac{3}{4}\sqrt{x^2 - 8x} \qquad \text{Simplify}$$

To obtain the graph of the hyperbola, we graph the functions

$$y = -1 + 0.75\sqrt{x^2 - 8x} \quad \text{and} \quad y = -1 - 0.75\sqrt{x^2 - 8x}$$

as shown in Figure 6(b).

■ The General Equation of a Shifted Conic

If we expand and simplify the equations of any of the shifted conics illustrated in Figures 1, 3, and 5, then we will always obtain an equation of the form

$$Ax^2 + Cy^2 + Dx + Ey + F = 0$$

where A and C are not both 0. Conversely, if we begin with an equation of this form, then we can complete the square in x and y to see which type of conic section the equation represents. In some cases the graph of the equation turns out to be just a pair of lines, a single point, or there may be no graph at all. These cases are called **degenerate conics**. If the equation is not degenerate, then we can tell whether it represents a parabola, an ellipse, or a hyperbola simply by examining the signs of A and C, as described in the following box.

GENERAL EQUATION OF A SHIFTED CONIC

The graph of the equation

$$Ax^2 + Cy^2 + Dx + Ey + F = 0$$

where A and C are not both 0, is a conic or a degenerate conic. In the nondegenerate cases the graph is

1. a parabola if A or C is 0,
2. an ellipse if A and C have the same sign (or a circle if $A = C$),
3. a hyperbola if A and C have opposite signs.

▶ **EXAMPLE 4** | An Equation That Leads to a Degenerate Conic

Sketch the graph of the equation

$$9x^2 - y^2 + 18x + 6y = 0$$

▼ **SOLUTION**   Because the coefficients of x^2 and y^2 are of opposite sign, this equation looks as if it should represent a hyperbola (like the equation of Example 3). To see whether this is in fact the case, we complete the squares:

$$9(x^2 + 2x \qquad) - (y^2 - 6y \qquad) = 0 \qquad \text{Group terms and factor 9}$$

$$9(x^2 + 2x + 1) - (y^2 - 6y + 9) = 0 + 9 \cdot 1 - 9 \qquad \text{Complete the squares}$$

$$9(x + 1)^2 - (y - 3)^2 = 0 \qquad \text{Factor}$$

$$(x + 1)^2 - \frac{(y - 3)^2}{9} = 0 \qquad \text{Divide by 9}$$

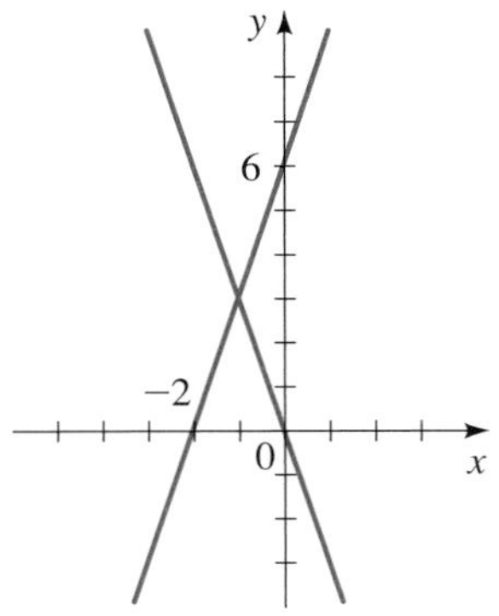

For this to fit the form of the equation of a hyperbola, we would need a nonzero constant to the right of the equal sign. In fact, further analysis shows that this is the equation of a pair of intersecting lines:

$$(y - 3)^2 = 9(x + 1)^2$$

$$y - 3 = \pm 3(x + 1) \qquad \text{Take square roots}$$

$$y = 3(x + 1) + 3 \qquad \text{or} \qquad y = -3(x + 1) + 3$$

$$y = 3x + 6 \qquad\qquad\qquad y = -3x$$

These lines are graphed in Figure 7.   ▲

FIGURE 7
$9x^2 - y^2 + 18x + 6y = 0$

Because the equation in Example 4 looked at first glance like the equation of a hyperbola but, in fact, turned out to represent simply a pair of lines, we refer to its graph as a **degenerate hyperbola**. Degenerate ellipses and parabolas can also arise when we complete the square(s) in an equation that seems to represent a conic. For example, the equation

$$4x^2 + y^2 - 8x + 2y + 6 = 0$$

looks as if it should represent an ellipse, because the coefficients of x^2 and y^2 have the same sign. But completing the squares leads to

$$(x - 1)^2 + \frac{(y + 1)^2}{4} = -\frac{1}{4}$$

which has no solution at all (since the sum of two squares cannot be negative). This equation is therefore degenerate.

▶ CHAPTER 8 | REVIEW

▼ PROPERTIES AND FORMULAS

Geometric Definition of a Parabola (p. 364)

A **parabola** is the set of points in the plane that are equidistant from a fixed point F (the **focus**) and a fixed line l (the **directrix**).

Graphs of Parabolas with Vertex at the Origin (p. 367)

A parabola with vertex at the origin has an equation of the form $x^2 = 4py$ if its axis is vertical and an equation of the form $y^2 = 4px$ if its axis is horizontal.

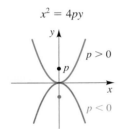

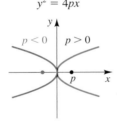

Focus $(0, p)$, directrix $y = -p$ Focus $(p, 0)$, directrix $x = -p$

Geometric Definition of an Ellipse (p. 370)

An **ellipse** is the set of all the points in the plane for which the sum of the distances to each of two given points F_1 and F_2 (the **foci**) is a fixed constant.

Graphs of Ellipses with Center at the Origin (p. 372)

An ellipse with center at the origin has an equation of the form $\frac{x^2}{a^2} + \frac{y^2}{b^2} = 1$ if its axis is horizontal and an equation of the form $\frac{x^2}{b^2} + \frac{y^2}{a^2} = 1$ if its axis is vertical (where in each case $a > b > 0$).

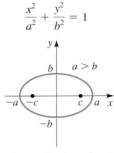

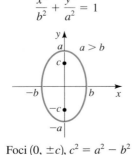

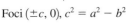

Foci $(\pm c, 0)$, $c^2 = a^2 - b^2$ Foci $(0, \pm c)$, $c^2 = a^2 - b^2$

Eccentricity of an Ellipse (p. 375)

The **eccentricity** of an ellipse with equation $\frac{x^2}{a^2} + \frac{y^2}{b^2} = 1$ or $\frac{x^2}{b^2} + \frac{y^2}{a^2} = 1$ (where $a > b > 0$) is the number

$$e = \frac{c}{a}$$

where $c = \sqrt{a^2 - b^2}$. The eccentricity e of any ellipse is a number between 0 and 1. If e is close to 0, then the ellipse is nearly circular; the closer e gets to 1, the more elongated it becomes.

Geometric Definition of a Hyperbola (p. 376)

A **hyperbola** is the set of all those points in the plane for which the absolute value of the difference of the distances to each of two given points F_1 and F_2 (the **foci**) is a fixed constant.

Graphs of Hyperbolas with Center at the Origin (p. 377)

A **hyperbola** with center at the origin has an equation of the form $\frac{x^2}{a^2} - \frac{y^2}{b^2} = 1$ if its axis is horizontal and an equation of the form $-\frac{x^2}{b^2} + \frac{y^2}{a^2} = 1$ if its axis is vertical.

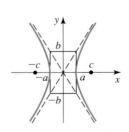

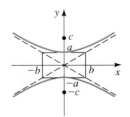

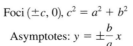

Foci $(\pm c, 0)$, $c^2 = a^2 + b^2$ Foci $(0, \pm c)$, $c^2 = a^2 + b^2$

Asymptotes: $y = \pm\frac{b}{a}x$ Asymptotes: $y = \pm\frac{a}{b}x$

Shifted Conics (pp. 382–386)

If the vertex of a parabola or the center of an ellipse or a hyperbola does not lie at the origin but rather at the point (h, k), then we refer to the curve as a **shifted conic**. To find the equation of the shifted conic, we use the "unshifted" form for the appropriate curve and replace x by $x-h$ and y by $y-k$.

General Equation of a Shifted Conic (p. 387)

The graph of the equation

$$Ax^2 + Cy^2 + Dx + Ey + F = 0$$

(where A and C are not both 0) is either a conic or a degenerate conic. In the nondegenerate cases the graph is:

1. A **parabola** if $A = 0$ or $C = 0$.

2. An **ellipse** if A and C have the same sign (or a circle if $A = C$).

3. A **hyperbola** if A and C have opposite sign.

To graph a conic whose equation is given in general form, **complete the square** in x and y to put the equation in standard form for a parabola, an ellipse, or a hyperbola.

▼ CONCEPT SUMMARY

	Review Exercises
Section 8.1	
▪ Find geometric properties of a parabola from its equation	1–4, 56
▪ Find the equation of a parabola from some of its geometric properties	25, 43–44
Section 8.2	Review Exercises
▪ Find geometric properties of an ellipse from its equation	9–12, 55
▪ Find the equation of an ellipse from some of its geometric properties	26, 45
Section 8.3	Review Exercises
▪ Find geometric properties of a hyperbola from its equation	17–20
▪ Find the equation of a hyperbola from some of its geometric properties	27, 47
Section 8.4	Review Exercises
▪ Find geometric properties of a shifted conic from its equation	5–8, 13–16, 21–24, 31–42
▪ Find the equation of a shifted conic from some of its geometric properties	28–30, 47, 48–54

▼ EXERCISES

1–8 ▪ Find the vertex, focus, and directrix of the parabola, and sketch the graph.

1. $y^2 = 4x$

2. $x = \frac{1}{12}y^2$

3. $x^2 + 8y = 0$

4. $2x - y^2 = 0$

5. $x - y^2 + 4y - 2 = 0$

6. $2x^2 + 6x + 5y + 10 = 0$

7. $\frac{1}{2}x^2 + 2x = 2y + 4$

8. $x^2 = 3(x + y)$

9–16 ▪ Find the center, vertices, foci, eccentricity, and the lengths of the major and minor axes of the ellipse, and sketch the graph.

9. $\dfrac{x^2}{9} + \dfrac{y^2}{25} = 1$

10. $\dfrac{x^2}{49} + \dfrac{y^2}{9} = 1$

11. $x^2 + 4y^2 = 16$

12. $9x^2 + 4y^2 = 1$

13. $\dfrac{(x-3)^2}{9} + \dfrac{y^2}{16} = 1$

14. $\dfrac{(x-2)^2}{25} + \dfrac{(y+3)^2}{16} = 1$

15. $4x^2 + 9y^2 = 36y$

16. $2x^2 + y^2 = 2 + 4(x - y)$

17–24 ▪ Find the center, vertices, foci, and asymptotes of the hyperbola, and sketch the graph.

17. $-\dfrac{x^2}{9} + \dfrac{y^2}{16} = 1$

18. $\dfrac{x^2}{49} - \dfrac{y^2}{32} = 1$

19. $x^2 - 2y^2 = 16$

20. $x^2 - 4y^2 + 16 = 0$

21. $\dfrac{(x+4)^2}{16} - \dfrac{y^2}{16} = 1$

22. $\dfrac{(x-2)^2}{8} - \dfrac{(y+2)^2}{8} = 1$

23. $9y^2 + 18y = x^2 + 6x + 18$ **24.** $y^2 = x^2 + 6y$

25–30 ▪ Find an equation for the conic whose graph is shown.

25.

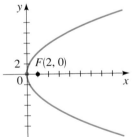

26.

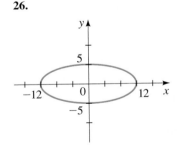

27.

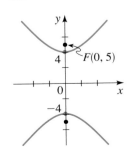

28.

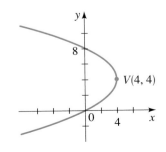

29.

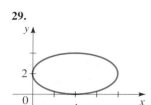

30.

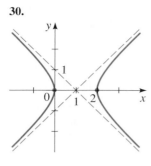

31–42 ▪ Determine the type of curve represented by the equation. Find the foci and vertices (if any), and sketch the graph.

31. $\dfrac{x^2}{12} + y = 1$

32. $\dfrac{x^2}{12} + \dfrac{y^2}{144} = \dfrac{y}{12}$

33. $x^2 - y^2 + 144 = 0$

34. $x^2 + 6x = 9y^2$

35. $4x^2 + y^2 = 8(x + y)$

36. $3x^2 - 6(x + y) = 10$

37. $x = y^2 - 16y$

38. $2x^2 + 4 = 4x + y^2$

39. $2x^2 - 12x + y^2 + 6y + 26 = 0$

40. $36x^2 - 4y^2 - 36x - 8y = 31$

41. $9x^2 + 8y^2 - 15x + 8y + 27 = 0$

42. $x^2 + 4y^2 = 4x + 8$

43–52 ▪ Find an equation for the conic section with the given properties.

43. The parabola with focus $F(0, 1)$ and directrix $y = -1$

44. The parabola with vertex at the origin and focus $F(5, 0)$

45. The ellipse with center at the origin and with x-intercepts ± 2 and y-intercepts ± 5

46. The ellipse with center $C(0, 4)$, foci $F_1(0, 0)$ and $F_2(0, 8)$, and major axis of length 10

47. The hyperbola with vertices $V(0, \pm 2)$ and asymptotes $y = \pm \frac{1}{2}x$

48. The hyperbola with center $C(2, 4)$, foci $F_1(2, 1)$ and $F_2(2, 7)$, and vertices $V_1(2, 6)$ and $V_2(2, 2)$

49. The ellipse with foci $F_1(1, 1)$ and $F_2(1, 3)$ and with one vertex on the x-axis

50. The parabola with vertex $V(5, 5)$ and directrix the y-axis

51. The ellipse with vertices $V_1(7, 12)$ and $V_2(7, -8)$ and passing through the point $P(1, 8)$

52. The parabola with vertex $V(-1, 0)$ and horizontal axis of symmetry and crossing the y-axis at $y = 2$

53. The path of the earth around the sun is an ellipse with the sun at one focus. The ellipse has major axis 186,000,000 mi and eccentricity 0.017. Find the distance between the earth and the sun when the earth is **(a)** closest to the sun and **(b)** farthest from the sun.

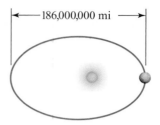

54. A ship is located 40 mi from a straight shoreline. LORAN stations A and B are located on the shoreline, 300 mi apart. From the LORAN signals, the captain determines that the ship is 80 mi closer to A than to B. Find the location of the ship. (Place A and B on the y-axis with the x-axis halfway between them. Find the x- and y-coordinates of the ship.)

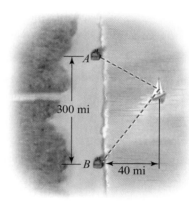

55. (a) Draw graphs of the following family of ellipses for $k = 1, 2, 4$, and 8.

$$\frac{x^2}{16 + k^2} + \frac{y^2}{k^2} = 1$$

(b) Prove that all the ellipses in part (a) have the same foci.

 56. (a) Draw graphs of the following family of parabolas for $k = \frac{1}{2}, 1, 2$, and 4.

$$y = kx^2$$

(b) Find the foci of the parabolas in part (a).
(c) How does the location of the focus change as k increases?

1. Find the focus and directrix of the parabola $x^2 = -12y$, and sketch its graph.

2. Find the vertices, foci, and the lengths of the major and minor axes for the ellipse
   $\dfrac{x^2}{16} + \dfrac{y^2}{4} = 1$. Then sketch its graph.

3. Find the vertices, foci, and asymptotes of the hyperbola $\dfrac{y^2}{9} - \dfrac{x^2}{16} = 1$. Then sketch its graph.

4–6 ■ Find an equation for the conic whose graph is shown.

4.

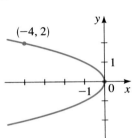

5.

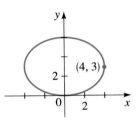

6.

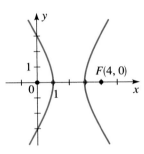

7–9 ■ Sketch the graph of the equation.

7. $16x^2 + 36y^2 - 96x + 36y + 9 = 0$

8. $9x^2 - 8y^2 + 36x + 64y = 164$

9. $2x + y^2 + 8y + 8 = 0$

10. Find an equation for the hyperbola with foci $(0, \pm 5)$ and with asymptotes $y = \pm\frac{3}{4}x$.

11. Find an equation for the parabola with focus $(2, 4)$ and directrix the x-axis.

12. Find an equation for the ellipse with foci $(3, \pm 4)$ and with x-intercepts 0 and 6.

13. A parabolic reflector for a car headlight forms a bowl shape that is 6 in. wide at its opening and 3 in. deep, as shown in the figure. How far from the vertex should the filament of the bulb be placed if it is to be located at the focus?

6 in.

◄— 3 in. —►

1. Consider the following system of equations.

$$\begin{cases} x^2 + y^2 = 4y \\ x^2 - 2y = 0 \end{cases}$$

(a) Is the system linear or nonlinear? Explain.

(b) Find all solutions of the system.

(c) The graph of each equation is a conic section. Name the type of conic section in each case.

(d) Graph both equations on the same set of axes.

(e) On your graph, shade the region that corresponds to the solution of the system of inequalities.

$$\begin{cases} x^2 + y^2 \le 4y \\ x^2 - 2y \le 0 \end{cases}$$

2. Find the complete solution of each linear system, or show that no solution exists.

(a) $\begin{cases} x + y - z = 2 \\ 2x + 3y - z = 5 \\ 3x + 5y + 2z = 11 \end{cases}$

(b) $\begin{cases} y - z = 2 \\ x + 2y - 3z = 3 \\ 3x + 5y - 8z = 7 \end{cases}$

3. Xavier, Yolanda, and Zachary go fishing. Yolanda catches as many fish as Xavier and Zachary put together. Zachary catches 2 more fish than Xavier. The total catch for all three people is 20 fish. How many did each person catch?

4. Let

$$A = \begin{bmatrix} 1 & 5 \\ 2 & 0 \end{bmatrix} \qquad B = \begin{bmatrix} -2 & 1 & 0 \\ -\frac{1}{2} & 0 & 1 \end{bmatrix}$$

$$C = \begin{bmatrix} 1 & 0 & 1 \\ 0 & 2 & 1 \\ -1 & 0 & 0 \end{bmatrix} \qquad D = \begin{bmatrix} 1 & 4 & 3 \\ 1 & 6 & 5 \\ 0 & 1 & 1 \end{bmatrix}$$

(a) Calculate each of the following, or explain why the calculation can't be done.

$$A + B, \quad C - D, \quad AB, \quad CB, \quad BD, \quad \det(B), \quad \det(C), \quad \det(D)$$

(b) Based on the values you calculated for $\det(C)$ and $\det(D)$, which matrix, C or D, has an inverse? Find the inverse of the invertible one.

5. Consider the following system of equations.

$$\begin{cases} 5x - 3y = 5 \\ 6x - 4y = 0 \end{cases}$$

(a) Write a matrix equation of the form $AX = B$ that is equivalent to this system.

(b) Find A^{-1}, the inverse of the coefficient matrix.

(c) Solve the matrix equation by multiplying each side by A^{-1}.

(d) Now solve the system using Cramer's Rule. Did you get the same solution as in part (b)?

6. Find the partial fraction decomposition of the rational function $r(x) = \dfrac{4x + 8}{x^4 + 4x^2}$.

7. Find the focus and directrix of each parabola, and sketch its graph.

(a) $x^2 + 6y = 0$

(b) $x - 2y^2 + 4y = 2$

8. Determine whether the equation represents an ellipse or a hyperbola. If it is an ellipse, find the coordinates of its vertices and foci, and sketch its graph. If it is a hyperbola, find the coordinates of its vertices and foci, find the equations of its asymptotes, and sketch its graph.

   (a) $\dfrac{x^2}{9} - y^2 = 1$

   (b) $\dfrac{x^2}{9} + y^2 = 1$

   (c) $-\dfrac{x^2}{9} + y^2 = 1$

9. Sketch the graph of each conic section, and find the coordinates of its foci. What type of conic section does each equation represent?

   (a) $9x^2 + 4y^2 = 24y$

   (b) $x^2 + 6x - y^2 + 8y = 16$

10. Find an equation for the conic whose graph is shown.

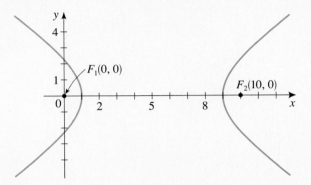

Sequences and Series

In this chapter we study sequences and series of numbers. A sequence is simply a list of numbers, and a series is the sum of a sequence. We learn how to find patterns in sequences, and how to write algebraic formulas that model them. Among other applications, we use series to calculate the value of an annuity and the amount of a mortgage payment.

The following Warm-Up Exercises will help you review the basic algebra skills that you will need for this chapter.

WARM-UP EXERCISES

1. Let $f(n) = \dfrac{(-1)^n}{n^2}$.

   (a) Compute $f(1), f(5),$ and $f(10)$.   (b) Is $f(5420)$ positive or negative?

2. Fill in the blank.

   (a) 4, 8, 12, __, 20, 24, . . .

   (b) 1, 4, 9, 16, __, 49, . . .

   (c) 1, 2, 4, 8, 16, __, . . .

3. Find the quotient.

   (a) $\dfrac{3(2^{10})}{3(2^9)}$ (b) $\dfrac{a(b^{k+1})}{a(b^k)}$

4. Factor the following expression:

   $$\left(\frac{a}{3^2} + \frac{a}{3^3} + \frac{a}{3^4} + \frac{a}{3^5} + \frac{a}{3^6} + \frac{a}{3^7} + \frac{a}{3^8} + \frac{a}{3^9} + \frac{a}{3^{10}} \right)$$

5. If a person invests \$3000 at an interest rate of 5% compounded semiannually for 2 years, how much does that person have at the end of the 2-year period?

6. Compute the following:

   (a) $(2x + 1)^2$ (b) $(2x + 1)^5$

9.1 | Sequences and Summation Notation

LEARNING OBJECTIVES
After completing this section, you will be able to:

■ Find the terms of a sequence
■ Find the terms of a recursive sequence
■ Find the partial sums of a sequence
■ Use sigma notation

In this chapter we study sequences. Roughly speaking, a sequence is an infinite list of numbers. The numbers in the sequence are often written as $a_1, a_2, a_3, \ldots$. The dots mean that the list continues forever. A simple example is the sequence

$$5, \quad 10, \quad 15, \quad 20, \quad 25, \ldots$$
$$\uparrow \quad \uparrow \quad \uparrow \quad \uparrow \quad \uparrow$$
$$a_1 \quad a_2 \quad a_3 \quad a_4 \quad a_5 \ldots$$

Sequences arise in many real-world situations. For example, if you deposit a sum of money into an interest-bearing account, the interest that is earned each month forms a sequence. If you drop a ball and let it bounce, the height the ball reaches at each successive bounce is a sequence.

We can describe the pattern of the sequence displayed above by the following *formula*:

$$a_n = 5n$$

You may have already thought of a different way to describe the pattern—namely, "you go from one number to the next by adding 5." This natural way of describing the sequence is expressed by the *recursive formula*:

$$a_n = a_{n-1} + 5$$

starting with $a_1 = 5$. Try substituting $n = 1, 2, 3, \ldots$ in each of these formulas to see how they produce the numbers in the sequence. In this section we see how these different ways are used to describe specific sequences.

■ Sequences

A *sequence* is a set of numbers written in a specific order:

$$a_1, a_2, a_3, a_4, \ldots, a_n, \ldots$$

The number a_1 is called the *first term*, a_2 is the *second term*, and in general a_n is the *nth term*. Since for every natural number n there is a corresponding number a_n, we can define a sequence as a function.

DEFINITION OF A SEQUENCE

A **sequence** is a function f whose domain is the set of natural numbers. The values $f(1), f(2), f(3), \ldots$ are called the **terms** of the sequence.

We usually write a_n instead of the function notation $f(n)$ for the value of the function at the number n.

Here is a simple example of a sequence:

$$2, 4, 6, 8, 10, \ldots$$

The dots indicate that the sequence continues indefinitely. We can write a sequence in this way when it's clear what the subsequent terms of the sequence are. This sequence consists of even numbers. To be more accurate, however, we need to specify a procedure for finding *all* the terms of the sequence. This can be done by giving a formula for the *n*th term a_n of the sequence. In this case,

Another way to write this sequence is to use function notation:

$$a(n) = 2n$$

so $a(1) = 2, a(2) = 4, a(3) = 6, \ldots$

$$a_n = 2n$$

and the sequence can be written as

2,   4,   6,   8,   . . . ,   2n,   . . .

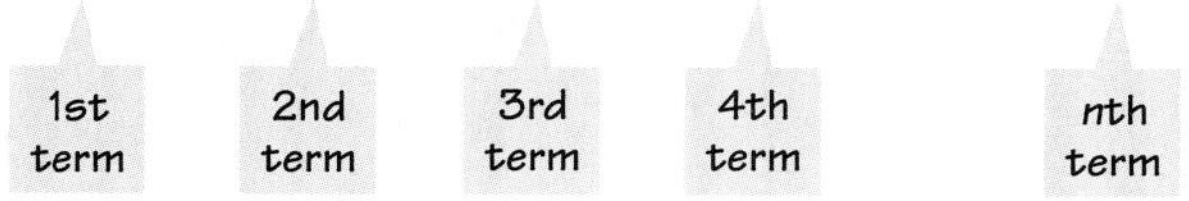

1st term   2nd term   3rd term   4th term   *n*th term

Notice how the formula $a_n = 2n$ gives all the terms of the sequence. For instance, substituting 1, 2, 3, and 4 for *n* gives the first four terms:

$$a_1 = 2 \cdot 1 = 2 \qquad a_2 = 2 \cdot 2 = 4$$
$$a_3 = 2 \cdot 3 = 6 \qquad a_4 = 2 \cdot 4 = 8$$

To find the 103rd term of this sequence, we use $n = 103$ to get

$$a_{103} = 2 \cdot 103 = 206$$

▶ **EXAMPLE 1** | Finding the Terms of a Sequence

Find the first five terms and the 100th term of the sequence defined by each formula.

(a) $a_n = 2n - 1$

(b) $c_n = n^2 - 1$

(c) $t_n = \dfrac{n}{n+1}$

(d) $r_n = \dfrac{(-1)^n}{2^n}$

▼ **SOLUTION**   To find the first five terms, we substitute $n = 1, 2, 3, 4$, and 5 in the formula for the *n*th term. To find the 100th term, we substitute $n = 100$. This gives the following.

*n*th term	First five terms	100th term
(a) $2n - 1$	1, 3, 5, 7, 9	199
(b) $n^2 - 1$	0, 3, 8, 15, 24	9999
(c) $\dfrac{n}{n+1}$	$\dfrac{1}{2}, \dfrac{2}{3}, \dfrac{3}{4}, \dfrac{4}{5}, \dfrac{5}{6}$	$\dfrac{100}{101}$
(d) $\dfrac{(-1)^n}{2^n}$	$-\dfrac{1}{2}, \dfrac{1}{4}, -\dfrac{1}{8}, \dfrac{1}{16}, -\dfrac{1}{32}$	$\dfrac{1}{2^{100}}$

In Example 1(d) the presence of $(-1)^n$ in the sequence has the effect of making successive terms alternately negative and positive.

It is often useful to picture a sequence by sketching its graph. Since a sequence is a function whose domain is the natural numbers, we can draw its graph in the Cartesian plane. For instance, the graph of the sequence

$$1, \frac{1}{2}, \frac{1}{3}, \frac{1}{4}, \frac{1}{5}, \frac{1}{6}, \ldots, \frac{1}{n}, \ldots$$

is shown in Figure 1.

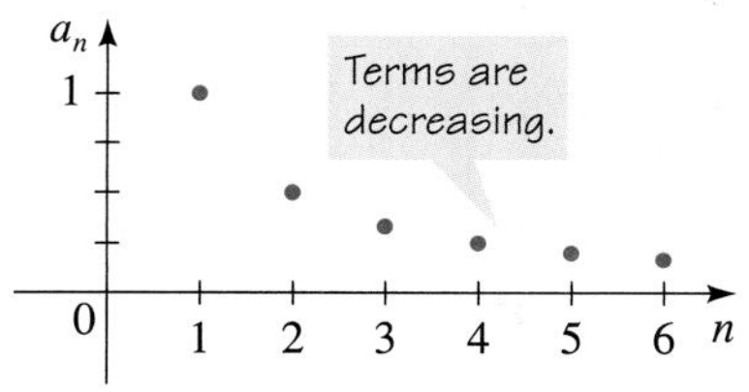

Terms are decreasing.

FIGURE 1

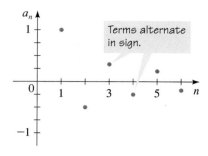

FIGURE 2

Compare the graph of the sequence shown in Figure 1 to the graph of

$$1, -\frac{1}{2}, \frac{1}{3}, -\frac{1}{4}, \frac{1}{5}, -\frac{1}{6}, \ldots, \frac{(-1)^{n+1}}{n}, \ldots$$

shown in Figure 2. The graph of every sequence consists of isolated points that are *not* connected.

Graphing calculators are useful in analyzing sequences. To work with sequences on a TI-83, we put the calculator in SEQ mode ("sequence" mode) as in Figure 3(a). If we enter the sequence $u(n) = n/(n + 1)$ of Example 1(c), we can display the terms using the TABLE command as shown in Figure 3(b). We can also graph the sequence as shown in Figure 3(c).

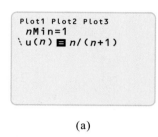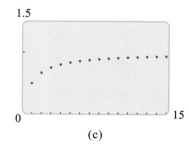

FIGURE 3
$u(n) = n/(n + 1)$

(a)   (b)   (c)

Finding patterns is an important part of mathematics. Consider a sequence that begins

$$1, 4, 9, 16, \ldots$$

Can you detect a pattern in these numbers? In other words, can you define a sequence whose first four terms are these numbers? The answer to this question seems easy; these numbers are the squares of the numbers 1, 2, 3, 4. Thus, the sequence we are looking for is defined by $a_n = n^2$. However, this is not the *only* sequence whose first four terms are 1, 4, 9, 16. In other words, the answer to our problem is not unique. In the next example we are interested in finding an *obvious* sequence whose first few terms agree with the given ones.

Not all sequences can be defined by a formula. For example, there is no known formula for the sequence of prime numbers:

$$2, 3, 5, 7, 11, 13, 17, 19, 23, \ldots$$

▶ **EXAMPLE 2** | Finding the *n*th Term of a Sequence

Find the *n*th term of a sequence whose first several terms are given.

(a) $\frac{1}{2}, \frac{3}{4}, \frac{5}{6}, \frac{7}{8}, \ldots$ **(b)** $-2, 4, -8, 16, -32, \ldots$

▼ **SOLUTION**

(a) We notice that the numerators of these fractions are the odd numbers and the denominators are the even numbers. Even numbers are of the form $2n$, and odd numbers are of the form $2n - 1$ (an odd number differs from an even number by 1). So a sequence that has these numbers for its first four terms is given by

$$a_n = \frac{2n - 1}{2n}$$

(b) These numbers are powers of 2, and they alternate in sign, so a sequence that agrees with these terms is given by

$$a_n = (-1)^n 2^n$$

You should check that these formulas do indeed generate the given terms.   ▲

■ Recursively Defined Sequences

Some sequences do not have simple defining formulas like those of the preceding example. The nth term of a sequence may depend on some or all of the terms preceding it. A sequence defined in this way is called **recursive**. Here are two examples.

▶ **EXAMPLE 3** | Finding the Terms of a Recursively Defined Sequence

Find the first five terms of the sequence defined recursively by $a_1 = 1$ and

$$a_n = 3(a_{n-1} + 2)$$

▼ **SOLUTION**   The defining formula for this sequence is recursive. It allows us to find the nth term a_n if we know the preceding term a_{n-1}. Thus, we can find the second term from the first term, the third term from the second term, the fourth term from the third term, and so on. Since we are given the first term $a_1 = 1$, we can proceed as follows.

$$a_2 = 3(a_1 + 2) = 3(1 + 2) = 9$$

$$a_3 = 3(a_2 + 2) = 3(9 + 2) = 33$$

$$a_4 = 3(a_3 + 2) = 3(33 + 2) = 105$$

$$a_5 = 3(a_4 + 2) = 3(105 + 2) = 321$$

Thus, the first five terms of this sequence are

$$1, 9, 33, 105, 321, \ldots$$

Note that to find the 20th term of the sequence in Example 3, we must first find all 19 preceding terms. This is most easily done by using a graphing calculator. Figure 4(a) shows how to enter this sequence on the TI-83 calculator. From Figure 4(b) we see that the 20th term of the sequence is $a_{20} = 4{,}649{,}045{,}865$.

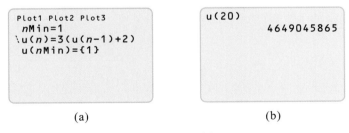

```
Plot1 Plot2 Plot3
 nMin=1
\u(n)=3(u(n-1)+2)
 u(nMin)={1}
```

(a)

```
u(20)
             4649045865
```

(b)

FIGURE 4 $u(n) = 3(u(n-1) + 2), u(1) = 1$

▶ **EXAMPLE 4** | The Fibonacci Sequence

Find the first 11 terms of the sequence defined recursively by $F_1 = 1$, $F_2 = 1$ and

$$F_n = F_{n-1} + F_{n-2}$$

▼ **SOLUTION**   To find F_n, we need to find the two preceding terms F_{n-1} and F_{n-2}. Since we are given F_1 and F_2, we proceed as follows.

$$F_3 = F_2 + F_1 = 1 + 1 = 2$$

$$F_4 = F_3 + F_2 = 2 + 1 = 3$$

$$F_5 = F_4 + F_3 = 3 + 2 = 5$$

It's clear what is happening here. Each term is simply the sum of the two terms that precede it, so we can easily write down as many terms as we please. Here are the first 11 terms:

$$1, 1, 2, 3, 5, 8, 13, 21, 34, 55, 89, \ldots$$ ▲

The sequence in Example 4 is called the **Fibonacci sequence**, named after the 13th century Italian mathematician who used it to solve a problem about the breeding of rabbits. The sequence also occurs in numerous other applications in nature. (See Figures 5 and 6.) In fact, so many phenomena behave like the Fibonacci sequence that one mathematical journal, the *Fibonacci Quarterly*, is devoted entirely to its properties.

FIGURE 5 The Fibonacci sequence in the branching of a tree

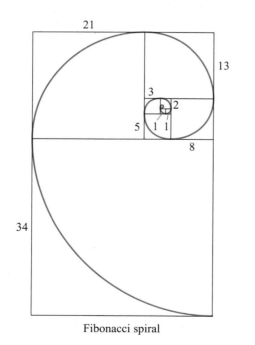

FIGURE 6 Fibonacci spiral   Nautilus shell

■ The Partial Sums of a Sequence

In calculus we are often interested in adding the terms of a sequence. This leads to the following definition on the next page.

THE PARTIAL SUMS OF A SEQUENCE

For the sequence

$$a_1, a_2, a_3, a_4, \ldots, a_n, \ldots$$

the **partial sums** are

$$S_1 = a_1$$
$$S_2 = a_1 + a_2$$
$$S_3 = a_1 + a_2 + a_3$$
$$S_4 = a_1 + a_2 + a_3 + a_4$$
$$\vdots$$
$$S_n = a_1 + a_2 + a_3 + \cdots + a_n$$
$$\vdots$$

S_1 is called the **first partial sum**, S_2 is the **second partial sum**, and so on. S_n is called the **nth partial sum**. The sequence $S_1, S_2, S_3, \ldots, S_n, \ldots$ is called the **sequence of partial sums**.

▶ **EXAMPLE 5** | Finding the Partial Sums of a Sequence

Find the first four partial sums and the nth partial sum of the sequence given by $a_n = 1/2^n$.

▼ **SOLUTION**   The terms of the sequence are

$$\frac{1}{2}, \frac{1}{4}, \frac{1}{8}, \ldots$$

The first four partial sums are

$$S_1 = \frac{1}{2} \qquad\qquad = \frac{1}{2}$$

$$S_2 = \frac{1}{2} + \frac{1}{4} \qquad\qquad = \frac{3}{4}$$

$$S_3 = \frac{1}{2} + \frac{1}{4} + \frac{1}{8} \qquad = \frac{7}{8}$$

$$S_4 = \frac{1}{2} + \frac{1}{4} + \frac{1}{8} + \frac{1}{16} = \frac{15}{16}$$

Notice that in the value of each partial sum the denominator is a power of 2 and the numerator is one less than the denominator. In general, the nth partial sum is

$$S_n = \frac{2^n - 1}{2^n} = 1 - \frac{1}{2^n}$$

The first five terms of a_n and S_n are graphed in Figure 7.

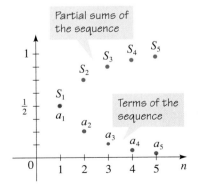

FIGURE 7 Graph of the sequence a_n and the sequence of partial sums S_n

▶ **EXAMPLE 6** | Finding the Partial Sums of a Sequence

Find the first four partial sums and the nth partial sum of the sequence given by

$$a_n = \frac{1}{n} - \frac{1}{n+1}$$

▼ **SOLUTION**   The first four partial sums are

$$S_1 = \left(1 - \frac{1}{2}\right) \qquad\qquad\qquad\qquad\qquad = 1 - \frac{1}{2}$$

$$S_2 = \left(1 - \frac{1}{2}\right) + \left(\frac{1}{2} - \frac{1}{3}\right) \qquad\qquad\qquad = 1 - \frac{1}{3}$$

$$S_3 = \left(1 - \frac{1}{2}\right) + \left(\frac{1}{2} - \frac{1}{3}\right) + \left(\frac{1}{3} - \frac{1}{4}\right) \qquad = 1 - \frac{1}{4}$$

$$S_4 = \left(1 - \frac{1}{2}\right) + \left(\frac{1}{2} - \frac{1}{3}\right) + \left(\frac{1}{3} - \frac{1}{4}\right) + \left(\frac{1}{4} - \frac{1}{5}\right) = 1 - \frac{1}{5}$$

Do you detect a pattern here? Of course. The nth partial sum is

$$S_n = 1 - \frac{1}{n+1}$$

▲

■ Sigma Notation

Given a sequence

$$a_1, a_2, a_3, a_4, \ldots$$

we can write the sum of the first n terms using **summation notation**, or **sigma notation**. This notation derives its name from the Greek letter Σ (capital sigma, corresponding to our S for "sum"). Sigma notation is used as follows:

> This tells us to end with $k = n$.

> This tells us to add. $\displaystyle\sum_{k=1}^{n} a_k$

> This tells us to start with $k = 1$.

$$\sum_{k=1}^{n} a_k = a_1 + a_2 + a_3 + a_4 + \cdots + a_n$$

The left side of this expression is read "The sum of a_k from $k = 1$ to $k = n$." The letter k is called the **index of summation**, or the **summation variable**, and the idea is to replace k in the expression after the sigma by the integers $1, 2, 3, \ldots, n$, and add the resulting expressions, arriving at the right side of the equation.

▶ **EXAMPLE 7** | Sigma Notation

Find each sum.

(a) $\displaystyle\sum_{k=1}^{5} k^2$ **(b)** $\displaystyle\sum_{j=3}^{5} \frac{1}{j}$ **(c)** $\displaystyle\sum_{i=5}^{10} i$ **(d)** $\displaystyle\sum_{i=1}^{6} 2$

▼ **SOLUTION**

(a) $\displaystyle\sum_{k=1}^{5} k^2 = 1^2 + 2^2 + 3^2 + 4^2 + 5^2 = 55$

(b) $\displaystyle\sum_{j=3}^{5} \frac{1}{j} = \frac{1}{3} + \frac{1}{4} + \frac{1}{5} = \frac{47}{60}$

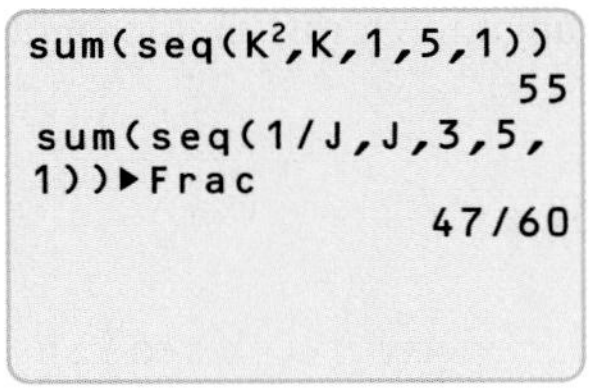

```
sum(seq(K²,K,1,5,1))
                   55
sum(seq(1/J,J,3,5,
1))►Frac
                47/60
```

FIGURE 8

(c) $\displaystyle\sum_{i=5}^{10} i = 5 + 6 + 7 + 8 + 9 + 10 = 45$

(d) $\displaystyle\sum_{i=1}^{6} 2 = 2 + 2 + 2 + 2 + 2 + 2 = 12$

We can use a graphing calculator to evaluate sums. For instance, Figure 8 shows how the TI-83 can be used to evaluate the sums in parts (a) and (b) of Example 7.   ▲

▶ **EXAMPLE 8** | Writing Sums in Sigma Notation

Write each sum using sigma notation.

(a) $1^3 + 2^3 + 3^3 + 4^3 + 5^3 + 6^3 + 7^3$

(b) $\sqrt{3} + \sqrt{4} + \sqrt{5} + \cdots + \sqrt{77}$

▼ **SOLUTION**

(a) We can write

$$1^3 + 2^3 + 3^3 + 4^3 + 5^3 + 6^3 + 7^3 = \sum_{k=1}^{7} k^3$$

(b) A natural way to write this sum is

$$\sqrt{3} + \sqrt{4} + \sqrt{5} + \cdots + \sqrt{77} = \sum_{k=3}^{77} \sqrt{k}$$

However, there is no unique way of writing a sum in sigma notation. We could also write this sum as

$$\sqrt{3} + \sqrt{4} + \sqrt{5} + \cdots + \sqrt{77} = \sum_{k=0}^{74} \sqrt{k+3}$$

or

$$\sqrt{3} + \sqrt{4} + \sqrt{5} + \cdots + \sqrt{77} = \sum_{k=1}^{75} \sqrt{k+2}$$ ▲

The following properties of sums are natural consequences of properties of the real numbers.

PROPERTIES OF SUMS

Let $a_1, a_2, a_3, a_4, \ldots$ and $b_1, b_2, b_3, b_4, \ldots$ be sequences. Then for every positive integer n and any real number c, the following properties hold.

1. $\displaystyle\sum_{k=1}^{n} (a_k + b_k) = \sum_{k=1}^{n} a_k + \sum_{k=1}^{n} b_k$

2. $\displaystyle\sum_{k=1}^{n} (a_k - b_k) = \sum_{k=1}^{n} a_k - \sum_{k=1}^{n} b_k$

3. $\displaystyle\sum_{k=1}^{n} ca_k = c\left(\sum_{k=1}^{n} a_k \right)$

▼ **PROOF**   To prove Property 1, we write out the left side of the equation to get

$$\sum_{k=1}^{n} (a_k + b_k) = (a_1 + b_1) + (a_2 + b_2) + (a_3 + b_3) + \cdots + (a_n + b_n)$$

Because addition is commutative and associative, we can rearrange the terms on the right side to read

$$\sum_{k=1}^{n} (a_k + b_k) = (a_1 + a_2 + a_3 + \cdots + a_n) + (b_1 + b_2 + b_3 + \cdots + b_n)$$

Rewriting the right side using sigma notation gives Property 1. Property 2 is proved in a similar manner. To prove Property 3, we use the Distributive Property:

$$\sum_{k=1}^{n} ca_k = ca_1 + ca_2 + ca_3 + \cdots + ca_n$$

$$= c(a_1 + a_2 + a_3 + \cdots + a_n) = c\left(\sum_{k=1}^{n} a_k\right)$$

9.2 | Arithmetic Sequences

LEARNING OBJECTIVES
After completing this section, you will be able to:

- ▪ Find the terms of an arithmetic sequence
- ▪ Find the partial sums of an arithmetic sequence

In this section we study a special type of sequence, called an arithmetic sequence.

▪ Arithmetic Sequences

Perhaps the simplest way to generate a sequence is to start with a number a and add to it a fixed constant d, over and over again.

DEFINITION OF AN ARITHMETIC SEQUENCE

An **arithmetic sequence** is a sequence of the form

$$a, a + d, a + 2d, a + 3d, a + 4d, \ldots$$

The number a is the **first term**, and d is the **common difference** of the sequence. The **nth term** of an arithmetic sequence is given by

$$a_n = a + (n - 1)d$$

The number d is called the common difference because any two consecutive terms of an arithmetic sequence differ by d.

▶ **EXAMPLE 1** | Arithmetic Sequences

(a) If $a = 2$ and $d = 3$, then we have the arithmetic sequence

$$2, 2 + 3, 2 + 6, 2 + 9, \ldots$$

or $2, 5, 8, 11, \ldots$

Any two consecutive terms of this sequence differ by $d = 3$. The nth term is $a_n = 2 + 3(n - 1)$.

(b) Consider the arithmetic sequence

$$9, 4, -1, -6, -11, \ldots$$

Here the common difference is $d = -5$. The terms of an arithmetic sequence decrease if the common difference is negative. The nth term is $a_n = 9 - 5(n - 1)$.

(c) The graph of the arithmetic sequence $a_n = 1 + 2(n - 1)$ is shown in Figure 1. Notice that the points in the graph lie on a straight line with slope $d = 2$.

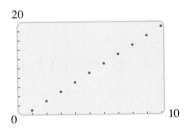

FIGURE 1

An arithmetic sequence is determined completely by the first term a and the common difference d. Thus, if we know the first two terms of an arithmetic sequence, then we can find a formula for the nth term, as the next example shows.

▶ **EXAMPLE 2** | Finding Terms of an Arithmetic Sequence

Find the first six terms and the 300th term of the arithmetic sequence

$$13, 7, \ldots$$

▼ **SOLUTION**   Since the first term is 13, we have $a = 13$. The common difference is $d = 7 - 13 = -6$. Thus, the nth term of this sequence is

$$a_n = 13 - 6(n - 1)$$

From this we find the first six terms:

$$13, 7, 1, -5, -11, -17, \ldots$$

The 300th term is $a_{300} = 13 - 6(299) = -1781$.   ▲

The next example shows that an arithmetic sequence is determined completely by *any* two of its terms.

▶ **EXAMPLE 3** | Finding Terms of an Arithmetic Sequence

The 11th term of an arithmetic sequence is 52, and the 19th term is 92. Find the 1000th term.

▼ **SOLUTION**   To find the nth term of this sequence, we need to find a and d in the formula

$$a_n = a + (n - 1)d$$

From this formula we get

$$a_{11} = a + (11 - 1)d = a + 10d$$
$$a_{19} = a + (19 - 1)d = a + 18d$$

Since $a_{11} = 52$ and $a_{19} = 92$, we get the two equations:

$$\begin{cases} 52 = a + 10d \\ 92 = a + 18d \end{cases}$$

Solving this system for a and d, we get $a = 2$ and $d = 5$. (Verify this.) Thus, the nth term of this sequence is

$$a_n = 2 + 5(n - 1)$$

The 1000th term is $a_{1000} = 2 + 5(999) = 4997$.   ▲

Partial Sums of Arithmetic Sequences

Suppose we want to find the sum of the numbers $1, 2, 3, 4, \ldots, 100$, that is,

$$\sum_{k=1}^{100} k$$

When the famous mathematician C. F. Gauss was a schoolboy, his teacher posed this problem to the class and expected that it would keep the students busy for a long time. But Gauss answered the question almost immediately. His idea was this: Since we are adding numbers produced according to a fixed pattern, there must also be a pattern (or formula) for finding the sum. He started by writing the numbers from 1 to 100 and then below them wrote the same numbers in reverse order. Writing S for the sum and adding corresponding terms gives

$$\begin{array}{rcccccccc} S = & 1 + & 2 + & 3 + \cdots + & 98 + & 99 + & 100 \\ S = & 100 + & 99 + & 98 + \cdots + & 3 + & 2 + & 1 \\ \hline 2S = & 101 + & 101 + & 101 + \cdots + & 101 + & 101 + & 101 \end{array}$$

It follows that $2S = 100(101) = 10{,}100$ and so $S = 5050$.

Of course, the sequence of natural numbers $1, 2, 3, \ldots$ is an arithmetic sequence (with $a = 1$ and $d = 1$), and the method for summing the first 100 terms of this sequence can be used to find a formula for the nth partial sum of any arithmetic sequence. We want to find the sum of the first n terms of the arithmetic sequence whose terms are $a_k = a + (k - 1)d$; that is, we want to find

$$S_n = \sum_{k=1}^{n} [a + (k - 1)d]$$

$$= a + (a + d) + (a + 2d) + (a + 3d) + \cdots + [a + (n - 1)d]$$

Using Gauss's method, we write

$$\begin{array}{rcccccc} S_n = & a & + & (a + d) & + \cdots + & [a + (n - 2)d] + [a + (n - 1)d] \\ S_n = & [a + (n - 1)d] & + & [a + (n - 2)d] & + \cdots + & (a + d) & + & a \\ \hline 2S_n = & [2a + (n - 1)d] & + & [2a + (n - 1)d] & + \cdots + & [2a + (n - 1)d] + [2a + (n - 1)d] \end{array}$$

There are n identical terms on the right side of this equation, so

$$2S_n = n[2a + (n - 1)d]$$

$$S_n = \frac{n}{2}[2a + (n - 1)d]$$

Notice that $a_n = a + (n - 1)d$ is the nth term of this sequence. So, we can write

$$S_n = \frac{n}{2}[a + a + (n - 1)d] = n\left(\frac{a + a_n}{2}\right)$$

This last formula says that the sum of the first n terms of an arithmetic sequence is the average of the first and nth terms multiplied by n, the number of terms in the sum. We now summarize this result.

PARTIAL SUMS OF AN ARITHMETIC SEQUENCE

For the arithmetic sequence $a_n = a + (n - 1)d$ the **nth partial sum**

$$S_n = a + (a + d) + (a + 2d) + (a + 3d) + \cdots + [a + (n - 1)d]$$

is given by either of the following formulas.

1. $S_n = \dfrac{n}{2}[2a + (n - 1)d]$ **2.** $S_n = n\left(\dfrac{a + a_n}{2}\right)$

▶ **EXAMPLE 4** | Finding a Partial Sum of an Arithmetic Sequence

Find the sum of the first 40 terms of the arithmetic sequence

$$3, 7, 11, 15, \ldots$$

▼ **SOLUTION** For this arithmetic sequence, $a = 3$ and $d = 4$. Using Formula 1 for the partial sum of an arithmetic sequence, we get

$$S_{40} = \tfrac{40}{2}[2(3) + (40 - 1)4] = 20(6 + 156) = 3240$$

▲

▶ **EXAMPLE 5** | Finding a Partial Sum of an Arithmetic Sequence

Find the sum of the first 50 odd numbers.

▼ **SOLUTION** The odd numbers form an arithmetic sequence with $a = 1$ and $d = 2$. The nth term is $a_n = 1 + 2(n - 1) = 2n - 1$, so the 50th odd number is $a_{50} = 2(50) - 1 = 99$. Substituting in Formula 2 for the partial sum of an arithmetic sequence, we get

$$S_{50} = 50\left(\frac{a + a_{50}}{2}\right) = 50\left(\frac{1 + 99}{2}\right) = 50 \cdot 50 = 2500$$

▲

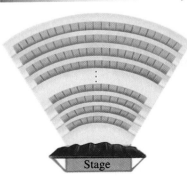

▶ **EXAMPLE 6** | Finding the Seating Capacity of an Amphitheater

An amphitheater has 50 rows of seats with 30 seats in the first row, 32 in the second, 34 in the third, and so on. Find the total number of seats.

▼ **SOLUTION** The numbers of seats in the rows form an arithmetic sequence with $a = 30$ and $d = 2$. Since there are 50 rows, the total number of seats is the sum

$$S_{50} = \tfrac{50}{2}[2(30) + 49(2)] \qquad S_n = \frac{n}{2}[2a + (n - 1)d]$$

$$= 3950$$

Thus, the amphitheater has 3950 seats.

▲

▶ **EXAMPLE 7** | Finding the Number of Terms in a Partial Sum

How many terms of the arithmetic sequences 5, 7, 9, . . . must be added to get 572?

▼ **SOLUTION** We are asked to find n when $S_n = 572$. Substituting $a = 5$, $d = 2$, and $S_n = 572$ in Formula 1 for the partial sum of an arithmetic sequence, we get

$$572 = \frac{n}{2}[2 \cdot 5 + (n - 1)2] \qquad S_n = \frac{n}{2}[2a + (n-1)d]$$

$$572 = 5n + n(n - 1)$$

$$0 = n^2 + 4n - 572$$

$$0 = (n - 22)(n + 26)$$

This gives $n = 22$ or $n = -26$. But since n is the *number* of terms in this partial sum, we must have $n = 22$. ▲

9.3 | Geometric Sequences

LEARNING OBJECTIVES

After completing this section, you will be able to:

▣ Find the terms of a geometric sequence
▣ Find the partial sums of a geometric sequence
▣ Find the sum of an infinite geometric series

In this section we study geometric sequences. This type of sequence occurs frequently in applications to finance, population growth, and other fields.

▣ Geometric Sequences

Recall that an arithmetic sequence is generated when we repeatedly add a number d to an initial term a. A *geometric* sequence is generated when we start with a number a and repeatedly *multiply* by a fixed nonzero constant r.

DEFINITION OF A GEOMETRIC SEQUENCE

A **geometric sequence** is a sequence of the form

$$a, ar, ar^2, ar^3, ar^4, \ldots$$

The number a is the **first term**, and r is the **common ratio** of the sequence. The **nth term** of a geometric sequence is given by

$$a_n = ar^{n-1}$$

The number r is called the common ratio because the ratio of any two consecutive terms of the sequence is r.

▶ **EXAMPLE 1** | Geometric Sequences

(a) If $a = 3$ and $r = 2$, then we have the geometric sequence

$$3, \quad 3 \cdot 2, \quad 3 \cdot 2^2, \quad 3 \cdot 2^3, \quad 3 \cdot 2^4, \quad \ldots$$

or $3, 6, 12, 24, 48, \ldots$

Notice that the ratio of any two consecutive terms is $r = 2$. The nth term is $a_n = 3(2)^{n-1}$.

(b) The sequence

$$2, -10, 50, -250, 1250, \ldots$$

is a geometric sequence with $a = 2$ and $r = -5$. When r is negative, the terms of the sequence alternate in sign. The nth term is $a_n = 2(-5)^{n-1}$.

(c) The sequence

$$1, \frac{1}{3}, \frac{1}{9}, \frac{1}{27}, \frac{1}{81}, \ldots$$

is a geometric sequence with $a = 1$ and $r = \frac{1}{3}$. The nth term is $a_n = 1\left(\frac{1}{3}\right)^{n-1}$.

(d) The graph of the geometric sequence $a_n = \frac{1}{5} \cdot 2^{n-1}$ is shown in Figure 1. Notice that the points in the graph lie on the graph of the exponential function $y = \frac{1}{5} \cdot 2^{x-1}$.

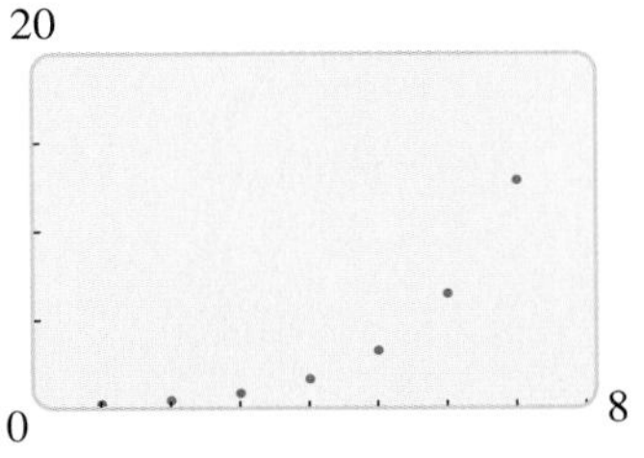

If $0 < r < 1$, then the terms of the geometric sequence ar^{n-1} decrease, but if $r > 1$, then the terms increase. (What happens if $r = 1$?) ▲

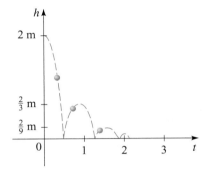

FIGURE 1

Geometric sequences occur naturally. Here is a simple example. Suppose a ball has elasticity such that when it is dropped, it bounces up one-third of the distance it has fallen. If this ball is dropped from a height of 2 m, then it bounces up to a height of $2\left(\frac{1}{3}\right) = \frac{2}{3}$ m. On its second bounce, it returns to a height of $\left(\frac{2}{3}\right)\left(\frac{1}{3}\right) = \frac{2}{9}$ m, and so on (see Figure 2). Thus, the height h_n that the ball reaches on its nth bounce is given by the geometric sequence

$$h_n = \frac{2}{3}\left(\frac{1}{3}\right)^{n-1} = 2\left(\frac{1}{3}\right)^n$$

We can find the nth term of a geometric sequence if we know any two terms, as the following examples show.

FIGURE 2

▶ **EXAMPLE 2** | Finding Terms of a Geometric Sequence

Find the eighth term of the geometric sequence $5, 15, 45, \ldots$.

▼ **SOLUTION**   To find a formula for the nth term of this sequence, we need to find a and r. Clearly, $a = 5$. To find r, we find the ratio of any two consecutive terms. For instance, $r = \frac{45}{15} = 3$. Thus,

$$a_n = 5(3)^{n-1}$$

The eighth term is $a_8 = 5(3)^{8-1} = 5(3)^7 = 10{,}935$. ▲

▶ **EXAMPLE 3** | Finding Terms of a Geometric Sequence

The third term of a geometric sequence is $\frac{63}{4}$, and the sixth term is $\frac{1701}{32}$. Find the fifth term.

▼ **SOLUTION**   Since this sequence is geometric, its nth term is given by the formula $a_n = ar^{n-1}$. Thus,

$$a_3 = ar^{3-1} = ar^2$$

$$a_6 = ar^{6-1} = ar^5$$

From the values we are given for these two terms, we get the following system of equations:

$$\begin{cases} \frac{63}{4} = ar^2 \\ \frac{1701}{32} = ar^5 \end{cases}$$

We solve this system by dividing.

$$\frac{ar^5}{ar^2} = \frac{\frac{1701}{32}}{\frac{63}{4}}$$

$$r^3 = \tfrac{27}{8} \qquad \text{Simplify}$$

$$r = \tfrac{3}{2} \qquad \text{Take cube root of each side}$$

Substituting for r in the first equation, $\frac{63}{4} = ar^2$, gives

$$\tfrac{63}{4} = a\left(\tfrac{3}{2}\right)^2$$

$$a = 7 \qquad \text{Solve for } a$$

It follows that the nth term of this sequence is

$$a_n = 7\left(\tfrac{3}{2}\right)^{n-1}$$

Thus, the fifth term is

$$a_5 = 7\left(\tfrac{3}{2}\right)^{5-1} = 7\left(\tfrac{3}{2}\right)^4 = \tfrac{567}{16}$$

▲

■ Partial Sums of Geometric Sequences

For the geometric sequence $a, ar, ar^2, ar^3, ar^4, \ldots, ar^{n-1}, \ldots$, the nth partial sum is

$$S_n = \sum_{k=1}^{n} ar^{k-1} = a + ar + ar^2 + ar^3 + ar^4 + \cdots + ar^{n-1}$$

To find a formula for S_n, we multiply S_n by r and subtract from S_n.

$$S_n = a + ar + ar^2 + ar^3 + ar^4 + \cdots + ar^{n-1}$$

$$\underline{rS_n = \qquad ar + ar^2 + ar^3 + ar^4 + \cdots + ar^{n-1} + ar^n}$$

$$S_n - rS_n = a - ar^n$$

So $\qquad S_n(1 - r) = a(1 - r^n)$

$$S_n = \frac{a(1 - r^n)}{1 - r} \qquad (r \neq 1)$$

We summarize this result.

PARTIAL SUMS OF A GEOMETRIC SEQUENCE

For the geometric sequence $a_n = ar^{n-1}$, the **nth partial sum**

$$S_n = a + ar + ar^2 + ar^3 + ar^4 + \cdots + ar^{n-1} \qquad (r \neq 1)$$

is given by

$$S_n = a\frac{1 - r^n}{1 - r}$$

▶ **EXAMPLE 4** | Finding a Partial Sum of a Geometric Sequence

Find the sum of the first five terms of the geometric sequence

$$1, 0.7, 0.49, 0.343, \ldots$$

▼ **SOLUTION**   The required sum is the sum of the first five terms of a geometric sequence with $a = 1$ and $r = 0.7$. Using the formula for S_n with $n = 5$, we get

$$S_5 = 1 \cdot \frac{1 - (0.7)^5}{1 - 0.7} = 2.7731$$

Thus, the sum of the first five terms of this sequence is 2.7731.   ▲

▶ **EXAMPLE 5** | Finding a Partial Sum of a Geometric Sequence

Find the sum $\displaystyle\sum_{k=1}^{5} 7\left(-\frac{2}{3}\right)^k$.

▼ **SOLUTION**   The given sum is the fifth partial sum of a geometric sequence with first term $a = 7\left(-\frac{2}{3}\right)^1 = -\frac{14}{3}$ and common ratio $r = -\frac{2}{3}$. Thus, by the formula for S_n we have

$$S_5 = -\frac{14}{3} \cdot \frac{1 - \left(-\frac{2}{3}\right)^5}{1 - \left(-\frac{2}{3}\right)} = -\frac{14}{3} \cdot \frac{1 + \frac{32}{243}}{\frac{5}{3}} = -\frac{770}{243}$$ ▲

■ What Is an Infinite Series?

An expression of the form

$$a_1 + a_2 + a_3 + a_4 + \cdots$$

is called an **infinite series**. The dots mean that we are to continue the addition indefinitely. What meaning can we attach to the sum of infinitely many numbers? It seems at first that it is not possible to add infinitely many numbers and arrive at a finite number. But consider the following problem. You have a cake, and you want to eat it by first eating half the cake, then eating half of what remains, then again eating half of what remains. This process can continue indefinitely because at each stage some of the cake remains. (See Figure 3.)

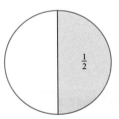 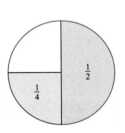 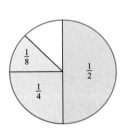 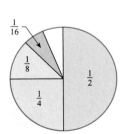 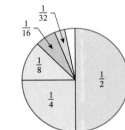

FIGURE 3

Does this mean that it's impossible to eat all of the cake? Of course not. Let's write down what you have eaten from this cake:

$$\frac{1}{2} + \frac{1}{4} + \frac{1}{8} + \frac{1}{16} + \cdots + \frac{1}{2^n} + \cdots$$

This is an infinite series, and we note two things about it: First, from Figure 3 it's clear that no matter how many terms of this series we add, the total will never exceed 1. Second, the more terms of this series we add, the closer the sum is to 1 (see Figure 3). This suggests that the number 1 can be written as the sum of infinitely many smaller numbers:

$$1 = \frac{1}{2} + \frac{1}{4} + \frac{1}{8} + \frac{1}{16} + \cdots + \frac{1}{2^n} + \cdots$$

To make this more precise, let's look at the partial sums of this series:

$$S_1 = \frac{1}{2} \qquad\qquad\qquad = \frac{1}{2}$$

$$S_2 = \frac{1}{2} + \frac{1}{4} \qquad\qquad = \frac{3}{4}$$

$$S_3 = \frac{1}{2} + \frac{1}{4} + \frac{1}{8} \qquad = \frac{7}{8}$$

$$S_4 = \frac{1}{2} + \frac{1}{4} + \frac{1}{8} + \frac{1}{16} = \frac{15}{16}$$

and, in general (see Example 5 of Section 9.1),

$$S_n = 1 - \frac{1}{2^n}$$

As n gets larger and larger, we are adding more and more of the terms of this series. Intuitively, as n gets larger, S_n gets closer to the sum of the series. Now notice that as n gets large, $1/2^n$ gets closer and closer to 0. Thus, S_n gets close to $1 - 0 = 1$. Using the notation of Section 4.6, we can write

$$S_n \to 1 \quad \text{as} \quad n \to \infty$$

In general, if S_n gets close to a finite number S as n gets large, we say that S is the **sum of the infinite series**.

■ Infinite Geometric Series

Here is another way to arrive at the formula for the sum of an infinite geometric series:

$$S = a + ar + ar^2 + ar^3 + \cdots$$
$$= a + r(a + ar + ar^2 + \cdots)$$
$$= a + rS$$

Solve the equation $S = a + rS$ for S to get

$$S - rS = a$$
$$(1 - r)S = a$$
$$S = \frac{a}{1 - r}$$

An **infinite geometric series** is a series of the form

$$a + ar + ar^2 + ar^3 + ar^4 + \cdots + ar^{n-1} + \cdots$$

We can apply the reasoning used earlier to find the sum of an infinite geometric series. The nth partial sum of such a series is given by the formula

$$S_n = a\frac{1 - r^n}{1 - r} \qquad (r \neq 1)$$

It can be shown that if $|r| < 1$, then r^n gets close to 0 as n gets large (you can easily convince yourself of this using a calculator). It follows that S_n gets close to $a/(1 - r)$ as n gets large, or

$$S_n \to \frac{a}{1 - r} \quad \text{as} \quad n \to \infty$$

Thus, the sum of this infinite geometric series is $a/(1 - r)$.

SUM OF AN INFINITE GEOMETRIC SERIES

If $|r| < 1$, then the infinite geometric series

$$a + ar + ar^2 + ar^3 + ar^4 + \cdots + ar^{n-1} + \cdots$$

has the sum

$$S = \frac{a}{1 - r}$$

▶ **EXAMPLE 6** | Finding the Sum of an Infinite Geometric Series

Find the sum of the infinite geometric series

$$2 + \frac{2}{5} + \frac{2}{25} + \frac{2}{125} + \cdots + \frac{2}{5^n} + \cdots$$

▼ **SOLUTION** We use the formula for the sum of an infinite geometric series. In this case $a = 2$ and $r = \frac{1}{5}$. Thus, the sum of this infinite series is

$$S = \frac{2}{1 - \frac{1}{5}} = \frac{5}{2}$$ ▲

▶ **EXAMPLE 7** | Writing a Repeated Decimal as a Fraction

Find the fraction that represents the rational number $2.3\overline{51}$.

▼ **SOLUTION** This repeating decimal can be written as a series:

$$\frac{23}{10} + \frac{51}{1000} + \frac{51}{100,000} + \frac{51}{10,000,000} + \frac{51}{1,000,000,000} + \cdots$$

After the first term, the terms of this series form an infinite geometric series with

$$a = \frac{51}{1000} \qquad \text{and} \qquad r = \frac{1}{100}$$

Thus, the sum of this part of the series is

$$S = \frac{\frac{51}{1000}}{1 - \frac{1}{100}} = \frac{\frac{51}{1000}}{\frac{99}{100}} = \frac{51}{1000} \cdot \frac{100}{99} = \frac{51}{990}$$

So

$$2.3\overline{51} = \frac{23}{10} + \frac{51}{990} = \frac{2328}{990} = \frac{388}{165}$$ ▲

9.4 | Mathematics of Finance

LEARNING OBJECTIVES

After completing this section, you will be able to:

■ Find the amount of an annuity
■ Find the present value of an annuity
■ Find the amount of the installment payments on a loan

Many financial transactions involve payments that are made at regular intervals. For example, if you deposit $100 each month in an interest-bearing account, what will the value of your account be at the end of 5 years? If you borrow $100,000 to buy a house, how much must your monthly payments be in order to pay off the loan in 30 years? Each of these questions involves the sum of a sequence of numbers; we use the results of the preceding section to answer them here.

■ The Amount of an Annuity

An **annuity** is a sum of money that is paid in regular equal payments. Although the word *annuity* suggests annual (or yearly) payments, they can be made semiannually, quarterly, monthly, or at some other regular interval. Payments are usually made at the end of the payment interval. The **amount of an annuity** is the sum of all the individual payments from the time of the first payment until the last payment is made, together with all the interest. We denote this sum by A_f (the subscript f here is used to denote *final* amount).

▶ **EXAMPLE 1** | Calculating the Amount of an Annuity

An investor deposits $400 every December 15 and June 15 for 10 years in an account that earns interest at the rate of 8% per year, compounded semiannually. How much will be in the account immediately after the last payment?

▼ **SOLUTION** We need to find the amount of an annuity consisting of 20 semiannual payments of $400 each. Since the interest rate is 8% per year, compounded semiannually, the interest rate per time period is $i = 0.08/2 = 0.04$. The first payment is in the account for 19 time periods, the second for 18 time periods, and so on.

The last payment receives no interest. The situation can be illustrated by the time line in Figure 1.

⊘ When using interest rates in calculators, remember to convert percentages to decimals. For example, 8% is 0.08.

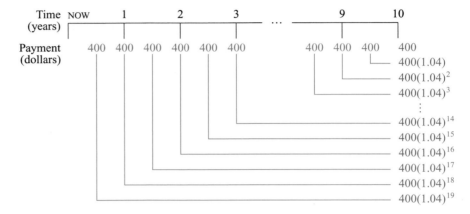

FIGURE 1

The amount A_f of the annuity is the sum of these 20 amounts. Thus,

$$A_f = 400 + 400(1.04) + 400(1.04)^2 + \cdots + 400(1.04)^{19}$$

But this is a geometric series with $a = 400$, $r = 1.04$, and $n = 20$, so

$$A_f = 400 \, \frac{1 - (1.04)^{20}}{1 - 1.04} \approx 11{,}911.23$$

Thus, the amount in the account after the last payment is $11,911.23. ▲

In general, the regular annuity payment is called the **periodic rent** and is denoted by R. We also let i denote the interest rate per time period and n the number of payments. *We always assume that the time period in which interest is compounded is equal to the time between payments.* By the same reasoning as in Example 1, we see that the amount A_f of an annuity is

$$A_f = R + R(1 + i) + R(1 + i)^2 + \cdots + R(1 + i)^{n-1}$$

Since this is the nth partial sum of a geometric sequence with $a = R$ and $r = 1 + i$, the formula for the partial sum gives

$$A_f = R \, \frac{1 - (1 + i)^n}{1 - (1 + i)} = R \, \frac{1 - (1 + i)^n}{-i} = R \, \frac{(1 + i)^n - 1}{i}$$

AMOUNT OF AN ANNUITY

The amount A_f of an annuity consisting of n regular equal payments of size R with interest rate i per time period is given by

$$A_f = R \, \frac{(1 + i)^n - 1}{i}$$

▶ **EXAMPLE 2** | Calculating the Amount of an Annuity

How much money should be invested every month at 12% per year, compounded monthly, in order to have $4000 in 18 months?

▼ **SOLUTION**   In this problem $i = 0.12/12 = 0.01$, $A_f = 4000$, and $n = 18$. We need to find the amount R of each payment. By the formula for the amount of an annuity,

$$4000 = R\,\frac{(1 + 0.01)^{18} - 1}{0.01}$$

Solving for R, we get

$$R = \frac{4000(0.01)}{(1 + 0.01)^{18} - 1} \approx 203.928$$

Thus, the monthly investment should be $203.93.   ▲

■ The Present Value of an Annuity

If you were to receive $10,000 five years from now, it would be worth much less than if you got $10,000 right now. This is because of the interest you could accumulate during the next five years if you invested the money now. What smaller amount would you be willing to accept *now* instead of receiving $10,000 in five years? This is the amount of money that, together with interest, would be worth $10,000 in five years. The amount that we are looking for here is called the *discounted value* or *present value*. If the interest rate is 8% per year, compounded quarterly, then the interest per time period is $i = 0.08/4 = 0.02$, and there are $4 \times 5 = 20$ time periods. If we let PV denote the present value, then by the formula for compound interest (Section 5.1), we have

$$10,000 = PV(1 + i)^n = PV(1 + 0.02)^{20}$$

so

$$PV = 10,000(1 + 0.02)^{-20} \approx 6729.713$$

Thus, in this situation the present value of $10,000 is $6729.71. This reasoning leads to a general formula for present value. If an amount A_f is to be paid in a lump sum n time periods from now and the interest rate per time period is i, then its **present value** A_p is given by

$$A_p = A_f(1 + i)^{-n}$$

Similarly, the **present value of an annuity** is the amount A_p that must be invested now at the interest rate i per time period to provide n payments, each of amount R. Clearly, A_p is the sum of the present values of each individual payment. Another way of finding A_p is to note that A_p is the present value of A_f:

$$A_p = A_f(1 + i)^{-n} = R\,\frac{(1 + i)^n - 1}{i}\,(1 + i)^{-n} = R\,\frac{1 - (1 + i)^{-n}}{i}$$

THE PRESENT VALUE OF AN ANNUITY

The **present value** A_p of an annuity consisting of n regular equal payments of size R and interest rate i per time period is given by

$$A_p = R\,\frac{1 - (1 + i)^{-n}}{i}$$

▶ **EXAMPLE 3** | Calculating the Present Value of an Annuity

A person wins $10,000,000 in the California lottery, and the amount is paid in yearly installments of half a million dollars each for 20 years. What is the present value of his winnings? Assume that he can earn 10% interest, compounded annually.

▼ **SOLUTION**   Since the amount won is paid as an annuity, we need to find its present value. Here $i = 0.1$, $R = \$500,000$, and $n = 20$. Thus,

$$A_p = 500,000 \, \frac{1 - (1 + 0.1)^{-20}}{0.1} \approx 4,256,781.859$$

This means that the winner really won only $4,256,781.86 if it were paid immediately.   ▲

■ Installment Buying

When you buy a house or a car by installment, the payments that you make are an annuity whose present value is the amount of the loan.

▶ **EXAMPLE 4** | The Amount of a Loan

A student wishes to buy a car. She can afford to pay $200 per month but has no money for a down payment. If she can make these payments for four years and the interest rate is 12%, what purchase price can she afford?

▼ **SOLUTION**   The payments that the student makes constitute an annuity whose present value is the price of the car (which is also the amount of the loan, in this case). Here we have $i = 0.12/12 = 0.01$, $R = 200$, and $n = 12 \times 4 = 48$, so

$$A_p = R \, \frac{1 - (1 + i)^{-n}}{i} = 200 \, \frac{1 - (1 + 0.01)^{-48}}{0.01} \approx 7594.792$$

Thus, the student can buy a car priced at $7594.79.   ▲

When a bank makes a loan that is to be repaid with regular equal payments R, then the payments form an annuity whose present value A_p is the amount of the loan. So to find the size of the payments, we solve for R in the formula for the amount of an annuity. This gives the following formula for R.

INSTALLMENT BUYING

If a loan A_p is to be repaid in n regular equal payments with interest rate i per time period, then the size R of each payment is given by

$$R = \frac{iA_p}{1 - (1 + i)^{-n}}$$

▶ **EXAMPLE 5** | Calculating Monthly Mortgage Payments

A couple borrows $100,000 at 9% interest as a mortgage loan on a house. They expect to make monthly payments for 30 years to repay the loan. What is the size of each payment?

▼ **SOLUTION** The mortgage payments form an annuity whose present value is $A_p = \$100{,}000$. Also, $i = 0.09/12 = 0.0075$, and $n = 12 \times 30 = 360$. We are looking for the amount R of each payment.

From the formula for installment buying, we get

$$R = \frac{iA_p}{1 - (1 + i)^{-n}} = \frac{(0.0075)(100{,}000)}{1 - (1 + 0.0075)^{-360}} \approx 804.623$$

Thus, the monthly payments are \$804.62. ▲

We now illustrate the use of graphing devices in solving problems related to installment buying.

▶ **EXAMPLE 6** | Calculating the Interest Rate from the Size of Monthly Payments

A car dealer sells a new car for \$18,000. He offers the buyer payments of \$405 per month for 5 years. What interest rate is this car dealer charging?

▼ **SOLUTION** The payments form an annuity with present value $A_p = \$18{,}000$, $R = 405$, and $n = 12 \times 5 = 60$. To find the interest rate, we must solve for i in the equation

$$R = \frac{iA_p}{1 - (1 + i)^{-n}}$$

A little experimentation will convince you that it is not possible to solve this equation for i algebraically. So to find i, we use a graphing device to graph R as a function of the interest rate x, and we then use the graph to find the interest rate corresponding to the value of R we want (\$405 in this case). Since $i = x/12$, we graph the function

$$R(x) = \frac{\dfrac{x}{12}(18{,}000)}{1 - \left(1 + \dfrac{x}{12}\right)^{-60}}$$

in the viewing rectangle $[0.06, 0.16] \times [350, 450]$, as shown in Figure 2. We also graph the horizontal line $R(x) = 405$ in the same viewing rectangle. Then, by moving the cursor to the point of intersection of the two graphs, we find that the corresponding x-value is approximately 0.125. Thus, the interest rate is about $12\frac{1}{2}\%$. ▲

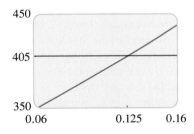

FIGURE 2

9.5 | Mathematical Induction

LEARNING OBJECTIVE

After completing this section, you will be able to:

■ Prove a statement using the Principle of Mathematical Induction

There are two aspects to mathematics—discovery and proof—and they are of equal importance. We must discover something before we can attempt to prove it, and we cannot be certain of its truth until it has been proved. In this section we examine the relationship between these two key components of mathematics more closely.

▦ Conjecture and Proof

Let's try a simple experiment. We add more and more of the odd numbers as follows:

$$1 = 1$$
$$1 + 3 = 4$$
$$1 + 3 + 5 = 9$$
$$1 + 3 + 5 + 7 = 16$$
$$1 + 3 + 5 + 7 + 9 = 25$$

What do you notice about the numbers on the right side of these equations? They are, in fact, all perfect squares. These equations say the following:

The sum of the first 1 odd number is 1^2.

The sum of the first 2 odd numbers is 2^2.

The sum of the first 3 odd numbers is 3^2.

The sum of the first 4 odd numbers is 4^2.

The sum of the first 5 odd numbers is 5^2.

Consider the polynomial

$$p(n) = n^2 - n + 41$$

Here are some values of $p(n)$:

$$p(1) = 41 \quad p(2) = 43$$
$$p(3) = 47 \quad p(4) = 53$$
$$p(5) = 61 \quad p(6) = 71$$
$$p(7) = 83 \quad p(8) = 97$$

All the values so far are prime numbers. In fact, if you keep going, you will find that $p(n)$ is prime for all natural numbers up to $n = 40$. It might seem reasonable at this point to conjecture that $p(n)$ is prime for *every* natural number n. But that conjecture would be too hasty, because it is easily seen that $p(41)$ is *not* prime. This illustrates that we cannot be certain of the truth of a statement no matter how many special cases we check. We need a convincing argument—a *proof*—to determine the truth of a statement.

This leads naturally to the following question: Is it true that for every natural number n, the sum of the first n odd numbers is n^2? Could this remarkable property be true? We could try a few more numbers and find that the pattern persists for the first 6, 7, 8, 9, and 10 odd numbers. At this point we feel quite sure that this is always true, so we make a *conjecture*:

The sum of the first n odd numbers is n^2.

Since we know that the nth odd number is $2n - 1$, we can write this statement more precisely as

$$1 + 3 + 5 + \cdots + (2n - 1) = n^2$$

It is important to realize that this is still a conjecture. We cannot conclude by checking a finite number of cases that a property is true for all numbers (there are infinitely many). To see this more clearly, suppose someone tells us that he has added up the first trillion odd numbers and found that they do *not* add up to 1 trillion squared. What would you tell this person? It would be silly to say that you're sure it's true because you have already checked the first five cases. You could, however, take out paper and pencil and start checking it yourself, but this task would probably take the rest of your life. The tragedy would be that after completing this task, you would still not be sure of the truth of the conjecture! Do you see why?

Herein lies the power of mathematical proof. A **proof** is a clear argument that demonstrates the truth of a statement beyond doubt.

▦ Mathematical Induction

Let's consider a special kind of proof called **mathematical induction**. Here is how it works: Suppose we have a statement that says something about all natural numbers n. Let's call this statement P. For example, we could consider the statement

P: For every natural number n, the sum of the first n odd numbers is n^2.

Since this statement is about *all* natural numbers, it contains infinitely many statements; we will call them $P(1), P(2), \ldots$.

$P(1)$:   The sum of the first 1 odd number is 1^2.

$P(2)$:   The sum of the first 2 odd numbers is 2^2.

$P(3)$:   The sum of the first 3 odd numbers is 3^2.

$$\vdots \qquad\qquad \vdots$$

How can we prove all of these statements at once? Mathematical induction is a clever way of doing just that.

The crux of the idea is this: Suppose we can prove that whenever one of these statements is true, then the one following it in the list is also true. In other words,

$$\text{For every } k, \text{ if } P(k) \text{ is true, then } P(k + 1) \text{ is true.}$$

This is called the **induction step** because it leads us from the truth of one statement to the next. Now, suppose that we can also prove that

$$P(1) \text{ is true.}$$

The induction step now leads us through the following chain of statements:

$$P(1) \text{ is true, so } P(2) \text{ is true.}$$
$$P(2) \text{ is true, so } P(3) \text{ is true.}$$
$$P(3) \text{ is true, so } P(4) \text{ is true.}$$
$$\vdots \qquad \vdots$$

So we see that if both the induction step and $P(1)$ are proved, then statement P is proved for all n. Here is a summary of this important method of proof.

PRINCIPLE OF MATHEMATICAL INDUCTION

For each natural number n, let $P(n)$ be a statement depending on n. Suppose that the following two conditions are satisfied.

1. $P(1)$ is true.

2. For every natural number k, if $P(k)$ is true then $P(k + 1)$ is true.

Then $P(n)$ is true for all natural numbers n.

To apply this principle, there are two steps:

Step 1 Prove that $P(1)$ is true.

Step 2 Assume that $P(k)$ is true, and use this assumption to prove that $P(k + 1)$ is true.

Notice that in Step 2 we do not prove that $P(k)$ is true. We only show that *if* $P(k)$ is true, *then* $P(k + 1)$ is also true. The assumption that $P(k)$ is true is called the **induction hypothesis**.

©1979 National Council of Teachers of Mathematics.
Used by permission. Courtesy of Andrejs Dunkels, Sweden.

We now use mathematical induction to prove that the conjecture that we made at the beginning of this section is true.

▶ **EXAMPLE 1** | A Proof by Mathematical Induction

Prove that for all natural numbers n,

$$1 + 3 + 5 + \cdots + (2n - 1) = n^2$$

▼ **SOLUTION** Let $P(n)$ denote the statement $1 + 3 + 5 + \cdots + (2n - 1) = n^2$.

Step 1 We need to show that $P(1)$ is true. But $P(1)$ is simply the statement that $1 = 1^2$, which is of course true.

Step 2 We assume that $P(k)$ is true. Thus, our induction hypothesis is

$$1 + 3 + 5 + \cdots + (2k - 1) = k^2$$

We want to use this to show that $P(k + 1)$ is true, that is,

$$1 + 3 + 5 + \cdots + (2k - 1) + [2(k + 1) - 1] = (k + 1)^2$$

[Note that we get $P(k + 1)$ by substituting $k + 1$ for each n in the statement $P(n)$.] We start with the left side and use the induction hypothesis to obtain the right side of the equation:

$$1 + 3 + 5 + \cdots + (2k - 1) + [2(k + 1) - 1]$$

$= [1 + 3 + 5 + \cdots + (2k - 1)] + [2(k + 1) - 1]$	Group the first k terms
$= k^2 + [2(k + 1) - 1]$	Induction hypothesis
$= k^2 + [2k + 2 - 1]$	Distributive Property
$= k^2 + 2k + 1$	Simplify
$= (k + 1)^2$	Factor

This equals k^2 by the induction hypothesis.

Thus, $P(k + 1)$ follows from $P(k)$, and this completes the induction step. ▲

Having proved Steps 1 and 2, we conclude by the Principle of Mathematical Induction that $P(n)$ is true for all natural numbers n.

▶ **EXAMPLE 2** | A Proof by Mathematical Induction

Prove that for every natural number n,

$$1 + 2 + 3 + \cdots + n = \frac{n(n + 1)}{2}$$

▼ **SOLUTION** Let $P(n)$ be the statement $1 + 2 + 3 + \cdots + n = n(n + 1)/2$. We want to show that $P(n)$ is true for all natural numbers n.

Step 1 We need to show that $P(1)$ is true. But $P(1)$ says that

$$1 = \frac{1(1 + 1)}{2}$$

and this statement is clearly true.

Step 2 Assume that $P(k)$ is true. Thus, our induction hypothesis is

$$1 + 2 + 3 + \cdots + k = \frac{k(k + 1)}{2}$$

We want to use this to show that $P(k + 1)$ is true, that is,

$$1 + 2 + 3 + \cdots + k + (k + 1) = \frac{(k + 1)[(k + 1) + 1]}{2}$$

So we start with the left side and use the induction hypothesis to obtain the right side:

$1 + 2 + 3 + \cdots + k + (k + 1)$

$= [1 + 2 + 3 + \cdots + k] + (k + 1)$ Group the first k terms

$= \dfrac{k(k + 1)}{2} + (k + 1)$ Induction hypothesis

$= (k + 1)\left(\dfrac{k}{2} + 1\right)$ Factor $k + 1$

$= (k + 1)\left(\dfrac{k + 2}{2}\right)$ Common denominator

$= \dfrac{(k + 1)[(k + 1) + 1]}{2}$ Write $k + 2$ as $k + 1 + 1$

This equals $\dfrac{k(k + 1)}{2}$ by the induction process.

Thus, $P(k + 1)$ follows from $P(k)$ and this completes the induction step.

Having proved Steps 1 and 2, we conclude by the Principle of Mathematical Induction that $P(n)$ is true for all natural numbers n.

Formulas for the sums of powers of the first n natural numbers are important in calculus. Formula 1 in the following box is proved in Example 2. The other formulas are also proved by using mathematical induction.

SUMS OF POWERS

0. $\sum_{k=1}^{n} 1 = n$ **1.** $\sum_{k=1}^{n} k = \dfrac{n(n + 1)}{2}$

2. $\sum_{k=1}^{n} k^2 = \dfrac{n(n + 1)(2n + 1)}{6}$ **3.** $\sum_{k=1}^{n} k^3 = \dfrac{n^2(n + 1)^2}{4}$

It might happen that a statement $P(n)$ is false for the first few natural numbers but true from some number on. For example, we might want to prove that $P(n)$ is true for $n \geq 5$. Notice that if we prove that $P(5)$ is true, then this fact, together with the induction step, would imply the truth of $P(5), P(6), P(7), \ldots$. The next example illustrates this point.

▶ **EXAMPLE 3** | Proving an Inequality by Mathematical Induction

Prove that $4n < 2^n$ for all $n \geq 5$.

▼ **SOLUTION** Let $P(n)$ denote the statement $4n < 2^n$.

Step 1 $P(5)$ is the statement that $4 \cdot 5 < 2^5$, or $20 < 32$, which is true.

Step 2 Assume that $P(k)$ is true. Thus, our induction hypothesis is

$$4k < 2^k$$

We get $P(k + 1)$ by replacing k by $k + 1$ in the statement $P(k)$.

We want to use this to show that $P(k + 1)$ is true, that is,

$$4(k + 1) < 2^{k+1}$$

So we start with the left-hand side of the inequality and use the induction hypothesis to show that it is less than the right-hand side. For $k \geq 5$ we have

$$
\begin{aligned}
4(k + 1) &= 4k + 4 \\
&< 2^k + 4 && \text{Induction hypothesis} \\
&< 2^k + 4k && \text{Because } 4 < 4k \\
&< 2^k + 2^k && \text{Induction hypothesis} \\
&= 2 \cdot 2^k \\
&= 2^{k+1} && \text{Property of exponents}
\end{aligned}
$$

Thus, $P(k + 1)$ follows from $P(k)$, and this completes the induction step.

Having proved Steps 1 and 2, we conclude by the Principle of Mathematical Induction that $P(n)$ is true for all natural numbers $n \geq 5$. ▲

9.6 | The Binomial Theorem

LEARNING OBJECTIVES
After completing this section, you will be able to:

- Expand powers of binomials using Pascal's triangle
- Find binomial coefficients
- Expand powers of binomials using the Binomial Theorem
- Find a particular term in a binomial expansion

An expression of the form $a + b$ is called a **binomial**. Although in principle it's easy to raise $a + b$ to any power, raising it to a very high power would be tedious. In this section we find a formula that gives the expansion of $(a + b)^n$ for any natural number n and then prove it using mathematical induction.

▪ Expanding $(a + b)^n$

To find a pattern in the expansion of $(a + b)^n$, we first look at some special cases:

$$
\begin{aligned}
(a + b)^1 &= a + b \\
(a + b)^2 &= a^2 + 2ab + b^2 \\
(a + b)^3 &= a^3 + 3a^2b + 3ab^2 + b^3 \\
(a + b)^4 &= a^4 + 4a^3b + 6a^2b^2 + 4ab^3 + b^4 \\
(a + b)^5 &= a^5 + 5a^4b + 10a^3b^2 + 10a^2b^3 + 5ab^4 + b^5
\end{aligned}
$$

$$\vdots$$

The following simple patterns emerge for the expansion of $(a + b)^n$:

1. There are $n + 1$ terms, the first being a^n and the last being b^n.
2. The exponents of a decrease by 1 from term to term, while the exponents of b increase by 1.
3. The sum of the exponents of a and b in each term is n.

For instance, notice how the exponents of a and b behave in the expansion of $(a + b)^5$.

The exponents of *a* decrease:

$$(a + b)^5 = \overset{⑤}{a} + 5\overset{④}{a}b^1 + 10\overset{③}{a}b^2 + 10\overset{②}{a}b^3 + 5\overset{①}{a}b^4 + b^5$$

The exponents of *b* increase:

$$(a + b)^5 = a^5 + 5a^4\overset{①}{b} + 10a^3\overset{②}{b} + 10a^2\overset{③}{b} + 5a^1\overset{④}{b} + \overset{⑤}{b}$$

With these observations we can write the form of the expansion of $(a + b)^n$ for any natural number n. For example, writing a question mark for the missing coefficients, we have

$$(a + b)^8 = a^8 + ?\,a^7b + ?\,a^6b^2 + ?\,a^5b^3 + ?\,a^4b^4 + ?\,a^3b^5 + ?\,a^2b^6 + ?\,ab^7 + b^8$$

To complete the expansion, we need to determine these coefficients. To find a pattern, let's write the coefficients in the expansion of $(a + b)^n$ for the first few values of n in a triangular array as shown in the following array, which is called **Pascal's triangle**.

$$
\begin{array}{ll}
(a + b)^0 & \qquad\qquad\qquad 1 \\
(a + b)^1 & \qquad\qquad\quad 1 \quad 1 \\
(a + b)^2 & \qquad\qquad 1 \quad 2 \quad 1 \\
(a + b)^3 & \qquad\quad ① \quad ③ \quad 3 \quad 1 \\
(a + b)^4 & \qquad 1 \quad ④ \quad ⑥ \quad ④ \quad 1 \\
(a + b)^5 & \quad 1 \quad 5 \quad 10 \quad ⑩ \quad 5 \quad 1 \\
\end{array}
$$

The row corresponding to $(a + b)^0$ is called the zeroth row and is included to show the symmetry of the array. The key observation about Pascal's triangle is the following property.

KEY PROPERTY OF PASCAL'S TRIANGLE

Every entry (other than a 1) is the sum of the two entries diagonally above it.

From this property it is easy to find any row of Pascal's triangle from the row above it. For instance, we find the sixth and seventh rows, starting with the fifth row:

$$
\begin{array}{lllllllll}
(a + b)^5 & & 1 & 5 & 10 & 10 & 5 & 1 \\
(a + b)^6 & & 1 & 6 & 15 & 20 & 15 & 6 & 1 \\
(a + b)^7 & 1 & 7 & 21 & 35 & 35 & 21 & 7 & 1 \\
\end{array}
$$

To see why this property holds, let's consider the following expansions:

$$(a + b)^5 = a^5 + 5a^4b + 10a^3b^2 + \boxed{10a^2b^3} + \boxed{5ab^4} + b^5$$

$$(a + b)^6 = a^6 + 6a^5b + 15a^4b^2 + 20a^3b^3 + \boxed{15a^2b^4} + 6ab^5 + b^6$$

We arrive at the expansion of $(a + b)^6$ by multiplying $(a + b)^5$ by $(a + b)$. Notice, for instance, that the circled term in the expansion of $(a + b)^6$ is obtained via this multiplication from the two circled terms above it. We get this term when the two terms above it are multiplied by b and a, respectively. Thus, its coefficient is the sum of the coefficients of these two terms. We will use this observation at the end of this section when we prove the Binomial Theorem.

Having found these patterns, we can now easily obtain the expansion of any binomial, at least to relatively small powers.

▶ **EXAMPLE 1** | Expanding a Binomial Using Pascal's Triangle

Find the expansion of $(a + b)^7$ using Pascal's triangle.

▼ **SOLUTION** The first term in the expansion is a^7, and the last term is b^7. Using the fact that the exponent of a decreases by 1 from term to term and that of b increases by 1 from term to term, we have

$$(a + b)^7 = a^7 + ?\, a^6 b + ?\, a^5 b^2 + ?\, a^4 b^3 + ?\, a^3 b^4 + ?\, a^2 b^5 + ?\, ab^6 + b^7$$

The appropriate coefficients appear in the seventh row of Pascal's triangle. Thus,

$$(a + b)^7 = a^7 + 7a^6 b + 21a^5 b^2 + 35a^4 b^3 + 35a^3 b^4 + 21a^2 b^5 + 7ab^6 + b^7 \qquad \blacktriangle$$

▶ **EXAMPLE 2** | Expanding a Binomial Using Pascal's Triangle

Use Pascal's triangle to expand $(2 - 3x)^5$.

▼ **SOLUTION** We find the expansion of $(a + b)^5$ and then substitute 2 for a and $-3x$ for b. Using Pascal's triangle for the coefficients, we get

$$(a + b)^5 = a^5 + 5a^4 b + 10a^3 b^2 + 10a^2 b^3 + 5ab^4 + b^5$$

Substituting $a = 2$ and $b = -3x$ gives

$$(2 - 3x)^5 = (2)^5 + 5(2)^4(-3x) + 10(2)^3(-3x)^2 + 10(2)^2(-3x)^3 + 5(2)(-3x)^4 + (-3x)^5$$

$$= 32 - 240x + 720x^2 - 1080x^3 + 810x^4 - 243x^5 \qquad \blacktriangle$$

■ The Binomial Coefficients

Although Pascal's triangle is useful in finding the binomial expansion for reasonably small values of n, it isn't practical for finding $(a + b)^n$ for large values of n. The reason is that the method we use for finding the successive rows of Pascal's triangle is recursive. Thus, to find the 100th row of this triangle, we must first find the preceding 99 rows.

We need to examine the pattern in the coefficients more carefully to develop a formula that allows us to calculate directly any coefficient in the binomial expansion. Such a formula exists, and the rest of this section is devoted to finding and proving it. However, to state this formula, we need some notation.

The product of the first n natural numbers is denoted by $n!$ and is called n **factorial**:

$4! = 1 \cdot 2 \cdot 3 \cdot 4 = 24$

$7! = 1 \cdot 2 \cdot 3 \cdot 4 \cdot 5 \cdot 6 \cdot 7 = 5040$

$10! = 1 \cdot 2 \cdot 3 \cdot 4 \cdot 5 \cdot 6 \cdot 7 \cdot 8 \cdot 9 \cdot 10$

$\qquad = 3,628,800$

$$n! = 1 \cdot 2 \cdot 3 \cdot \cdots \cdot (n - 1) \cdot n$$

We also define 0! as follows:

$$0! = 1$$

This definition of 0! makes many formulas involving factorials shorter and easier to write.

THE BINOMIAL COEFFICIENT

Let n and r be nonnegative integers with $r \le n$. The **binomial coefficient** is denoted by $\binom{n}{r}$ and is defined by

$$\binom{n}{r} = \frac{n!}{r!(n - r)!}$$

▶ **EXAMPLE 3** | Calculating Binomial Coefficients

(a) $\displaystyle \binom{9}{4} = \frac{9!}{4!(9-4)!} = \frac{9!}{4!5!} = \frac{1 \cdot 2 \cdot 3 \cdot 4 \cdot 5 \cdot 6 \cdot 7 \cdot 8 \cdot 9}{(1 \cdot 2 \cdot 3 \cdot 4)(1 \cdot 2 \cdot 3 \cdot 4 \cdot 5)}$

$$= \frac{6 \cdot 7 \cdot 8 \cdot 9}{1 \cdot 2 \cdot 3 \cdot 4} = 126$$

(b) $\displaystyle \binom{100}{3} = \frac{100!}{3!(100-3)!} = \frac{1 \cdot 2 \cdot 3 \cdots 97 \cdot 98 \cdot 99 \cdot 100}{(1 \cdot 2 \cdot 3)(1 \cdot 2 \cdot 3 \cdots 97)}$

$$= \frac{98 \cdot 99 \cdot 100}{1 \cdot 2 \cdot 3} = 161{,}700$$

(c) $\displaystyle \binom{100}{97} = \frac{100!}{97!(100-97)!} = \frac{1 \cdot 2 \cdot 3 \cdots 97 \cdot 98 \cdot 99 \cdot 100}{(1 \cdot 2 \cdot 3 \cdots 97)(1 \cdot 2 \cdot 3)}$

$$= \frac{98 \cdot 99 \cdot 100}{1 \cdot 2 \cdot 3} = 161{,}700$$

▲

Although the binomial coefficient $\binom{n}{r}$ is defined in terms of a fraction, all the results of Example 3 are natural numbers. In fact, $\binom{n}{r}$ is always a natural number. Notice that the binomial coefficients in parts (b) and (c) of Example 3 are equal. This is a special case of the following relation, which you are asked to prove in the exercise set.

$$\binom{n}{r} = \binom{n}{n-r}$$

To see the connection between the binomial coefficients and the binomial expansion of $(a + b)^n$, let's calculate the following binomial coefficients:

$$\binom{5}{2} = \frac{5!}{2!(5-2)!} = 10$$

$$\binom{5}{0} = 1 \qquad \binom{5}{1} = 5 \qquad \binom{5}{2} = 10 \qquad \binom{5}{3} = 10 \qquad \binom{5}{4} = 5 \qquad \binom{5}{5} = 1$$

These are precisely the entries in the fifth row of Pascal's triangle. In fact, we can write Pascal's triangle as follows.

$$\binom{0}{0}$$

$$\binom{1}{0} \quad \binom{1}{1}$$

$$\binom{2}{0} \quad \binom{2}{1} \quad \binom{2}{2}$$

$$\binom{3}{0} \quad \binom{3}{1} \quad \binom{3}{2} \quad \binom{3}{3}$$

$$\binom{4}{0} \quad \binom{4}{1} \quad \binom{4}{2} \quad \binom{4}{3} \quad \binom{4}{4}$$

$$\binom{5}{0} \quad \binom{5}{1} \quad \binom{5}{2} \quad \binom{5}{3} \quad \binom{5}{4} \quad \binom{5}{5}$$

$$\cdot \quad \cdot \quad \cdot \quad \cdot \quad \cdot \quad \cdot \quad \cdot \quad \cdot$$

$$\binom{n}{0} \quad \binom{n}{1} \quad \binom{n}{2} \quad \cdot \quad \cdot \quad \cdot \quad \binom{n}{n-1} \quad \binom{n}{n}$$

To demonstrate that this pattern holds, we need to show that any entry in this version of Pascal's triangle is the sum of the two entries diagonally above it. In other words, we must show that each entry satisfies the key property of Pascal's triangle. We now state this property in terms of the binomial coefficients.

KEY PROPERTY OF THE BINOMIAL COEFFICIENTS

For any nonnegative integers r and k with $r \le k$,

$$\binom{k}{r-1} + \binom{k}{r} = \binom{k+1}{r}$$

Notice that the two terms on the left side of this equation are adjacent entries in the kth row of Pascal's triangle and the term on the right side is the entry diagonally below them, in the $(k+1)$st row. Thus, this equation is a restatement of the key property of Pascal's triangle in terms of the binomial coefficients. A proof of this formula is outlined in the exercise set.

■ The Binomial Theorem

We are now ready to state the Binomial Theorem.

THE BINOMIAL THEOREM

$$(a+b)^n = \binom{n}{0}a^n + \binom{n}{1}a^{n-1}b + \binom{n}{2}a^{n-2}b^2 + \cdots + \binom{n}{n-1}ab^{n-1} + \binom{n}{n}b^n$$

We prove this theorem at the end of this section. First, let's look at some of its applications.

▶ **EXAMPLE 4** │ Expanding a Binomial Using the Binomial Theorem

Use the Binomial Theorem to expand $(x + y)^4$.

▼ **SOLUTION**   By the Binomial Theorem,

$$(x+y)^4 = \binom{4}{0}x^4 + \binom{4}{1}x^3y + \binom{4}{2}x^2y^2 + \binom{4}{3}xy^3 + \binom{4}{4}y^4$$

Verify that

$$\binom{4}{0} = 1 \qquad \binom{4}{1} = 4 \qquad \binom{4}{2} = 6 \qquad \binom{4}{3} = 4 \qquad \binom{4}{4} = 1$$

It follows that

$$(x+y)^4 = x^4 + 4x^3y + 6x^2y^2 + 4xy^3 + y^4 \qquad \blacktriangle$$

▶ **EXAMPLE 5** │ Expanding a Binomial Using the Binomial Theorem

Use the Binomial Theorem to expand $\left(\sqrt{x} - 1\right)^8$.

▼ **SOLUTION**   We first find the expansion of $(a+b)^8$ and then substitute $\sqrt{x}$ for a and -1 for b. Using the Binomial Theorem, we have

$$(a + b)^8 = \binom{8}{0}a^8 + \binom{8}{1}a^7b + \binom{8}{2}a^6b^2 + \binom{8}{3}a^5b^3 + \binom{8}{4}a^4b^4$$
$$+ \binom{8}{5}a^3b^5 + \binom{8}{6}a^2b^6 + \binom{8}{7}ab^7 + \binom{8}{8}b^8$$

Verify that

$$\binom{8}{0} = 1 \qquad \binom{8}{1} = 8 \qquad \binom{8}{2} = 28 \qquad \binom{8}{3} = 56 \qquad \binom{8}{4} = 70$$

$$\binom{8}{5} = 56 \qquad \binom{8}{6} = 28 \qquad \binom{8}{7} = 8 \qquad \binom{8}{8} = 1$$

So

$$(a + b)^8 = a^8 + 8a^7b + 28a^6b^2 + 56a^5b^3 + 70a^4b^4 + 56a^3b^5$$
$$+ 28a^2b^6 + 8ab^7 + b^8$$

Performing the substitutions $a = x^{1/2}$ and $b = -1$ gives

$$(\sqrt{x} - 1)^8 = (x^{1/2})^8 + 8(x^{1/2})^7(-1) + 28(x^{1/2})^6(-1)^2 + 56(x^{1/2})^5(-1)^3$$
$$+ 70(x^{1/2})^4(-1)^4 + 56(x^{1/2})^3(-1)^5 + 28(x^{1/2})^2(-1)^6$$
$$+ 8(x^{1/2})(-1)^7 + (-1)^8$$

This simplifies to

$$(\sqrt{x} - 1)^8 = x^4 - 8x^{7/2} + 28x^3 - 56x^{5/2} + 70x^2 - 56x^{3/2} + 28x - 8x^{1/2} + 1 \quad \blacktriangle$$

The Binomial Theorem can be used to find a particular term of a binomial expansion without having to find the entire expansion.

GENERAL TERM OF THE BINOMIAL EXPANSION

The term that contains a^r in the expansion of $(a + b)^n$ is

$$\binom{n}{n-r}a^rb^{n-r}$$

▶ **EXAMPLE 6** | Finding a Particular Term in a Binomial Expansion

Find the term that contains x^5 in the expansion of $(2x + y)^{20}$.

▼ **SOLUTION**   The term that contains x^5 is given by the formula for the general term with $a = 2x$, $b = y$, $n = 20$, and $r = 5$. So this term is

$$\binom{20}{15}a^5b^{15} = \frac{20!}{15!(20-15)!}(2x)^5y^{15} = \frac{20!}{15!5!}32x^5y^{15} = 496{,}128x^5y^{15} \quad \blacktriangle$$

▶ **EXAMPLE 7** | Finding a Particular Term in a Binomial Expansion

Find the coefficient of x^8 in the expansion of $\left(x^2 + \dfrac{1}{x}\right)^{10}$.

▼ **SOLUTION** Both x^2 and $1/x$ are powers of x, so the power of x in each term of the expansion is determined by both terms of the binomial. To find the required coefficient, we first find the general term in the expansion. By the formula we have $a = x^2$, $b = 1/x$, and $n = 10$, so the general term is

$$\binom{10}{10-r}(x^2)^r\left(\frac{1}{x}\right)^{10-r} = \binom{10}{10-r}x^{2r}(x^{-1})^{10-r} = \binom{10}{10-r}x^{3r-10}$$

Thus, the term that contains x^8 is the term in which

$$3r - 10 = 8$$
$$r = 6$$

So the required coefficient is

$$\binom{10}{10-6} = \binom{10}{4} = 210$$

▲

■ Proof of the Binomial Theorem

We now give a proof of the Binomial Theorem using mathematical induction.

▼ **PROOF** Let $P(n)$ denote the statement

$$(a + b)^n = \binom{n}{0}a^n + \binom{n}{1}a^{n-1}b + \binom{n}{2}a^{n-2}b^2 + \cdots + \binom{n}{n-1}ab^{n-1} + \binom{n}{n}b^n$$

Step 1 We show that $P(1)$ is true. But $P(1)$ is just the statement

$$(a + b)^1 = \binom{1}{0}a^1 + \binom{1}{1}b^1 = 1a + 1b = a + b$$

which is certainly true.

Step 2 We assume that $P(k)$ is true. Thus, our induction hypothesis is

$$(a + b)^k = \binom{k}{0}a^k + \binom{k}{1}a^{k-1}b + \binom{k}{2}a^{k-2}b^2 + \cdots + \binom{k}{k-1}ab^{k-1} + \binom{k}{k}b^k$$

We use this to show that $P(k + 1)$ is true.

$$(a + b)^{k+1} = (a + b)[(a + b)^k]$$

$$= (a + b)\left[\binom{k}{0}a^k + \binom{k}{1}a^{k-1}b + \binom{k}{2}a^{k-2}b^2 + \cdots + \binom{k}{k-1}ab^{k-1} + \binom{k}{k}b^k\right] \quad \text{Induction hypothesis}$$

$$= a\left[\binom{k}{0}a^k + \binom{k}{1}a^{k-1}b + \binom{k}{2}a^{k-2}b^2 + \cdots + \binom{k}{k-1}ab^{k-1} + \binom{k}{k}b^k\right]$$

$$\quad + b\left[\binom{k}{0}a^k + \binom{k}{1}a^{k-1}b + \binom{k}{2}a^{k-2}b^2 + \cdots + \binom{k}{k-1}ab^{k-1} + \binom{k}{k}b^k\right] \quad \text{Distributive Property}$$

$$= \binom{k}{0}a^{k+1} + \binom{k}{1}a^kb + \binom{k}{2}a^{k-1}b^2 + \cdots + \binom{k}{k-1}a^2b^{k-1} + \binom{k}{k}ab^k$$

$$\quad + \binom{k}{0}a^kb + \binom{k}{1}a^{k-1}b^2 + \binom{k}{2}a^{k-2}b^3 + \cdots + \binom{k}{k-1}ab^k + \binom{k}{k}b^{k+1} \quad \text{Distributive Property}$$

$$= \binom{k}{0}a^{k+1} + \left[\binom{k}{0} + \binom{k}{1}\right]a^kb + \left[\binom{k}{1} + \binom{k}{2}\right]a^{k-1}b^2$$

$$\quad + \cdots + \left[\binom{k}{k-1} + \binom{k}{k}\right]ab^k + \binom{k}{k}b^{k+1} \quad \text{Group like terms}$$

Using the key property of the binomial coefficients, we can write each of the expressions in square brackets as a single binomial coefficient. Also, writing the first and last coefficients as $\binom{k+1}{0}$ and $\binom{k+1}{k+1}$ (these are equal to 1 as shown in the exercise set) gives

$$(a + b)^{k+1} = \binom{k+1}{0}a^{k+1} + \binom{k+1}{1}a^k b + \binom{k+1}{2}a^{k-1}b^2 + \cdots + \binom{k+1}{k}ab^k + \binom{k+1}{k+1}b^{k+1}$$

But this last equation is precisely $P(k+1)$, and this completes the induction step.

Having proved Steps 1 and 2, we conclude by the Principle of Mathematical Induction that the theorem is true for all natural numbers n. ▲

▶ CHAPTER 9 | REVIEW

▼ PROPERTIES AND FORMULAS

Sequences (p. 396)

A **sequence** is a function whose domain is the set of natural numbers. Instead of writing $a(n)$ for the value of the sequence at n, we generally write a_n, and we refer to this value as the **nth term** of the sequence. Sequences are often described in list form:

$$a_1, a_2, a_3, \ldots$$

Partial Sums of a Sequence (pp. 401–402)

For the sequence $a_1, a_2, a_3, \ldots$ the **nth partial sum** S_n is the sum of the first n terms of the sequence:

$$S_n = a_1 + a_2 + a_3 + \cdots + a_n$$

The nth partial sum of a sequence can also be expressed by using **sigma notation**:

$$S_n = \sum_{k=1}^{n} a_k$$

Arithmetic Sequences (p. 404)

An **arithmetic sequence** is a sequence whose terms are obtained by adding the same fixed constant d to each term to get the next term. Thus, an arithmetic sequence has the form

$$a, a + d, a + 2d, a + 3d, \ldots$$

The number a is the **first term** of the sequence, and the number d is the **common difference**. The nth term of the sequence is

$$a_n = a + (n - 1)d$$

Partial Sums of an Arithmetic Sequence (p. 407)

For the arithmetic sequence $a_n = a + (n - 1)d$ the nth partial sum

$S_n = \sum_{k=1}^{n} [a + (k - 1)d]$ is given by either of the following equivalent formulas:

1. $S_n = \dfrac{n}{2}[2a + (n - 1)d]$

2. $S_n = n\left(\dfrac{a + a_n}{2}\right)$

Geometric Sequences (p. 408)

A **geometric sequence** is a sequence whose terms are obtained by multiplying each term by the same fixed constant r to get the next term. Thus, a geometric sequence has the form

$$a, ar, ar^2, ar^3, \ldots$$

The number a is the **first term** of the sequence, and the number r is the **common ratio**. The nth term of the sequence is

$$a_n = ar^{n-1}$$

Partial Sums of a Geometric Sequence (p. 410)

For the geometric sequence $a_n = ar^{n-1}$ the nth partial sum

$S_n = \sum_{k=1}^{n} ar^{k-1}$ (where $r \neq 1$) is given by

$$S_n = a\frac{1 - r^n}{1 - r}$$

Infinite Geometric Series (p. 412)

An **infinite geometric series** is a series of the form

$$a + ar + ar^2 + ar^3 + \cdots + ar^{n-1} + \cdots$$

An infinite series for which $|r| < 1$ has the sum

$$S = \frac{a}{1 - r}$$

Amount of an Annuity (p. 414)

The amount A_f of an **annuity** consisting of n regular equal payments of size R with interest rate i per time period is given by

$$A_f = R\frac{(1 + i)^n - 1}{i}$$

Present Value of an Annuity (p. 415)

The **present value** A_p of an annuity consisting of n regular equal payments of size R with interest rate i per time period is given by

$$A_p = R\frac{1 - (1 + i)^{-n}}{i}$$

Present Value of a Future Amount (p. 415)

If an amount A_f is to be paid in one lump sum, n time periods from now, and the interest rate per time period is i, then its **present value** A_p is given by

$$A_p = A_f(1 + i)^{-n}$$

Installment Buying (p. 416)

If a loan A_p is to be repaid in n regular equal payments with interest rate i per time period, then the size R of each payment is given by

$$R = \frac{iA_p}{1 - (1 + i)^{-n}}$$

Principle of Mathematical Induction (p. 419)

For each natural number n, let $P(n)$ be a statement that depends on n. Suppose that each of the following conditions is satisfied.

1. $P(1)$ is true.
2. For every natural number k, if $P(k)$ is true, then $P(k + 1)$ is true.

Then $P(n)$ is true for all natural numbers n.

Sums of Powers (p. 421)

0. $\sum_{k=1}^{n} 1 = n$ **1.** $\sum_{k=1}^{n} k = \frac{n(n + 1)}{2}$

2. $\sum_{k=1}^{n} k^2 = \frac{n(n + 1)(2n + 1)}{6}$ **3.** $\sum_{k=1}^{n} k^3 = \frac{n^2(n + 1)^2}{4}$

Binomial Coefficients (pp. 424–426)

If n and r are positive integers with $n \geq r$, then the **binomial coefficient** $\binom{n}{r}$ is defined by

$$\binom{n}{r} = \frac{n!}{r!(n - r)!}$$

Binomial coefficients satisfy the following properties:

$$\binom{n}{r} = \binom{n}{n - r}$$

$$\binom{k}{r - 1} + \binom{k}{r} = \binom{k + 1}{r}$$

The Binomial Theorem (p. 426)

$$(a + b)^n = \binom{n}{0}a^n + \binom{n}{1}a^{n-1}b + \binom{n}{2}a^{n-2}b^2 + \cdots + \binom{n}{n}b^n$$

▼ CONCEPT SUMMARY

	Review Exercises
Section 9.1	
■ Find the terms of a sequence	1–6, 11(a)–14(a)
■ Find the terms of a recursive sequence	7–10
■ Find the partial sums of a sequence	11(c)–14(c), 37–40
■ Use sigma notation	37–48
Section 9.2	
■ Find the terms of an arithmetic sequence	11, 14, 15–17, 25, 26, 30
■ Find the partial sums of an arithmetic sequence	50–52, 59
Section 9.3	
■ Find the terms of a geometric sequence	12, 13, 18, 19, 21–24, 27–29, 31
■ Find the partial sums of a geometric sequence	49, 53–54, 60, 61
■ Find the sum of an infinite geometric series	55–58
Section 9.4	
■ Find the amount of an annuity	62
■ Find the present value of an annuity	63
■ Find the amount of the installment payments on a loan	64
Section 9.5	
■ Prove a statement using the Principle of Mathematical Induction	65–70
Section 9.6	
■ Expand a binomial using Pascal's triangle	75–76
■ Find binomial coefficients	71–74
■ Expand a binomial using the Binomial Theorem	77–78
■ Find a particular term in a binomial expansion	79–81

▼ EXERCISES _____

1–6 ▪ Find the first four terms as well as the tenth term of the sequence with the given nth term.

1. $a_n = \dfrac{n^2}{n+1}$

2. $a_n = (-1)^n \dfrac{2^n}{n}$

3. $a_n = \dfrac{(-1)^n + 1}{n^3}$

4. $a_n = \dfrac{n(n+1)}{2}$

5. $a_n = \dfrac{(2n)!}{2^n n!}$

6. $a_n = \dbinom{n+1}{2}$

7–10 ▪ A sequence is defined recursively. Find the first seven terms of the sequence.

7. $a_n = a_{n-1} + 2n - 1, \quad a_1 = 1$

8. $a_n = \dfrac{a_{n-1}}{n}, \quad a_1 = 1$

9. $a_n = a_{n-1} + 2a_{n-2}, \quad a_1 = 1, a_2 = 3$

10. $a_n = \sqrt{3a_{n-1}}, \quad a_1 = \sqrt{3}$

11–14 ▪ The nth term of a sequence is given.
(a) Find the first five terms of the sequence.
(b) Graph the terms you found in part (a).
(c) Find the fifth partial sum of the sequence.
(d) Determine whether the series is arithmetic or geometric. Find the common difference or the common ratio.

11. $a_n = 2n + 5$

12. $a_n = \dfrac{5}{2^n}$

13. $a_n = \dfrac{3^n}{2^{n+1}}$

14. $a_n = 4 - \dfrac{n}{2}$

15–22 ▪ The first four terms of a sequence are given. Determine whether they can be the terms of an arithmetic sequence, a geometric sequence, or neither. If the sequence is arithmetic or geometric, find the fifth term.

15. $5, 5.5, 6, 6.5, \ldots$

16. $\sqrt{2}, 2\sqrt{2}, 3\sqrt{2}, 4\sqrt{2}, \ldots$

17. $t - 3, t - 2, t - 1, t, \ldots$

18. $\sqrt{2}, 2, 2\sqrt{2}, 4, \ldots$

19. $t^3, t^2, t, 1, \ldots$

20. $1, -\dfrac{3}{2}, 2, -\dfrac{5}{2}, \ldots$

21. $\dfrac{3}{4}, \dfrac{1}{2}, \dfrac{1}{3}, \dfrac{2}{9}, \ldots$

22. $a, 1, \dfrac{1}{a}, \dfrac{1}{a^2}, \ldots$

23. Show that $3, 6i, -12, -24i, \ldots$ is a geometric sequence, and find the common ratio. (Here $i = \sqrt{-1}$.)

24. Find the nth term of the geometric sequence $2, 2 + 2i, 4i, -4 + 4i, -8, \ldots$ (Here $i = \sqrt{-1}$.)

25. The sixth term of an arithmetic sequence is 17, and the fourth term is 11. Find the second term.

26. The 20th term of an arithmetic sequence is 96, and the common difference is 5. Find the nth term.

27. The third term of a geometric sequence is 9, and the common ratio is $\frac{3}{2}$. Find the fifth term.

28. The second term of a geometric sequence is 10, and the fifth term is $\frac{1250}{27}$. Find the nth term.

29. A teacher makes $32,000 in his first year at Lakeside School and gets a 5% raise each year.
(a) Find a formula for his salary A_n in his nth year at this school.
(b) List his salaries for his first 8 years at this school.

30. A colleague of the teacher in Exercise 29, hired at the same time, makes $35,000 in her first year, and gets a $1200 raise each year.
(a) What is her salary A_n in her nth year at this school?
(b) Find her salary in her eighth year at this school, and compare it to the salary of the teacher in Exercise 29 in his eighth year.

31. A certain type of bacteria divides every 5 s. If three of these bacteria are put into a petri dish, how many bacteria are in the dish at the end of 1 min?

32. If $a_1, a_2, a_3, \ldots$ and $b_1, b_2, b_3, \ldots$ are arithmetic sequences, show that $a_1 + b_1, a_2 + b_2, a_3 + b_3, \ldots$ is also an arithmetic sequence.

33. If $a_1, a_2, a_3, \ldots$ and $b_1, b_2, b_3, \ldots$ are geometric sequences, show that $a_1 b_1, a_2 b_2, a_3 b_3, \ldots$ is also a geometric sequence.

34. (a) If $a_1, a_2, a_3, \ldots$ is an arithmetic sequence, is the sequence $a_1 + 2, a_2 + 2, a_3 + 2, \ldots$ arithmetic?
(b) If $a_1, a_2, a_3, \ldots$ is a geometric sequence, is the sequence $5a_1, 5a_2, 5a_3, \ldots$ geometric?

35. Find the values of x for which the sequence $6, x, 12, \ldots$ is
(a) arithmetic   **(b)** geometric

36. Find the values of x and y for which the sequence $2, x, y, 17, \ldots$ is
(a) arithmetic   **(b)** geometric

37–40 ▪ Find the sum.

37. $\displaystyle\sum_{k=3}^{6} (k+1)^2$

38. $\displaystyle\sum_{i=1}^{4} \dfrac{2i}{2i-1}$

39. $\displaystyle\sum_{k=1}^{6} (k+1)2^{k-1}$

40. $\displaystyle\sum_{m=1}^{5} 3^{m-2}$

41–44 ▪ Write the sum without using sigma notation. Do not evaluate.

41. $\displaystyle\sum_{k=1}^{10} (k-1)^2$

42. $\displaystyle\sum_{j=2}^{100} \dfrac{1}{j-1}$

43. $\displaystyle\sum_{k=1}^{50} \dfrac{3^k}{2^{k+1}}$

44. $\displaystyle\sum_{n=1}^{10} n^2 2^n$

45–48 ▪ Write the sum using sigma notation. Do not evaluate.

45. $3 + 6 + 9 + 12 + \cdots + 99$

46. $1^2 + 2^2 + 3^2 + \cdots + 100^2$

47. $1 \cdot 2^3 + 2 \cdot 2^4 + 3 \cdot 2^5 + 4 \cdot 2^6 + \cdots + 100 \cdot 2^{102}$

48. $\dfrac{1}{1 \cdot 2} + \dfrac{1}{2 \cdot 3} + \dfrac{1}{3 \cdot 4} + \cdots + \dfrac{1}{999 \cdot 1000}$

49–54 ▪ Determine whether the expression is a partial sum of an arithmetic or geometric sequence. Then find the sum.

49. $1 + 0.9 + (0.9)^2 + \cdots + (0.9)^5$

50. $3 + 3.7 + 4.4 + \cdots + 10$

51. $\sqrt{5} + 2\sqrt{5} + 3\sqrt{5} + \cdots + 100\sqrt{5}$

52. $\frac{1}{3} + \frac{2}{3} + 1 + \frac{4}{3} + \cdots + 33$

53. $\displaystyle\sum_{n=0}^{6} 3(-4)^n$ **54.** $\displaystyle\sum_{k=0}^{8} 7(5)^{k/2}$

55–58 ▪ Find the sum of the infinite geometric series.

55. $1 - \frac{2}{5} + \frac{4}{25} - \frac{8}{125} + \cdots$

56. $0.1 + 0.01 + 0.001 + 0.0001 + \cdots$

57. $1 + \dfrac{1}{3^{1/2}} + \dfrac{1}{3} + \dfrac{1}{3^{3/2}} + \cdots$

58. $a + ab^2 + ab^4 + ab^6 + \cdots$

59. The first term of an arithmetic sequence is $a = 7$, and the common difference is $d = 3$. How many terms of this sequence must be added to obtain 325?

60. The sum of the first three terms of a geometric series is 52, and the common ratio is $r = 3$. Find the first term.

61. A person has two parents, four grandparents, eight great-grandparents, and so on. What is the total number of a person's ancestors in 15 generations?

62. Find the amount of an annuity consisting of 16 annual payments of $1000 each into an account that pays 8% interest per year, compounded annually.

63. How much money should be invested every quarter at 12% per year, compounded quarterly, in order to have $10,000 in one year?

64. What are the monthly payments on a mortgage of $60,000 at 9% interest if the loan is to be repaid in
  (a) 30 years?   **(b)** 15 years?

65–67 ▪ Use mathematical induction to prove that the formula is true for all natural numbers n.

65. $1 + 4 + 7 + \cdots + (3n - 2) = \dfrac{n(3n - 1)}{2}$

66. $\dfrac{1}{1 \cdot 3} + \dfrac{1}{3 \cdot 5} + \dfrac{1}{5 \cdot 7} + \cdots + \dfrac{1}{(2n - 1)(2n + 1)}$

$= \dfrac{n}{2n + 1}$

67. $\left(1 + \dfrac{1}{1}\right)\left(1 + \dfrac{1}{2}\right)\left(1 + \dfrac{1}{3}\right) \cdots \left(1 + \dfrac{1}{n}\right) = n + 1$

68. Show that $7^n - 1$ is divisible by 6 for all natural numbers n.

69. Let $a_{n+1} = 3a_n + 4$ and $a_1 = 4$. Show that $a_n = 2 \cdot 3^n - 2$ for all natural numbers n.

70. Prove that the Fibonacci number F_{4n} is divisible by 3 for all natural numbers n.

71–74 ▪ Evaluate the expression.

71. $\dbinom{5}{2}\dbinom{5}{3}$ **72.** $\dbinom{10}{2} + \dbinom{10}{6}$

73. $\displaystyle\sum_{k=0}^{5} \binom{5}{k}$ **74.** $\displaystyle\sum_{k=0}^{8} \binom{8}{k}\binom{8}{8 - k}$

75–78 ▪ Expand the expression.

75. $(A - B)^3$ **76.** $(x + 2)^5$

77. $(1 - x^2)^6$ **78.** $(2x + y)^4$

79. Find the 20th term in the expansion of $(a + b)^{22}$.

80. Find the first three terms in the expansion of $(b^{-2/3} + b^{1/3})^{20}$.

81. Find the term containing A^6 in the expansion of $(A + 3B)^{10}$.

1. Find the first six terms and the sixth partial sum of the sequence whose nth term is $a_n = 2n^2 - n$.

2. A sequence is defined recursively by $a_{n+1} = 3a_n - n$, $a_1 = 2$. Find the first six terms of the sequence.

3. An arithmetic sequence begins $2, 5, 8, 11, 14, \ldots$.
   (a) Find the common difference d for this sequence.
   (b) Find a formula for the nth term a_n of the sequence.
   (c) Find the 35th term of the sequence.

4. A geometric sequence begins $12, 3, 3/4, 3/16, 3/64, \ldots$.
   (a) Find the common ratio r for this sequence.
   (b) Find a formula for the nth term a_n of the sequence.
   (c) Find the tenth term of the sequence.

5. The first term of a geometric sequence is 25, and the fourth term is $\frac{1}{5}$.
   (a) Find the common ratio r and the fifth term.
   (b) Find the partial sum of the first eight terms.

6. The first term of an arithmetic sequence is 10 and the tenth term is 2.
   (a) Find the common difference and the 100th term of the sequence.
   (b) Find the partial sum of the first ten terms.

7. Let $a_1, a_2, a_3, \ldots$ be a geometric sequence with initial term a and common ratio r. Show that $a_1^2, a_2^2, a_3^2, \ldots$ is also a geometric sequence by finding its common ratio.

8. Write the expression without using sigma notation, and then find the sum.
   (a) $\displaystyle\sum_{n=1}^{5} (1 - n^2)$
   (b) $\displaystyle\sum_{n=3}^{6} (-1)^n 2^{n-2}$

9. Find the sum.
   (a) $\dfrac{1}{3} + \dfrac{2}{3^2} + \dfrac{2^2}{3^3} + \dfrac{2^3}{3^4} + \cdots + \dfrac{2^9}{3^{10}}$
   (b) $1 + \dfrac{1}{2^{1/2}} + \dfrac{1}{2} + \dfrac{1}{2^{3/2}} + \cdots$

10. Use mathematical induction to prove that for all natural numbers n,
$$1^2 + 2^2 + 3^2 + \cdots + n^2 = \frac{n(n + 1)(2n + 1)}{6}$$

11. Expand $(2x + y^2)^5$.

12. Find the term containing x^3 in the binomial expansion of $(3x - 2)^{10}$.

13. A puppy weighs 0.85 lb at birth, and each week he gains 24% in weight. Let a_n be his weight in pounds at the end of his nth week of life.
    (a) Find a formula for a_n.
    (b) How much does the puppy weigh when he is six weeks old?
    (c) Is the sequence $a_1, a_2, a_3, \ldots$ arithmetic, geometric, or neither?

CHAPTER P

SOLUTIONS TO WARM-UP EXERCISES ▪ page 1

1. (a) Every tick mark represents half a unit. The number $\frac{5}{2}$ is equal to 2.5, i.e., 1 half-unit to the right of 2. The number -1.75 is halfway between -1.5 and -2.

$$-1.75 \quad x \qquad\qquad 2.5$$

(b) The number x is halfway between $-\frac{1}{2}$ and -1, i.e., $x = -\frac{3}{4}$.

2. (a) 20 **(b)** -50 (a negative number times a positive number gives a negative result) **(c)** 12 (a negative number times a negative number gives a positive result)

3. (a) 27 **(b)** $52 - 25 = 27$. When we consider $25 - 52$, we are subtracting the larger number from the smaller number, so the answer is the opposite of 27, i.e., -27. **(c)** Because -52 and -25 are both negative, we add 52 to 25 to get 77, and our answer is negative, i.e., -77.

4. (a) In the absence of parenthesis, we do multiplication before addition: $4 \times 3 = 12$, and $12 + 6 = 18$. **(b)** The division bar acts as a grouping symbol, so we compute the numerator and the denominator separately, and then do the division: $\frac{-3}{27} = -\frac{1}{9}$. **(c)** We do the operation in the parenthesis first: $6 - 3 = 3$, and $5 \times 3 = 15$.

5. (a) The common denominator is 30. We multiply the first fraction by $\frac{5}{5}$ and the second by $\frac{3}{3}$: $\frac{5}{30} - \frac{4}{30} = \frac{1}{30}$. **(b)** We first compute the numerator and the denominator separately. Numerator: $\frac{2}{2} + \frac{1}{2} = \frac{3}{2}$. Denominator: $\frac{8}{4} - \frac{1}{4} = \frac{7}{4}$. We have the compound fraction $\dfrac{\frac{3}{2}}{\frac{7}{4}}$. We invert and multiply to obtain $\frac{3}{2} \cdot \frac{4}{7} = \frac{6}{7}$.

6. The quantity $\frac{2+5}{6-6}$ has a value of zero in the denominator, so it is undefined. The quantity $\sqrt{-6}$ is the square root of a negative number, and therefore is undefined.

7. Substitution yields $P(10) = (7.5)(8.3)(294)$, or $10P = 18{,}301.50$. Dividing both sides by 10 gives the answer 1830.15.

8. Make a table that gives the value of x^2, for each integer x from 1 through 6. Each column of the table will consist of an integer, followed by its square, as follows:

x	1	2	3	4	5	6
x^2	1	4	9	16	25	36

CHAPTER P REVIEW ▪ page 39

1. (a) $T = 250 - 2x$ **(b)** 190 **(c)** 125 **3. (a)** rational, natural number, integer **(b)** rational, integer **(c)** rational, natural number, integer **(d)** irrational **(e)** rational, neither **(f)** rational, integer **5.** Commutative Property for Addition

7. Distributive Property **9. (a)** $\frac{3}{2}$ **(b)** $\frac{1}{6}$ **11. (a)** $\frac{9}{2}$ **(b)** $\frac{25}{32}$

13. $-2 \le x < 6$

15. $x \le 4$

17. $[5, \infty)$

19. $(-1, 5]$

21. (a) $\{-1, 0, \frac{1}{2}, 1, 2, 3, 4\}$ **(b)** $\{1\}$ **23. (a)** $\{1, 2\}$ **(b)** $\{\frac{1}{72}, 1\}$ **25.** 3 **27.** 6 **29.** $\frac{1}{72}$ **31.** $\frac{1}{6}$ **33.** 11 **35.** -5 **37. (a)** $|3 - 5| = 2$ **(b)** $|3 - (-5)| = 8$ **39.** x^{-2} **41.** x^{4m+2} **43.** x^{a+b+c} **45.** x^{5c-1} **47. (a)** $7^{1/3}$ **(b)** $7^{4/5}$ **49. (a)** $x^{5/6}$ **(b)** $x^{9/2}$ **51.** $12x^5y^4$ **53.** $9x^3$ **55.** x^2y^2 **57.** $\dfrac{x(2 - \sqrt{x})}{4 - x}$ **59.** $\dfrac{4r^{5/2}}{s^7}$ **61.** 7.825×10^{10} **63.** 1.65×10^{-32} **65.** $2xy(x - 3y)$ **67.** $(x - 6)(x - 3)$ **69.** $(3x + 1)(x - 1)$ **71.** $(4t + 3)(t - 4)$ **73.** $(5 - 4t)(5 + 4t)$ **75.** $(x - 1)(x^2 + x + 1)(x + 1)(x^2 - x + 1)$ **77.** $x^{-1/2}(x - 1)^2$ **79.** $(x - 2)(4x^2 + 3)$ **81.** $\sqrt{x^2 + 2}(x^2 + x + 2)^2$ **83.** $y(a + b)(a - b)$ **85.** x^2 **87.** $6x^2 - 21x + 3$ **89.** $4a^4 - 4a^2b + b^2$ **91.** $x^3 - 6x^2 + 11x - 6$ **93.** $2x^{3/2} + x - x^{1/2}$ **95.** $2x^3 - 6x^2 + 4x$ **97.** $\dfrac{x - 3}{2x + 3}$ **99.** $\dfrac{3(x + 3)}{x + 4}$ **101.** $\dfrac{x + 1}{x - 4}$ **103.** $\dfrac{x + 1}{(x - 1)(x^2 + 1)}$ **105.** $\dfrac{1}{x + 1}$ **107.** $-\dfrac{1}{2x}$ **109.** $6x + 3h - 5$ **111.** $\{x \mid x \ne -10\}$ **113.** $\{x \mid x \ge 0 \text{ and } x \ne 4\}$ **115.** No **117.** Yes **119.** No **121.** No

CHAPTER P TEST ▪ page 42

1. (a) $C = 9 + 1.50x$ **(b)** \$15 **2. (a)** Rational, natural number, integer **(b)** Irrational **(c)** Rational, integer **(d)** Rational, integer **3. (a)** $\{0, 1, 5\}$ **(b)** $\{-2, 0, \frac{1}{2}, 1, 3, 5, 7\}$

4. (a)

(b) Intersection $[0, 2)$

Union $[-4, 3]$ **(c)** $|-4 - 2| = 6$

5. (a) -64 **(b)** 64 **(c)** $\frac{1}{64}$ **(d)** $\frac{1}{49}$ **(e)** $\frac{4}{9}$ **(f)** $\frac{1}{2}$ **(g)** $\frac{9}{16}$

(h) $\frac{1}{27}$ **6. (a)** 1.86×10^{11} **(b)** 3.965×10^{-7} **7. (a)** $6\sqrt{2}$

(b) $48a^5b^7$ **(c)** $5x^3$ **(d)** $\dfrac{x}{9y^7}$ **8. (a)** $11x - 2$

(b) $4x^2 + 7x - 15$ **(c)** $a - b$ **(d)** $4x^2 + 12x + 9$

(e) $x^3 + 6x^2 + 12x + 8$ **(f)** $x^4 - 9x^2$ **9. (a)** $(2x - 5)(2x + 5)$

(b) $(2x - 3)(x + 4)$ **(c)** $(x - 3)(x - 2)(x + 2)$

(d) $x(x + 3)(x^2 - 3x + 9)$ **(e)** $3x^{-1/2}(x - 1)(x - 2)$

(f) $xy(x - 2)(x + 2)$ **10. (a)** $\dfrac{x + 2}{x - 2}$ **(b)** $\dfrac{x - 1}{x - 3}$ **(c)** $\dfrac{1}{x - 2}$

(d) $-(x + y)$ **11. (a)** $3\sqrt[3]{2}$ **(b)** $2\sqrt{6} - 3\sqrt{2}$

FOCUS ON PROBLEM SOLVING · page 46

1. 37.5 mi/h **3.** 150 mi **5.** 427 **7.** 75 s
9. The same amount **11.** 2π
15. 15,999,999,999,992,000,000,000,001

CHAPTER 1

SOLUTIONS TO WARM-UP EXERCISES · page 49

1. (a) We combine like terms:
$(3x^2 + x^2) + (-4x + 5x) + 2 = 4x^2 + x + 2$
(b) We distribute the negative sign and combine like terms:
$3y - 3 - 2y - 1 = (3y - 2y) + (-3 - 1) = y - 4$
(c) We use FOIL to obtain $2r^2 - 7r - 4$.
2. (a) The LCD will be a multiple of both 60 and 90.
The multiples of 60 are 60, 120, 180, 240, 300, 360, . . .
The multiples of 90 are 90, 180, 270, 360, 450, 540, . . .
The lowest number in both lists is 180.
(This can also be done by prime factorization. $60 = (2^2)(3)(5)$
and $90 = (2)(3^2)(5)$ so the LCD is $(2^2)(3^2)(5) = 180$.)
(b) To obtain the smallest polynomial that is a multiple
of both $2x + 1$ and $x + 2$ we multiply them to obtain
$(2x + 1)(x + 2) = 2x^2 + 5x + 2$. (This method works because
both of the denominators have no common factors.)
3. (a) 4, because $4^3 = 64$ **(b)** 3, because $3^4 = 81$ **(c)** $3x^2$, because
$(3x^2)^3 = 27x^6$ **(d)** $|2x + 1|$, because we want the *positive* square
root of $(2x+1)^2$ **4.** We multiply 247 by 17%, recalling that $17\% =$
0.17. $(247)(0.17) = 41.99$ **5. (a)** We use $d = rt$. 1.25 (miles) $=$
r 0.25 (hours). $r = 5$ miles/hour **(b)** We use $d = rt$. 1.5 (miles) $=$
6 (miles/hour) t (hours). $t = 0.25$ hours, or 15 minutes
6. We substitute to obtain

$$\frac{+7 - \sqrt{(-7)^2 - 4(2)(3)}}{2(2)} = \frac{7 - 5}{4} = \frac{1}{2}.$$

7. We factor by grouping, and then using difference of two squares.

$$(x^3 + 2x^2) + (-9x - 18) = x^2(x + 2) - 9(x + 2)$$
$$= (x^2 - 9)(x + 2)$$
$$= (x + 3)(x - 3)(x + 2)$$

8. (a) 63 **(b)** $\sqrt{2}$ **(c)** If $x > 10$, then $2x + 1$ is positive, and
therefore $|2x + 1| = 2x + 1$. **(d)** If $x < 1$ then $x - 2$ is negative,
and therefore $|x - 2| = -(x - 2) = -x + 2$.

CHAPTER 1 REVIEW · page 93

1. 5 **3.** 4 **5.** 5 **7.** $\frac{15}{2}$ **9.** -6 **11.** 0 **13.** ±12 **15.** $\pm\dfrac{2}{\sqrt[4]{5}}$

17. $\sqrt[3]{3}$ **19.** $x = 2A - y$ **21.** $t = \dfrac{11}{6J}$ **23.** -5 **25.** -27

27. 625 **29.** No solution **31.** 2, 7 **33.** $-1, \frac{1}{2}$ **35.** $0, \pm\frac{5}{2}$

37. $\dfrac{-2 \pm \sqrt{7}}{3}$ **39.** $\dfrac{3 \pm \sqrt{6}}{3}$ **41.** ±3 **43.** 1 **45.** 3, 11

47. $-2, 7$ **49.** 20 lb raisins, 30 lb nuts **51.** $\frac{1}{4}(\sqrt{329} - 3)$

≈ 3.78 mi/h **53.** 12 cm, 16 cm **55.** 23 ft by 46 ft by 8 ft

57. $-3 - 9i$ **59.** $19 + 40i$ **61.** $\dfrac{-5 - 12i}{13}$ **63.** i

65. $(2 + 2\sqrt{3}) + (2 - 2\sqrt{3})i$ **67.** $\pm4i$ **69.** $-3 \pm i$

71. $\pm4, \pm4i$ **73.** $\pm\dfrac{\sqrt{3}}{3}i$

75. $(-3, \infty)$

77. $(-3, -1]$

79. $(-\infty, -6) \cup (2, \infty)$

81. $[-4, -1)$

83. $(-\infty, -2) \cup (2, 4]$

85. $[2, 8]$

87. $(-\infty, -1] \cup [0, \infty)$

89. (a) $[-3, \frac{8}{3}]$ **(b)** $(0, 1)$

CHAPTER 1 TEST · page 95

1. (a) 5 **(b)** $-\frac{5}{2}$ **(c)** 512 **(d)** $\frac{15}{2}$ **2.** $c = \sqrt{\dfrac{E}{m}}$ **3.** 150 km

4. (a) $5 - 4i$ **(b)** $1 - 6i$ **(c)** $7 + 3i$ **(d)** $-\frac{1}{5} + \frac{2}{5}i$ **(e)** i

(f) $2\sqrt{2} - 8i$ **5. (a)** $-3, 4$ **(b)** $\dfrac{-2 \pm i\sqrt{2}}{2}$ **(c)** No solution

(d) 1, 16 **(e)** $0, \pm4$ **(f)** $\frac{2}{3}, \frac{22}{3}$ **6.** 50 ft by 120 ft

7. (a) $(-\frac{5}{2}, 3]$ **(b)** $(0, 1) \cup (2, \infty)$ **(c)** $(1, 5)$ **(d)** $[-4, -1)$
8. 41°F to 50°F **9.** $0 \le x \le 4$

CHAPTER 2

SOLUTIONS TO WARM-UP EXERCISES · page 97

1. (a) We want all x-values to the right of zero, including zero.

(b) If $|x| = 2$, then x can be positive or negative 2.

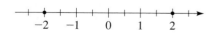

(c) We want all x-values whose distance from 0 is less than or
equal to 2.

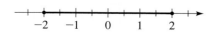

(d) We want all x-values whose distance from 1 is greater than 1.

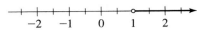

2. (a) We use the quadratic formula with $a = 3$, $b = 5$, $c = -2$:

$$\frac{-5 \pm \sqrt{5^2 - 4(3)(-2)}}{2(3)} = \frac{-5 \pm 7}{6} = \frac{1}{3}, -2$$

(b) We use the quadratic formula with $a = 1$, $b = 2$, $c = 5$:

$$\frac{-2 \pm \sqrt{2^2 - 4(1)(5)}}{2(1)} = \frac{-2 \pm \sqrt{-16}}{2} \quad \text{no real solutions}$$

3. $2x^2 - 4x - 16 = 0$

$$2x^2 - 4x = 16$$
$$2(x^2 - 2x) = 16$$
$$2(x^2 - 2x + 1) = 16 + 2$$
$$(x^2 - 2x + 1) = 9$$
$$(x - 1)^2 = 9$$
$$x - 1 = \pm 3$$
$$x = 4, -2$$

4. We use the fact that -1 to an even power is 1, and -1 to an odd power is -1. We obtain $3x^2y^4 - 6x - x^2y^3$.

5. We use our rules for working with inequalities:

$$3x - 2 > 0$$
$$3x > 2$$
$$x > \frac{2}{3}$$

6. We use the method described in Section 1.6:

$$x^2 + x - 2 \leq 0$$
$$(x + 2)(x - 1) \leq 0$$

The relevant intervals are $(-\infty, -2)$, $(-2, 1)$, $(1, \infty)$.

We make a table of values to test the intervals:

Interval	$(-\infty, -2)$	$(-2, 1)$	$(1, \infty)$
Sign of $(x + 2)$	$-$	$+$	$+$
Sign of $(x - 1)$	$-$	$-$	$+$
Sign of $(x + 2)(x - 1)$	$+$	$-$	$+$

Therefore, the solution to the inequality $x^2 + x - 2 \leq 0$ is $[-2, 1]$ (square brackets because we are including the endpoints).

7. Solve the following equations for b:
(a) Multiplying both sides by $\frac{1}{5}$ gives $b = -\frac{1}{5}$.   **(b)** If $a = 0$, then there are no solutions, because $0 \neq -1$. If $a \neq 0$, then multiplying both sides by $\frac{1}{a}$ gives $b = -\frac{1}{a}$.   **8.** The fraction will be undefined precisely when $2 + a$ is zero, or $a = -2$.

CHAPTER 2 REVIEW ▪ page 130

1. (a)

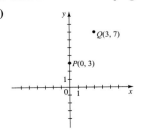

(b) 5   **(c)** $\left(\frac{3}{2}, 5\right)$ **(d)** $m = \frac{4}{3}$; point-slope: $y - 7 = \frac{4}{3}(x - 3)$; slope-intercept: $y = \frac{4}{3}x + 3$;   **(e)** $x^2 + (y - 3)^2 = 25$

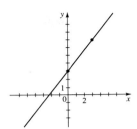
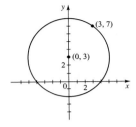

3. (a)

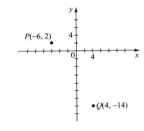

(b) $2\sqrt{89}$ **(c)** $(-1, -6)$

(d) $m = -\frac{8}{5}$;   point-slope: $y + 14 = -\frac{8}{5}(x - 4)$; slope-intercept: $y = -\frac{8}{5}x - \frac{38}{5}$;   **(e)** $(x + 6)^2 + (y - 2)^2 = 356$

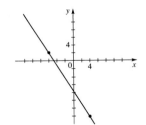
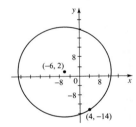

5.

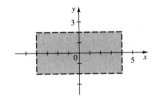

7. B **9.** $(x + 5)^2 + (y + 1)^2 = 26$

11. (a) Circle   **(b)** Center $(-1, 3)$, radius 1

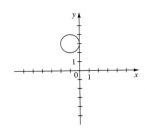

13. (a) No graph

15.

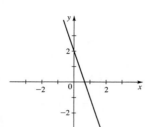

17.

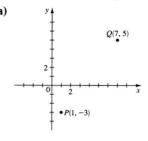

19.

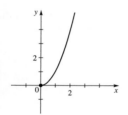

21.

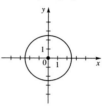

23.

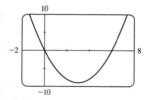

25. (a) Symmetry about y-axis   **(b)** x-intercepts $-3, 3$; y-intercept 9   **27. (a)** Symmetry about y-axis   **(b)** x-intercept 0; y-intercepts 0, 2   **29. (a)** Symmetry about x- and y-axes and the origin   **(b)** x-intercepts $-4, 4$; no y-intercept

31.

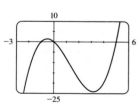

33.

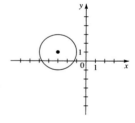

35. $-1, 7$ **37.** $-2.72, -1.15, 1.00, 2.87$ **39.** $[1, 3]$
41. $(-1.85, -0.60) \cup (0.45, 2.00)$ **43. (a)** $y = 2x + 6$
(b) $2x - y + 6 = 0$ **45. (a)** $y = \frac{2}{3}x - \frac{16}{3}$
(b) $2x - 3y - 16 = 0$ **47. (a)** $x = 3$ **(b)** $x - 3 = 0$
49. (a) $y = \frac{2}{5}x + \frac{3}{5}$ **(b)** $2x - 5y + 3 = 0$
51. (a) $y = -2x$ **(b)** $2x + y = 0$ **53. (a)** The slope represents the amount the spring lengthens for a one-pound increase in weight. The S-intercept represents the unstretched length of the spring.
(b) 4 in.   **55.** $M = 8z$ **57. (a)** $I = k/d^2$ **(b)** 64,000

(c) 160 candles   **59.** 11.0 mi/h   **61.** $F = 0.000125q_1q_2$
63. $x^2 + y^2 = 169, 5x - 12y + 169 = 0$

CHAPTER 2 TEST ▪ page 132

1. (a)

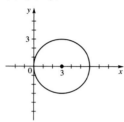

(b) 10   **(c)** $(4, 1)$ **(d)** $\frac{4}{3}$ **(e)** $y = -\frac{3}{4}x + 4$
(f) $(x - 4)^2 + (y - 1)^2 = 25$

2. (a) $(0, 0); \frac{5}{2}$ **(b)** $(3, 0); 3$

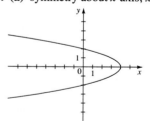

(c) $(-3, 1); 2$

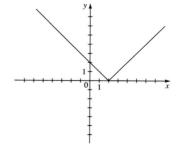

3. (a) symmetry about x-axis; x-intercept 4; y-intercepts $-2, 2$

(b) No symmetry; x-intercept 2; y-intercept 2

4. (a) x-intercept 5; y-intercept -3

(b)

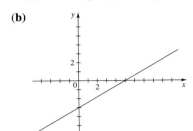

(c) $y = \frac{3}{5}x - 3$ **(d)** $\frac{3}{5}$ **(e)** $-\frac{5}{3}$

5. (a) $3x + y - 3 = 0$ **(b)** $2x + 3y - 12 = 0$

6. (a) $4°C$ **(b)**

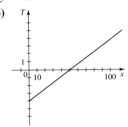

(c) The slope is the rate of change in temperature, the x-intercept is the depth at which the temperature is $0°C$, and the T-intercept is the temperature at ground level.   **7. (a)** $-2.94, -0.11, 3.05$
(b) $[-1.07, 3.74]$ **8. (a)** $M = kwh^2/L$ **(b)** 400
(c) 12,000 lb

ANSWERS TO CUMULATIVE REVIEW TEST FOR CHAPTERS 1 AND 2 ▪ page 133

1. $8000   **2. (a)** $-\frac{1}{3} + 3i$ **(b)** 10   **(c)** $-\frac{9}{5} - \frac{12}{5}i$
(d) $2 + i\frac{\sqrt{3}}{2}$
3. (a) -2 **(b)** -16 **(c)** $-1, 6$ **(d)** $\frac{2}{3} \pm i\frac{\sqrt{2}}{3}$
(e) $\pm 2, \pm 2i$ **(f)** $-1.75, 2.75$
4. (a) $[3, \infty)$ **(b)** $(-\infty, -4) \cup (2, \infty)$

(c) $\left[-\frac{1}{2}, 1\right]$ **(d)** $(3, \infty)$

5. (a)

(b) $-2, -1, 2, 3$ **(c)** $-2 \le x \le -1$ and $2 \le x \le 3$
6. (a) $\sqrt{194} \approx 13.93$ **(b)** $y = -\frac{5}{13}x + \frac{63}{13}$
(c) $\left(x - \frac{7}{2}\right)^2 + \left(y - \frac{7}{2}\right)^2 = \frac{97}{2}$ **(d)** $y = \frac{13}{5}x + \frac{69}{5}$
(e) $y = -\frac{5}{13}x$

7. (a)   **(b)** $C(3, 0), 3$

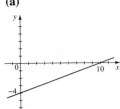   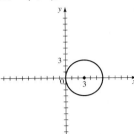

8. (a) $y = -2x + 6$ **(b)** $(x + 3)^2 + (y - 4)^2 = 25$
9. 8 mi/h   **10. (a)** $S = 187.5x^2$ **(b)** $18,750.00
(c) 18 years

CHAPTER 3

SOLUTIONS TO WARM-UP EXERCISES ▪ page 135

1. (a) We distribute the negative sign and combine terms with the same powers. $x^2 + 3x + 2 - x^2 + 2x - 3 = 5x - 1$.
(b) We use FOIL to obtain $x^3 + 2x + 4x^2 + 8 = x^3 + 4x^2 + 2x + 8$.
2. (a) We use FOIL on the first term, and combine terms with the same powers. $x^2 - 4x + 4 + x - 2 + 4 = x^2 - 3x + 6$
(b) We only have to combine the -3 and the 4: $x^2 + x + 1$.
3. We use FOIL and a little bit of factoring. Notice that we are assuming $h \ne 0$:

$$\frac{(x + h)^2 - x^2}{h} = \frac{x^2 + 2xh + h^2 - x^2}{h}$$

$$= \frac{2xh + h^2}{h}$$

$$= \frac{h(2x + h)}{h}$$

$$= 2x + h$$

4. (a) It is impossible to divide by zero, or to take the square root of a negative number. So we need $x - 2 > 0$, i.e., $x > 2$.
(b) It is impossible to divide by zero, so we need $x - 2 \ne 0$, or $x \ne 2$. (It IS possible to take the cube root of a negative number, so $x - 2 > 0$ is NOT a requirement.)

5. We use $m = \dfrac{y_2 - y_1}{x_2 - x_1}$ to obtain a slope of $\dfrac{-3 - 2}{5 - 3} = -\dfrac{5}{2}$.
6. We pay careful attention to order of operations to obtain

$$\frac{[(4x - 5)^{1/7}]^7 + 5}{4} = \frac{(4x - 5) + 5}{4} = \frac{4x}{4} = x$$

7. We wish to isolate y on one side of the equation:

$$x = \frac{6y - 2}{3}$$

$$3x = 6y - 2$$

$$3x + 2 = 6y$$

$$\frac{3x + 2}{6} = y$$

So our final answer is $y = \dfrac{3x + 2}{6}$.

8. We use our graphing calculator, and use trial-and-error to obtain a viewing window that makes sense.

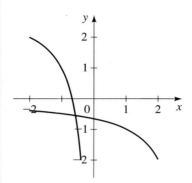

CHAPTER 3 REVIEW ▪ page 180

1. $f(x) = x^2 - 5$ **3.** Add 10, then multiply the result by 3.

5.

x	$g(x)$
-1	5
0	0
1	-3
2	-4
3	-3

7. **(a)** $C(1000) = 34{,}000$, $C(10{,}000) = 205{,}000$
(b) The costs of printing 1000 and 10,000 copies of the book
(c) $C(0) = 5000$; fixed costs **9.** $6, 2, 18, a^2 - 4a + 6$,
$a^2 + 4a + 6, x^2 - 2x + 3, 4x^2 - 8x + 6, 2x^2 - 8x + 10$
11. **(a)** Not a function **(b)** Function **(c)** Function, one-to-one
(d) Not a function **13.** Domain $[-3, \infty)$, range $[0, \infty)$
15. $(-\infty, \infty)$ **17.** $[-4, \infty)$ **19.** $\{x \mid x \neq -2, -1, 0\}$
21. $(-\infty, -1] \cup [1, 4]$

23.

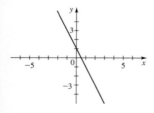

25.

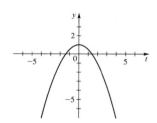

27.

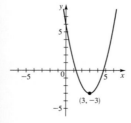

29.

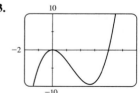

31.

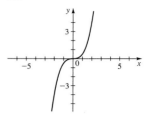

33.

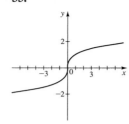

35.

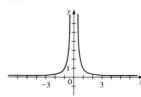

37.

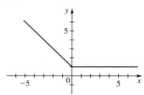

39.

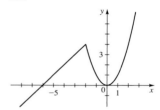

41. No **43.** Yes **45.** (iii)
47.

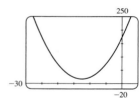

49.

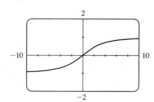

51. $[-2.1, 0.2] \cup [1.9, \infty)$

53.

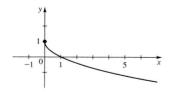

Increasing on $(-\infty, 0]$,
$[2.67, \infty)$; decreasing on
$[0, 2.67]$

55. 5 **57.** $\dfrac{-1}{3(3 + h)}$ **59.** **(a)** $P(10) = 5010$, $P(20) = 7040$; the
populations in 1995 and 2005 **(b)** 203 people/yr; average annual
population increase **61.** **(a)** $\frac{1}{2}, \frac{1}{2}$ **(b)** Yes, because it is a
linear function **63.** **(a)** Shift upward 8 units **(b)** Shift to the left
8 units **(c)** Stretch vertically by a factor of 2, then shift upward
1 unit **(d)** Shift to the right 2 units and downward 2 units
(e) Reflect in y-axis **(f)** Reflect in y-axis, then in x-axis
(g) Reflect in x-axis **(h)** Reflect in line $y = x$ **65.** **(a)** Neither
(b) Odd **(c)** Even **(d)** Neither **67.** $g(-1) = -7$ **69.** 68 ft

71. Local maximum ≈ 3.79 when $x \approx 0.46$; local minimum ≈ 2.81 when $x \approx -0.46$

73.

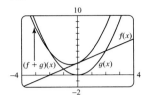

75. (a) $(f + g)(x) = x^2 - 6x + 6$ (b) $(f - g)(x) = x^2 - 2$
(c) $(fg)(x) = -3x^3 + 13x^2 - 18x + 8$
(d) $(f/g)(x) = (x^2 - 3x + 2)/(4 - 3x)$
(e) $(f \circ g)(x) = 9x^2 - 15x + 6$
(f) $(g \circ f)(x) = -3x^2 + 9x - 2$
77. $(f \circ g)(x) = -3x^2 + 6x - 1, (-\infty, \infty)$;
$(g \circ f)(x) = -9x^2 + 12x - 3, (-\infty, \infty); (f \circ f)(x) = 9x - 4,$
$(-\infty, \infty); (g \circ g)(x) = -x^4 + 4x^3 - 6x^2 + 4x, (-\infty, \infty)$
79. $(f \circ g \circ h)(x) = 1 + \sqrt{x}$ **81.** Yes   **83.** No

85. No   **87.** $f^{-1}(x) = \dfrac{x + 2}{3}$ **89.** $f^{-1}(x) = \sqrt[3]{x} - 1$

91. (a), (b)

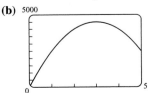

(c) $f^{-1}(x) = \sqrt{x + 4}$

CHAPTER 3 TEST ▪ page 184

1. (a) and (b) are graphs of functions, (a) is one-to-one
2. (a) $2/3, \sqrt{6}/5, \sqrt{a}/(a - 1)$ (b) $[-1, 0) \cup (0, \infty)$
3. (a) $f(x) = (x - 2)^3$
(b)

x	$f(x)$
-1	-27
0	-8
1	-1
2	0
3	1
4	8

(c)

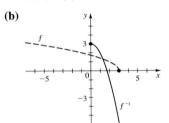

(d) By the Horizontal Line Test; take the cube root, then add 2
(e) $f^{-1}(x) = x^{1/3} + 2$ **4.** (a) $R(2) = 4000, R(4) = 4000$; total sales revenue with prices of $2 and $4
(b)

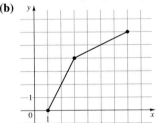

revenue increases until price reaches $3, then decreases

(c) $4500; $3   **5.** 5

6. (a)

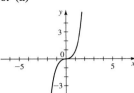

(b)

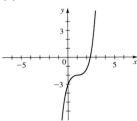

7. (a) Shift to the right 3 units, then shift upward 2 units
(b) Reflect in y-axis
8. (a) $3, 0$ (b)

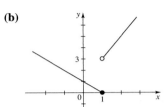

9. (a) $(f \circ g)(x) = (x - 3)^2 + 1$ (b) $(g \circ f)(x) = x^2 - 2$
(c) 2 (d) 2 (e) $(g \circ g \circ g)(x) = x - 9$
10. (a) $f^{-1}(x) = 3 - x^2, x \geq 0$
(b)

11. (a) Domain $[0, 6]$, range $[1, 7]$
(b)   (c) $\frac{5}{4}$

12. (a)   (b) No

(c) Local minimum ≈ -27.18 when $x \approx -1.61$; local maximum ≈ -2.55 when $x \approx 0.18$; local minimum ≈ -11.93 when $x \approx 1.43$
(d) $[-27.18, \infty)$ (e) Increasing on $[-1.61, 0.18] \cup [1.43, \infty)$;
decreasing on $(-\infty, -1.61] \cup [0.18, 1.43]$

CHAPTER 4

SOLUTIONS TO WARM-UP EXERCISES ▪ page 187

1.
$$2x^2 - x - 1 = 0$$
$$2x^2 - x = 1$$
$$2\left(x^2 - \frac{1}{2}x\right) = 1$$
$$2\left(x^2 - \frac{1}{2}x + \frac{1}{16}\right) = 1 + \frac{1}{8}$$
$$2\left(x - \frac{1}{4}\right)^2 = \frac{9}{8}$$
$$\left(x - \frac{1}{4}\right)^2 = \frac{9}{16}$$
$$x - \frac{1}{4} = \pm\frac{3}{4}$$
$$x = \frac{1 \pm 3}{4}$$
$$x = 1, -\frac{1}{2}$$

2. $f(x) = (x - 30)(x + 1)^2(x - 2)$. Substituting in $x = 30$, -1, or 2 gives a zero term, causing the product to be zero. Substituting in $x = 0$ gives the simple computation $-30 \cdot 1 \cdot -2 = 60$.

3. Consider the arithmetic problem $21,198 \div 15$. Use long division to compute the quotient and remainder.

```
          1  4  1  3
   1 5 | 2  1  1  9  8
        -1  5
         6  1
        -6  0
            1  9
           -1  5
               4  8
              -4  5
                  3
```

Quotient: 1413 Remainder: 3

4. We use a factor tree or similar method to find the prime factors: $90 = 2 \cdot 3^2 \cdot 5$. Then we take all combinations of the prime factors:

1 =	1
2 =	2
3 =	3
5 =	5
6 =	$2 \cdot 3$
9 =	$3 \cdot 3$
10 =	$2 \cdot 5$
15 =	$3 \cdot 5$
18 =	$2 \cdot 3^2$
30 =	$2 \cdot 3 \cdot 5$
45 =	$3^2 \cdot 5$
90 =	$2 \cdot 3^2 \cdot 5$

Including the negatives, we obtain 24 factors:
$\pm 1, \pm 2, \pm 3, \pm 5, \pm 6, \pm 9, \pm 10, \pm 15, \pm 18, \pm 30, \pm 45, \pm 90$

5. We can first factor out an x: $x(x^2 - x - 6)$. Now we are left with a quadratic to factor, using any method: $x(x - 3)(x + 2)$.

6. (a) $f(0) = \dfrac{-5}{-8} = \dfrac{5}{8}$ (b) 2.01108871, 2.000011, 2.000000011; $f(x)$ is getting closer and closer to 2.
(c) -915.125153, 91667.90277, -916664.67; $f(x)$ is getting larger and larger . . . large and negative for x-values less than 2, and large and positive for x-values greater than 2.

CHAPTER 4 REVIEW ▪ page 237

1. (a) $f(x) = (x + 2)^2 - 3$
(b)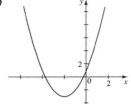

3. (a) $g(x) = -(x - 4)^2 + 17$
(b)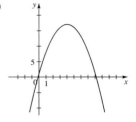

5. Minimum $f(-1) = -7$ **7.** 68 feet

9.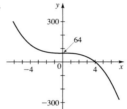

11.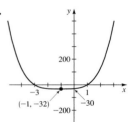

13.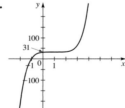

15. (a) 0 (multiplicity 3), 2 (multiplicity 2)
(b)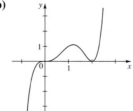

17.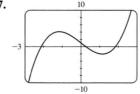

x-intercepts $-2.1, 0.3, 1.9$
y-intercept 1
local maximum $(-1.2, 4.1)$
local minimum $(1.2, -2.1)$

$y \to \infty$ as $x \to \infty$
$y \to -\infty$ as $x \to -\infty$

19. 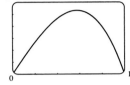 x-intercepts $-0.1, 2.1$
y-intercept -1
local minimum $(1.4, -14.5)$
$y \to \infty$ as $x \to \infty$
$y \to \infty$ as $x \to -\infty$

21. (a) $S = 13.8x(100 - x^2)$ **(b)** $0 \le x \le 10$
(c) 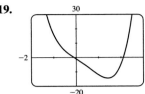   **(d)** 5.8 in.

In answers 23–29 the first polynomial given is the quotient, and the second is the remainder.
23. $x - 1, 3$ **25.** $x^2 + 3x + 23, 94$ **27.** $x^3 - 5x^2 + 17x - 83,$
422 **29.** $2x - 3, 12$ **31.** 3   **35.** 8   **37. (a)** $\pm 1, \pm 2, \pm 3, \pm 6,$
$\pm 9, \pm 18$ **(b)** 2 or 0 positive, 3 or 1 negative
39. (a) $-4, 0, 4$ **41. (a)** $-2, 0$ (multiplicity 2), 1
(b) 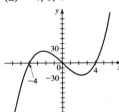   **(b)**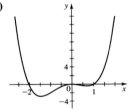

43. (a) $-2, -1, 2, 3$ **45. (a)** $-\frac{1}{2}, 1$
(b) 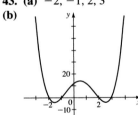   **(b)**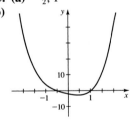

47. $4x^3 - 18x^2 + 14x + 12$ **49.** No; since the complex conjugates of imaginary zeros will also be zeros, the polynomial would have 8 zeros, contradicting the requirement that it have degree 4.
51. $-3, 1, 5$ **53.** $-1 \pm 2i, -2$ (multiplicity 2)   **55.** $\pm 2,$
1 (multiplicity 3)   **57.** $\pm 2, \pm 1 \pm i\sqrt{3}$ **59.** $1, 3, \dfrac{-1 \pm i\sqrt{7}}{2}$
61. $x = -0.5, 3$ **63.** $x \approx -0.24, 4.24$
65. $2, P(x) = (x - 2)(x^2 + 2x + 2)$
67. 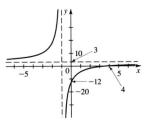   **69.**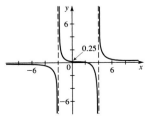

71.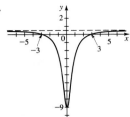

73. 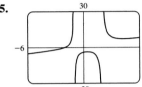 x-intercept 3
y-intercept -0.5
vertical $x = -3$
horizontal $y = 0.5$
no local extrema

75. 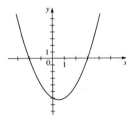 x-intercept -2
y-intercept -4
vertical $x = -1, x = 2$
slant $y = x + 1$
local maximum $(0.425, -3.599)$
local minimum $(4.216, 7.175)$

77. $(-2, -28), (1, 26), (2, 68), (5, 770)$

CHAPTER 4 TEST ▪ page 239
1. $f(x) = \left(x - \frac{1}{2}\right)^2 - \frac{25}{4}$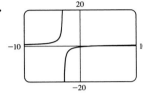

2. Minimum $f\left(-\frac{3}{2}\right) = -\frac{3}{2}$ **3. (a)** 2500 ft   **(b)** 1000 ft
4.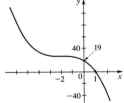

5. (a) $x^3 + 2x^2 + 2, 9$ **(b)** $x^3 + 2x^2 + \frac{1}{2}, \frac{15}{2}$
6. (a) $\pm 1, \pm 3, \pm \frac{1}{2}, \pm \frac{3}{2}$ **(b)** $2(x - 3)\left(x - \frac{1}{2}\right)(x + 1)$
(c) $-1, \frac{1}{2}, 3$ **(d)**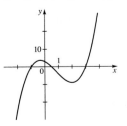

7. $3, -1 \pm i$ **8.** $(x - 1)^2(x - 2i)(x + 2i)$
9. $x^4 + 2x^3 + 10x^2 + 18x + 9$

10. (a) 4, 2, or 0 positive; 0 negative
(c) 0.17, 3.93

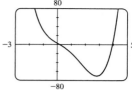

(d) Local minimum $(2.8, -70.3)$
11. (a) r, u **(b)** s **(c)** s
(d)

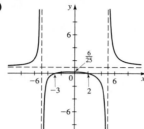

(e) $x^2 - 2x - 5$

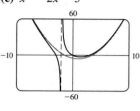

CHAPTER 5

SOLUTIONS TO WARM-UP EXERCISES ▪ page 241

1. (a) $2 \cdot 2 \cdot 2 \cdot 2 \cdot 2 = 32$ **(b)** $(-1) \cdot (-1) \cdot (-1) \cdot (-1) \cdot (-1) \cdot (-1) \cdot (-1) = 1$ **(c)** $\frac{1}{\sqrt[3]{27}} = \frac{1}{3}$ **(d)** $\left(\sqrt{16}\right)^3 = 64$

2. (a) We shift the original graph left by 4 units:

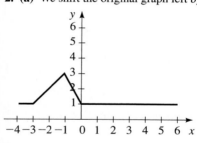

(b) We shift the original graph up by 2 units:

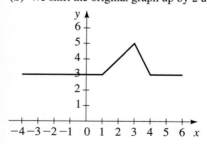

(c) We vertically stretch the original graph by a factor of 2:

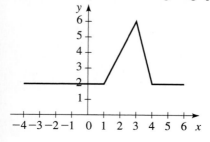

(d) We flip the original graph across the y-axis:

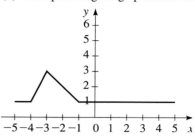

3. We obtain, directly from the table: $f(1) = 1, f(4) = 7, f(5) = 8$.
Since $x = 14$ is not in the table, we cannot compute $f(14)$.
For the inverse functions such as $f^{-1}(4)$, we look for a value of x such that $f(x) = 4$.
Since $f(3) = 4$, we know $f^{-1}(4) = 3$.
Similarly, since $f(1) = 1$, we know $f^{-1}(1) = 1$, and since $f(7) = 14$, we know $f^{-1}(14) = 7$. Since there is no value of x in the table for which $f(x) = 5$, we cannot compute $f^{-1}(5)$.

4.
$$\frac{4x^{1/3} - 3}{5} = 1$$
$$4x^{1/3} - 3 = 5$$
$$4x^{1/3} = 8$$
$$x^{1/3} = 2$$
$$x = 8$$

5.
$$x^2 - 8x + 12 = 0$$
$$(x - 2)(x - 6) = 0$$
$$x = 2, 6$$

6. (a) A preliminary graph does not give us enough detail to find the intercepts:

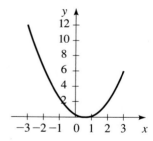

We zoom in to find where the graph touches the x-axis:

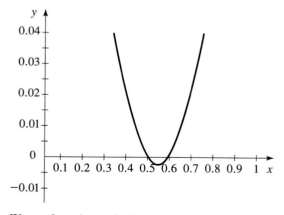

We see, from the graph, that the x-intercepts are at $x = 0.5, 0.6$.

(b) The solutions to this equation are precisely the *x*-intercepts—the values of *x* for which $f(x) = 0$. $x = 0.5, 0.6$

7. We solve $6 = 3(1) + 5 + a$ to obtain $a = -2$.

CHAPTER 5 REVIEW ▪ page 281

1. 0.089, 9.739, 55.902 **3.** 11.954, 2.989, 2.518

5. $\mathbb{R}, (0, \infty), y = 0$ **7.** $\mathbb{R}, (3, \infty), y = 3$

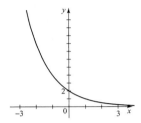 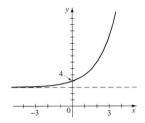

9. $(1, \infty), \mathbb{R}, x = 1$ **11.** $(0, \infty), \mathbb{R}, x = 0$

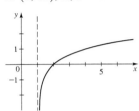 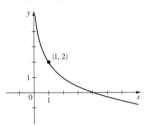

13. $\mathbb{R}, (-1, \infty), y = -1$ **15.** $(0, \infty), \mathbb{R}, x = 0$

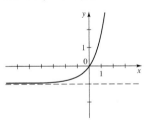 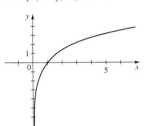

17. $\left(-\infty, \frac{1}{2}\right)$ **19.** $(-\infty, -2) \cup (2, \infty)$ **21.** $2^{10} = 1024$

23. $10^y = x$ **25.** $\log_2 64 = 6$ **27.** $\log 74 = x$ **29.** 7 **31.** 45

33. 6 **35.** -3 **37.** $\frac{1}{2}$ **39.** 2 **41.** 92 **43.** $\frac{2}{3}$

45. $\log A + 2 \log B + 3 \log C$ **47.** $\frac{1}{2}[\ln(x^2 - 1) - \ln(x^2 + 1)]$

49. $2 \log_5 x + \frac{3}{2} \log_5(1 - 5x) - \frac{1}{2} \log_5(x^3 - x)$ **51.** $\log 96$

53. $\log_2\left(\dfrac{(x - y)^{3/2}}{(x^2 + y^2)^2}\right)$ **55.** $\log\left(\dfrac{x^2 - 4}{\sqrt{x^2 + 4}}\right)$ **57.** 5

59. 2.60 **61.** -1.15 **63.** $-4, 2$ **65.** -15 **67.** 3

69. 0.430618 **71.** 2.303600

73.

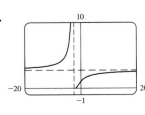

vertical asymptote
$x = -2$
horizontal asymptote
$y = 2.72$
no maximum or minimum

75.

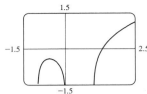

vertical asymptotes
$x = -1, x = 0, x = 1$
local maximum
$\approx (-0.58, -0.41)$

77. 2.42 **79.** $0.16 < x < 3.15$

81. Increasing on $(-\infty, 0]$ and $[1.10, \infty)$, decreasing on $[0, 1.10]$

83. 1.953445 **85.** -0.579352 **87.** $\log_4 258$

89. (a) \$16,081.15 **(b)** \$16,178.18 **(c)** \$16,197.64

(d) \$16,198.31 **91.** 1.83 yr **93.** 4.341\%

95. (a) $n(t) = 30e^{0.15t}$ **(b)** 55 **(c)** 19 yr **97. (a)** 9.97 mg

(b) 1.39×10^5 yr **99. (a)** $n(t) = 150e^{-0.0004359t}$ **(b)** 97.0 mg

(c) 2520 yr **101. (a)** $n(t) = 1500e^{0.1515t}$ **(b)** 7940

103. 7.9, basic **105.** 8.0

CHAPTER 5 TEST ▪ page 284

1. (a) $\mathbb{R}, (4, \infty), y = 4$ **(b)** $(-3, \infty), \mathbb{R}, x = -3$

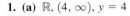 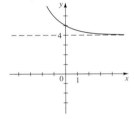 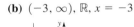 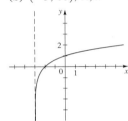

2. (a) $\log_6 25 = 2x$ **(b)** $e^3 = A$ **3. (a)** 36 **(b)** 3 **(c)** $\frac{3}{2}$

(d) 3 **(e)** $\frac{2}{3}$ **(f)** 2 **4.** $\frac{1}{3}[\log(x + 2) - 4 \log x - \log(x^2 + 4)]$

5. $\ln\left(\dfrac{x\sqrt{3 - x^4}}{(x^2 + 1)^2}\right)$ **6. (a)** 4.32 **(b)** 0.77 **(c)** 5.39 **(d)** 2

7. (a) $n(t) = 1000e^{2.07944t}$ **(b)** 22,627 **(c)** 1.3 h

(d)

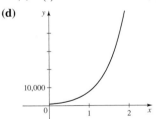

8. (a) $A(t) = 12,000\left(1 + \dfrac{0.056}{12}\right)^{12t}$ **(b)** \$14,195.06

(c) 9.249 yr **9. (a)** $A(t) = 3e^{-0.069t}$ **(b)** 0.048 g

(c) after 3.6 min **10.** 1995 times more intense

CUMULATIVE REVIEW TEST FOR CHAPTERS 3, 4, AND 5 ▪ page 285

1. (a) $(-\infty, \infty)$ **(b)** $[-4, \infty)$ **(c)** 12, 0, 0, 2, $2\sqrt{3}$, undefined

(d) $x^2 - 4, \sqrt{x + 6}, -4 + h^2$ **(e)** $\frac{1}{8}$

(f) $f \circ g = x + 4 - \sqrt{x + 4}, g \circ f = |x - 2|, f(g(12)) = 0$,

$g(f(12)) = 10$ **(g)** $g^{-1}(x) = x^2 - 4, x \geq 0$

2. (a) 4, 4, 4, 0, 1   **(b)**

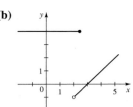

3. (a) $f(x) = -2(x - 2)^2 + 13$ **(b)** Maximum 13
(c)

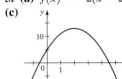

   (d) Increasing on $(-\infty, 2]$;
decreasing on $[2, \infty)$
   (e) Shift upward 5 units
   (f) Shift to the left 3 units

4. f, D; g, C; r, A; s, F; h, B; k, E
5. (a) $\pm 1, \pm 2, \pm 4, \pm 8, \pm\frac{1}{2}$ **(b)** $2, 4, -\frac{1}{2}$
(c) $P(x) = 2(x - 2)(x - 4)\left(x + \frac{1}{2}\right)$ **(d)**

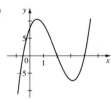

6. (a) 1 (multiplicity 2); $-1, 1 + i, 1 - i$ (multiplicity 1)
(b) $Q(x) = (x - 1)^2(x + 1)(x - 1 - i)(x - 1 + i)$
(c) $Q(x) = (x - 1)^2(x + 1)(x^2 - 2x + 2)$

7. x-intercepts 0, -2; y-intercept 0; horizontal
asymptote $y = 3$; vertical asymptotes $x = 2$
and $x = -1$

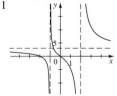   **8.**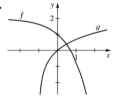

9. (a) -4 **(b)** $5 \log x + \frac{1}{2} \log(x - 1) - \log(2x - 3)$
10. (a) 4   **(b)** $\ln 2, \ln 4$ **11. (a)** \$29,396.15
(b) After 6.23 years   **(c)** 12.837 years
12. (a) $P(t) = 120e^{0.0565t}$ **(b)** 917   **(c)** After 49.8 months

CHAPTER 6

SOLUTIONS TO WARM-UP EXERCISES ▪ page 287

1. Rather than going through the effort to solve the equation, we can
just check the proposed solutions by substituting them into the expression $x^2 e^x - x e^x - 2e^x - x^2 + x$ and seeing if the result is -2.

-2: $(-2)^2 e^{-2} - (-2)e^{-2} - 2e^{-2} - (-2)^2 + (-2) = 4e^{-2} - 4$
-1: $(-1)^2 e^{-1} - (-1)e^{-1} - 2e^{-1} - (-1)^2 + (-1) = -2$
0: $(0)^2 e^0 - (0)e^0 - 2e^0 - (0)^2 + (0) = -2$
1: $(1)^2 e^1 - (1)e^1 - 2e^1 - (1)^2 + (1) = -2e$
2: $(2)^2 e^2 - (2)e^2 - 2e^2 - (2)^2 + (2) = -2$

We see that the solutions are $-1, 0, 2$.

2. $3x + 2y = 4$
$3x = 4 - 2y$
$x = \dfrac{4 - 2y}{3}$

3. The slopes of all but the fourth line can be read directly from
their equations. We put the fourth equation in slope-intercept form:
$y = \frac{3}{2}x + 3$. We now see the slopes are $\frac{3}{2}, 4, -\frac{2}{3}, \frac{3}{2}$, and 2. The first
and fourth line have the same slope and are therefore parallel.
4. 3 quarts of lemonade $\cdot$ 0.15 (15%) = 0.45 quarts of lemon juice.
Now if we add a quart of water, we have 0.45 quarts of lemonade/
4 quarts of lemonade = 0.1125. So the new solution is 11.25%
lemon juice.

5. If we rewrite as $\dfrac{12n}{3n - 2} - \dfrac{8}{3n - 2}$, the rational expressions have a
common denominator and we can add them directly.
$\dfrac{12n - 8}{3n - 2} = \dfrac{4(3n - 2)}{3n - 2} = 4$, provided $3n - 2$ is not zero.

6. (a) The common denominator is $(x + 2)(x - 2) = x^2 - 4$. So we
put both fractions over the common denominator and add them together.
$$\dfrac{3}{x + 2} + \dfrac{4}{x - 2} = \dfrac{3(x - 2)}{x^2 - 4} + \dfrac{4(x + 2)}{x^2 - 4}$$
$$= \dfrac{3(x - 2) + 4(x + 2)}{x^2 - 4}$$
$$= \dfrac{3x - 6 + 4x + 8}{x^2 - 4}$$
$$= \dfrac{7x + 2}{x^2 - 4}$$

(b) In this case, the common denominator is $(x - 1)^2$. Again, we put
both fractions over the common denominator and add them together.
$$\dfrac{6}{(x - 1)} + \dfrac{7}{(x - 1)^2} = \dfrac{6(x - 1)}{(x - 1)^2} + \dfrac{7}{(x - 1)^2}$$
$$= \dfrac{6(x - 1) + 7}{(x - 1)^2}$$
$$= \dfrac{6x - 6 + 7}{(x - 1)^2}$$
$$= \dfrac{6x + 1}{(x - 1)^2}$$

7. (a) We may express our final answer as an inequality, or as an
interval.
$3x - 6 \le -9$
$3x \le -3$
$x \le -1$
$(-\infty, -1]$
(b) Again, we may express our final answer as an inequality, or as
an interval.
$-2x + 4 > 10$
$-2x > 6$
$x < -3$
$(-\infty, -3)$
Notice that when we multiplied or divided both sides of the inequality by a negative number, we remembered to reverse the direction of
the inequality.
8. We first put the equation of the circle into standard form.
$$x^2 - 4x + y^2 - 2y = 4$$
$$(x^2 - 4x + 4) + (y^2 - 2y + 1) = 4 + 4 + 1$$
$$(x - 2)^2 + (y - 1)^2 = 3^2$$

So we have a circle with center $(2, 1)$ and radius 3.

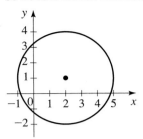

CHAPTER 6 REVIEW ▪ page 315

1. $(2, 1)$ **3.** $\left(-\frac{1}{2}, \frac{7}{4}\right), (2, -2)$

5. $(2, 1)$

7. $x =$ any number
$y = \frac{2}{7}x - 4$

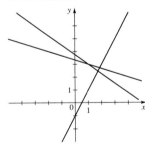

9. No solution

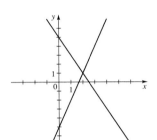

11. $(-3, 3), (2, 8)$ **13.** $\left(\frac{16}{7}, -\frac{14}{3}\right)$ **15.** $(21.41, -15.93)$

17. $(11.94, -1.39), (12.07, 1.44)$ **19.** $(1, 1, 2)$ **21.** No solution

23. $x = -4t + 1, y = -t - 1, z = t$

25. $x = 6 - 5t, y = \frac{1}{2}(7 - 3t), z = t$ **27.** Siobhan is 9 years old; Kieran is 13 years old **29.** 12 nickels, 30 dimes, 8 quarters

31. $\dfrac{2}{x - 5} + \dfrac{1}{x + 3}$ **33.** $\dfrac{-4}{x} + \dfrac{4}{x - 1} + \dfrac{-2}{(x - 1)^2}$

35. $\dfrac{-1}{x} + \dfrac{x + 2}{x^2 + 1}$ **37.** $\dfrac{3}{x^2 + 2} - \dfrac{x}{(x^2 + 2)^2}$ **39.** $x + y^2 \le 4$

41.

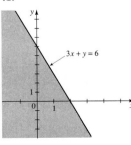

43.

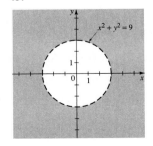

45.

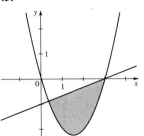

47.

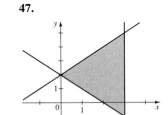

49.

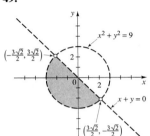

bounded

51.

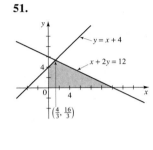

bounded

53. $x = \dfrac{b + c}{2}, y = \dfrac{a + c}{2}, z = \dfrac{a + b}{2}$ **55.** 2, 3

CHAPTER 6 TEST ▪ page 317

1. (a) linear **(b)** $(3, -1)$ **2. (a)** nonlinear **(b)** $\left(\frac{1}{2}, 1\right)$, $(4, -6)$ **3.** $(-0.55, -0.78), (0.43, -0.29), (2.12, 0.56)$

4. Wind 60 km/h, airplane 300 km/h **5. (a)** $(2, 1, -1)$

(b) Neither **6. (a)** No solution **(b)** Inconsistent

7. (a) $x = \frac{1}{7}(t + 1), y = \frac{1}{7}(9t + 2), z = t$ **(b)** Dependent

8. (a) $(10, 0, 1)$ **(b)** neither **9.** Coffee $1.50, juice $1.75, donut $0.75

10.

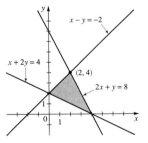

11.

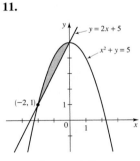

12. $\dfrac{1}{x - 1} + \dfrac{1}{(x - 1)^2} - \dfrac{1}{x + 2}$ **13.** $-\dfrac{1}{x} + \dfrac{x + 2}{x^2 + 3}$

CHAPTER 7

SOLUTIONS TO WARM-UP EXERCISES ▪ page 319

1.
$$\begin{aligned}
x + y + z &= 0 \\
2x - 2y + 3z &= -1 \\
x + 2y + 2z &= 4
\end{aligned}$$

$$\begin{aligned}
x + y + z &= 0 \\
-x + y - \frac{3}{2}z &= \frac{1}{2} \qquad \text{Multiply Equation 2 by } -\frac{1}{2}. \\
-x - 2y - 2z &= -4 \qquad \text{Multiply Equation 3 by } -1.
\end{aligned}$$

$$x + y + z = 0$$
$$2y - \frac{1}{2}z = \frac{1}{2}$$ We get Equation 2 by adding Equations 1
$$-y - z = -4$$ and 2 from the previous step.
We get Equation 3 by adding Equations 1
and 3 from the previous step.

$$x + y + z = 0$$
$$y - \frac{1}{4}z = \frac{1}{4}$$ Multiply Equation 2 by $\frac{1}{2}$.
$$-y - z = -4$$
$$x + y + z = 0$$
$$y - \frac{1}{4}z = \frac{1}{4}$$
$$-\frac{5}{4}z = -\frac{15}{4}$$ We get Equation 3 by adding Equations 2
and 3 from the previous step.

$$x + y + z = 0$$
$$y - \frac{1}{4}z = \frac{1}{4}$$
$$z = 3$$ Multiply Equation 3 by $-\frac{4}{5}$.
$$x = -4$$
$$y = 1$$ Substitute $z = 3$ back into Equation 2 to
$$z = 3$$ get $y = 1$ and substitute both z and y into
Equation 1 to get $x = -4$.

2. We solve the first equation for x to obtain $x = 3y + 2$. We then substitute into the second equation, and solve.
$$3(3y + 2) - 9y = 4$$
$$9y + 6 - 9y = 4$$
$$6 = 4$$
Since 6 does not equal 4, we have shown that this system has no solution. (We could have solved this system using other methods, of course, but all of them would lead to the same result.)

3. (a) The commutative property of equality $a + b = b + a$
(b) The associative property of equality $a(bc) = (ab)c$
(c) The distributive property of equality $a(b + c) = ab + ac$
4. We notice each term has a common factor of bx^2 and use the distributive property of equality to obtain $bx^2(a - 3)$.
5. The multiplicative inverse of a is the number that, when multiplied by a, gives a result of 1 (the multiplicative identity). In the case of real numbers, this is the reciprocal.
(a) The reciprocal of 3 is $\frac{1}{3}$, because $3 \cdot \frac{1}{3} = 1$.
(b) The reciprocal of $\dfrac{x + 3}{x - 2}$ is $\dfrac{x - 2}{x + 3}$, because $\dfrac{x + 3}{x - 2} \cdot \dfrac{x - 2}{x + 3} = 1$, provided $x \neq 2$, $x \neq -3$. If $x = 2$, then the original expression is not defined, and if $x = -3$, then the reciprocal is not defined.
(c) Notice that the numerator of $\dfrac{a^{1/3} - \sqrt[3]{a}}{a + 5}$ is equal to zero, making the whole fraction zero. The number zero has no inverse, because $\frac{1}{0}$ is not defined.

CHAPTER 7 REVIEW ▪ **page 358**

1. (a) 2×3 **(b)** Yes   **(c)** No   **(d)** $\begin{cases} x + 2y = -5 \\ \quad\quad y = 3 \end{cases}$

3. (a) 3×4 **(b)** Yes   **(c)** Yes   **(d)** $\begin{cases} x + 8z = 0 \\ y + 5z = -1 \\ \quad\quad 0 = 0 \end{cases}$

5. (a) 3×4 **(b)** No   **(c)** No   **(d)** $\begin{cases} \quad\quad y - 3z = 4 \\ x + y \quad\quad = 7 \\ x + 2y + z = 2 \end{cases}$

7. $(0, 1, 2)$ **9.** No solution   **11.** $(1, 0, 1, -2)$
13. $\left(-\frac{4}{3}t + \frac{4}{3}, \frac{5}{3}t - \frac{2}{3}, t\right)$ **15.** $(s + 1, 2s - t + 1, s, t)$
17. No solution   **19.** $(1, t + 1, t, 0)$ **21.** Not equal
23. Impossible

25. $\begin{bmatrix} 4 & 18 \\ 4 & 0 \\ 2 & 2 \end{bmatrix}$ **27.** $\begin{bmatrix} 10 & 0 & -5 \end{bmatrix}$ **29.** $\begin{bmatrix} -\frac{7}{2} & 10 \\ 1 & -\frac{9}{2} \end{bmatrix}$

31. $\begin{bmatrix} 30 & 22 & 2 \\ -9 & 1 & -4 \end{bmatrix}$ **33.** $\begin{bmatrix} -\frac{1}{2} & \frac{11}{2} \\ \frac{15}{4} & -\frac{3}{2} \\ -\frac{1}{2} & 1 \end{bmatrix}$ **37.** $\frac{1}{3}\begin{bmatrix} -1 & -3 \\ -5 & 2 \end{bmatrix}$

39. $\begin{bmatrix} \frac{7}{2} & -2 \\ 0 & 8 \end{bmatrix}$ **41.** $\begin{bmatrix} 2 & -2 & 6 \\ -4 & 5 & -9 \end{bmatrix}$ **43.** $1, \begin{bmatrix} 9 & -4 \\ -2 & 1 \end{bmatrix}$

45. 0, no inverse   **47.** $-1, \begin{bmatrix} 3 & 2 & -3 \\ 2 & 1 & -2 \\ -8 & -6 & 9 \end{bmatrix}$

49. $24, \begin{bmatrix} 1 & 0 & 0 & -\frac{1}{4} \\ 0 & \frac{1}{2} & 0 & -\frac{1}{4} \\ 0 & 0 & \frac{1}{3} & -\frac{1}{4} \\ 0 & 0 & 0 & \frac{1}{4} \end{bmatrix}$ **51.** $(65, 154)$ **53.** $\left(-\frac{1}{12}, \frac{1}{12}, \frac{1}{12}\right)$

55. (a) Matrix A describes the number of pounds of each vegetable sold on each day; matrix B lists the price per pound of each vegetable.
(b) $AB = \begin{bmatrix} 68.5 \\ 41.0 \end{bmatrix}$; \$68.50 was the total made on Saturday, and \$41.00 was the total made on Sunday.   **57.** $\left(\frac{1}{5}, \frac{9}{5}\right)$ **59.** $\left(-\frac{87}{26}, \frac{21}{26}, \frac{3}{2}\right)$
61. 11 **63.** \$2,500 in bank A, \$40,000 in bank B, \$17,500 in bank C

CHAPTER 7 TEST ▪ **page 361**

1. Row-echelon form   **2.** Neither   **3.** Reduced row-echelon form
4. Reduced row-echelon form   **5.** $\left(\frac{5}{2}, \frac{5}{2}, 0\right)$ **6.** No solution
7. $\left(-\frac{3}{5} + \frac{2}{5}t, \frac{1}{5} + \frac{1}{5}t, t\right)$ **8.** Incompatible dimensions
9. Incompatible dimensions   **10.** $\begin{bmatrix} 6 & 10 \\ 3 & -2 \\ -3 & 9 \end{bmatrix}$ **11.** $\begin{bmatrix} 36 & 58 \\ 0 & -3 \\ 18 & 28 \end{bmatrix}$

12. $\begin{bmatrix} 2 & -\frac{3}{2} \\ -1 & 1 \end{bmatrix}$ **13.** B is not square   **14.** B is not square

15. -3 **16. (a)** $\begin{bmatrix} 4 & -3 \\ 3 & -2 \end{bmatrix}\begin{bmatrix} x \\ y \end{bmatrix} = \begin{bmatrix} 10 \\ 30 \end{bmatrix}$ **(b)** $(70, 90)$

17. $|A| = 0$, $|B| = 2$, $B^{-1} = \begin{bmatrix} 1 & -2 & 0 \\ 0 & \frac{1}{2} & 0 \\ 3 & -6 & 1 \end{bmatrix}$ **18.** $(5, -5, -4)$

19. 1.2 lb almonds, 1.8 lb walnuts

CHAPTER 8

SOLUTIONS TO WARM-UP EXERCISES ▪ **page 363**

1. We put the equation in standard form, by completing the square:
$$y = -2x^2 + 12x - 16$$
$$y = -2(x^2 - 6x) - 16$$
$$y = -2(x^2 - 6x + 9) - 16 + 18$$
$$y = -2(x - 3)^2 + 2$$

So we have a parabola that opens downward, with vertex $(3, 2)$. The y-intercept is $-2(0 - 3)^2 + 2 = -16$.

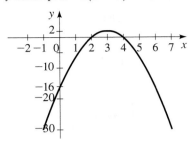

2. We use the distance formula:
$$\sqrt{(2 - 3)^2 + (-4 - 1)^2} = \sqrt{1 + 25} = \sqrt{26}.$$

3. We multiply both sides of the equation by 9 to get the equation of a circle in standard form: $x^2 + y^2 = 3^2$. So we graph a circle of radius 3, centered at the origin.

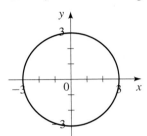

4.
$$\frac{x^2}{a^2} - \frac{y^2}{b^2} = 1$$
$$\frac{b^2 x^2}{a^2} - y^2 = b^2$$
$$\frac{b^2 x^2}{a^2} - b^2 = y^2$$
$$\pm \sqrt{\frac{b^2 x^2}{a^2} - b^2} = y$$
$$\pm \sqrt{\frac{b^2 x^2 - b^2 a^2}{a^2}} = y$$
$$\pm \sqrt{\frac{b^2 (x^2 - a^2)}{a^2}} = y$$
$$y = \pm \frac{b}{a} \sqrt{x^2 - a^2}$$

5. We write $y = f(x - 2) + 1$, which is the original function shifted 2 units to the right and 1 unit up.

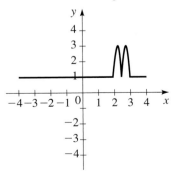

6. The center of the circle will be the midpoint of $(1, 3)$ and $(9, 9)$.
$$\left(\frac{1 + 9}{2}, \frac{3 + 9}{2} \right) = (5, 6).$$

The radius of the circle will be half the diameter of the circle, i.e., half the distance between $(1, 3)$ and $(9, 9)$.
$$\frac{\sqrt{(9 - 1)^2 + (9 - 3)^2}}{2} = \frac{\sqrt{64 + 36}}{2} = 5$$
Therefore, the equation of the circle is $(x - 5)^2 + (y - 6)^2 = 25$.

CHAPTER 8 REVIEW ▪ page 389

1. $V(0, 0)$; $F(1, 0)$; $x = -1$ **3.** $V(0, 0)$; $F(0, -2)$; $y = 2$

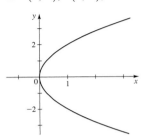   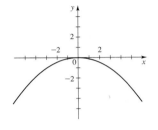

5. $V(-2, 2)$; $F\left(-\frac{7}{4}, 2\right)$; $x = -\frac{9}{4}$ **7.** $V(-2, -3)$; $F(-2, -2)$; $y = -4$

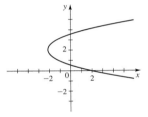   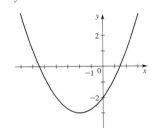

9. $C(0, 0)$; $V(0, \pm 5)$; $F(0, \pm 4)$; $e = \frac{4}{5}$; axes 10, 6   **11.** $C(0, 0)$; $V(\pm 4, 0)$; $F(\pm 2\sqrt{3}, 0)$; $e = \frac{\sqrt{3}}{2}$; axes 8, 4

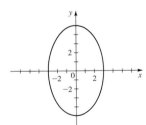   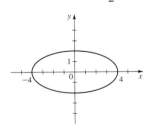

13. $C(3, 0)$; $V(3, \pm 4)$; $F(3, \pm \sqrt{7})$; $e = \frac{\sqrt{7}}{4}$; axes 8, 6   **15.** $C(0, 2)$; $V(\pm 3, 2)$; $F(\pm \sqrt{5}, 2)$; $e = \frac{\sqrt{5}}{3}$; axes 6, 4

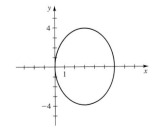   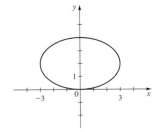

17. $C(0, 0)$; $V(0, \pm 4)$; $F(0, \pm 5)$; asymptotes $y = \pm\frac{4}{3}x$

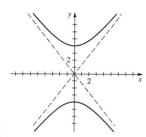

19. $C(0, 0)$; $V(\pm 4, 0)$; $F(\pm 2\sqrt{6}, 0)$; asymptotes $y = \pm\frac{1}{\sqrt{2}}x$

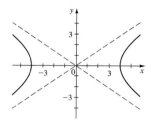

39. Ellipse; $F(3, -3 \pm 1/\sqrt{2})$; $V_1(3, -4)$, $V_2(3, -2)$

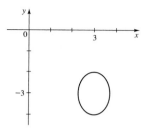

21. $C(-4, 0)$; $V_1(-8, 0)$, $V_2(0, 0)$; $F(-4 \pm 4\sqrt{2}, 0)$; asymptotes $y = \pm(x + 4)$

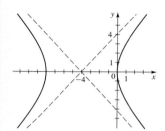

23. $C(-3, -1)$; $V(-3, -1 \pm \sqrt{2})$; $F(-3, -1 \pm 2\sqrt{5})$; asymptotes $y = \frac{1}{3}x$, $y = -\frac{1}{3}x - 2$

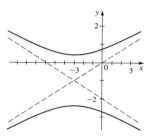

41. Has no graph   **43.** $x^2 = 4y$ **45.** $\dfrac{x^2}{4} + \dfrac{y^2}{25} = 1$

47. $\dfrac{y^2}{4} - \dfrac{x^2}{16} = 1$ **49.** $\dfrac{(x-1)^2}{3} + \dfrac{(y-2)^2}{4} = 1$

51. $\dfrac{4(x-7)^2}{225} + \dfrac{(y-2)^2}{100} = 1$ **53.** (a) 91,419,000 mi

(b) 94,581,000 mi

55. (a)

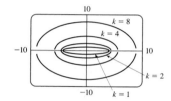

25. $y^2 = 8x$ **27.** $\dfrac{y^2}{16} - \dfrac{x^2}{9} = 1$

29. $\dfrac{(x-4)^2}{16} + \dfrac{(y-2)^2}{4} = 1$

31. Parabola; $F(0, -2)$; $V(0, 1)$

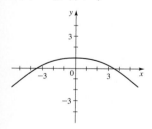

33. Hyperbola; $F(0, \pm 12\sqrt{2})$; $V(0, \pm 12)$

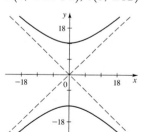

CHAPTER 8 TEST ▪ page 391

1. $F(0, -3)$, $y = 3$

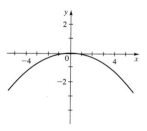

2. $V(\pm 4, 0)$; $F(\pm 2\sqrt{3}, 0)$; 8, 4

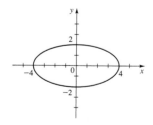

35. Ellipse; $F(1, 4 \pm \sqrt{15})$; $V(1, 4 \pm 2\sqrt{5})$

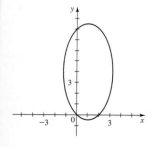

37. Parabola; $F(-\frac{255}{4}, 8)$; $V(-64, 8)$

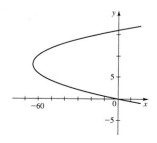

3. $V(0, \pm 3)$; $F(0, \pm 5)$; $y = \pm\frac{3}{4}x$

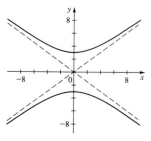

4. $y^2 = -x$ **5.** $\dfrac{x^2}{16} + \dfrac{(y-3)^2}{9} = 1$ **6.** $(x-2)^2 - \dfrac{y^2}{3} = 1$

6. $\dfrac{1}{x} + \dfrac{2}{x^2} - \dfrac{x+2}{x^2+4}$

7. $\dfrac{(x-3)^2}{9} + \dfrac{(y+\frac{1}{2})^2}{4} = 1$ **8.** $\dfrac{(x+2)^2}{8} - \dfrac{(y-4)^2}{9} = 1$

7. **(a)** $F(0, -\frac{3}{2})$, $y = \frac{3}{2}$ **(b)** $F(\frac{1}{8}, 1)$, $x = -\frac{1}{8}$

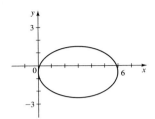

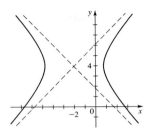

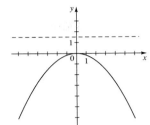

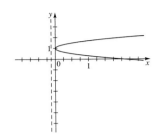

9. $(y+4)^2 = -2(x-4)$

8. **(a)** Hyperbola; $V(\pm 3, 0)$, $F(\pm\sqrt{10}, 0)$; $y = \pm\frac{1}{3}x$

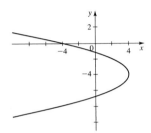

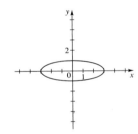

10. $\dfrac{y^2}{9} - \dfrac{x^2}{16} = 1$ **11.** $x^2 - 4x - 8y + 20 = 0$

(b) Ellipse; $V(\pm 3, 0)$, $F(\pm 2\sqrt{2}, 0)$

12. $\dfrac{(x-3)^2}{9} + \dfrac{y^2}{25} = 1$ **13.** $\frac{3}{4}$ in.

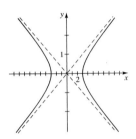

CUMULATIVE REVIEW TEST FOR
CHAPTERS 6, 7, AND 8 ▪ page 392

1. **(a)** Nonlinear   **(b)** $(0, 0), (2, 2), (-2, 2)$
(c) Circle, parabola   **(d), (e)**

(c) Hyperbola; $V(0, \pm 1)$, $F(0, \pm\sqrt{10})$; $y = \pm\frac{1}{3}x$

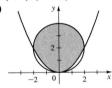

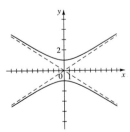

2. **(a)** $(3, 0, 1)$ **(b)** $x = t-1, y = t+2, z = t$ **3.** Xavier 4,
Yolanda 10, Zachary 6
4. **(a)** $A + B$ impossible; $C - D =$
$\begin{bmatrix} 0 & -4 & -2 \\ -1 & -4 & -4 \\ -1 & -1 & -1 \end{bmatrix}$; $AB = \begin{bmatrix} -\frac{9}{2} & 1 & 5 \\ -4 & 2 & 0 \end{bmatrix}$; CB impossible;

$BD = \begin{bmatrix} -1 & -2 & -1 \\ -\frac{1}{2} & -1 & -\frac{1}{2} \end{bmatrix}$; $\det(B)$ impossible; $\det(C) = 2$; $\det(D) = 0$

(b) $C^{-1} = \begin{bmatrix} 0 & 0 & -1 \\ -\frac{1}{2} & \frac{1}{2} & -\frac{1}{2} \\ 1 & 0 & 1 \end{bmatrix}$ **5.** **(a)** $\begin{bmatrix} 5 & -3 \\ 6 & -4 \end{bmatrix}\begin{bmatrix} x \\ y \end{bmatrix} = \begin{bmatrix} 5 \\ 0 \end{bmatrix}$

(b) $\begin{bmatrix} 2 & -\frac{3}{2} \\ 3 & -\frac{5}{2} \end{bmatrix}$ **(c)** $X = \begin{bmatrix} 10 \\ 15 \end{bmatrix}$ **(d)** $x = 10, y = 15$

9. **(a)**

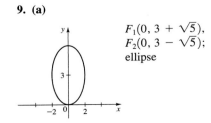

$F_1(0, 3 + \sqrt{5})$,
$F_2(0, 3 - \sqrt{5})$;
ellipse

(b)

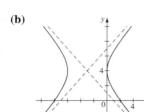

$F_1(-3 + 3\sqrt{2}, 4)$,
$F_2(-3 - 3\sqrt{2}, 4)$;
hyperbola

10. $\dfrac{(x-5)^2}{16} - \dfrac{y^2}{9} = 1$

CHAPTER 9

SOLUTIONS TO WARM-UP EXERCISES • page 395

1. (a) $\dfrac{(-1)^1}{1^2} = -1$, $\dfrac{(-1)^5}{5^2} = -\dfrac{1}{25}$, $\dfrac{(-1)^{10}}{10^2} = \dfrac{1}{100}$
(b) 5420 is even, so $(-1)^{5420} = 1$. Therefore, $f(5420)$ is positive.
2. (a) We observe that this is a list of multiples of 4, so the missing entry is 16. **(b)** We observe that this is a list of perfect squares of integers, so the missing entry is 25. **(c)** We observe that this is a list of powers of 2, so the missing entry is 32. **3. (a)** We cancel the 3's and use laws of exponents to get the answer of 2. **(b)** We cancel the a's and use laws of exponents to get the answer of b.

4. Every term contains a factor of $\dfrac{2}{3^2}$. We use the distributive law to obtain:

$$\dfrac{a}{3^2}\left(1 + \dfrac{1}{3} + \dfrac{1}{3^2} + \dfrac{1}{3^3} + \dfrac{1}{3^4} + \dfrac{1}{3^5} + \dfrac{1}{3^6} + \dfrac{1}{3^7} + \dfrac{1}{3^8}\right)$$

5. We use the compound interest formula $A = P\left(1 + \dfrac{r}{n}\right)^{nt}$, with $P = 3000$, $r = 0.05$, $n = 2$ and $t = 2$.

$$3000\left(1 + \dfrac{0.05}{2}\right)^4 = 3311.44$$

So the total amount is $3311.44.
6. (a) We use FOIL to obtain $4x^2 + 4x + 1$.
(b) We use the result of 6a to obtain $(2x + 1)^3 =$
$(2x + 1)(4x^2 + 4x + 1)$. Now we use the distributive property:
$(2x + 1)(4x^2 + 4x + 1) = 8x^3 + 8x^2 + 2x + 4x^2 + 4x + 1$
$$= 8x^3 + 12x^2 + 6x + 1.$$

CHAPTER 9 REVIEW • page 431

1. $\dfrac{1}{2}, \dfrac{4}{3}, \dfrac{9}{4}, \dfrac{16}{5}; \dfrac{100}{11}$ **3.** $0, \dfrac{1}{4}, 0, \dfrac{1}{32}; \dfrac{1}{500}$ **5.** 1, 3, 15, 105; 654,729,075
7. 1, 4, 9, 16, 25, 36, 49   **9.** 1, 3, 5, 11, 21, 43, 85
11. (a) 7, 9, 11, 13, 15

(b)

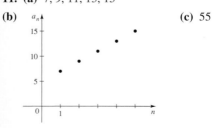

(c) 55

(d) Arithmetic, common difference 2
13. (a) $\dfrac{3}{4}, \dfrac{9}{8}, \dfrac{27}{16}, \dfrac{81}{32}, \dfrac{243}{64}$

(b)

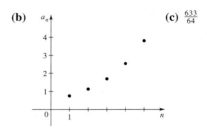

(c) $\dfrac{633}{64}$

(d) Geometric, common ratio $\dfrac{3}{2}$ **15.** Arithmetic, 7

17. Arithmetic, $t + 1$ **19.** Geometric, $\dfrac{1}{t}$

21. Geometric, $\dfrac{4}{27}$ **23.** $2i$ **25.** 5   **27.** $\dfrac{81}{4}$

29. (a) $A_n = 32,000(1.05)^{n-1}$ **(b)** $32,000, $33,600, $35,280,
$37,044, $38,896.20, $40,841.01, $42,883.06, $45,027.21
31. 12,288   **35. (a)** 9   **(b)** $\pm 6\sqrt{2}$ **37.** 126   **39.** 384

41. $0^2 + 1^2 + 2^2 + \cdots + 9^2$ **43.** $\dfrac{3}{2^2} + \dfrac{3^2}{2^3} + \dfrac{3^3}{2^4} + \cdots + \dfrac{3^{50}}{2^{51}}$

45. $\displaystyle\sum_{k=1}^{33} 3k$ **47.** $\displaystyle\sum_{k=1}^{100} k2^{k+2}$ **49.** Geometric; 4.68559

51. Arithmetic, $5050\sqrt{5}$ **53.** Geometric, 9831   **55.** $\dfrac{5}{7}$

57. $\dfrac{1}{2}(3 + \sqrt{3})$ **59.** 13   **61.** 65,534   **63.** $2390.27

65. Let $P(n)$ denote the statement

$$1 + 4 + 7 + \cdots + (3n - 2) = \dfrac{n(3n - 1)}{2}.$$

Step 1 $P(1)$ is true, since $1 = \dfrac{1(3 \cdot 1 - 1)}{2}$.

Step 2 Suppose $P(k)$ is true. Then

$$1 + 4 + 7 + \cdots + (3k - 2) + [3(k + 1) - 2]$$
$$= \dfrac{k(3k - 1)}{2} + [3k + 1] \qquad \text{Induction hypothesis}$$
$$= \dfrac{3k^2 - k + 6k + 2}{2}$$
$$= \dfrac{(k + 1)(3k + 2)}{2}$$
$$= \dfrac{(k + 1)[3(k + 1) - 1]}{2}$$

So $P(k + 1)$ follows from $P(k)$. Thus, by the Principle of Mathematical Induction $P(n)$ holds for all n.

67. Let $P(n)$ denote the statement
$\left(1 + \dfrac{1}{1}\right)\left(1 + \dfrac{1}{2}\right)\cdots\left(1 + \dfrac{1}{n}\right) = n + 1$.

Step 1 $P(1)$ is true, since $\left(1 + \dfrac{1}{1}\right) = 1 + 1$.
Step 2 Suppose $P(k)$ is true. Then

$$\left(1 + \dfrac{1}{1}\right)\left(1 + \dfrac{1}{2}\right)\cdots\left(1 + \dfrac{1}{k}\right)\left(1 + \dfrac{1}{k + 1}\right)$$
$$= (k + 1)\left(1 + \dfrac{1}{k + 1}\right) \qquad \text{Induction hypothesis}$$
$$= (k + 1) + 1$$

So $P(k + 1)$ follows from $P(k)$. Thus, by the Principle of Mathematical Induction $P(n)$ holds for all n.

69. Let $P(n)$ denote the statement $a_n = 2 \cdot 3^n - 2$.

Step 1 $P(1)$ is true, since $a_1 = 2 \cdot 3^1 - 2 = 4$.
Step 2 Suppose $P(k)$ is true. Then

$$a_{k+1} = 3a_k + 4$$
$$= 3(2 \cdot 3^k - 2) + 4 \quad \text{Induction hypothesis}$$
$$= 2 \cdot 3^{k+1} - 2$$

So $P(k + 1)$ follows from $P(k)$. Thus, by the Principle of Mathematical Induction $P(n)$ holds for all n.

71. 100   **73.** 32   **75.** $A^3 - 3A^2B + 3AB^2 - B^3$
77. $1 - 6x^2 + 15x^4 - 20x^6 + 15x^8 - 6x^{10} + x^{12}$
79. $1540a^3b^{19}$ **81.** $17{,}010A^6B^4$

CHAPTER 9 TEST ▪ page 433

1. 1, 6, 15, 28, 45, 66; 161   **2.** 2, 5, 13, 36, 104, 307   **3. (a)** 3
(b) $a_n = 2 + (n - 1)3$ **(c)** 104   **4. (a)** $\frac{1}{4}$ **(b)** $a_n = 12\left(\frac{1}{4}\right)^{n-1}$
(c) $3/4^8$ **5. (a)** $\frac{1}{5}, \frac{1}{25}$ **(b)** $\dfrac{5^8 - 1}{12{,}500}$ **6. (a)** $-\frac{8}{9}, -78$ **(b)** 60
8. (a) $(1 - 1^2) + (1 - 2^2) + (1 - 3^2) + (1 - 4^2) +$
$(1 - 5^2) = -50$
(b) $(-1)^3 2^1 + (-1)^4 2^2 + (-1)^5 2^3 + (-1)^6 2^4 = 10$
9. (a) $\frac{58{,}025}{59{,}049}$ **(b)** $2 + \sqrt{2}$

10. Let $P(n)$ denote the statement

$$1^2 + 2^2 + \cdots + n^2 = \frac{n(n + 1)(2n + 1)}{6}.$$

Step 1 $P(1)$ is true, since $1^2 = \dfrac{1(1 + 1)(2 \cdot 1 + 1)}{6}$.

Step 2 Suppose $P(k)$ is true. Then

$$1^2 + 2^2 + \cdots + k^2 + (k + 1)^2$$
$$= \frac{k(k + 1)(2k + 1)}{6} + (k + 1)^2 \quad \text{Induction hypothesis}$$
$$= \frac{k(k + 1)(2k + 1) + 6(k + 1)^2}{6}$$
$$= \frac{(k + 1)[k(2k + 1) + 6(k + 1)]}{6}$$
$$= \frac{(k + 1)(2k^2 + 7k + 6)}{6}$$
$$= \frac{(k + 1)[(k + 1) + 1][2(k + 1) + 1]}{6}$$

So $P(k + 1)$ follows from $P(k)$. Thus, by the Principle of Mathematical Induction $P(n)$ holds for all n.

11. $32x^5 + 80x^4y^2 + 80x^3y^4 + 40x^2y^6 + 10xy^8 + y^{10}$

12. $\dbinom{10}{3}(3x)^3(-2)^7 = -414{,}720x^3$

13. (a) $a_n = (0.85)(1.24)^n$ **(b)** 3.09 lb   **(c)** Geometric

INDEX

▼ **PROPERTIES AND FORMULAS** _____

Properties of Real Numbers (p. 7)

Commutative: $a + b = b + a$
                $ab = ba$
Associative: $(a + b) + c = a + (b + c)$
                $(ab)c = a(bc)$
Distributive: $a(b + c) = ab + ac$

Absolute Value (p. 13)

$$|a| = \begin{cases} a & \text{if } a \geq 0 \\ -a & \text{if } a < 0 \end{cases}$$

$$|ab| = |a||b|$$

$$\left|\frac{a}{b}\right| = \frac{|a|}{|b|}$$

Distance between a and b is

$$d(a, b) = |b - a|$$

Exponents (p. 16)

$$a^m a^n = a^{m+n}$$
$$\frac{a^m}{a^n} = a^{m-n}$$
$$(a^m)^n = a^{mn}$$
$$(ab)^n = a^n b^n$$
$$\left(\frac{a}{b}\right)^n = \frac{a^n}{b^n}$$

Radicals (p. 20)

$\sqrt[n]{a} = b$ means $b^n = a$

$\sqrt[n]{ab} = \sqrt[n]{a}\sqrt[n]{b}$

$\sqrt[n]{\frac{a}{b}} = \frac{\sqrt[n]{a}}{\sqrt[n]{b}}$

$\sqrt[m]{\sqrt[n]{a}} = \sqrt[mn]{a}$

If n is odd, then $\sqrt[n]{a^n} = a$

If n is even, then $\sqrt[n]{a^n} = |a|$

$a^{m/n} = \sqrt[n]{a^m}$

Special Product Formulas (p. 26)

Sum and difference of same terms:
$$(A + B)(A - B) = A^2 - B^2$$

Square of a sum or difference:
$$(A + B)^2 = A^2 + 2AB + B^2$$
$$(A - B)^2 = A^2 - 2AB + B^2$$

Cube of a sum or difference:
$$(A + B)^3 = A^3 + 3A^2B + 3AB^2 + B^3$$
$$(A - B)^3 = A^3 - 3A^2B + 3AB^2 - B^3$$

Factoring Formulas (p. 29)

Difference of squares:
$$A^2 - B^2 = (A + B)(A - B)$$

Perfect squares:
$$A^2 + 2AB + B^2 = (A + B)^2$$
$$A^2 - 2AB + B^2 = (A - B)^2$$

Sum or difference of cubes:
$$A^3 - B^3 = (A - B)(A^2 + AB + B^2)$$
$$A^3 + B^3 = (A + B)(A^2 - AB + B^2)$$

Rational Expressions (pp. 33–34)

We can cancel common factors:
$$\frac{AC}{BC} = \frac{A}{B}$$

To multiply two fractions, we multiply their numerators together and their denominators together:
$$\frac{A}{B} \times \frac{C}{D} = \frac{AC}{BD}$$

To divide fractions, we invert the divisor and multiply:
$$\frac{A}{B} \div \frac{C}{D} = \frac{A}{B} \times \frac{D}{C}$$

To add fractions, we find a common denominator:
$$\frac{A}{C} + \frac{B}{C} = \frac{A + B}{C}$$

▼ PROPERTIES AND FORMULAS

Properties of Equality (p. 50)

$$A = B \iff A + C = B + C$$
$$A = B \iff CA = CB \quad (C \neq 0)$$

Linear Equations (p. 50)

A **linear equation** is an equation of the form $ax + b = 0$

Power Equations (p. 53)

A **power equation** is an equation of the form $X^n = a$. Its solutions are

$$X = \sqrt[n]{a} \quad \text{if } n \text{ is odd}$$
$$X = \pm\sqrt[n]{a} \quad \text{if } n \text{ is even}$$

If n is even and $a < 0$, the equation has no real solution.

Simple Interest (p. 57)

If a principal P is invested at simple annual interest rate r for t years, then the interest I is given by

$$I = Prt$$

Zero-Product Property (p. 64)

If $AB = 0$ then $A = 0$ or $B = 0$.

Completing the Square (p. 65)

To make $x^2 + bx$ a perfect square, add $\left(\dfrac{b}{2}\right)^2$. This gives the perfect square

$$x^2 + bx + \left(\frac{b}{2}\right)^2 = \left(x + \frac{b}{2}\right)^2$$

Quadratic Formula (pp. 66–68)

A **quadratic equation** is an equation of the form

$$ax^2 + bx + c = 0$$

Its solutions are given by the **Quadratic Formula:**

$$x = \frac{-b \pm \sqrt{b^2 - 4ac}}{2a}$$

The **discriminant** is $D = b^2 - 4ac$.

If $D > 0$, the equation has two real solutions.

If $D = 0$, the equation has one solution.

If $D < 0$, the equation has two complex solutions.

Complex Numbers (pp. 71–74)

A **complex number** is a number of the form $a + bi$, where $i = \sqrt{-1}$.

The **complex conjugate** of $a + bi$ is

$$\overline{a + bi} = a - bi$$

To **multiply** complex numbers, treat them as binomials and use $i^2 = -1$ to simplify the result.

To **divide** complex numbers, multiply numerator and denominator by the complex conjugate of the denominator:

$$\frac{a + bi}{c + di} = \left(\frac{a + bi}{c + di}\right) \cdot \left(\frac{c - di}{c - di}\right) = \frac{(a + bi)(c - di)}{c^2 + d^2}$$

Inequalities (p. 82)

Adding the same quantity to each side of an inequality gives an equivalent inequality:

$$A < B \iff A + C < B + C$$

Multiplying each side of an inequality by the same *positive* quantity gives an equivalent inequality. Multiplying each side by the same *negative* quantity reverses the direction of the inequality:

$$A < B \iff CA < CB \quad \text{if } C > 0$$
$$A < B \iff CA > CB \quad \text{if } C < 0$$

Absolute Value Equations (p. 89)

To solve an absolute value equation, we use

$$|x| = C \iff x = C \quad \text{or} \quad x = -C$$

Absolute Value Inequalities (p. 90)

To solve an absolute value inequality, we use

$$|x| < C \iff -C < x < C$$
$$|x| > C \iff x < -C \quad \text{or} \quad x > C$$

▼ PROPERTIES AND FORMULAS

The Distance Formula (p. 99)

The distance between the points $A(x_1, y_1)$ and $B(x_2, y_2)$ is

$$d(A, B) = \sqrt{(x_1 - x_2)^2 + (y_1 - y_2)^2}$$

The Midpoint Formula (p. 101)

The midpoint of the line segment from $A(x_1, y_1)$ to $B(x_2, y_2)$ is

$$\left(\frac{x_1 + x_2}{2}, \frac{y_1 + y_2}{2} \right)$$

Intercepts (p. 103)

To find the **x-intercepts** of the graph of an equation, set $y = 0$ and solve for x.

To find the **y-intercepts** of the graph of an equation, set $x = 0$ and solve for y.

Circles (p. 104)

The circle with center $(0, 0)$ and radius r has equation

$$x^2 + y^2 = r^2$$

The circle with center (h, k) and radius r has equation

$$(x - h)^2 + (y - k)^2 = r^2$$

Symmetry (p. 107)

The graph of an equation is **symmetric with respect to the x-axis** if the equation remains unchanged when you replace y by $-y$.

The graph of an equation is **symmetric with respect to the y-axis** if the equation remains unchanged when you replace x by $-x$.

The graph of an equation is **symmetric with respect to the origin** if the equation remains unchanged when you replace x by $-x$ and y by $-y$.

Slope of a Line (p. 117)

The slope of the nonvertical line that contains the points $A(x_1, y_1)$ and $B(x_2, y_2)$ is

$$m = \frac{\text{rise}}{\text{run}} = \frac{y_2 - y_1}{x_2 - x_1}$$

Equations of Lines (pp. 118–120)

If a line has slope m, has y-intercept b, and contains the point (x_1, y_1), then:

the **point-slope form** of its equation is

$$y - y_1 = m(x - x_1)$$

the **slope-intercept form** of its equation is

$$y = mx + b$$

The equation of any line can be expressed in the **general form**

$$Ax + By + C = 0$$

(where A and B can't both be 0).

Vertical and Horizontal Lines (p. 119)

The **vertical** line containing the point (a, b) has the equation $x = a$.

The **horizontal** line containing the point (a, b) has the equation $y = b$.

Parallel and Perpendicular Lines (pp. 121–122)

Two lines with slopes m_1 and m_2 are

parallel if and only if $m_1 = m_2$

perpendicular if and only if $m_1 m_2 = -1$

Variation (pp. 125–127)

If y is **directly proportional** to x, then

$$y = kx$$

If y is **inversely proportional** to x, then

$$y = \frac{k}{x}$$

If z is **jointly proportional** to x and y, then

$$z = kxy$$

In each case, k is the **constant of proportionality**.

▼ PROPERTIES AND FORMULAS

Function Notation (p. 137)

If a function is given by the formula $y = f(x)$, then x is the independent variable and denotes the **input**; y is the dependent variable and denotes the **output**; the **domain** is the set of all possible inputs x; the **range** is the set of all possible outputs y.

The Graph of a Function (p. 143)

The graph of a function f is the graph of the equation $y = f(x)$ that defines f.

The Vertical Line Test (p. 148)

A curve in the coordinate plane is the graph of a function if and only if no vertical line intersects the graph more than once.

Increasing and Decreasing Functions (p. 152)

A function f is **increasing** on an interval if $f(x_1) < f(x_2)$ whenever $x_1 < x_2$ in the interval.

A function f is **decreasing** on an interval if $f(x_1) > f(x_2)$ whenever $x_1 < x_2$ in the interval.

Local Maximum and Minimum Values (p. 154)

The function value $f(a)$ is a **local maximum value** of the function f if $f(a) \geq f(x)$ for all x near a. In this case we also say that f has a **local maximum** at $x = a$.

The function value $f(b)$ is a **local minimum value** of the function f if $f(b) \leq f(x)$ for all x near b. In this case we also say that f has a **local minimum** at $x = b$.

Average Rate of Change (p. 157)

The **average rate of change** of the function f between $x = a$ and $x = b$ is the slope of the **secant** line between $(a, f(a))$ and $(b, f(b))$:

$$\text{average rate of change} = \frac{f(b) - f(a)}{b - a}$$

Vertical and Horizontal Shifts of Graphs (pp. 161–162)

Let c be a positive constant.

To graph $y = f(x) + c$, shift the graph of $y = f(x)$ **upward** by c units.

To graph $y = f(x) - c$, shift the graph of $y = f(x)$ **downward** by c units.

To graph $y = f(x - c)$, shift the graph of $y = f(x)$ **to the right** by c units.

To graph $y = f(x + c)$, shift the graph of $y = f(x)$ **to the left** by c units.

Reflecting Graphs (p. 163)

To graph $y = -f(x)$, **reflect** the graph of $y = f(x)$ in the **x-axis**.

To graph $y = f(-x)$, **reflect** the graph of $y = f(x)$ in the **y-axis**.

Vertical and Horizontal Stretching and Shrinking of Graphs (pp. 164–165)

If $c > 1$, then to graph $y = cf(x)$, **stretch** the graph of $y = f(x)$ **vertically** by a factor of c.

If $0 < c < 1$, then to graph $y = cf(x)$, **shrink** the graph of $y = f(x)$ **vertically** by a factor of c.

If $c > 1$, then to graph $y = f(cx)$, **shrink** the graph of $y = f(x)$ **horizontally** by a factor of $1/c$.

If $0 < c < 1$, then to graph $y = f(cx)$, **stretch** the graph of $y = f(x)$ **horizontally** by a factor of $1/c$.

Even and Odd Functions (p. 167)

A function f is

even if $f(-x) = f(x)$

odd if $f(-x) = -f(x)$

for every x in the domain of f.

Composition of Functions (p. 170)

Given two functions f and g, the **composition** of f and g is the function $f \circ g$ defined by

$$(f \circ g)(x) = f(g(x))$$

The **domain** of $f \circ g$ is the set of all x for which both $g(x)$ and $f(g(x))$ are defined.

One-to-One Functions (p. 174)

A function f is **one-to-one** if $f(x_1) \neq f(x_2)$ whenever x_1 and x_2 are *different* elements of the domain of f.

To see from its graph whether a function is one-to-one, use the **Horizontal Line Test**: A function is one-to-one if and only if no horizontal line intersects its graph more than once.

Inverse of a Function (pp. 175–176)

Let f be a one-to-one function with domain A and range B.

The **inverse** of f is the function f^{-1} defined by

$$f^{-1}(y) = x \iff f(x) = y$$

The inverse function f^{-1} has domain B and range A.

The functions f and f^{-1} satisfy the following **cancellation properties**:

$$f^{-1}(f(x)) = x \quad \text{for every } x \text{ in } A$$
$$f(f^{-1}(x)) = x \quad \text{for every } x \text{ in } B$$

▼ PROPERTIES AND FORMULAS

Quadratic Functions (pp. 188–191)

A **quadratic function** is a function of the form

$$f(x) = ax^2 + bx + c$$

It can be expressed in the **standard form**

$$f(x) = a(x - h)^2 + k$$

by completing the square.

The graph of a quadratic function in standard form is a **parabola** with **vertex** (h, k).

If $a > 0$, then the quadratic function f has the **minimum value** k at $x = h$.

If $a < 0$, then the quadratic function f has the **maximum value** k at $x = h$.

Polynomial Functions (p. 194)

A **polynomial function** of **degree** n is a function P of the form

$$P(x) = a_n x^n + a_{n-1} x^{n-1} + \cdots + a_1 x + a_0$$

The numbers a_i are the **coefficients** of the polynomial; a_n is the **leading coefficient**, and a_0 is the **constant coefficient** (or **constant term**).

The graph of a polynomial function is a smooth, continuous curve.

Real Zeros of Polynomials (p. 198)

A **zero** of a polynomial P is a number c for which $P(c) = 0$. The following are equivalent ways of describing real zeros of polynomials:

1. c is a real zero of P.

2. $x = c$ is a solution of the equation $P(x) = 0$.

3. $x - c$ is a factor of $P(x)$.

4. c is an x-intercept of the graph of P.

Multiplicity of a Zero (p. 202)

A zero c of a polynomial P has multiplicity m if m is the highest power for which $(x - c)^m$ is a factor of $P(x)$.

Local Maxima and Minima (p. 203)

A polynomial function P of degree n has $n - 1$ or fewer **local extrema** (i.e., local maxima and minima).

Division of Polynomials (p. 205)

If P and D are any polynomials (with $D(x) \neq 0$), then we can divide P by D using either **long division** or (if D is linear) **synthetic division**. The result of the division can be expressed in one of the following equivalent forms:

$$P(x) = D(x) \cdot Q(x) + R(x)$$
$$\frac{P(x)}{D(x)} = Q(x) + \frac{R(x)}{D(x)}$$

In this division, P is the **dividend**, D is the **divisor**, Q is the **quotient**, and R is the **remainder**. When the division is continued to its completion, the degree of R will be less than the degree of D (or $R(x) = 0$).

Remainder Theorem (p. 207)

When $P(x)$ is divided by the linear divisor $D(x) = x - c$, the **remainder** is the constant $P(c)$. So one way to **evaluate** a polynomial function P at c is to use synthetic division to divide $P(x)$ by $x - c$ and observe the value of the remainder.

Rational Zeros of Polynomials (pp. 210–211)

If the polynomial P given by

$$P(x) = a_n x^n + a_{n-1} x^{n-1} + \cdots + a_1 x + a_0$$

has integer coefficients, then all the **rational zeros** of P have the form

$$x = \pm \frac{p}{q}$$

where p is a divisor of the constant term a_0 and q is a divisor of the leading coefficient a_n.

So to find all the rational zeros of a polynomial, we list all the *possible* rational zeros given by this principle and then check to see which *actually* are zeros by using synthetic division.

Descartes' Rule of Signs (p. 213)

Let P be a polynomial with real coefficients. Then:

The number of positive real zeros of P either is the number of **changes of sign** in the coefficients of $P(x)$ or is less than that by an even number.

The number of negative real zeros of P either is the number of **changes of sign** in the coefficients of $P(-x)$ or is less than that by an even number.

Upper and Lower Bounds Theorem (p. 213)

Suppose we divide the polynomial P by the linear expression $x - c$ and arrive at the result

$$P(x) = (x - c) \cdot Q(x) + r$$

If $c > 0$ and the coefficients of Q, followed by r, are all nonnegative, then c is an **upper bound** for the zeros of P.

If $c < 0$ and the coefficients of Q, followed by r (including zero coefficients), are alternately nonnegative and nonpositive, then c is a **lower bound** for the zeros of P.

The Fundamental Theorem of Algebra, Complete Factorization, and the Zeros Theorem (pp. 217–219)

Every polynomial P of degree n with complex coefficients has exactly n complex zeros, provided that each zero of multiplicity m is counted m times. P factors into n linear factors as follows:

$$P(x) = a(x - c_1)(x - c_2) \cdots (x - c_n)$$

where a is the leading coefficient of P and $c_1, c_1, \ldots, c_n$ are the zeros of P.

Conjugate Zeros Theorem (p. 221)

If the polynomial P has real coefficients and if $a + bi$ is a zero of P, then its complex conjugate $a - bi$ is also a zero of P.

Linear and Quadratic Factors Theorem (p. 222)

Every polynomial with real coefficients can be factored into linear and irreducible quadratic factors with real coefficients.

Rational Functions (p. 223)

A **rational function** r is a quotient of polynomial functions:

$$r(x) = \frac{P(x)}{Q(x)}$$

We generally assume that the polynomials P and Q have no factors in common.

Asymptotes (pp. 223–225)

The line $x = a$ is a **vertical asymptote** of the function $y = f(x)$ if

$$y \to \infty \quad \text{or} \quad y \to -\infty \quad \text{as} \quad x \to a^+ \quad \text{or} \quad x \to a^-$$

The line $y = b$ is a **horizontal asymptote** of the function $y = f(x)$ if

$$y \to b \quad \text{as} \quad x \to \infty \quad \text{or} \quad x \to -\infty$$

Asymptotes of Rational Functions (pp. 225–228)

Let $r(x) = \dfrac{P(x)}{Q(x)}$ be a rational function.

The vertical asymptotes of r are the lines $x = a$ where a is a zero of Q.

If the degree of P is less than the degree of Q, then the horizontal asymptote of r is the line $y = 0$.

If the degrees of P and Q are the same, then the horizontal asymptote of r is the line $y = b$, where

$$b = \frac{\text{leading coefficient of } P}{\text{leading coefficient of } Q}$$

If the degree of P is greater than the degree of Q, then r has no horizontal asymptote.

▼ PROPERTIES AND FORMULAS

Exponential Functions (pp. 242–245)

The **exponential function** f with base a (where $a > 0$, $a \neq 1$) is defined for all real numbers x by

$$f(x) = a^x$$

The domain of f is $\mathbb{R}$, and the range of f is $(0, \infty)$ The graph of f has one of the following shapes, depending on the value of a:

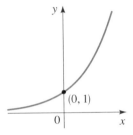

$f(x) = a^x$ for $a > 1$

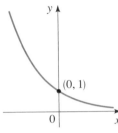

$f(x) = a^x$ for $0 < a < 1$

The Natural Exponential Function (p. 247)

The **natural exponential function** is the exponential function with base e:

$$f(x) = e^x$$

The number e is defined to be the number that the expression $(1 + 1/n)^n$ approaches as $n \to \infty$. An approximate value for the irrational number e is

$$e \approx 2.7182818284590\ldots$$

Compound Interest (pp. 249–250)

If a principal P is invested in an account paying an annual interest rate r, compounded n times a year, then after t years the **amount** $A(t)$ in the account is

$$A(t) = P\left(1 + \frac{r}{n}\right)^{nt}$$

If the interest is compounded **continuously**, then the amount is

$$A(t) = Pe^{rt}$$

Logarithmic Functions (p. 251)

The **logarithmic function** $\log_a$ with base a (where $a > 0$, $a \neq 1$) is defined for $x > 0$ by

$$\log_a x = y \iff a^y = x$$

So $\log_a x$ is the exponent to which the base a must be raised to give y.

The domain of $\log_a$ is $(0, \infty)$, and the range is $\mathbb{R}$. For $a > 1$, the graph of the function $\log_a$ has the following shape:

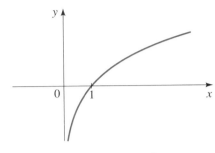

$y = \log_a x$, $a > 1$

Properties of Logarithms (p. 252)

1. $\log_a 1 = 0$ **2.** $\log_a a = 1$

3. $\log_a a^x = x$ **4.** $a^{\log_a x} = x$

Common and Natural Logarithms (pp. 255–257)

The logarithm function with base 10 is called the **common logarithm** and is denoted **log**. So

$$\log x = \log_{10} x$$

The logarithm function with base e is called the **natural logarithm** and is denoted **ln**. So

$$\ln x = \log_e x$$

Laws of Logarithms (p. 258)

Let a be a logarithm base ($a > 0$, $a \neq 1$), and let A, B, and C be any real numbers or algebraic expressions that represent real numbers, with $A > 0$ and $B > 0$. Then:

1. $\log_a(AB) = \log_a A + \log_a B$

2. $\log_a(A/B) = \log_a A - \log_a B$

3. $\log_a(A^C) = C \log_a A$

Change of Base Formula (p. 261)

$$\log_b x = \frac{\log_a x}{\log_a b}$$

Guidelines for Solving Exponential Equations (p. 263)

1. Isolate the exponential term on one side of the equation.

2. Take the logarithm of each side, and use the Laws of Logarithms to "bring down the exponent."

3. Solve for the variable.

Guidelines for Solving Logarithmic Equations (p. 265)

1. Isolate the logarithmic term(s) on one side of the equation, and use the Laws of Logarithms to combine logarithmic terms if necessary.

2. Rewrite the equation in exponential form.

3. Solve for the variable.

Exponential Growth Model (p. 271)

A population experiences **exponential growth** if it can be modeled by the exponential function

$$n(t) = n_0 e^{rt}$$

where $n(t)$ is the population at time t, n_0 is the initial population (at time $t = 0$), and r is the relative growth rate (expressed as a proportion of the population).

Radioactive Decay Model (p. 274)

If a **radioactive substance** with half-life h has initial mass m_0, then at time t the mass $m(t)$ of the substance that remains is modeled by the exponential function

$$m(t) = m_0 e^{-rt}$$

where $r = \dfrac{\ln 2}{h}$.

Newton's Law of Cooling (p. 275)

If an object has an initial temperature that is D_0 degrees warmer than the surrounding temperature T_s, then at time t the temperature $T(t)$ of the object is modeled by the function

$$T(t) = T_s + D_0 e^{-kt}$$

where the constant $k > 0$ depends on the size and type of the object.

Logarithmic Scales (pp. 276–279)

The **pH scale** measures the acidity of a solution:

$$\text{pH} = -\log[\text{H}^+]$$

The **Richter scale** measures the intensity of earthquakes:

$$M = \log \frac{I}{S}$$

The **decibel scale** measures the intensity of sound:

$$B = 10 \log \frac{I}{I_0}$$

▼ PROPERTIES AND FORMULAS

Substitution Method (p. 289)

To solve a pair of equations in two variables by substitution:

1. Solve for one variable in terms of the other variable in one equation.

2. **Substitute** into the other equation to get an equation in one variable, and solve for this variable.

3. Substitute the value(s) of the variable you have found into either original equation, and solve for the remaining variable.

Elimination Method (p. 290)

To solve a pair of equations in two variables by elimination:

1. Multiply the equations by appropriate constants so that the term(s) involving one of the variables are of opposite sign in the equations.

2. Add the equations to **eliminate** that one variable; this gives an equation in the other variable. Solve for this variable.

3. Substitute the value(s) of the variable that you have found into either original equation, and solve for the remaining variable.

Graphical Method (p. 292)

To solve a pair of equations in two variables graphically:

1. Put each equation in function form, $y = f(x)$

2. Use a graphing calculator to **graph** the equations on a common screen.

3. Find the points of intersection of the graphs. The solutions are the x- and y-coordinates of the points of intersection.

Gaussian Elimination (p. 300)

When we use **Gaussian elimination** to solve a system of linear equations, we use the following operations to change the system to an **equivalent** simpler system:

1. Add a nonzero multiple of one equation to another.

2. Multiply an equation by a nonzero constant.

3. Interchange the position of two equations in the system.

Number of Solutions of a System of Linear Equations (pp. 294, 301)

A system of linear equations can have:

1. A unique solution for each variable.

2. No solution, in which case the system is **inconsistent**.

3. Infinitely many solutions, in which case the system is **dependent**.

How to Determine the Number of Solutions of a Linear System (p. 301)

When we use **Gaussian elimination** to solve a system of linear equations, then we can tell that the system has:

1. **No solution** (is *inconsistent*) if we arrive at a false equation of the form $0 = c$, where c is nonzero.

2. **Infinitely many solutions** (is *dependent*) if the system is consistent but we end up with fewer equations than variables (after discarding redundant equations of the form $0 = 0$).

Partial Fractions (pp. 304–308)

The *partial fraction decomposition* of a rational function

$$r(x) = \frac{P(x)}{Q(x)}$$

(where the degree of P is less than the degree of Q) is a sum of simpler fractional expressions that equal $r(x)$ when brought to a common denominator. The denominator of each simpler fraction is either a linear or quadratic factor of $Q(x)$ or a power of such a linear or quadratic factor. So to find the terms of the partial fraction decomposition, we first factor $Q(x)$ into linear and irreducible quadratic factors. The terms then have the following forms, depending on the factors of $Q(x)$.

1. For every **distinct linear factor** $ax + b$, there is a term of the form

$$\frac{A}{ax + b}$$

2. For every **repeated linear factor** $(ax + b)^m$, there are terms of the form

$$\frac{A_1}{ax + b} + \frac{A_2}{(ax + b)^2} + \cdots + \frac{A_m}{(ax + b)^m}$$

3. For every **distinct quadratic factor** $ax^2 + bx + c$, there is a term of the form

$$\frac{Ax + B}{ax^2 + bx + c}$$

4. For every **repeated quadratic factor** $(ax^2 + bx + c)^m$, there are terms of the form

$$\frac{A_1x + B_1}{ax^2 + bx + c} + \frac{A_2x + B_2}{(ax^2 + bx + c)^2} + \cdots + \frac{A_mx + B_m}{(ax^2 + bx + c)^m}$$

Graphing Inequalities (p. 309)

To graph an inequality:

1. Graph the equation that corresponds to the inequality. This "boundary curve" divides the coordinate plane into separate regions.

2. Use **test points** to determine which region(s) satisfy the inequality.

3. Shade the region(s) that satisfy the inequality, and use a solid line for the boundary curve if it satisfies the inequality ($\leq$ or $\geq$) and a dashed line if it does not ($<$ or $>$).

Graphing Systems of Inequalities (pp. 310–311)

To graph the solution of a system of inequalities (or **feasible region** determined by the inequalities):

1. Graph all the inequalities on the same coordinate plane.

2. The solution is the intersection of the solutions of all the inequalities, so shade the region that satisfies all the inequalities.

3. Determine the coordinates of the intersection points of all the boundary curves that touch the solution set of the system. These points are the **vertices** of the solution.

▼ PROPERTIES AND FORMULAS _____

Matrices (p. 321)

A **matrix** A of **dimension** $m \times n$ is a rectangular array of numbers with m **rows** and n **columns**:

$$A = \begin{bmatrix} a_{11} & a_{12} & \cdots & a_{1n} \\ a_{21} & a_{22} & \cdots & a_{2n} \\ \vdots & \vdots & \ddots & \vdots \\ a_{m1} & a_{m2} & \cdots & a_{mn} \end{bmatrix}$$

Augmented Matrix of a System (p. 321)

The **augmented matrix** of a system of linear equations is the matrix consisting of the coefficients and the constant terms. For example, for the two-variable system

$$a_{11}x + a_{12}x = b_1$$
$$a_{21}x + a_{22}x = b_2$$

the augmented matrix is

$$\begin{bmatrix} a_{11} & a_{12} & b_1 \\ a_{21} & a_{22} & b_2 \end{bmatrix}$$

Elementary Row Operations (p. 322)

To solve a system of linear equations using the augmented matrix of the system, the following operations can be used to transform the rows of the matrix:

1. Add a nonzero multiple of one row to another.
2. Multiply a row by a nonzero constant.
3. Interchange two rows.

Row-Echelon Form of a Matrix (p. 323)

A matrix is in **row-echelon form** if its entries satisfy the following conditions:

1. The first nonzero entry in each row (the **leading entry**) is the number 1.
2. The leading entry of each row is to the right of the leading entry in the row above it.
3. All rows consisting entirely of zeros are at the bottom of the matrix.

If the matrix also satisfies the following condition, it is in **reduced row-echelon** form:

4. If a column contains a leading entry, then every other entry in that column is a 0.

Number of Solutions of a Linear System (p. 327)

If the augmented matrix of a system of linear equations has been reduced to row-echelon form using elementary row operations, then the system has:

1. **No solution** if the row-echelon form contains a row that represents the equation $0 = 1$. In this case the system is **inconsistent**.
2. **One solution** if each variable in the row-echelon form is a **leading variable**.

3. **Infinitely many solutions** if the system is not inconsistent but not every variable is a leading variable. In this case the system is **dependent**.

Operations on Matrices (p. 332)

If A and B are $m \times n$ matrices and c is a scalar (real number), then:

1. The **sum** $A + B$ is the $m \times n$ matrix that is obtained by adding corresponding entries of A and B.
2. The **difference** $A - B$ is the $m \times n$ matrix that is obtained by subtracting corresponding entries of A and B.
3. The **scalar product** cA is the $m \times n$ matrix that is obtained by multiplying each entry of A by c.

Multiplication of Matrices (p. 334)

If A is an $m \times n$ matrix and B is an $n \times k$ matrix (so the number of columns of A is the same as the number of rows of B), then the **matrix product** AB is the $m \times k$ matrix whose ij-entry is the inner product of the ith row of A and the jth column of B.

Properties of Matrix Operations (pp. 333, 335)

If A, B, and C are matrices of compatible dimensions then the following properties hold:

1. **Commutativity of addition:**
$$A + B = B + A$$

2. **Associativity:**
$$(A + B) + C = A + (B + C)$$
$$(AB)C = A(BC)$$

3. **Distributivity:**
$$A(B + C) = AB + AC$$
$$(B + C)A = BA + CA$$

(Note that matrix *multiplication* is *not* commutative.)

Identity Matrix (p. 339)

The **identity matrix** I_n is the $n \times n$ matrix whose main diagonal entries are all 1 and whose other entries are all 0:

$$I_n = \begin{bmatrix} 1 & 0 & \cdots & 0 \\ 0 & 1 & \cdots & 0 \\ \vdots & \vdots & \ddots & \vdots \\ 0 & 0 & \cdots & 1 \end{bmatrix}$$

If A is an $m \times n$ matrix, then

$$AI_n = A \qquad \text{and} \qquad I_mA = A$$

Inverse of a Matrix (p. 340)

If A is an $n \times n$ matrix, then the inverse of A is the $n \times n$ matrix A^{-1} with the following properties:

$$A^{-1}A = I_n \qquad \text{and} \qquad AA^{-1} = I_n$$

To find the inverse of a matrix, we use a procedure involving elementary row operations (explained on page 322). (Note that *some* square matrices do not have an inverse.)

Inverse of a 2 × 2 Matrix (p. 340)

For 2 × 2 matrices the following special rule provides a shortcut for finding the inverse:

$$A = \begin{bmatrix} a & b \\ c & d \end{bmatrix} \;\Rightarrow\; A^{-1} = \frac{1}{ad - bc} \begin{bmatrix} d & -b \\ -c & a \end{bmatrix}$$

Writing a Linear System as a Matrix Equation (pp. 343–344)

A system of n linear equations in n variables can be written as a single matrix equation

$$AX = B$$

where A is the $n \times n$ matrix of coefficients, X is the $n \times 1$ matrix of the variables, and B is the $n \times 1$ matrix of the constants. For example, the linear system of two equations in two variables

$$a_{11}x + a_{12}x = b_1$$
$$a_{21}x + a_{22}x = b_2$$

can be expressed as

$$\begin{bmatrix} a_{11} & a_{12} \\ a_{21} & a_{22} \end{bmatrix} \begin{bmatrix} x \\ y \end{bmatrix} = \begin{bmatrix} b_1 \\ b_2 \end{bmatrix}$$

Solving Matrix Equations (p. 344)

If A is an invertible $n \times n$ matrix, X is an $n \times 1$ variable matrix, and B is an $n \times 1$ constant matrix, then the matrix equation

$$AX = B$$

has the unique solution

$$X = A^{-1}B$$

Determinant of a 2 × 2 Matrix (p. 347)

The **determinant** of the matrix

$$A = \begin{bmatrix} a & b \\ c & d \end{bmatrix}$$

is the *number*

$$\det(A) = |A| = ad - bc$$

Minors and Cofactors (p. 348)

If $A = |a_{ij}|$ is an $n \times n$ matrix, then the **minor** M_{ij} of the entry a_{ij} is the determinant of the matrix obtained by deleting the ith row and the jth column of A.

The **cofactor** A_{ij} of the entry a_{ij} is

$$A_{ij} = (-1)^{i+j}M_{ij}$$

(Thus, the minor and the cofactor of each entry either are the same or are negatives of each other.)

Determinant of an $n \times n$ Matrix (p. 348)

To find the **determinant** of the $n \times n$ matrix

$$A = \begin{bmatrix} a_{11} & a_{12} & \cdots & a_{1n} \\ a_{21} & a_{22} & \cdots & a_{2n} \\ \vdots & \vdots & \ddots & \vdots \\ a_{n1} & a_{n2} & \cdots & a_{nn} \end{bmatrix}$$

we choose a row or column to **expand**, and then we calculate the number that is obtained by multiplying each element of that row or column by its cofactor and then adding the resulting products. For example, if we choose to expand about the first row, we get

$$\det(A) = |A| = a_{11}A_{11} + a_{12}A_{12} + \cdots + a_{1n}A_{1n}$$

Invertibility Criterion (p. 350)

A square matrix has an inverse if and only if its determinant is not 0.

Row and Column Transformations (p. 351)

If we add a nonzero multiple of one row to another row in a square matrix or a nonzero multiple of one column to another column, then the determinant of the matrix is unchanged.

Cramer's Rule (pp. 352–354)

If a system of n linear equations in the n variables $x_1, x_2, \ldots, x_n$ is equivalent to the matrix equation $DX = B$ and if $|D| \neq 0$, then the solutions of the system are

$$x_1 = \frac{|D_{x_1}|}{|D|} \qquad x_2 = \frac{|D_{x_2}|}{|D|} \qquad \cdots \qquad x_n = \frac{|D_{x_n}|}{|D|}$$

where D_{x_i} is the matrix that is obtained from D by replacing its ith column by the constant matrix B.

Area of a Triangle Using Determinants (p. 355)

If a triangle in the coordinate plane has vertices (a_1, b_1), (a_2, b_2), and (a_3, b_3), then the area of the triangle is given by

$$\text{area} = \pm\frac{1}{2} \begin{vmatrix} a_1 & b_1 & 1 \\ a_2 & b_2 & 1 \\ a_3 & b_3 & 1 \end{vmatrix}$$

where the sign is chosen to make the area positive.

▼ PROPERTIES AND FORMULAS

Geometric Definition of a Parabola (p. 364)

A **parabola** is the set of points in the plane that are equidistant from a fixed point F (the **focus**) and a fixed line l (the **directrix**).

Graphs of Parabolas with Vertex at the Origin (p. 367)

A parabola with vertex at the origin has an equation of the form $x^2 = 4py$ if its axis is vertical and an equation of the form $y^2 = 4px$ if its axis is horizontal.

$$x^2 = 4py \qquad\qquad y^2 = 4px$$

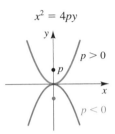 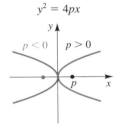

Focus $(0, p)$, directrix $y = -p$ Focus $(p, 0)$, directrix $x = -p$

Geometric Definition of an Ellipse (p. 370)

An **ellipse** is the set of all the points in the plane for which the sum of the distances to each of two given points F_1 and F_2 (the **foci**) is a fixed constant.

Graphs of Ellipses with Center at the Origin (p. 372)

An ellipse with center at the origin has an equation of the form $\frac{x^2}{a^2} + \frac{y^2}{b^2} = 1$ if its axis is horizontal and an equation of the form $\frac{x^2}{b^2} + \frac{y^2}{a^2} = 1$ if its axis is vertical (where in each case $a > b > 0$).

$$\frac{x^2}{a^2} + \frac{y^2}{b^2} = 1 \qquad\qquad \frac{x^2}{b^2} + \frac{y^2}{a^2} = 1$$

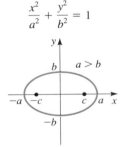 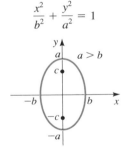

Foci $(\pm c, 0)$, $c^2 = a^2 - b^2$ Foci $(0, \pm c)$, $c^2 = a^2 - b^2$

Eccentricity of an Ellipse (p. 375)

The **eccentricity** of an ellipse with equation $\frac{x^2}{a^2} + \frac{y^2}{b^2} = 1$ or $\frac{x^2}{b^2} + \frac{y^2}{a^2} = 1$ (where $a > b > 0$) is the number

$$e = \frac{c}{a}$$

where $c = \sqrt{a^2 - b^2}$. The eccentricity e of any ellipse is a number between 0 and 1. If e is close to 0, then the ellipse is nearly circular; the closer e gets to 1, the more elongated it becomes.

Geometric Definition of a Hyperbola (p. 376)

A **hyperbola** is the set of all those points in the plane for which the absolute value of the difference of the distances to each of two given points F_1 and F_2 (the **foci**) is a fixed constant.

Graphs of Hyperbolas with Center at the Origin (p. 377)

A **hyperbola** with center at the origin has an equation of the form $\frac{x^2}{a^2} - \frac{y^2}{b^2} = 1$ if its axis is horizontal and an equation of the form $-\frac{x^2}{b^2} + \frac{y^2}{a^2} = 1$ if its axis is vertical.

$$\frac{x^2}{a^2} - \frac{y^2}{b^2} = 1 \qquad\qquad -\frac{x^2}{b^2} + \frac{y^2}{a^2} = 1$$

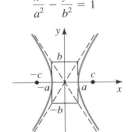 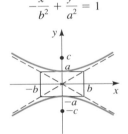

Foci $(\pm c, 0)$, $c^2 = a^2 + b^2$ Foci $(0, \pm c)$, $c^2 = a^2 + b^2$

Asymptotes: $y = \pm\frac{b}{a}x$ Asymptotes: $y = \pm\frac{a}{b}x$

Shifted Conics (pp. 382–386)

If the vertex of a parabola or the center of an ellipse or a hyperbola does not lie at the origin but rather at the point (h, k), then we refer to the curve as a **shifted conic**. To find the equation of the shifted conic, we use the "unshifted" form for the appropriate curve and replace x by $x-h$ and y by $y-k$.

General Equation of a Shifted Conic (p. 387)

The graph of the equation

$$Ax^2 + Cy^2 + Dx + Ey + F = 0$$

(where A and C are not both 0) is either a conic or a degenerate conic. In the nondegenerate cases the graph is:

1. A **parabola** if $A = 0$ or $C = 0$.

2. An **ellipse** if A and C have the same sign (or a circle if $A = C$).

3. A **hyperbola** if A and C have opposite sign.

To graph a conic whose equation is given in general form, **complete the square** in x and y to put the equation in standard form for a parabola, an ellipse, or a hyperbola.

▼ PROPERTIES AND FORMULAS

Sequences (p. 396)

A **sequence** is a function whose domain is the set of natural numbers. Instead of writing $a(n)$ for the value of the sequence at n, we generally write a_n, and we refer to this value as the **nth term** of the sequence. Sequences are often described in list form:

$$a_1, a_2, a_3, \ldots$$

Partial Sums of a Sequence (pp. 401–402)

For the sequence $a_1, a_2, a_3, \ldots$ the **nth partial sum** S_n is the sum of the first n terms of the sequence:

$$S_n = a_1 + a_2 + a_3 + \cdots + a_n$$

The nth partial sum of a sequence can also be expressed by using **sigma notation**:

$$S_n = \sum_{k=1}^{n} a_k$$

Arithmetic Sequences (p. 404)

An **arithmetic sequence** is a sequence whose terms are obtained by adding the same fixed constant d to each term to get the next term. Thus, an arithmetic sequence has the form

$$a, a + d, a + 2d, a + 3d, \ldots$$

The number a is the **first term** of the sequence, and the number d is the **common difference**. The nth term of the sequence is

$$a_n = a + (n - 1)d$$

Partial Sums of an Arithmetic Sequence (p. 407)

For the arithmetic sequence $a_n = a + (n - 1)d$ the nth partial sum $S_n = \sum_{k=1}^{n} [a + (k - 1)d]$ is given by either of the following equivalent formulas:

1. $\qquad S_n = \dfrac{n}{2}[2a + (n - 1)d]$

2. $\qquad S_n = n\left(\dfrac{a + a_n}{2}\right)$

Geometric Sequences (p. 408)

A **geometric sequence** is a sequence whose terms are obtained by multiplying each term by the same fixed constant r to get the next term. Thus, a geometric sequence has the form

$$a, ar, ar^2, ar^3, \ldots$$

The number a is the **first term** of the sequence, and the number r is the **common ratio**. The nth term of the sequence is

$$a_n = ar^{n-1}$$

Partial Sums of a Geometric Sequence (p. 410)

For the geometric sequence $a_n = ar^{n-1}$ the nth partial sum $S_n = \sum_{k=1}^{n} ar^{k-1}$ (where $r \neq 1$) is given by

$$S_n = a\frac{1 - r^n}{1 - r}$$

Infinite Geometric Series (p. 412)

An **infinite geometric series** is a series of the form

$$a + ar + ar^2 + ar^3 + \cdots + ar^{n-1} + \cdots$$

An infinite series for which $|r| < 1$ has the sum

$$S = \frac{a}{1 - r}$$

Amount of an Annuity (p. 414)

The amount A_f of an **annuity** consisting of n regular equal payments of size R with interest rate i per time period is given by

$$A_f = R\frac{(1 + i)^n - 1}{i}$$

Present Value of an Annuity (p. 415)

The **present value** A_p of an annuity consisting of n regular equal payments of size R with interest rate i per time period is given by

$$A_p = R\frac{1 - (1 + i)^{-n}}{i}$$

Present Value of a Future Amount (p. 415)

If an amount A_f is to be paid in one lump sum, n time periods from now, and the interest rate per time period is i, then its **present value** A_p is given by

$$A_p = A_f(1 + i)^{-n}$$

Installment Buying (p. 416)

If a loan A_p is to be repaid in n regular equal payments with interest rate i per time period, then the size R of each payment is given by

$$R = \frac{iA_p}{1 - (1 + i)^{-n}}$$

Principle of Mathematical Induction (p. 419)

For each natural number n, let $P(n)$ be a statement that depends on n. Suppose that each of the following conditions is satisfied.

1. $P(1)$ is true.
2. For every natural number k, if $P(k)$ is true, then $P(k + 1)$ is true.

Then $P(n)$ is true for all natural numbers n.

Sums of Powers (p. 421)

0. $\displaystyle\sum_{k=1}^{n} 1 = n$

1. $\displaystyle\sum_{k=1}^{n} k = \frac{n(n+1)}{2}$

2. $\displaystyle\sum_{k=1}^{n} k^2 = \frac{n(n+1)(2n+1)}{6}$

3. $\displaystyle\sum_{k=1}^{n} k^3 = \frac{n^2(n+1)^2}{4}$

Binomial Coefficients (pp. 424–426)

If n and r are positive integers with $n \geq r$, then the **binomial coefficient** $\dbinom{n}{r}$ is defined by

$$\binom{n}{r} = \frac{n!}{r!(n-r)!}$$

Binomial coefficients satisfy the following properties:

$$\binom{n}{r} = \binom{n}{n-r}$$

$$\binom{k}{r-1} + \binom{k}{r} = \binom{k+1}{r}$$

The Binomial Theorem (p. 426)

$$(a + b)^n = \binom{n}{0}a^n + \binom{n}{1}a^{n-1}b + \binom{n}{2}a^{n-2}b^2 + \cdots + \binom{n}{n}b^n$$

GEOMETRIC FORMULAS

Formulas for area A, perimeter P, circumference C, volume V:

Rectangle

$A = lw$ $P = 2l + 2w$

Box

$V = lwh$

Triangle

$A = \frac{1}{2}bh$

Pyramid

$V = \frac{1}{3}ha^2$

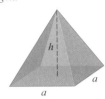

Circle

$A = \pi r^2$ $C = 2\pi r$

Sphere

$V = \frac{4}{3}\pi r^3$ $A = 4\pi r^2$

Cylinder

$V = \pi r^2 h$

Cone

$V = \frac{1}{3}\pi r^2 h$

PYTHAGOREAN THEOREM

In a right triangle, the square on the hypotenuse is equal to the sum of the squares on the other two sides.

$$a^2 + b^2 = c^2$$

SIMILAR TRIANGLES

Two triangles are similar if corresponding angles are equal.

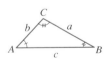

If $\triangle ABC$ is similar to $\triangle A'B'C'$, then ratios of corresponding sides are equal:

$$\frac{a}{a'} = \frac{b}{b'} = \frac{c}{c'}$$

CONIC SECTIONS

Circles

$(x - h)^2 + (y - k)^2 = r^2$

Parabolas

$x^2 = 4py$ $y^2 = 4px$

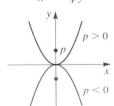

Focus $(0, p)$, directrix $y = -p$ Focus $(p, 0)$, directrix $x = -p$

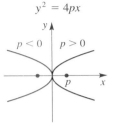

$y = a(x - h)^2 + k,$
$a < 0,\quad h > 0,\quad k > 0$

$y = a(x - h)^2 + k,$
$a > 0,\quad h > 0,\quad k > 0$

Ellipses

$$\frac{x^2}{a^2} + \frac{y^2}{b^2} = 1 \qquad\qquad \frac{x^2}{b^2} + \frac{y^2}{a^2} = 1$$

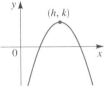
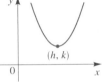

Foci $(\pm c, 0)$, $c^2 = a^2 - b^2$ Foci $(0, \pm c)$, $c^2 = a^2 - b^2$

Hyperbolas

$$\frac{x^2}{a^2} - \frac{y^2}{b^2} = 1 \qquad\qquad -\frac{x^2}{b^2} + \frac{y^2}{a^2} = 1$$

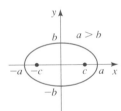
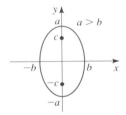

Foci $(\pm c, 0)$, $c^2 = a^2 + b^2$ Foci $(0, \pm c)$, $c^2 = a^2 + b^2$

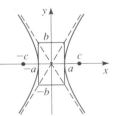
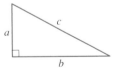
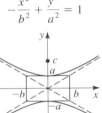

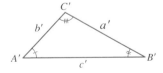

SEQUENCES AND SERIES

Arithmetic

$$a, a + d, a + 2d, a + 3d, a + 4d, \ldots$$

$$a_n = a + (n - 1)d$$

$$S_n = \sum_{k=1}^{n} a_k = \frac{n}{2}[2a + (n-1)d] = n\left(\frac{a + a_n}{2}\right)$$

Geometric

$$a, ar, ar^2, ar^3, ar^4, \ldots$$

$$a_n = ar^{n-1}$$

$$S_n = \sum_{k=1}^{n} a_k = a\frac{1 - r^n}{1 - r}$$

If $|r| < 1$, then the sum of an infinite geometric series is

$$S = \frac{a}{1 - r}$$

THE BINOMIAL THEOREM

$$(a + b)^n = \binom{n}{0}a^n + \binom{n}{1}a^{n-1}b + \cdots$$

$$+ \binom{n}{n-1}ab^{n-1} + \binom{n}{n}b^n$$

FINANCE

Compound interest

$$A = P\left(1 + \frac{r}{n}\right)^{nt}$$

where A is the amount after t years, P is the principal, r is the interest rate, and the interest is compounded n times per year.

Amount of an annuity

$$A_f = R\frac{(1 + i)^n - 1}{i}$$

where A_f is the final amount, i is the interest rate per time period, and there are n payments of size R.

Present value of an annuity

$$A_p = R\frac{1 - (1 + i)^{-n}}{i}$$

where A_p is the present value, i is the interest rate per time period, and there are n payments of size R.

Installment buying

$$R = \frac{iA_p}{1 - (1 + i)^{-n}}$$

where R is the size of each payment, i is the interest rate per time period, A_p is the amount of the loan, and n is the number of payments.

Getting the Most from Your Enhanced WebAssign Experience

You will likely be doing most if not all of your homework assignments within Enhanced WebAssign (EWA), an online homework system. With many of the online homework exercises in EWA you may see a variety of available tutorial links.

Each of these links provides you with the help you need to complete your homework. Below we'll explain the "Master It" tutorial and how best to get the most out of the tutorial help provided. Complete instructions on registration, login procedures, answer entry, and so on are available online at **www.webassign.com**, or in *Enhanced WebAssign: The Start Smart Guide for Students* (ISBN: 0-495-38479-8).

■ Master It Link

Click on this link and you'll be given a fill-in-the-blank complete solution quite similar to your actual homework exercise. This tutorial will look something like this

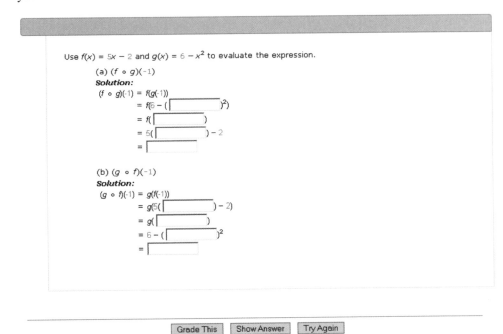

You'll notice an exercise similar to your homework problem, and a fill-in-the-blank solution of that exercise. Here you can see how to develop a solution, step by step.

1. Work out the solution on paper first. Now see if your steps are correct.

2. Begin to enter answers in the appropriate answer field.

3. At any point you can check your work. Go to the end of the solution and click the "Grade This" button. You don't have to fill in all the steps in order to see how you're doing. For each correct entry you'll get a green check mark, and for incorrect entries you'll get a red X. Unanswered entries get a red X, too.

Use $f(x) = 5x - 2$ and $g(x) = 6 - x^2$ to evaluate the expression.

(a) $(f \circ g)(-1)$

Solution:

$(f \circ g)(-1) = f(g(-1))$

$\qquad = f(6 - (\boxed{-1} \checkmark)^2)$

$\qquad = f(\boxed{5} \checkmark)$

$\qquad = 5(\boxed{} \times) - 2$

$\qquad = \boxed{} \times$

4. Continue building your solution by correcting wrong answers and/or entering answers into empty answer boxes. Whenever you want to check how you're doing, just click the "Grade This" button. You'll see green check marks replace the red X's for any of **your** new answers that are correct.

5. At any point you can click the "Show Answer" button to see all the correct answers **and** thus a completed solution.

Use $f(x) = 5x - 2$ and $g(x) = 6 - x^2$ to evaluate the expression.

(a) $(f \circ g)(-1)$

Solution:

$(f \circ g)(-1) = f(g(-1))$

$\qquad = f(6 - (\boxed{-1} \checkmark \boxed{\text{?} -1})^2)$

$\qquad = f(\boxed{5} \checkmark \boxed{\text{?} 5})$

$\qquad = 5(\boxed{5} \checkmark \boxed{\text{?} 5}) - 2$

$\qquad = \boxed{23} \checkmark \boxed{\text{?} 23}$

(b) $(g \circ f)(-1)$

Solution:

$(g \circ f)(-1) = g(f(-1))$

$\qquad = g(5(\boxed{} \times \boxed{\text{?} -1}) - 2)$

$\qquad = g(\boxed{} \times \boxed{\text{?} -7})$

$\qquad = 6 - (\boxed{} \times \boxed{\text{?} -7})^2$

$\qquad = \boxed{} \times \boxed{\text{?} -43}$

6. If you want to work on another solution just click the "Try Again" button and EWA will give you another similar exercise and solution-building tutorial.

EXPONENTS AND RADICALS

$$x^m x^n = x^{m+n} \qquad\qquad \frac{x^m}{x^n} = x^{m-n}$$

$$(x^m)^n = x^{mn} \qquad\qquad x^{-n} = \frac{1}{x^n}$$

$$(xy)^n = x^n y^n \qquad\qquad \left(\frac{x}{y}\right)^n = \frac{x^n}{y^n}$$

$$x^{1/n} = \sqrt[n]{x} \qquad\qquad x^{m/n} = \sqrt[n]{x^m} = \left(\sqrt[n]{x}\right)^m$$

$$\sqrt[n]{xy} = \sqrt[n]{x}\,\sqrt[n]{y} \qquad\qquad \sqrt[n]{\frac{x}{y}} = \frac{\sqrt[n]{x}}{\sqrt[n]{y}}$$

$$\sqrt[m]{\sqrt[n]{x}} = \sqrt[n]{\sqrt[m]{x}} = \sqrt[mn]{x}$$

SPECIAL PRODUCTS

$$(x+y)(x-y) = x^2 - y^2$$
$$(x+y)^2 = x^2 + 2xy + y^2$$
$$(x-y)^2 = x^2 - 2xy + y^2$$
$$(x+y)^3 = x^3 + 3x^2 y + 3xy^2 + y^3$$
$$(x-y)^3 = x^3 - 3x^2 y + 3xy^2 - y^3$$

FACTORING FORMULAS

$$x^2 - y^2 = (x+y)(x-y)$$
$$x^2 + 2xy + y^2 = (x+y)^2$$
$$x^2 - 2xy + y^2 = (x-y)^2$$
$$x^3 + y^3 = (x+y)(x^2 - xy + y^2)$$
$$x^3 - y^3 = (x-y)(x^2 + xy + y^2)$$

QUADRATIC FORMULA

If $ax^2 + bx + c = 0$, then

$$x = \frac{-b \pm \sqrt{b^2 - 4ac}}{2a}$$

INEQUALITIES AND ABSOLUTE VALUE

If $a < b$ and $b < c$, then $a < c$.

If $a < b$, then $a + c < b + c$.

If $a < b$ and $c > 0$, then $ca < cb$.

If $a < b$ and $c < 0$, then $ca > cb$.

If $a > 0$, then

$\quad |x| = a \quad$ means $\quad x = a \quad$ or $\quad x = -a$.

$\quad |x| < a \quad$ means $\quad -a < x < a$.

$\quad |x| > a \quad$ means $\quad x > a \quad$ or $\quad x < -a$.

DISTANCE AND MIDPOINT FORMULAS

Distance between $P_1(x_1, y_1)$ and $P_2(x_2, y_2)$:

$$d = \sqrt{(x_2 - x_1)^2 + (y_2 - y_1)^2}$$

Midpoint of $P_1 P_2$: $\quad \left(\dfrac{x_1 + x_2}{2}, \dfrac{y_1 + y_2}{2}\right)$

LINES

Slope of line through $P_1(x_1, y_1)$ and $P_2(x_2, y_2)$
$\qquad m = \dfrac{y_2 - y_1}{x_2 - x_1}$

Point-slope equation of line through $P_1(x_1, y_1)$ with slope m
$\qquad y - y_1 = m(x - x_1)$

Slope-intercept equation of line with slope m and y-intercept b
$\qquad y = mx + b$

Two-intercept equation of line with x-intercept a and y-intercept b
$\qquad \dfrac{x}{a} + \dfrac{y}{b} = 1$

The lines $y = m_1 x + b_1$ and $y = m_2 x + b_2$ are

$\quad$ **Parallel** if the slopes are the same $\qquad m_1 = m_2$

$\quad$ **Perpendicular** if the slopes are negative reciprocals $\qquad m_1 = -1/m_2$

LOGARITHMS

$$y = \log_a x \quad \text{means} \quad a^y = x$$

$$\log_a a^x = x \qquad\qquad a^{\log_a x} = x$$

$$\log_a 1 = 0 \qquad\qquad \log_a a = 1$$

Common and natural logarithms

$$\log x = \log_{10} x \qquad\qquad \ln x = \log_e x$$

Laws of logarithms

$$\log_a xy = \log_a x + \log_a y$$

$$\log_a\left(\frac{x}{y}\right) = \log_a x - \log_a y$$

$$\log_a x^b = b \log_a x$$

Change of base formula

$$\log_b x = \frac{\log_a x}{\log_a b}$$

GRAPHS OF FUNCTIONS

Linear functions: $f(x) = mx + b$

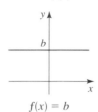

$f(x) = b$

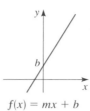

$f(x) = mx + b$

Power functions: $f(x) = x^n$

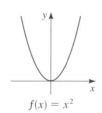

$f(x) = x^2$

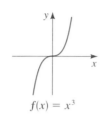

$f(x) = x^3$

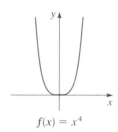

$f(x) = x^4$

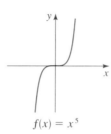

$f(x) = x^5$

Root functions: $f(x) = \sqrt[n]{x}$

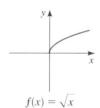

$f(x) = \sqrt{x}$

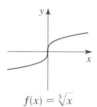

$f(x) = \sqrt[3]{x}$

Reciprocal functions: $f(x) = 1/x^n$

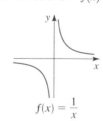

$f(x) = \dfrac{1}{x}$

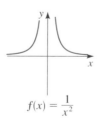

$f(x) = \dfrac{1}{x^2}$

Exponential functions: $f(x) = a^x$

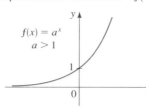

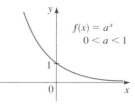

Logarithmic functions: $f(x) = \log_a x$

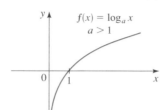

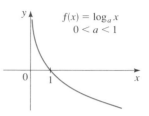

Absolute value function   Greatest integer function

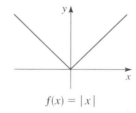

$f(x) = |x|$

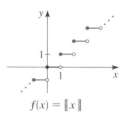

$f(x) = \lVert x \rVert$

SHIFTING OF FUNCTIONS

Vertical shifting

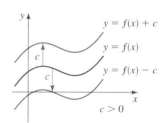

Horizontal shifting

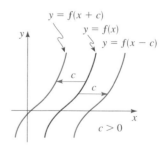

Getting the Most Out of Enhanced WebAssign Using the GraphPad

The Enhanced WebAssign GraphPad lets you graph one or more mathematical elements directly on a set of coordinate axes. Your graph is then scored automatically when you submit the assignment for grading.

The GraphPad currently supports points, rays, segments, lines, circles, and parabolas. Inequalities can also be indicated by filling one or more areas.

■ GraphPad Interface Overview

The middle of GraphPad is the drawing area. It contains labeled coordinate axes, which may have different axis scales and extents depending on the nature of the question you are working on.

On the left side of GraphPad is the list of Tools that let you create graph objects and select objects to edit.

The bottom of the GraphPad is the Object Properties Toolbar, which becomes active when you have a graph element selected. This toolbar shows you all the details about the selected graph object, and also lets you edit properties of that object.

On the right side of GraphPad is the list of Actions that lets you create fills and delete objects from your graph.

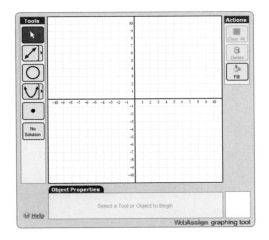

Drawing Tools	
Points:	Click the Point Tool, and then click where you want the point to appear.
Lines:	Click the Line Tool, and then place two points (which are on the desired line). The arrows on the end of the line indicate that the line goes off to infinity on both ends.
Rays:	Click the Ray Tool, place the endpoint, and then place another point on the ray.
Line Segments:	Click the Line Segment Tool, and then place the two endpoints of the line segment.
Circles:	Click the Circle Tool, place the point at the center first, and then place a point on the radius of the circle.
Parabolas:	Click the Parabola Tool (either vertical or horizontal), place the vertex first, and then place another point on the parabola.
No Solution:	If the question has no solution, simply click the No Solution Tool.

Note: Don't worry if you don't place the points exactly where you want them initially; you can move these points around before submitting for grading.

■ Selecting Graph Objects

To edit a graph object, that object must be "selected" as the active object. (When you first draw an object, it is created in the selected state.) When a graph element is "selected," the color of the line changes and two "handles" are visible. The handles are the two square points you clicked to create the object. To select an object, click the select tool, and then click on the object. To deselect the object, click on a blank area on the drawing area or on a drawing tool.

Not Selected   Selected

Once an object is selected, you can modify it by using your mouse or the keyboard. As you move it, you'll notice that you cannot move the handles off the drawing area. To move an object with the mouse, click and drag the object's line. Or, click and drag one of the handles to move just that handle. On the keyboard, the arrow keys also move the selected object around by one unit.

As you move the object or handle you'll see that the Object Properties toolbar changes to remain up to date.

Object Properties Toolbar

Coordinate Fields: ● Point 1 (-13 , -9) ■ Point 2 (15 , -6)	You can use the Coordinate Fields to edit the coordinates of the handles directly. Use this method to enter decimal or fractional coordinates.
Endpoint Controls: Endpoint ●○ Endpoint ●○	If the selected object is a segment or ray, the Endpoint Controls are used to define the endpoint as closed or open. As a shortcut, you can also define an endpoint by clicking on the endpoint when the ray or segment is in the unselected state.
Solid/Dash Controls: Solid Dash	For any selected object other than a point, the Solid/Dash Controls are used to define the object as solid or dashed. To change graph objects to solid or dashed, select the object and then click the solid or dash button.

■ Using Fractions or Decimals as Coordinates

To draw an object with handle coordinates that are fractions or decimals, you must use the Object Properties Toolbar. Draw the desired object anywhere on the drawing area, then use the Coordinate Fields to change the endpoint(s) to the desired value. For example, to enter a fraction, just type "3/4."

Note: The points and lines you draw must be exactly correct when you submit for grading. This means you should not round any of your values—if you want a point at 11/3, you should enter 11/3 in the coordinate box rather than 3.667. Also, mixed fractions are not acceptable entries. This means that 3 2/3 is an incorrect entry.

Actions

Fill Tool: Fill	To graph an inequality, you must specify a region on the graph. To do this, first draw the line(s), circle(s), or other object(s) that will define the region you want to represent your answer. Be sure to specify the objects as either solid or dashed, according to the inequality you are graphing! Then choose the Fill Tool and click inside the region that you want filled.
	  If you decide you wanted the fill in a different area, click the filled region that you want to unfill, and then click the region that you do want to fill.
Delete Tool: Delete	To erase a single graph object, first select that element in the drawing area then click the Delete Tool or press the Delete key on your keyboard.
Clear All Tool: Clear All	The Clear All Tool will erase all of your graph objects. (If the drawing area is already empty, the Clear All Tool is disabled.)

▶ **EXAMPLE:**
Let's suppose you're asked to graph the inequality $y > 5x + \frac{1}{5}$, and you want to use the points $(0, \frac{1}{5})$ and $(1, 5\frac{1}{5})$. You would first place any line on the drawing area.

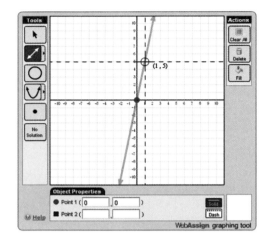

Then, adjust the points using the Coordinate Fields. Remember, you need to enter 5 1/5 as a decimal (5.2) or an improper fraction (26/5).

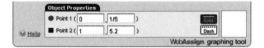

Next, you would define the line as dashed since the inequality does not include the values on the line.

Finally, you would select the Fill Tool and click on the desired region to complete the graph.

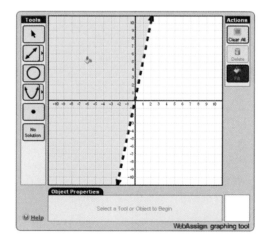

Using the MathPad in Enhanced WebAssign

For many of your online homework questions, Enhanced WebAssign provides you a **MathPad**. The **MathPad** allows you to easily input math notation and symbols, even the more complicated ones, into your answers. If your answer involves math notation or symbols the MathPad will become available when you click the answer box.

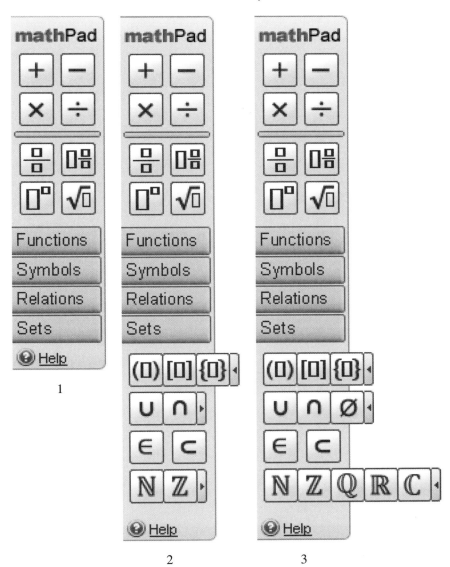

1

2   3

■ Top Symbols

The buttons on the top are single input buttons for frequently used operations.

■ Word Buttons

When you click one of the "word" buttons **Functions**, **Symbols**, **Relations**, or **Sets** you are given a drop-down menu with symbols or notation from which to choose. For example if you click the **Sets** button you get set notation (figure 2 above). If you then click a right arrow button additional symbols become available (figure 3 above).

To insert any available notation or symbol into your answer simply click it.